KV-388-116

Crystalline Plasticity and Solid State Flow in Metamorphic Rocks

---

SELECTED TOPICS IN GEOLOGICAL SCIENCES

Editor

M. H. P. Bott, *Department of Geology, University of Durham*

Crystalline Plasticity and Solid State Flow in Metamorphic Rocks:

A. Nicolas, Institut des Sciences de la Nature,
Département des Sciences de la Terre,
Université de Nantes
J. P. Poirier, Section de Recherches de Métallurgie,
Centre d'Études Nucléaires de Saclay

Crystalline Plasticity and Solid State Flow in Metamorphic Rocks

A. NICOLAS
Département des Sciences de la Terre,
Université de Nantes

J. P. POIRIER
Section de Recherches de Métallurgie Physique,
Centre d'Études Nucléaires de Saclay

A Wiley–Interscience Publication

JOHN WILEY & SONS
London · New York · Sydney · Toronto

Library of Congress Cataloging in Publication Data:

Nicolas, Adolphe, 1936–
Crystalline plasticity and solid state flow in metamorphic rocks.

'A Wiley–Interscience publication.'
Includes index.
1. Rocks, Metamorphic. 2. Rock deformation.
3. Dislocations in crystals. I. Poirier, Jean Paul.
II. Title.

QE475.N52 1976 552'.4 75-15981
ISBN 0 471 63791 2

Set on Linotron Filmsetter and printed in Great Britain by
J. W. Arrowsmith Ltd., Bristol. BS3 2NT.

Expliquer du visible compliqué par de l'invisible simple.

JEAN PERRIN

The general lesson . . . concerns not so much the problem of guessing quantitatively as the importance of qualitative observations—what we may describe as looking for the natural history of the phenomenon.

A. B. PIPPARD
Inaugural lecture, Cambridge, 1971.

Series Preface

Expansion in the geological sciences has taken place at an ever-increasing tempo throughout the twentieth century and reached a climax in the plate-tectonic revolution of the 1960's. Initially, the theory provided, for the first time, an underlying explanation of present-day tectonic activity, but with subsequent years of consolidation the new insight has penetrated deeply into many branches of geology and geophysics. Furthermore, the geological sciences have benefited greatly in recent years, not only from the application of the concepts of plate-tectonics but also from notable advances in other branches of physical and biological sciences. The present time is thus one of broad advance in almost all aspects of the subject.

Rapid advance brings with it difficulties for the teacher, the professional geologist and the student in keeping up to date with the new theories and findings. Few people can keep abreast of the avalanche of technical literature which has appeared in ever-increasing volume, and one way of helping scientists to achieve this is to supply reliable accounts epitomizing the advance of science. I hope this series of books will help fulfil this need.

Thus this series of books on selected topics in geological sciences has been launched at an appropriate time. The individual books cover topics in which there have been recent advances in knowledge. They have been written for teachers, undergraduate and postgraduate students and for professional earth scientists. I am sure that the series will make an important contribution to the advance of understanding in the geological sciences.

M. H. P. Bott

Department of Geological Sciences,
University of Durham,

May 1976.

Preface

In the last 15 years, kinematic and dynamic analyses in structural geology have rapidly progressed in two distinct directions. One, which will be referred to as the continuum mechanics approach, has progressed mainly from theoretical, experimental and geometrical results concerning folding and finite strain estimations; the other is based on a growing crop of data in the field of experimental deformation of rocks and rock-forming minerals. Those data have shown that there is no important difference between flow structures and processes in rocks and in metals. Therefore the way is now clear for geologists and physical metallurgists to collaborate. The former benefit from the longer and deeper experience of the latter who in turn can extend their interest to new materials deformed in different conditions. This is why we propose to call this direction of research in geology, the physical metallurgy approach.

In writing this book we had two purposes:

(1) To give a full account of the background of physical metallurgy which is necessary to describe correctly and interpret the structures of deformation in rocks. Our contribution differs from Spry's (1969) by a more theoretical and complete presentation of the basic data. However, Spry gives a comprehensive description of the textures in metamorphic rocks whereas we limit our study to those induced by flowage in metamorphic rocks;

(2) To investigate the interpretation in terms of flow of the metamorphic textures. To achieve this, we have felt that strain–stress analysis and description of various techniques which are not commonly used deserved extended treatment. In flow interpretation, dynamic considerations are discarded because we believe that they can be proposed only after the kinematic path has been clearly traced.

The use of physical metallurgy and geological concepts developed in our book will be illustrated by examining flow in mantle peridotites. Peridotites are probably the rocks most suited to this purpose, because olivine and enstatite which are dominant have a comparatively simple crystallographic structure and rather well-known mechanical properties; moreover the deformational structures indicate that the flowage was usually extensive and homogeneous. Such restrictive conditions show the limits of this method in the present state of knowledge.

At this point we should like to emphasize that we have accepted some risks in pushing the flow interpretation of deformed structures and textures as far as we could. Very little has yet been published on this subject and the interpretations are often personal and may be modified in the future. They have been suggested mainly by the study of flowing peridotites. Strictly speaking, they are valid only for those rocks but, as flow obeys general laws, we think this example is of general value. They challenge some interpretations which had been accredited by the continuum mechanics approach, for instance that concerning the kinematics of some folds.

It is with pleasure that we gratefully acknowledge the various help which made this book possible: discussions with H. W. Green; reviews for different parts of the book from M. Lelubre, C. Froidevaux, S. H. Kirby, G. Martin, F. Boudier, J. L. Bouchez, Y. Gueguen, M. Mattauer, E. den Tex, A. J. Ardell and J. M. Christie. We also express our thanks to S. Wightman who reviewed the English, L. Bureau who typed the manuscript, A. Cossard who drew the illustrations and J. Naulet who made the photographs.

A. NICOLAS
J. P. POIRIER

Nantes and Saclay,
March, 1975

Contents

CHAPTER 1

Introduction: Structural Analysis in Metamorphic Rocks

1.1. METAMORPHIC ROCKS UNDER CONSIDERATION

Structural analysis in metamorphic rocks is aimed at investigating the nature of the observed deformations and their relationship with the associated metamorphic events. For most petrologists, metamorphism applies to phase changes occurring mainly in the solid state and ascribed to changes in temperature, partial and/or total pressures or chemistry in the considered system. Sander (1930, 1970) in his study of the structures in rocks deriving from plastic flow (direct componental movements) and recrystallization (indirect componental movements) extends the metamorphism concept to those physical changes in crystals, even though there has been no phase change. This wider definition is accepted here. It means that no formal difference is made between flowage in the ice of a glacier,* in the salt of a diapir, in schists of a crystalline basement, or in peridotites of the mantle. They all belong to the metamorphic history of the rock in which landmarks would be phase changes.

The deformation may be evaluated with regard to those phase changes. It may pre-date them, post-date them or be contemporaneous. The latter relationship is common and can be explained in different ways: the deformation in crystals may increase the kinetics of otherwise metastable situations; the deformation may also result from a flow which has possibly contributed to transporting the system into a new physical environment, requiring phase changes. This is illustrated by the deformational metamorphic history of the intrusion of the Lanzo peridotite body from the mantle into the crust developed in Chapters 10 and 11 of this book.

The rocks and rock formations dealt with here are those to which the metallurgical knowledge best applies. They have been deformed in a range of temperature and strain rates where minerals could plastically deform or recrystallize mainly by solid state diffusion. Impact deformation, characterized

* Although ice is a geological material, we will not consider it in this book because in a review which is necessarily limited its importance seemed to us weaker than that of the materials finally retained.

by extremely high strain rates, is not considered here. Neither do we consider specifically deformations related to metamorphism in the presence of an extensive fluid phase, like dynamic metamorphism at low temperatures when dehydration reactions take place or metamorphism with a large amount of partial melt at higher temperatures. These deformations can be correlated in a broad way with the prograde metamorphism of initially water-rich sediments. The flow mechanisms can be compaction, pressure solution or sliding at grain boundaries in the presence of a sufficient amount of fluid. Such mechanisms are unknown in plastic deformation taking place in an essentially solid state. The conditions for plastic deformation are more commonly met during retrograde metamorphism. Finally, we will principally deal with rocks displaying evidence of extensive flow. Objective signs of flow are recorded by all sorts of strain markers and in the occurrence of strongly anisotropic structures, textures and mineral lattice orientations. The rocks or formations need not show macroscopically interpretable features like folded layers, in which case they are better adapted to other types of structural analyses discussed below.

1.2. OBJECTIVES IN STRUCTURAL ANALYSIS OF METAMORPHIC ROCKS

Structural analysis of metamorphic rocks may have two distinct objectives which are not mutually exclusive. They are not always envisaged, possibly because to a certain extent the specific geological conditions leave no choice in the matter.

1.2.1. The tectonic and metamorphic history objective

Many metamorphic formations have been repeatedly deformed and metamorphosed in relation with one or more orogeneses. In order to trace the orogenic history, it is important to analyse carefully the chronological relations between tectonic and metamorphic events (the latter being characterized by remarkable phase changes) and to be able to date these events.

The deformations in this kind of analysis are typically evidenced by folding, and the structural analysis consists then of a geometric study of superimposed folds and associated planar and linear elements according to methods exposed by Ramsay (1967). The relations with the metamorphic events are determined by examining minor structures and using microscopic criteria on minerals formed during these events. The minerals can grow with preferred orientations or dimensions in the planes or along the lines associated with a given folding. Microscopic criteria (Zwart, 1963; Spry, 1969) make it possible to establish the time relation of the deformation with the mineral growth. Another piece of evidence is provided by the study of deformation and metamorphism printed in dikes injected at different periods of the historical sequence. Some dikes can also be the result of metamorphic mobilization. This method has been illustrated by Wegmann and Schaer (1962) and by Persoz (1967).

The tectonic and metamorphic historical sequence can be dated by radiometric chronology either on minerals grown during a specific metamorphic event or on dikes, injected at a known stage of the sequence but not necessarily related to the metamorphic history. To illustrate the first case one can refer to the alpine history in the western Alps where a sequence of deformation and metamorphic pulses has been largely clarified, using the microscopic criteria evoked above (Ellenberger, 1958; Nicolas, 1969). The two main successive metamorphic events, a blueschist facies metamorphism followed by a greenschist facies one (Ellenberger, 1958; Van Der Plas, 1959; Bearth, 1959) have been dated by Rb–Sr and K–Ar methods on minerals, characteristic of each facies (Steiger, 1964; Bocquet et al., 1974; Hunziker, 1974). The structural and metamorphic history of basement rocks in Northern Scotland illustrates the second case with a chronological sequence largely based on dike dating (Moorbath and Park, 1971; Park and Tarney, 1973).

Generally speaking, this aim of structural analysis has been best served by the structural geology school working in Scotland.

1.2.2. The kinematic and dynamic objectives

In areas where the superimposed deformations are not too complex or where one deformation displays an overwhelming print in rocks, the structures can be analysed in view of their kinematic and dynamic origins. The kinematic analysis, purports to trace the successive internal and external movements which have contributed to the present structural state and position. The notions of internal and external movements, that is movement related to coordinate axes internal or external to the considered object, have often been mixed up resulting in a great deal of confusion, as pointed out by Ramsay (1969). This is why it has been deemed necessary to give a full physical treatment of this problem in Chapter 2. The dynamic analysis consists in determining the successive states of forces responsible for the movements and ultimately the structure and position of the considered object.

Turner and Weiss (1963), regarding kinematics and dynamics as the interpretative side of structural analysis, insist on a progressive degree of uncertainty in these interpretations. The structural properties of the deformed object can be known by an exhaustive geometrical study. This is what we refer to as **structural analysis** ***stricto sensu.*** *Next to the* ***geometrical analysis*** *come the* ***kinematic analysis*** *which relies directly on the preceding study and the* ***dynamic analysis****. The ultimate aim is to define the structure (position?) and physical properties of the initial object. The kinetic analysis, that is rate of deformation, depends on the dynamic and kinematic ones.*

This logical order is compulsory in an interpretative study of natural deformations unless extensive reference to experimental results is possible. For example, Ramsay (1960) has designed a method to infer the flow directions responsible for the development of a similar fold from the study of the deformation of an older lineation oblique to the fold axis. Even assuming that besides the knowledge of the flow directions, the finite strain has been deduced from strain markers, it is obvious that a large number of stress orientations with different magnitudes could be responsible for the case under consideration.

On the other hand, in deformation experiments conducted either on scale models or on a limited amount of natural material, depending on the approach (see next section), the complete sequence is known, starting from the structural state and the physical properties of the initial object onwards. There is then a strong inclination to correlate the deformed state directly with the stress conditions and to operate in the same way for natural cases, disregarding the kinematic history. This can be illustrated by the kind of studies initiated by Turner (1953) and recently reviewed by Carter and Raleigh (1969). In experimentally deformed rocks, it has been observed that certain substructures in minerals, like kink bands, twins, lamellae, tend to appear in crystals with a special orientation in relation to the principal directions of the applied stresses. Conversely, statistics of the orientations of crystals displaying such substructures can allow a reconstruction of the stress field. This has been applied with success to natural examples. If the rock submitted to those stresses has extensively flowed, the crystals in the privileged orientations will have changed their orientations and the results will be biased. A substructure very similar to the kink banding related to compression can also appear during a recovery process accompanying plastic flow; its statistical orientation has only a kinematic inference (see Chapter 10). This proves that this dynamic method is applicable only in the case of a weak deformation.

In a more general manner, it is believed that directly relating the final structure to the stress regime is unreliable when the rock has considerably flowed. The structure reflects then primarily the flow and the dynamic project depends on the kinematic one. Hence, the main interpretation drawn from the description of structures in this book will be of a kinematic nature.

1.3. THE METHODS IN KINEMATIC ANALYSIS

Our attention is now focused on kinematic interpretation in metamorphic rocks with flowage structures. Two distinct methods which are not mutually exclusive complement each other by combining information obtained at different scale levels. The principal one at the present time considers the structure as a continuum, which implies that it operates mainly at the mesoscopic–megascopic levels. It proposes a global kinematic sketch for complex deformed structures. It is akin to the *continuum mechanics* and will be designated accordingly. The other method is analytical and depends on the physical study at the optical and electron microscope levels of the elementary flow mechanisms in crystals and in rocks, using the physical metallurgy concepts and techniques. Here it is called the *physical metallurgy* method.

1.3.1. The continuum mechanics method

This method has been predominantly used until now to obtain kinematic information in deformed rocks and is familiar to structural geologists. There are two variants which have been defined by Sander (1930, 1970) and are summarized by Turner and Weiss (1963) in these words:

'1. The geometric features of a deformed body can be interpreted directly in terms of kinematic concepts on the empirical assumption that the nature of the geometric order of the body reflects the geometric order of the differential displacements, rotations, and strains that must be present during deformation of a real polycrystalline body. These relative motions Sander collectively designates the movement picture of the deformation. It is in the evaluation of the movement picture that symmetry principles are of greatest importance.

2. The observed final state of a deformed body is compared with some assumed initial state, and a path of kinematic development is proposed. But even from the same observations and the same assumptions regarding parent states more than one kinematic reconstruction is possible.'

Considering a final structural state can in no way give direct information on its origin or on the kinematic path which led to it. External information is necessary to reach these ends. This information can be relative to the initial state; for instance in the case of a deformed fossil it is the initial shape of this fossil. We are now dealing with the second approach which is in fact a method of evaluating the finite strain. Unless some new information is produced, the effective kinematic path is usually unknown, as already observed by Turner and Weiss (1963), although detailed observations, for instance pressure fringes (Plate 19b) or evidence from folding of layers followed by boudinage, may cast some light on the effective path (Ramsay, 1967b).

The true kinematic method is the first one (movement picture method). The external information consists of an assumption about the kinematic path and deformation mechanisms leading from a preceding reference state to the observed final state. Both states are suggested by the structure of the deformed body. This means that even if this body is statistically homogeneous (Turner and Weiss, 1963), on a smaller scale it will have a complex geometry and will be heterogeneous, some parts being equated with the reference state. This is necessary since it is the heterogeneity which contains the information about differential strain.

Before proceeding, it may be observed that kinematic interpretation depending on the physical metallurgy approach advocated here, is akin to the movement picture method. Again a deformation mechanism is suggested by the study of the deformed structure and the kinematic interpretation is deduced from applying this mechanism to a known or supposed preceding state. A major difference is that the final structure may be homogeneous on any scale; in fact it is better when it is so.

A rapid survey of folds probably provides the best way to illustrate the future possibilities and limitations in the continuum mechanics method. Folds are heterogeneous structures, highly suggestive of movement. This is probably one reason why they have always fascinated geologists and why the most recent and sophisticated studies have focused on refined analyses of their geometry, on their experimental generation or on theoretical paths for their development. They have brought forth general ideas about their relations to stress, the origin of associated cleavage and foliation and even the flow mechanisms.

The theory of folding of stratified viscoelastic media due to elastic and viscous instabilities has been presented by Biot (1961) and Ramberg (1963) and developed by these authors and co-workers in more recent publications. The finite element method recently exposed by Stephanson and Berner (1971) has allowed further progress (Chapple, 1968; Dieterich and Carter, 1969; Parrish, 1973). Along with experiments on scale models involving deformation of contrasted viscosity layers (Biot et al., 1961; Ramberg, 1963), and geometric analysis of natural folds (Ramsay, 1967), these studies have demonstrated that many folds have been formed in nature by buckling and flattening of competent layers with the maximum principal compressive stress component normal to the axial plane. The cleavage or foliation appearing in the less competent matrix parallel to the axial surface of the folds is then normal to the principal compressive stress. Local deviations observed in natural folds are remarkably reproduced in models, both theoretical (Dieterich, 1969) and experimental (Roberts and Stromgard, 1972). They follow the local stress trajectories and reinforce the general interpretation. Again, repeated observations and the experimental evidence mentioned above, both using strain markers, have contributed to reaching an almost general consensus on the nature of slaty cleavage and foliation. These are the planes normal to the shortest axis of the finite strain ellipsoid and they can in no way be equated with shear planes (see Siddans, 1972, Wood, 1974 and Chapters 7 and 12). The often deduced parallelism between the principal axis of compression and the shortest axis of the strain ellipsoid has encouraged the investigation of flow mechanisms under a direct stress control. In flow in the presence of water, responsible for the slaty cleavage in shales, solution transfer due to pressure solution (Riecke's principle) has been shown to be an adequate mechanism (Von Plessman, 1964; Durney, 1972). However, in solid state flow, in conjunction with the thermodynamic theory of Kamb (1959) and with the first experimental results (see next section), this has accredited the more disputable interpretation of stress controlled syntectonic recrystallization (Hartman and den Tex, 1964; Goguel, 1965; de Vore, 1966) (see § 4.4.).

1.3.2. The physical metallurgy method

In solid state physics, the basic relations between flow and crystal properties were first understood in metals. Studies carried out in the last 15 years on other materials like oxides or alkali halides have shown that the flow structures were very comparable and consequently that the general concepts were still valid for more complex crystallographic lattices. Silicates which are more complex still, have been known for a long time to deform in a similar way to metals. After the First World War, this has inspired the Austrian School, whose most outstanding scientists were B. Sander and W. Schmidt. Goguel (1965) has critically exposed how the considerable effort made to gather data on preferred orientations of minerals and textures in rocks has degenerated into speculative interpretations because of a lack of correct information on flow mechanisms and mechanical properties in the considered minerals. Griggs (1940) was the first to provide such data. Information on calcite, quartz, olivine, pyroxenes and a few other minerals has been produced for a large part at the Institute of Geophysics at the University of California at Los Angeles in the last 15 years. The contributions will be exposed in Chapters 5 and 6.

A remarkable discussion of how the available metallurgical knowledge can be used in structural and kinematic analysis of rocks was presented in 1960 by Voll. It is noticeable that his kinematic analysis relies only on strain markers and interpretation of cleavages and folds, as plastic properties of rock-forming minerals were not yet known. In their pioneering study of quartz, Christie et al. (1964) introduced in the case of a geological material the concept of dislocations along with an appropriate technique for their observation. A physical metallurgist, D. McLean (1965), in a paper significantly entitled 'The Science of metamorphism in metals' pointed out that the same physical mechanisms were responsible for phase transformation, recrystallization or shear deformation in metals and in rocks. Transmission electron microscopy observations in silicates (McLaren and Phakey, 1965; McLaren et al., 1967) and a better understanding of their effective flow mechanisms in aggregates (Nicolas et al., 1971, 1973; Tullis et al., 1973) have carried further the analogy with metallurgical materials. On the basis of this analogy, a detailed account of the structures described by metallurgists and how they contribute to the flow, is thought to open the way to similar interpretations in geological materials. This account is presented in Chapter 3 of this book; contribution to flowage and related processes is the aim of Chapter 4, whereas Chapters 8 and 9 deal with the necessary metallurgical and geological techniques.

The general method (Chapter 7) can be explained by describing the successive pieces of evidence needed to interprete flow in deformed rocks.

(1) Textures and preferred mineral orientations

The study of optical textures is of primary importance. The textural observations and measurements must be made in the remarkable structural planes determined by a structural study of the rock sample. The results of this study act as a guide for the petrofabric (minerals preferred orientation) measurements on the various categories of minerals belonging to the same or different species. These general studies may have to be complemented by a more detailed investigation on the same minerals with the optical or electron microscope.

(2) Physical properties of the rock-forming minerals

In a general study these are mainly the plastic flow properties. They will be presented for the principal minerals in Chapter 5. More specific studies require data on diffusion coefficients, elastic constants, etc.

(3) Flow mechanisms in the rocks

The flow mechanisms operating in the rocks are derived from experiments in a large range of physical conditions on single crystals (Chapter 5) and on monomineralic aggregates (Chapter 6).

(4) Relations between mineral aggregate fabrics and the geometrical nature of the flow

Provided sufficient information has been gathered from points (1), (2) and (3), the problem of deducing the geometry of the flow from textures and preferred orientations is largely that of a retrodiction from the knowledge accumulated on experimental and computed deformation of aggregates (Chapter 6). Deformations from a small scale (Chapter 10) to a large scale (Chapter 11) in peridotites will illustrate the method.

1.3.3. Relations between the continuum mechanics and the physical metallurgy methods

To conclude this presentation of the continuum mechanics and the physical metallurgy methods, it may be interesting to consider their aims from a methodological point of view. Their respective positions may then appear similar to those of thermodynamics compared with statistical mechanics, in the sense that the mechanical approach is phenomenological whereas physical metallurgy has the ambition—admittedly not yet entirely fulfilled—to start from microscopic first principles and work its way up to explain the macroscopic reality. The example of how the phenomenon of flow is treated by both approaches may help to make our point clear.

(i) Continuum mechanics takes a phenomenological approach, that is, its aim will be to describe the global *response* (displacement, strain, strain rate) of a body or structure in terms of the macroscopic *solicitations* (external forces, stresses, etc.) and of *material constants* (elastic constants, viscosity). The body is considered as a homogeneous continuum and the material constants are global averages of the local constants. (The method therefore clearly applies only at mesoscopic and megascopic levels when the local inhomogeneities of structure and composition of the body are smeared out.)

Mechanics makes use of previously established constitutive relations between response and solicitations, of the type:

$$f\,(\text{stress, strain, strain rate, etc.}) = 0$$

It is important to note that these relations are established and can be used, without any reference to the physical mechanism by which deformation or flowage takes place.

For instance, a constitutive relation of the type

$$\text{Flow rate} \propto (\text{shear stress})^n$$

which characterizes a viscous flow, can be applied to the flow of a hot lava bed, the creep of a glacier, the high temperature deformation of marble in a laboratory experiment or the creep of the asthenosphere.

The viscosity for instance, in the case where the flow is Newtonian ($n = 1$), will be defined simply as the ratio of shear stress over flow rate or $\eta = \sigma/\dot{\varepsilon}$. It can be measured and tabulated but nothing more.

(ii) Physical metallurgy, as is evidenced by its name, has a physical approach and starts from the microscopic end of the problem. Its purpose, admittedly ambitious, is to derive the constitutive relations from first principles, starting with the knowledge of the deformation processes taking place in the physically meaningful elementary units: the single crystalline grains constitutive of the solid. This approach starts with the study of the lattice defects (vacancies, dislocations, etc.) responsible for the deformation of single crystals and makes use of physical models relating the deformation or the flow rate to the applied stress and physical parameters such as pressure and temperature. These models are embodied in microscopic constitutive relations where the material constants can be expressed in terms of physical constants and microstructural parameters.

For instance, for the Newtonian viscous flow occurring in solids under certain conditions (Nabarro–Herring creep) the viscosity

$$\eta = \sigma/\dot{\varepsilon} = K\,Td^2/D\Omega$$

is expressed as a function of the self-diffusion coefficient D, the atomic volume Ω, the temperature T and the grain size d.

The next step consists of expressing the behaviour of a macroscopic aggregate in terms of the deformation of its grains. Although the deformation of single crystals is now, by and large, rather well understood, it is still difficult to bridge the gap between the deformation of a grain and the deformation of a macroscopic aggregate; however, progress is being made, especially for high temperature deformation.

Finally, it can be said that the interest of the physical metallurgy approach is twofold:

(1) If a rock is formed of minerals for whose modes of deformation physical models exist, and if the physical conditions are known (P, T, applied stress, etc.), it is possible to predict reasonably well the behaviour of the material within certain limits.

(2) If a rock has been deformed in conditions partly unknown, the physical examination of the microstructure allows a certain amount of retrodiction, at least for the last stages of deformation, whenever the microstructural imprint of the deformation has not been wiped out. Thus, the physical study of deformation modes has enabled to ascribe to certain microstructural features the role of a signature of the geometry of the flow and of temperature or strain rate ranges (this is true for instance of the dislocations or subgrain boundaries arrangement in a crystal). In certain limited cases an estimate of the applied stress can even be ventured.

CHAPTER 2

Elements of Solid Mechanics

2.1. GENERAL CONSIDERATIONS

As recently pointed out by Ramsay (1969), kinematic analysis in naturally deformed structures has been obscured by a confusion between the notions of movement (or motion) and flow in kinematic language and the notions of displacement and strain (or deformation) in geometrical language. A clear distinction must be drawn between deformation internal to the considered structure, due to internal flowage, and displacement relative to external coordinates, resulting from bodily movement (or motion). In this regard, it must be clear that the total displacement is never directly recorded in the deformed structure and can be known only through external information. Assuming that the direction of displacement is known, the motion of the deformed structure can have occurred:

(i) With an internal flowage directly participating in the motion. For example, at the base of a glacier moving downwards by its own flowage.

(ii) Without any internal flowage. The body is translated passively and records nothing. For example, the surface of a glacier in a plane strain area.

(iii) With an internal flowage indirectly connected with the motion. For example, the surface of a glacier in a non-plane strain area where there is lateral extension or contraction.

The general motion can be extrapolated from the internal flowage only in case (i), if the deformation throughout a large body is shown to be homogeneous.

This calls for correct definitions in strain analysis. Moreover, infinitesimal and finite strain regimes need to be distinguished. The first part of this chapter gives a detailed account of infinitesimal strain analysis and has been written in a way that should be understood by anyone. A more geometrical approach in finite strain analysis is developed at the end of the chapter.

2.2. DEFINITIONS: DISPLACEMENT, DISTORTION, STRAIN, ROTATION

Let us consider a small volume element of the undeformed solid and let us identify its points by their coordinates $x_i (i = 1, 2, 3)$ in a system of rectilinear orthogonal axes. The distance $\mathrm{d}s$ between two neighbouring points $P(x_i)$ and $P'(x_i + \mathrm{d}x_i)$ is given by:

$$\mathrm{d}s^2 = (\mathrm{d}x_1)^2 + (\mathrm{d}x_2)^2 + (\mathrm{d}x_3)^2$$

or

$$\mathrm{d}s^2 = \sum_i (\mathrm{d}x_i)^2 \qquad \text{(Fig. 2.1(a))}$$

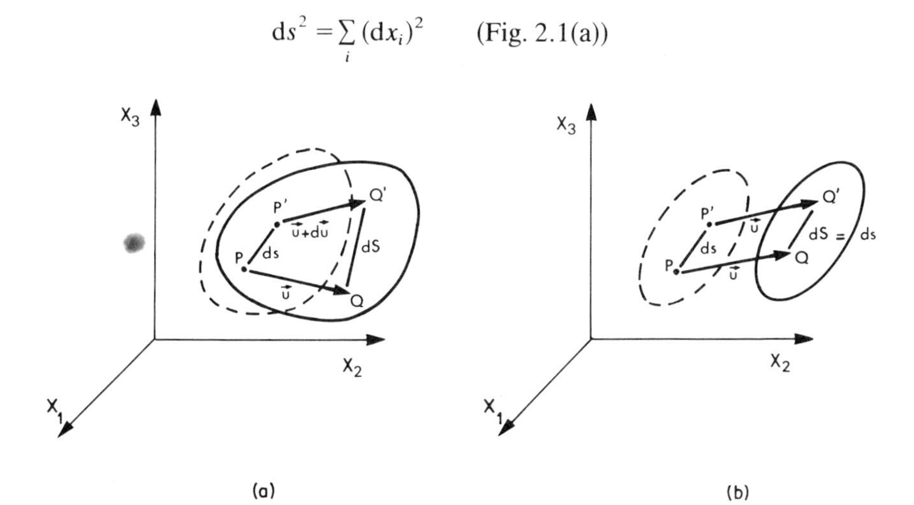

Fig. 2.1. (a) Deformation: two neighbouring points, P, P′ distant by ds are displaced by **u** and **u**+d**u** and come to Q and Q′ distant by d$S \neq$ ds; (b) rigid body translation: the distance between Q and Q′ is equal to the distance between P and P′

If the points corresponding to P and P' in the deformed solid, are Q and Q', their coordinates in the same axes will be respectively:

$$X_i = x_i + u_i$$

$$X_i + \mathrm{d}X_i = x_i + \mathrm{d}x_i + u_i + \mathrm{d}u_i$$

Their distance $\mathrm{d}S$ will be given by:

$$\mathrm{d}S^2 = \sum_i (\mathrm{d}X_i)^2$$

The vector $\mathbf{PQ} = \mathbf{u}$ is called the *displacement vector* at point P and its components u_i generally depend on the coordinates x_i of P.

(1) Let us first consider the special case where $\mathbf{u}$ is independent of x_i: all points will then be displaced by the same amount (i.e. $\mathrm{d}u_i \equiv 0$) and the distance

between neighbouring points will not change. The deformation is therefore reduced to a *rigid-body translation* (Fig. 2.1(b)).

(2) In the general case, we have $\mathbf{u}(x_i)$: the vector **PP′** linking two neighbouring points in the undeformed solid becomes **QQ′** after deformation. We can see (Fig. 2.1(a)) that its length has changed ($\mathrm{d}S \neq \mathrm{d}s$) as well as its orientation in the axes.

The local changes in position, orientation, dimensions and shape of a volume element by deformation will therefore be related to the variation of the displacement vector **u** as a function of the coordinates of the material points.

Let us then express d**u** as a function of the coordinates x_i.

Every component $\mathrm{d}u_i$ generally depends on all components $\mathrm{d}x_i$. We may write:

$$\begin{cases} \mathrm{d}u_1 = \dfrac{\partial u_1}{\partial x_1}\mathrm{d}x_1 + \dfrac{\partial u_1}{\partial x_2}\mathrm{d}x_2 + \dfrac{\partial u_1}{\partial x_3}\mathrm{d}x_3 \\ \mathrm{d}u_2 = \dfrac{\partial u_2}{\partial x_1}\mathrm{d}x_1 + \dfrac{\partial u_2}{\partial x_2}\mathrm{d}x_2 + \dfrac{\partial u_2}{\partial x_3}\mathrm{d}x_3 \\ \mathrm{d}u_3 = \dfrac{\partial u_3}{\partial x_1}\mathrm{d}x_1 + \dfrac{\partial u_3}{\partial x_2}\mathrm{d}x_2 + \dfrac{\partial u_3}{\partial x_3}\mathrm{d}x_3 \end{cases}$$

Or, in condensed notation,

$$\mathrm{d}u_i = \sum_j \frac{\partial u_i}{\partial x_j}\mathrm{d}x_j \qquad \begin{matrix} i = 1, 2, 3 \\ j = 1, 2, 3 \end{matrix} \tag{2.1}$$

the nine quantities $\partial u_i/\partial x_j$ may be written in a square array. They are called the components of the local *distortion* tensor.

$$\begin{pmatrix} \dfrac{\partial u_1}{\partial x_1} & \dfrac{\partial u_1}{\partial x_2} & \dfrac{\partial u_1}{\partial x_3} \\ \dfrac{\partial u_2}{\partial x_1} & \dfrac{\partial u_2}{\partial x_2} & \dfrac{\partial u_2}{\partial x_3} \\ \dfrac{\partial u_3}{\partial x_1} & \dfrac{\partial u_3}{\partial x_2} & \dfrac{\partial u_3}{\partial x_3} \end{pmatrix}$$

In everything that follows, we will assume that the $\partial u_i/\partial x_j$ components are infinitely small (i.e. **u** varies very slowly with x_i).

For the sake of simplicity we will analyse the physical significance of every component of the distortion tensor in the case of a two-dimensional deformation. The results will be easily generalized to the 3-D case.

Let us consider (Fig. 2.2) the deformation of an infinitesimal rectangle defined by three of its corners P, P', P'', whose coordinates in axes parallel to the edges are:

$$\begin{cases} P(0, 0) \\ P'(\mathrm{d}x_1, 0) \\ P''(0, \mathrm{d}x_2) \end{cases}$$

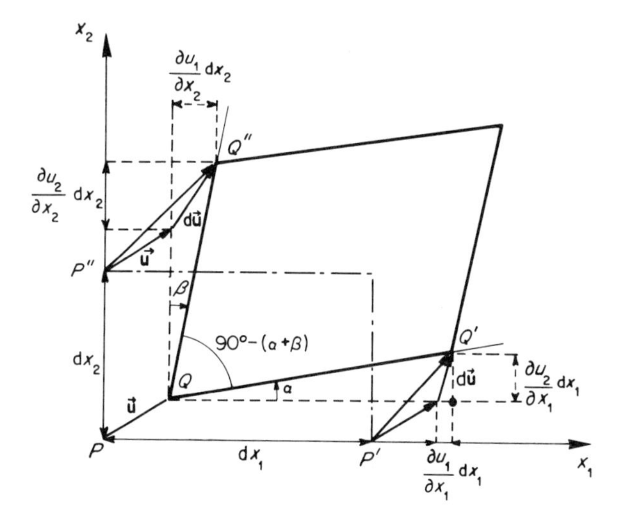

Fig. 2.2. Deformation of an infinitesimal rectangle: before deformation PP′P″ (broken line); after deformation QQ′Q″ (solid line). The length of the edges have changed (stretch) and the straight angle at P has changed to $90° - \gamma$, where $\gamma = \alpha + \beta$ is the angle of shear

The deformation brings P, P', P'' respectively to the positions Q, Q', Q'', whose coordinates are:

$$Q\begin{cases} u_1 \\ u_2 \end{cases}$$

$$Q'\begin{cases} \mathrm{d}x_1 + u_1 + \dfrac{\partial u_1}{\partial x_1}\mathrm{d}x_1 \\ u_2 + \dfrac{\partial u_2}{\partial x_1}\mathrm{d}x_1 \end{cases}$$

$$Q''\begin{cases} u_1 + \dfrac{\partial u_1}{\partial x_2}\mathrm{d}x_2 \\ \mathrm{d}x_2 + u_2 + \dfrac{\partial u_2}{\partial x_2}\mathrm{d}x_2 \end{cases}$$

The segment PP' whose initial length was dx_1, becomes QQ'. To the first order the length of QQ' is $dx_1[1+(\partial u_1/\partial x_1)]$, it follows that the quantity

$$\frac{\partial u_1}{\partial x_1} \simeq \frac{QQ' - PP'}{PP'}$$

is the *unit extension* (or stretch) undergone by a segment parallel to the x_1 axis. Similarly,

$$\frac{\partial u_2}{\partial x_2} \simeq \frac{QQ'' - PP''}{PP''}$$

is the unit extension undergone by a segment parallel to the x_2 axis.

The angle α that QQ' makes with the x_1 axis is given by

$$\alpha \simeq \tan \alpha \simeq \frac{\partial u_2}{\partial x_1}$$

In the undeformed solid, PP' was parallel to the x_1 axis. The quantity $\alpha \simeq \partial u_2/\partial x_1$ is therefore the angle through which the deformation rotates a segment initially parallel to the x_1 axis. Similarly, $\beta \simeq \partial u_1/\partial x_2$ is the angle through which the deformation rotates a segment initially parallel to the x_2 axis.

The initially right angle (PP', PP'') has therefore changed to $(QQ', QQ'') = 90° - (\alpha + \beta) = 90° - \gamma$.

The quantity

$$\frac{\partial u_2}{\partial x_1} + \frac{\partial u_1}{\partial x_2} = \gamma$$

is the angle by which a right angle is reduced.

The deformation of the rectangle into a parallelogram can therefore be decomposed into several stages (Fig. 2.3).

(a) The edges parallel to the x_1 and x_2 axes are respectively elongated by:

$$\frac{\Delta L_1}{L_1} = \frac{\partial u_1}{\partial x_1} \quad \text{and} \quad \frac{\Delta L_2}{L_2} = \frac{\partial u_2}{\partial x_2}$$

The shape is not altered but the dimensions are changed; the relative change in area of the rectangle is given to first order by

$$\frac{\Delta A}{A} \simeq \frac{\partial u_1}{\partial x_1} + \frac{\partial u_2}{\partial x_2}$$

(b) The rectangle is now *sheared* parallel to each axis by an angle

$$\frac{\alpha + \beta}{2} = \frac{1}{2}\left(\frac{\partial u_1}{\partial x_2} + \frac{\partial u_2}{\partial x_1}\right)$$

This *shear strain* transforms the rectangle into a parallelogram with the correct value of the angle at Q: $90° - (\alpha + \beta)$. But the angles the edges form with the axis are $(\alpha + \beta)/2$ instead of α and β.

(c) The correct orientation of the deformed parallelogram with respect to the axes, is given by a *rigid body rotation* through an angle $\omega = (\alpha - \beta)/2$ about the axis perpendicular at Q to the x_1x_2 plane.

We can now generalize this result to three dimensions and we arrive at the following conclusions:

The distortion tensor $\partial u_i/\partial x_j$ can be decomposed into a symmetrical and a skew-symmetrical part

$$\begin{pmatrix} \dfrac{\partial u_1}{\partial x_1} & \dfrac{\partial u_1}{\partial x_2} & \dfrac{\partial u_1}{\partial x_3} \\[2ex] \dfrac{\partial u_2}{\partial x_1} & \dfrac{\partial u_2}{\partial x_2} & \dfrac{\partial u_2}{\partial x_3} \\[2ex] \dfrac{\partial u_3}{\partial x_1} & \dfrac{\partial u_3}{\partial x_2} & \dfrac{\partial u_3}{\partial x_3} \end{pmatrix} \equiv \begin{pmatrix} \dfrac{\partial u_1}{\partial x_1} & \dfrac{1}{2}\left(\dfrac{\partial u_1}{\partial x_2}+\dfrac{\partial u_2}{\partial x_1}\right) & \dfrac{1}{2}\left(\dfrac{\partial u_1}{\partial x_3}+\dfrac{\partial u_3}{\partial x_1}\right) \\[2ex] \dfrac{1}{2}\left(\dfrac{\partial u_2}{\partial x_1}+\dfrac{\partial u_1}{\partial x_2}\right) & \dfrac{\partial u_2}{\partial x_2} & \dfrac{1}{2}\left(\dfrac{\partial u_2}{\partial x_3}+\dfrac{\partial u_3}{\partial x_2}\right) \\[2ex] \dfrac{1}{2}\left(\dfrac{\partial u_3}{\partial x_1}+\dfrac{\partial u_1}{\partial x_3}\right) & \dfrac{1}{2}\left(\dfrac{\partial u_3}{\partial x_2}+\dfrac{\partial u_2}{\partial x_3}\right) & \dfrac{\partial u_3}{\partial x_3} \end{pmatrix}$$

$$+ \begin{pmatrix} 0 & \dfrac{1}{2}\left(\dfrac{\partial u_1}{\partial x_2}-\dfrac{\partial u_2}{\partial x_1}\right) & \dfrac{1}{2}\left(\dfrac{\partial u_1}{\partial x_3}-\dfrac{\partial u_3}{\partial x_1}\right) \\[2ex] \dfrac{1}{2}\left(\dfrac{\partial u_2}{\partial x_1}-\dfrac{\partial u_1}{\partial x_2}\right) & 0 & \dfrac{1}{2}\left(\dfrac{\partial u_2}{\partial x_3}-\dfrac{\partial u_3}{\partial x_2}\right) \\[2ex] \dfrac{1}{2}\left(\dfrac{\partial u_3}{\partial x_1}-\dfrac{\partial u_1}{\partial x_3}\right) & \dfrac{1}{2}\left(\dfrac{\partial u_3}{\partial x_2}-\dfrac{\partial u_2}{\partial x_3}\right) & 0 \end{pmatrix}$$

or

$$\frac{\partial u_i}{\partial x_j} = \frac{1}{2}\left(\frac{\partial u_i}{\partial x_j}+\frac{\partial u_j}{\partial x_i}\right) + \frac{1}{2}\left(\frac{\partial u_i}{\partial x_j}-\frac{\partial u_j}{\partial x_i}\right)$$

or

$$\frac{\partial u_i}{\partial x_j} = \varepsilon_{ij} + \omega_{ij} \tag{2.2}$$

The quantities

$$\varepsilon_{ij} = \frac{1}{2}\left(\frac{\partial u_i}{\partial x_j}+\frac{\partial u_j}{\partial x_i}\right)$$

can be written in a square array

$$\begin{pmatrix} \varepsilon_{11} & \varepsilon_{12} & \varepsilon_{13} \\ \varepsilon_{12} & \varepsilon_{22} & \varepsilon_{23} \\ \varepsilon_{13} & \varepsilon_{23} & \varepsilon_{33} \end{pmatrix}$$

they are the components of the *strain* tensor. Since $\varepsilon_{ij} = \varepsilon_{ji}$, the number of non-identical components is reduced to 6. At a given point, in the solid, the 6

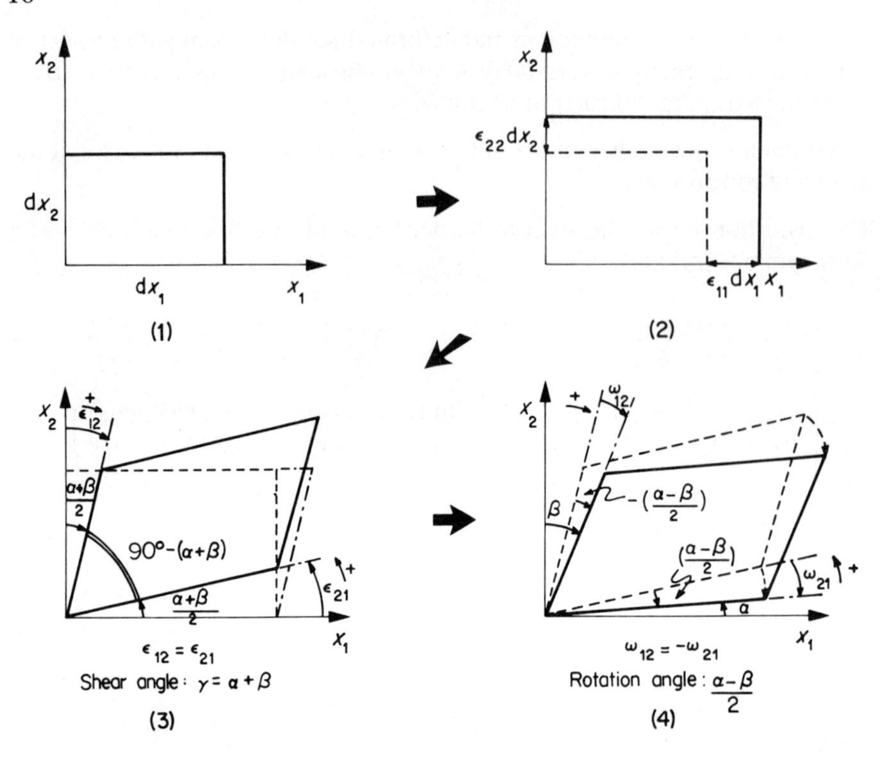

Fig. 2.3. Decomposition in successive stages of the deformation of an infinitesimal rectangle into a parallelogram. (1) initial undeformed state; (2) stretching of the edges without change of shape; (3) shear through an angle $\gamma = \alpha + \beta$; (4) rigid body rotation through an angle $\frac{1}{2}(\alpha - \beta)$

strain components contain all the information relative to the changes in dimensions and shape of a small volume element centred on this point.

A diagonal component ε_{ii} represents the *unit extension* (*tensile strain*) of a segment parallel to the axis x_i. An off-diagonal component $\varepsilon_{ij} (i \neq j)$ represents half the *angle of shear* γ by which a rectangle in the plane $x_i x_j$ is transformed into a parallelogram. The shear strains are responsible for the change in shape at constant volume of the initial volume element.

The volume change, or *dilatation*, is given by

$$\frac{\Delta V}{V} \simeq \sum_i \varepsilon_{ii} = \varepsilon_{11} + \varepsilon_{22} + \varepsilon_{33} \tag{2.3}$$

It is always possible to find a system of axes in which the strain tensor is diagonal (i.e. the ε_{ij} are equal to 0 if $i \neq j$). The directions of the axes are then called the principal directions and the corresponding strains the *principal strains*.

The quantities

$$\omega_{ij} = \frac{1}{2}\left(\frac{\partial u_i}{\partial x_j} - \frac{\partial u_j}{\partial x_i}\right)$$

are the components of the *rotation tensor* expressing the rigid-body rotation of the small volume element about the axis of the vector.

Uniform (homogeneous) strain. If all the $\partial u_i/\partial x_i$ quantities are constant throughout the volume (i.e. independent of the values of x_i), the components of the displacement vector $\mathbf{u}$ depend linearly on the coordinates x_i of the points. The components ε_{ij} of the strain tensor are the same at every point: the *strain is uniform.*

Rotational, irrotational strain. In a regime of uniform strain, if the components of the rotation tensor are all equal to zero at every point, the *strain is irrotational.* If this condition is not met, the strain is *rotational.*

For a uniform irrotational strain regime we have, therefore:

$$\begin{cases} \varepsilon_{ij} \text{ independent of } x_i \\ \omega_{ij} = 0 \end{cases}$$

Plane strain

Plane strain is a case of a simple regime of strain. All the displacements vectors $\mathbf{u}$ are parallel to a plane and therefore do not depend on the coordinate perpendicular to this plane (e.g., x_3):

We have

$$\begin{cases} u_3 = 0 \\ u_2, u_1 \text{ independent of } x_3 \end{cases}$$

this leads to:

$$\varepsilon_{13} = \varepsilon_{23} = \varepsilon_{33} = 0$$

the strains are two-dimensional.

Pure shear–simple shear

Two uniform plane strain regimes are of special interest:

(*a*) *Pure shear*

$$\frac{\partial u_1}{\partial x_2} = \frac{\partial u_2}{\partial x_1}$$

the rotation is equal to zero everywhere (irrotational strain) (Fig. 2.4(a))

$$\begin{cases} \varepsilon_{12} = \varepsilon_{21} = \dfrac{\gamma}{2} \\ \omega_{12} = -\omega_{21} = 0 \end{cases}$$

(*b*) *Simple shear*

$$\frac{\partial u_1}{\partial x_2} = \gamma \qquad \frac{\partial u_2}{\partial x_1} = 0$$

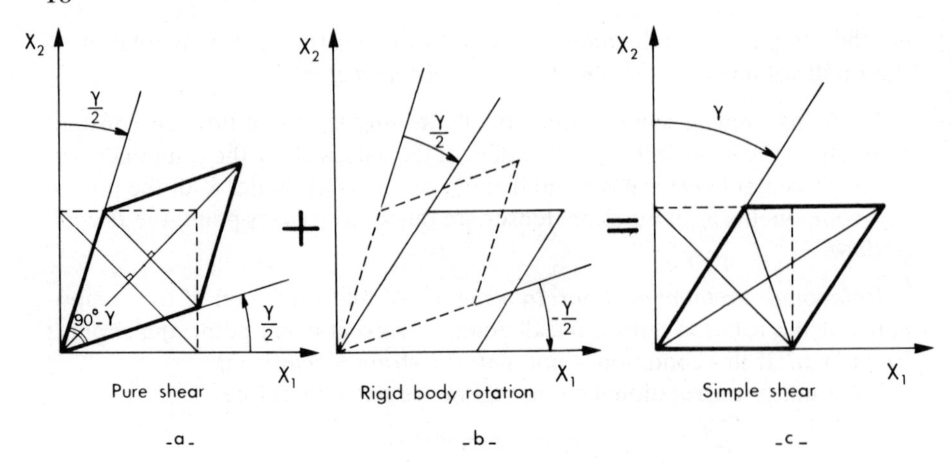

Fig. 2.4. Decomposition of simple shear through an angle γ into pure shear through an angle γ + rigid body rotation through an angle $\gamma/2$. (a) pure shear; (b) rigid body rotation; (c) simple shear

We have, therefore, (γ being the angle of shear) (Fig. 2.4(c))

$$\begin{cases} \varepsilon_{12} = \varepsilon_{21} = \dfrac{\gamma}{2} \\ \omega_{12} = \dfrac{\gamma}{2} \\ \omega_{21} = -\dfrac{\gamma}{2} \end{cases}$$

Every infinitesimal parallelogram undergoes a shear of angle γ and a rigid body rotation through an angle $\gamma/2$ (rotational strain).

Hydrostatic and deviatoric strain

It is possible to divide the strain tensor into two parts: the hydrostatic part, responsible for the *dilatation* or *compression* (isotropic volume change, positive or negative) and the deviatoric part responsible for the change in shape.

We have seen that the total dilatation is given by:

$$\frac{\Delta V}{V} \simeq \varepsilon_{11} + \varepsilon_{22} + \varepsilon_{33}$$

This dilatation can be achieved with a strain tensor, each diagonal term of which is equal to $\frac{1}{3}(\varepsilon_{11} + \varepsilon_{22} + \varepsilon_{33})$; the sum of its 3 diagonal terms will be equal to $\Delta V/V$.

The deviatoric part will be given as the difference between the strain tensor and its hydrostatic part; the sum of its diagonal term will then be equal to zero: the deviatoric strain is a strain at constant volume.

We have:

$$\begin{pmatrix} \varepsilon_{11}\varepsilon_{12}\varepsilon_{13} \\ \varepsilon_{12}\varepsilon_{22}\varepsilon_{23} \\ \varepsilon_{13}\varepsilon_{23}\varepsilon_{33} \end{pmatrix} \equiv \begin{pmatrix} \frac{1}{3}(\varepsilon_{11}+\varepsilon_{22}+\varepsilon_{33}) & 0 & 0 \\ 0 & \frac{1}{3}(\varepsilon_{11}+\varepsilon_{22}+\varepsilon_{33}) & 0 \\ 0 & 0 & \frac{1}{3}(\varepsilon_{11}+\varepsilon_{22}+\varepsilon_{33}) \end{pmatrix}$$

$$+ \begin{pmatrix} \varepsilon_{11}-\frac{1}{3}(\varepsilon_{11}+\varepsilon_{22}+\varepsilon_{33}) & \varepsilon_{12} & \varepsilon_{13} \\ \varepsilon_{12} & \varepsilon_{22}-\frac{1}{3}(\varepsilon_{11}+\varepsilon_{22}+\varepsilon_{33}) & \varepsilon_{23} \\ \varepsilon_{13} & \varepsilon_{13} & \varepsilon_{33}-\frac{1}{3}(\varepsilon_{11}+\varepsilon_{22}+\varepsilon_{33}) \end{pmatrix}$$

Strain ≡ Hydrostatic strain + Deviatoric strain
(change in volume) (change in shape)

Finite strains

Up to now we have assumed that the $\partial u_i/\partial x_j$ components were infinitely small, that is to say that we have considered only infinitely small strains. This approximation enabled us to define the strain by 6 independent components ε_{ij}. In other words, for an infinitely small strain, we can define the state of strain of a volume by the 6 ε_{ij}, and this state of strain does not depend on the order in which we have imposed the ε_{ij} on the solid, since they do not depend on each other. This is no longer true when the strain is great. However, for simple strain regimes, relevant components of strain can be defined but the difference between these components and the infinitesimal strain components must always be borne in mind, as can be seen in the following two examples of large plane strain.

(*a*) *Finite simple shear*

For infinitesimal shear, we had defined $\gamma = 2\varepsilon_{12} = \partial u_1/\partial x_2$. By assimilating the tangent to the angle we had taken γ to be the angular shear. This is no longer possible for a large simple shear.

We may now define a finite shear strain $\gamma = PP'/OP$ and an angular shear θ with

$$\gamma = \tan \theta \tag{2.4}$$

Note that if we take the deformation of a circle into an ellipse (Fig. 2.5(a)) as a strain marker, the angular shear θ is different from the complement of the angle α between the direction of maximum elongation of the ellipse and the shear direction. It can be shown (Nicolas, Boudier, and Boullier, 1973) that:

$$\cot 2\alpha = \tfrac{1}{2} \tan \theta \tag{2.5}$$

The curve θ vs α is shown in Fig. 2.5(b).

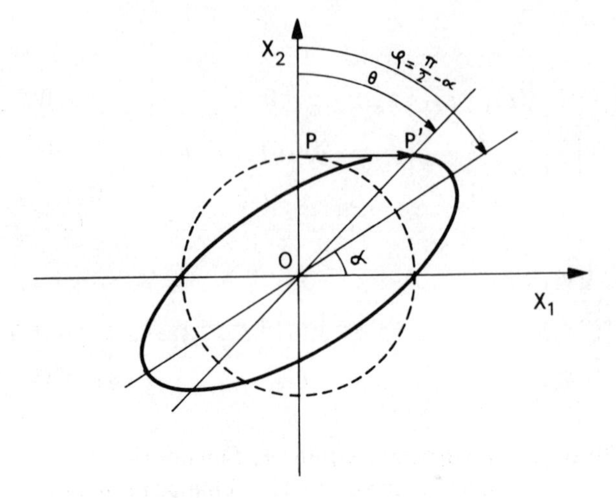

Fig. 2.5a. Deformation of a circle into an ellipse by finite simple shear. Finite shear strain $\gamma = PP'/OP$, θ: angular shear, α: angle between the direction of maximum elongation of the ellipse and the shear direction. $\varphi = \pi/2 - \alpha$. (Reproduced by permission of The American Journal of Science, from *Amer. J. Sci.*, **273**, 853, Figs 4 and 5 (1973))

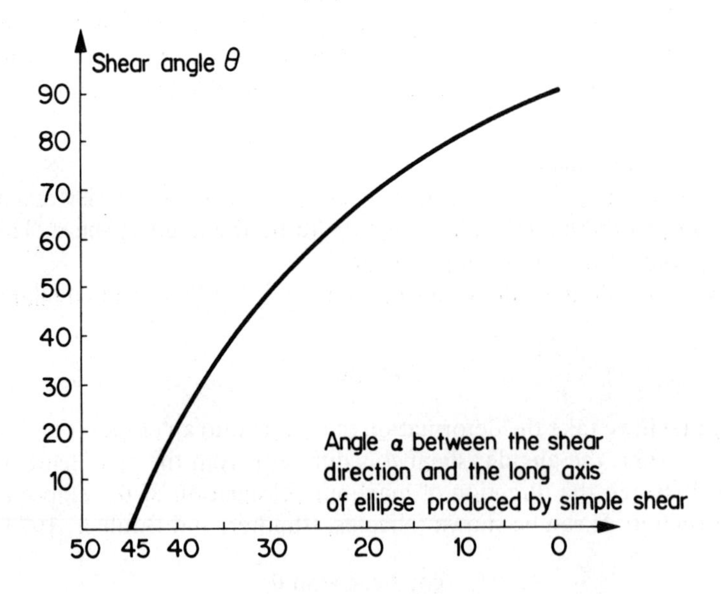

Fig. 2.5b. Relationship between angular shear θ and angle α for finite simple shear

Finite simple shear *cannot* be achieved by finite pure shear followed by finite rigid body rotation (Fig. 2.6).

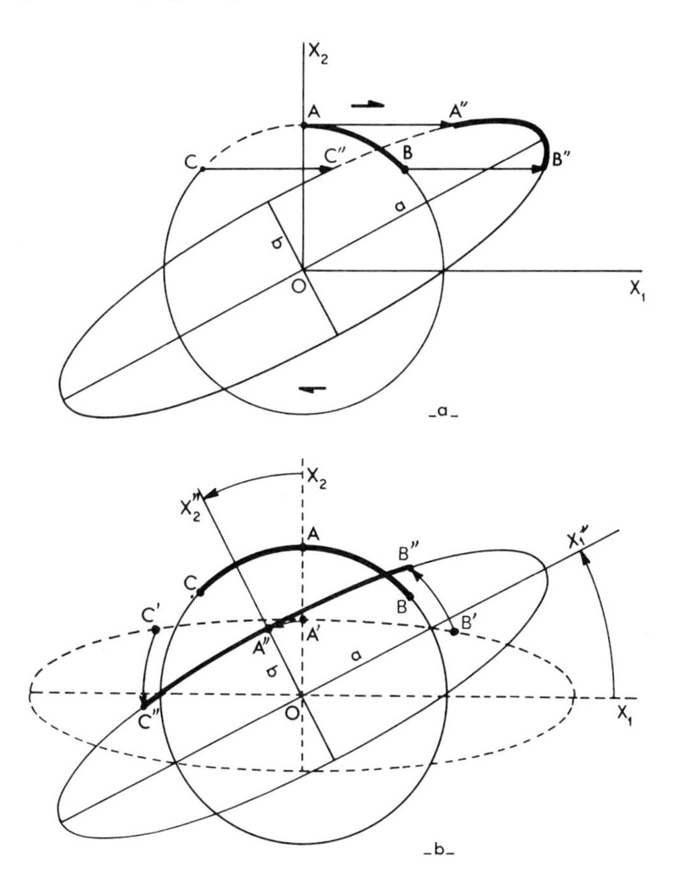

Fig. 2.6. Illustration of the non-equivalence of finite simple shear (a) and finite pure shear + appropriate rotation (b). The distortion of AB = AC lines on the initial circle is different in (a): A″B″ ≠ A″C″ and in (b): A″B″ = A″C″

(b) Finite uniaxial flattening (or extension) at constant volume

Let us again take the deformation of a circle as a strain marker. A uniaxial flattening along Ox_2 axis is equivalent to a shear in the axes $Ox'_1x'_2$ at 45° with respect to Ox_1x_2. The initial circle therefore deforms into an ellipse (Fig. 2.7). In the Ox_1x_2 axes, the finite strain can be defined by:

$$\varepsilon_{22} = \frac{PP'}{OP} = -\varepsilon_{11} = \frac{QQ'}{OQ}$$

since the deformation is at constant volume we have $\varepsilon_{11} + \varepsilon_{22} = 0$.

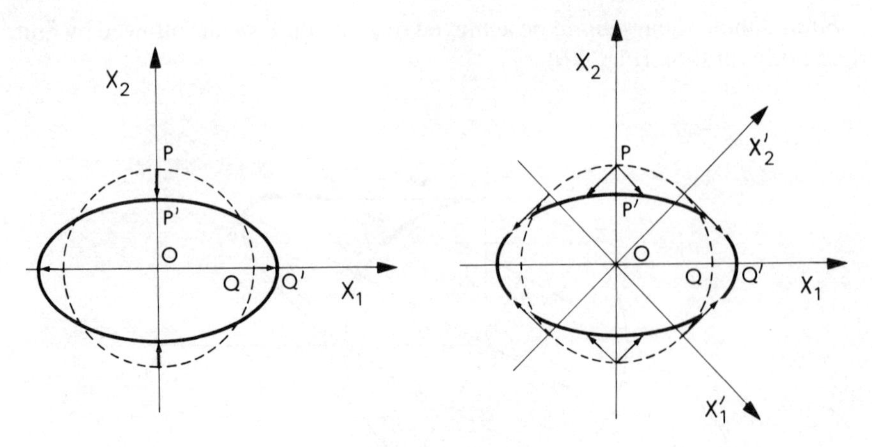

Fig. 2.7. Deformation of a circle into ellipse by uniaxial compression equivalent to finite pure shear

If the circle is drawn on a sample of initial length l_0, the uniform finite strain is given by:

$$\varepsilon_F = \frac{l_1 - l_0}{l_0}$$

($\varepsilon_F < 0$ for flattening; l_1 = length of the deformed sample).

Although ε_F can be immediately measured if l_0 and l_1 are known, it is not representative of the final state of strain.

This results from the fact that the true compression strain at any given time during the deformation is:

$$\varepsilon = \frac{dl}{l}$$

where $l < l_0$ is the *actual* length of the sample at this time. The *true strain* ε in the final state can therefore be considered as the sum of infinitesimal strains, the length of the specimen changing continuously:

$$\varepsilon = \int_{l_0}^{l_1} \frac{dl}{l} = \ln \frac{l_1}{l_0} = \ln(1 + \varepsilon_F)$$

It must be emphasized again that neither the finite strain, ε_F, nor the true strain ε for the final state can give any information about the deformation path followed by the specimen.

2.3. STRESSES

Let us consider a continuous solid in equilibrium under the action of applied forces.

We may divide these forces into two categories:

(a) Body forces: $\mathbf{\Phi}$ per unit volume, e.g. gravity.

(b) Surface forces applied to the external surface of the solid: **F** per unit area.

The notion of stress can be understood by means of the following thought experiment:

(1) Let us define inside the solid a volume of matter (V) bounded by a surface (S).

(2) Let us isolate the volume (V) by carving it out of the rest of the solid. If we want (V) to remain in equilibrium we will have to apply forces to (S). These forces distributed over (S) represent the action that the surrounding solid exerted upon (V) before the physical separation.

Let us focus our attention on an elementary area $\mathrm{d}A$ of (S) centred on a point O, and let the normal to (S) at O be oriented positively outward. The action that the surrounding part of the solid exerts on (V) at O, can be reduced to a force $\mathbf{T}\,\mathrm{dA}$ proportional to dA.

The *stress vector* **T** relative to an elementary area centred on O is defined as the force per unit area exerted by the part of the solid situated on the side of the positive normal, upon the part of the solid situated on the side of the negative normal.

Let us now consider an elementary tetrahedron OABC, of volume $\mathrm{d}V$, defined in the following way: O is the origin of the coordinates and the three orthogonal edges OA, OB, OC lie along the coordinate axes (Fig. 2.8).

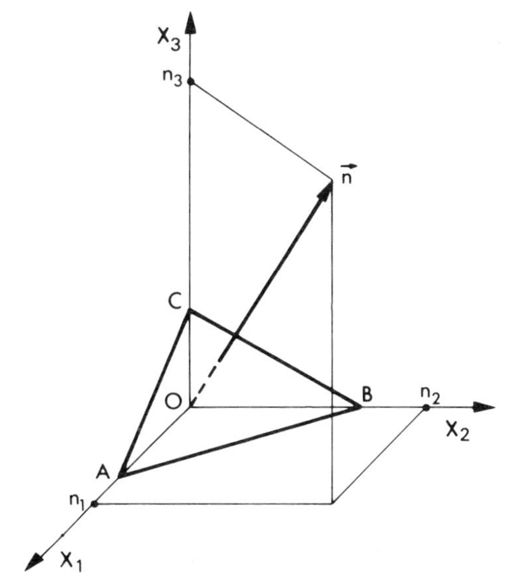

Fig. 2.8. Elementary tetrahedron OABC; n_1, n_2, n_3: projections along the axes of the vector **n** normal to the ABC face

If dA is the area of face ABC, the areas of the faces perpendicular to axes Ox_1, Ox_2, Ox_3, are respectively:

$$dA_1 = n_1\, dA$$
$$dA_2 = n_2\, dA$$
$$dA_3 = n_3\, dA$$

Where the n_i's ($i = 1, 2, 3$) are the components of $\mathbf{n}$, unit vector normal to face ABC.

Let us designate by $\mathbf{T}$ the stress vector relative to face ABC and by $\mathbf{T}'$, $\mathbf{T}''$, $\mathbf{T}'''$ the stress vectors relative to the faces normal respectively to Ox_1, Ox_2, Ox_3.

The tetrahedron, cut out of the solid, will be in equilibrium under the action of the surface forces:

$$+\mathbf{T}\, dA$$
$$-\mathbf{T}'\, dA_1,\ -\mathbf{T}''\, dA_2,\ -\mathbf{T}'''\, dA_3$$

and the volume force:

$$+\mathbf{\Phi}\, dV$$

The equilibrium condition is that the sum of the forces be equal to zero:

$$\mathbf{T}\, dA - \mathbf{T}' n_1\, dA - \mathbf{T}'' n_2\, dA - \mathbf{T}''' n_3\, dA + \mathbf{\Phi}\, dV = 0$$

or

$$\mathbf{T} - \mathbf{T}' n_1 - \mathbf{T}'' n_2 - \mathbf{T}''' n_3 + \mathbf{\Phi} \frac{dV}{dA} = 0$$

If the length of the tetrahedron edges is made to tend to zero, dV/dA tends to zero and the last term of the left-hand member drops out. Now if the components of the stress vectors are:

$$\mathbf{T}\begin{cases} T_1 \\ T_2 \\ T_3 \end{cases} \quad \mathbf{T}'\begin{cases} \sigma_{11} \\ \sigma_{21} \\ \sigma_{31} \end{cases} \quad \mathbf{T}''\begin{cases} \sigma_{12} \\ \sigma_{22} \\ \sigma_{32} \end{cases} \quad \mathbf{T}'''\begin{cases} \sigma_{13} \\ \sigma_{23} \\ \sigma_{33} \end{cases}$$

the equilibrium condition projected on to the axes can be written:

$$\begin{cases} T_1 = \sigma_{11} n_1 + \sigma_{12} n_2 + \sigma_{13} n_3 \\ T_2 = \sigma_{21} n_1 + \sigma_{22} n_2 + \sigma_{23} n_3 \\ T_3 = \sigma_{31} n_1 + \sigma_{32} n_2 + \sigma_{33} n_3 \end{cases}$$

These 3 equations can be written synthetically as:

$$T_i = \Sigma_j \sigma_{ij} n_j \tag{2.6}$$

We see that the *stress vector* $\mathbf{T}$ at point O (Fig. 2.9), relative to a surface element defined by $\mathbf{n}$, can be expressed as a function of 9 quantities σ_{ij} called the *stresses* at point O.

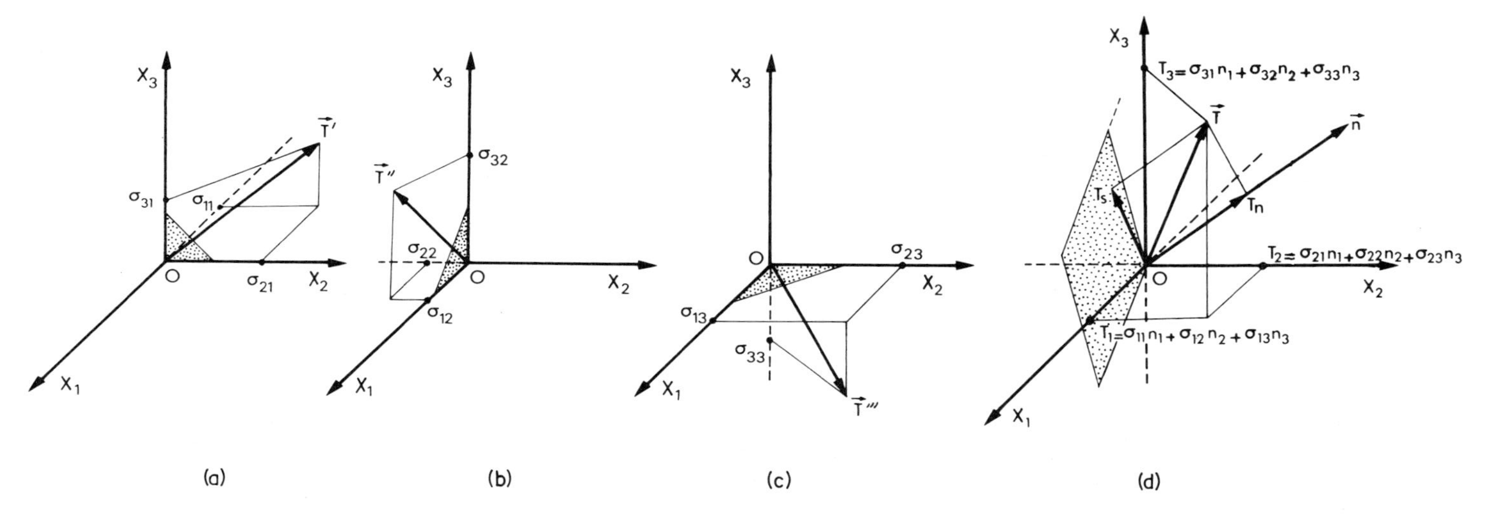

Fig. 2.9. Definition of the stresses: σ_{ij} is the projection along x_i axis of the stress vector relative to the face (stippled) of the elementary tetrahedron normal to x_j axis.

(a) **T′** stress vector relative to the face normal to ox_1
(b) **T″** stress vector relative to the face normal to ox_2
(c) **T‴** stress vector relative to the face normal to ox_3
(d) **T** stress vector relative to the face normal to **n**

These quantities entirely define the condition of stress at a given point in the solid.

The stress σ_{ij} is the component along x_i axis of the stress vector at O, *relative to the surface element normal to x_j axis.* It can be shown that, in general, $\sigma_{ij} = \sigma_{ji}$. In the general case, the 9 σ_{ij}, therefore, reduce to 6 independent stresses, which can be written in a square array representing the stress tensor.

$$\begin{pmatrix} \sigma_{11} & \sigma_{12} & \sigma_{13} \\ \sigma_{12} & \sigma_{22} & \sigma_{23} \\ \sigma_{13} & \sigma_{23} & \sigma_{33} \end{pmatrix}$$

The values of the σ_{ij} quantities obviously depend on the choice of the axes; however, the quantity $P = \sigma_{11} + \sigma_{22} + \sigma_{33}$ is independent of the choice of axes. $P/3$ represents the value of the *hydrostatic pressure* at point O, for the stress tensor can be divided into a hydrostatic part and a deviatoric part (as in the case of the strain tensor).

$$\begin{pmatrix} \sigma_{11} & \sigma_{12} & \sigma_{13} \\ \sigma_{12} & \sigma_{22} & \sigma_{23} \\ \sigma_{13} & \sigma_{23} & \sigma_{33} \end{pmatrix} = \begin{pmatrix} \frac{P}{3} & 0 & 0 \\ 0 & \frac{P}{3} & 0 \\ 0 & 0 & \frac{P}{3} \end{pmatrix} + \begin{pmatrix} \sigma_{11} - \frac{P}{3} & \sigma_{12} & \sigma_{13} \\ \sigma_{12} & \sigma_{22} - \frac{P}{3} & \sigma_{23} \\ \sigma_{13} & \sigma_{23} & \sigma_{33} - \frac{P}{3} \end{pmatrix}$$

We have seen how the components along the axes x_i of the stress vector can be expressed as a function of the stresses σ_{ij}; now the stress vector relative to a surface element can also be analysed in terms of its projections on the plane of the surface element and normal to it.

The projection of **T** on the plane of the surface element is called the *shear stress* at O on that plane: T_S and the projection of **T** on the normal **n** to the surface element is called the *normal stress*: T_n (Fig. 2.9).

T_S and T_n can of course be expressed as functions of the stresses σ_{ij}.

Thus, we can see that the diagonal stresses σ_{ij} are the normal stresses on the planes normal to the coordinate axes, whereas the off-diagonal stresses σ_{ij} $(i \neq j)$ are the shear stresses on these planes.

A few simple uniform stress regimes (see Table 2.1)

(1)

$$\begin{pmatrix} 0 & 0 & 0 \\ 0 & 0 & 0 \\ 0 & 0 & \sigma_{33} \end{pmatrix}$$

The stress vectors on the planes normal to the axes are:

$$\mathbf{T}'\begin{cases}0\\0\\0\end{cases}\quad \mathbf{T}''\begin{cases}0\\0\\0\end{cases}\quad \mathbf{T}'''\begin{cases}0\\0\\\sigma_{33}\end{cases}$$

This corresponds to a *uniaxial traction* along Ox_3 if σ_{33} is positive and a *uniaxial compression* if σ_{33} is negative.

On the plane x_1Ox_2, there is a normal stress $T_n = \sigma_{33}$ and no shear stress. The planes x_1Ox_3, x_2Ox_3 are entirely stress free: $T_n = T_S = 0$.

Let us consider any plane whose normal makes an angle of 45° with Ox_3, ($n_3 = \cos 45° = \sqrt{2}/2$), the stress vector on such a plane has as its components (Fig. 2.10):

$$\mathbf{T}\begin{cases}T_1 = 0\\T_2 = 0\\T_3 = n_3\sigma_{33} = \dfrac{\sqrt{2}}{2}\sigma_{33}\end{cases}$$

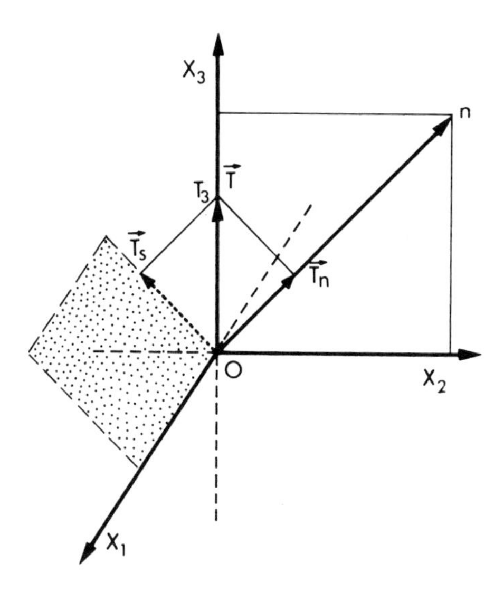

Fig. 2.10. Uniaxial traction along Ox_3. **T**: stress vector on plane at 45° from Ox_3 (stippled); T_n: normal stress; T_S: shear stress resolved on the plane

We can easily calculate the normal stress $T_n = T_3 \cos 45° = \sigma_{33}/2$. The shear stress (resolved on the plane) T_S is equal to T_n: $T_S = \sigma_{33}/2$. It is easily shown that any plane at 45° from the axis of tension is a *maximum resolved shear stress plane.*

(2)
$$\begin{pmatrix} \sigma_{11} & 0 & 0 \\ 0 & \sigma_{22} & 0 \\ 0 & 0 & \sigma_{33} \end{pmatrix}$$

(a)
$$\sigma_{11} = \sigma_{22} = \sigma_{33} = \sigma = P$$

The stress is purely hydrostatic: *hydrostatic compression* if σ is negative, *hydrostatic tension* if σ is positive.

The stress vector **T** on a plane defined by its normal

$$\mathbf{n}\begin{cases} n_1 \\ n_2 \\ n_3 \end{cases}$$

has as its components (see Fig. 2.9):

$$\mathbf{T}\begin{cases} T_1 = \sigma n_1 \\ T_2 = \sigma n_2 \\ T_3 = \sigma n_3 \end{cases}$$

The stress vector is always normal to the plane. There is *no shear stress.*

(b)
$$\begin{cases} \sigma_{11} = \sigma_{22} = \dfrac{P}{3} \\ \sigma_{33} = \dfrac{P}{3} + \sigma \end{cases}$$

We have

$$\begin{pmatrix} \dfrac{P}{3} & 0 & 0 \\ 0 & \dfrac{P}{3} & 0 \\ 0 & 0 & \dfrac{P}{3}+\sigma \end{pmatrix} = \begin{pmatrix} \dfrac{P}{3} & 0 & 0 \\ 0 & \dfrac{P}{3} & 0 \\ 0 & 0 & \dfrac{P}{3} \end{pmatrix} + \begin{pmatrix} 0 & 0 & 0 \\ 0 & 0 & 0 \\ 0 & 0 & +\sigma \end{pmatrix}$$

This case is therefore equivalent to a uniaxial tension or compression (according to the sign of σ) superimposed on a hydrostatic tension or compression (according to the sign of P).

The maximum shear stress is $\sigma/2$ on the 45° planes.

(c)
$$\begin{cases} \sigma_{11} = \sigma_{22} = -\dfrac{\sigma}{3} \\ \sigma_{33} = \dfrac{2\sigma}{3} \end{cases}$$

We have

$$P = \sigma_{11} + \sigma_{22} + \sigma_{33} = 0$$

The hydrostatic stress is equal to zero everywhere.

(d)
$$\begin{cases} \sigma_{11} = -\sigma_{22} = \sigma \\ \sigma_{33} = 0 \end{cases}$$

The planes containing Ox_3 and bisecting the angles x_1Ox_2 contain the stress vector **T** (Fig. 2.11):

$$\mathbf{T} \begin{cases} T_1 = \sigma n_1 \\ T_2 = -\sigma n_2 \\ T_3 = 0 \end{cases}$$

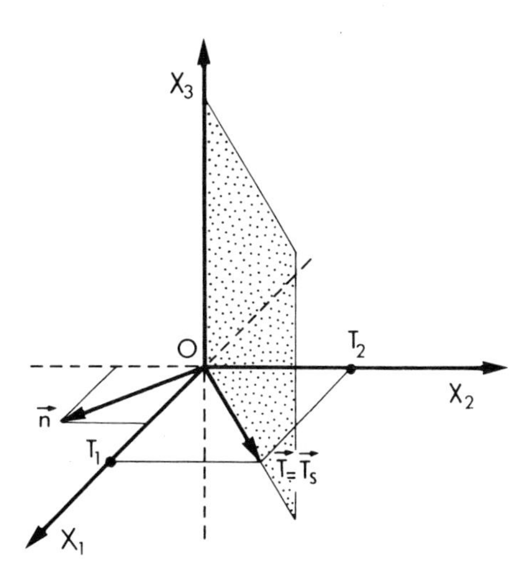

Fig. 2.11. Stress regime $\sigma_{11} = -\sigma_{22} = \sigma$, $\sigma_{33} = 0$. The stippled plane bears only a shear stress component T_S and no normal component

$n_1 = -n_2 = \sqrt{2}/2$ hence $T_1 = T_2 = \sigma\sqrt{2}/2$. There is *no normal stress but only a shear stress* on those planes.

Table 2.1. Axisymmetric Regimes of Stress

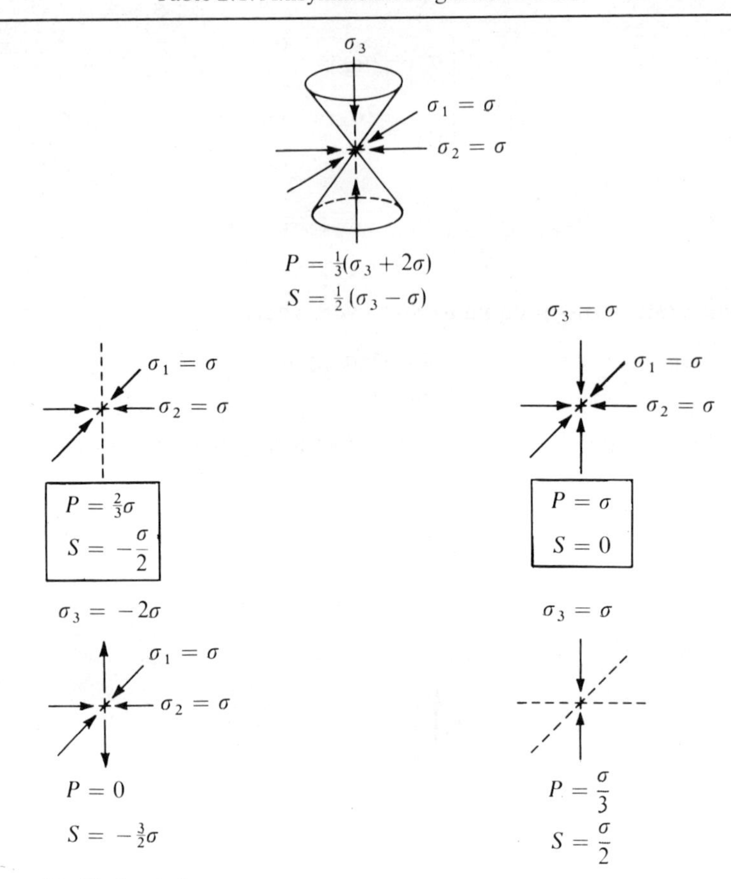

P = Hydrostatic pressure
S = Maximum shear stress on planes at 45° from σ_3 (tangent to cone)

2.4. RELATIONS BETWEEN STRESS AND STRAIN: ELASTIC DEFORMATION, PLASTIC DEFORMATION

We have defined the stresses by considering the forces that a surrounding solid exerts on the surface of an imbedded volume element. Of course, no solid can extend to infinity, and we must always consider an external surface to which forces are applied. By continuity, the distribution of applied forces on the surface of a solid, together with the body forces that may exist, defines the stress regime inside the solid.

In response to the solicitation represented by the stresses, all solids deform, and we may define strains at every point of the solid.

For the sake of simplicity we will, in what follows, consider regimes of uniform strain and stress, where, in an appropriate system of axes, we have only one component of stress σ and one component of strain ε; at every point in the solid, the stress σ is then equal to the applied stress and the strain ε is equal to the total strain of the solid.

Several important cases may exist:

(a) For small stresses, strain is proportional to stress (Hooke's law) and falls to zero if the stress is removed. The strain is uniquely determined once the stress is given. This is the domain of *elasticity* (Fig. 2.12).

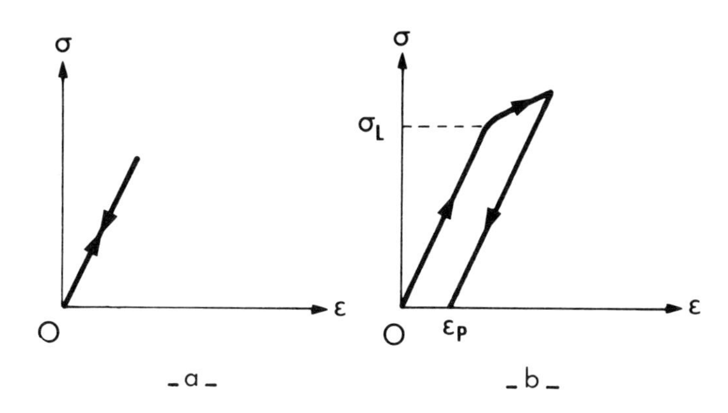

Fig. 2.12. Stress strain diagram, stress: σ, strain: ε. (a) Elastic regime: stress is proportional to strain; when the stress is removed there is no residual deformation; (b) For $\sigma < \sigma_L$ (elastic limit), the deformation is elastic. For $\sigma > \sigma_L$: the deformation is plastic; when the stress is removed a residual plastic strain ε_p remains

(b) For all stresses greater than a stress called *elastic limit*, strain is no longer proportional to stress. If the stress is removed, a permanent residual strain called *plastic strain* remains. This is the domain of *plasticity*. The plastic strain is not a function of stress only, but also depends on the whole mechanical and thermal history of the solid.

If the stress is applied for a time t, the strain increases with time: this is the phenomenon of *creep*. Creep is a special case of *plastic flow* under constant stress, and to characterize it at a given time we introduce the *strain rate* $\dot{\varepsilon} = \mathrm{d}\varepsilon/\mathrm{d}t$. If the solid is deformed at a constant strain rate, the stress or *flow stress* usually varies with time (i.e. with strain). As we will see later on, temperature is an important parameter of plastic flow and we can distinguish low temperature and high temperature regimes.

2.4.1. *Elastic constants*

The fundamental law of elasticity, Hooke's law, which expresses the proportionality between stress and strain, may be expressed for the most general regime of small stress and strain by writing that every component σ_{ij} of the stress tensor is a linear function of all the components ε_{ij} of the strain tensor:

$$\sigma_{11}=C_{1111}\varepsilon_{11}+C_{1112}\varepsilon_{12}+C_{1113}\varepsilon_{13}+C_{1121}\varepsilon_{21}+C_{1122}\varepsilon_{22}+C_{1123}\varepsilon_{23}+C_{1131}\varepsilon_{31}+C_{1132}\varepsilon_{32}+C_{1133}\varepsilon_{33}$$

$$\sigma_{12}=C_{1211}\varepsilon_{11}+C_{1212}\varepsilon_{12}+C_{1213}\varepsilon_{13}+C_{1221}\varepsilon_{21}+C_{1222}\varepsilon_{22}+C_{1223}\varepsilon_{23}+C_{1231}\varepsilon_{31}+C_{1232}\varepsilon_{32}+C_{1233}\varepsilon_{33}$$

$$\sigma_{13}=C_{1311}\varepsilon_{11}+C_{1312}\varepsilon_{12}+C_{1313}\varepsilon_{13}+C_{1321}\varepsilon_{21}+C_{1322}\varepsilon_{22}+C_{1323}\varepsilon_{23}+C_{1331}\varepsilon_{31}+C_{1332}\varepsilon_{32}+C_{1333}\varepsilon_{33}$$

$$\sigma_{22}=C_{2211}\varepsilon_{11}+C_{2212}\varepsilon_{12}+C_{2213}\varepsilon_{13}+C_{2221}\varepsilon_{21}+C_{2222}\varepsilon_{22}+C_{2223}\varepsilon_{23}+C_{2231}\varepsilon_{31}+C_{2232}\varepsilon_{32}+C_{2233}\varepsilon_{33}$$

$$\sigma_{23}=C_{2311}\varepsilon_{11}+C_{2312}\varepsilon_{12}+C_{2313}\varepsilon_{13}+C_{2321}\varepsilon_{21}+C_{2322}\varepsilon_{22}+C_{2323}\varepsilon_{23}+C_{2331}\varepsilon_{31}+C_{2332}\varepsilon_{32}+C_{2333}\varepsilon_{33}$$

$$\sigma_{33}=C_{3311}\varepsilon_{11}+C_{3312}\varepsilon_{12}+C_{3313}\varepsilon_{13}+C_{3321}\varepsilon_{21}+C_{3322}\varepsilon_{22}+C_{3323}\varepsilon_{23}+C_{3331}\varepsilon_{31}+C_{3332}\varepsilon_{32}+C_{3333}\varepsilon_{33}$$

The 6 equations may be written synthetically:

$$\sigma_{ij}=\sum_{kl} C_{ijkl}\varepsilon_{kl} \tag{2.7}$$

The C_{ijkl} coefficients are material constants depending only on the physical properties of the solid; they are called the *elastic constants.*

As we have seen, $\varepsilon_{ij}=\varepsilon_{ji}$ and $\sigma_{ij}=\sigma_{ji}$, consequently:

$$C_{ijkl}=C_{klij}$$

and

$$C_{ijkl}=C_{jilk}$$

hence in the most general case the 81 elastic constants are reduced to 21, which may be written in a square array of two index constants C_{ij} by adopting the following correspondence rule for each pair of indices

$$\begin{array}{ccc} 11\rightarrow 1 & 12\rightarrow 6 & 13\rightarrow 5 \\ 21\rightarrow 6 & 22\rightarrow 2 & 23\rightarrow 4 \\ 31\rightarrow 5 & 32\rightarrow 4 & 33\rightarrow 3 \end{array}$$

We then write the elastic constants symmetric matrix:

$$\begin{pmatrix} C_{11} & C_{12} & C_{13} & C_{14} & C_{15} & C_{16} \\ C_{12} & C_{22} & C_{23} & C_{24} & C_{25} & C_{26} \\ C_{13} & C_{23} & C_{33} & C_{34} & C_{35} & C_{36} \\ C_{14} & C_{24} & C_{34} & C_{44} & C_{45} & C_{46} \\ C_{15} & C_{25} & C_{35} & C_{45} & C_{55} & C_{56} \\ C_{16} & C_{26} & C_{36} & C_{46} & C_{56} & C_{66} \end{pmatrix}$$

For crystals with the lowest symmetry (triclinic), all 21 C_{ij} are non-zero. The number of independent C_{ij} decreases as the crystal symmetry increases and the cubic system has only 3 non-zero independent elastic constants: C_{11}, C_{12}, C_{44} (Nye, 1957)

$$\begin{pmatrix} C_{11} & C_{12} & C_{12} & 0 & 0 & 0 \\ C_{12} & C_{11} & C_{12} & 0 & 0 & 0 \\ C_{12} & C_{12} & C_{11} & 0 & 0 & 0 \\ 0 & 0 & 0 & C_{44} & 0 & 0 \\ 0 & 0 & 0 & 0 & C_{44} & 0 \\ 0 & 0 & 0 & 0 & 0 & C_{44} \end{pmatrix}$$

Finally, the most symmetric solids are the isotropic ones where all directions are equivalent. In this case there is a relation between the 3 elastic constants:

$$C_{44} = \tfrac{1}{2}(C_{11} - C_{12}) \tag{2.8}$$

and the number of independent constants falls to 2. A randomly oriented aggregate of crystals, or polycrystal, is statistically isotropic.

In the isotropic case, it is usual to write:

$$\begin{cases} C_{12} = \lambda \\ C_{44} = \mu \end{cases} \quad \text{with } C_{11} = \lambda + 2\mu \tag{2.9}$$

λ and μ are called *Lamé's coefficients.* From (2.7), (2.8) and (2.9), we have:

$$\begin{aligned} \sigma_{11} &= \lambda(\varepsilon_{11} + \varepsilon_{22} + \varepsilon_{33}) + 2\mu\varepsilon_{11} \\ \sigma_{12} &= 2\mu\varepsilon_{12} \\ \sigma_{13} &= 2\mu\varepsilon_{13} \\ \sigma_{22} &= \lambda(\varepsilon_{11} + \varepsilon_{22} + \varepsilon_{33}) + 2\mu\varepsilon_{22} \\ \sigma_{23} &= 2\mu\varepsilon_{23} \\ \sigma_{33} &= \lambda(\varepsilon_{11} + \varepsilon_{22} + \varepsilon_{33}) + 2\mu\varepsilon_{33} \end{aligned}$$

or

$$\sigma_{ij} = \lambda \delta_{ij} \sum_k \varepsilon_{kk} + 2\mu\varepsilon_{ij} \tag{2.10}$$

where

$$\delta_{ij} = 1 \qquad \text{if } i = j$$

$$\delta_{ij} = 0 \qquad \text{if } i \neq j$$

μ, the proportionality coefficient between the shear stresses and the shear strains, is called the *shear modulus.* (The shear modulus is often written G.)

The proportionality coefficient between the hydrostatic pressure $P = \frac{1}{3}(\sigma_{11} + \sigma_{22} + \sigma_{33})$ and the resulting dilatation $-\Delta V/V = \varepsilon_{11} + \varepsilon_{22} + \varepsilon_{33}$ is called the *Bulk modulus* B. It is easily seen that

$$B = -\frac{P}{\Delta V/V} = \frac{1}{3}(3\lambda + 2\mu) \tag{2.11}$$

The *compressibility* χ is defined as the inverse of B. It is interesting to define the elastic properties with the help of μ and B which determine the change in shape and volume for a given stress.

It is important to note that although cubic crystals are optically isotropic, they are not in general elastically isotropic, since there are three non-zero elastic constants. It follows in particular that the shear modulus depends on the plane and the direction of shear, thus:

$$\mu_{\{100\}} = C_{44} \tag{2.12}$$

is the shear modulus in the cube planes $\{100\}$ along the $\langle 010 \rangle$ directions

$$\mu_{\{110\}} = \tfrac{1}{2}(C_{11} - C_{12}) \tag{2.13}$$

is the shear modulus in the diagonal planes $\{110\}$ along the $\langle 1\bar{1}0 \rangle$ directions

$$A = \frac{(C_{11} - C_{12})}{2C_{44}} \tag{2.14}$$

is the anisotropy factor. The bulk modulus B clearly does not depend on directions.

Table 2.2 gives the values of the elastic constants for a few single crystalline minerals.

Average isotropic elastic constants can be calculated for polycrystalline material (Anderson, 1965).

Table 2.2

Cubic crystals	Elastic constants and moduli (kilobars)					
	T °K	C_{11}	C_{12}	C_{44}	$\frac{1}{2}(C_{11}-C_{12})$	A
NaCl	295	575	99	133	238	1·79
MgO	293	2860	870	1480	995	0·67
	1473	2091	787	1386	652	0·47
Diamond	300	10760	1250	5738	4755	0·83
Cu	293	1684	1214	754	235	0·31

Non-cubic crystals	Elastic constants (kilobars)										
	T °K	C_{11}	C_{12}	C_{13}	C_{14}	C_{22}	C_{23}	C_{33}	C_{44}	C_{55}	C_{66}
Trigonal											
calcite	293	1374	440	450	203	—	—	801	342		467
quartz	293	867	70	119	179	—	—	1072	579		398
tourmaline	293	2756	704	90	78	—	—	1638	680		1005
Orthorhombic											
olivine(1) (92% forsterite)	298	3237	664	716	—	1976	756	2351	646	780	790
orthopyroxene(2) (85% enstatite)		2299	701	573		1654	496	2057	831	764	785

(1) Kumazawa and Anderson, 1969.
(2) Kumazawa, 1969.

(After Anderson, 1965)

Physical meaning of the elastic constants

A crystal is a three-dimensional periodic array of atoms (or ions). In an unstrained crystal the atoms are maintained in equilibrium positions by interatomic forces. If a stress is applied to the crystal, the atoms are displaced from their equilibrium position by an amount which depends on the strength of the interatomic forces.

For instance, to shear an isotropic crystal through an angle γ, one must apply a shear stress σ, proportional to the shear modulus μ which expresses the resistance to shear arising from the interatomic forces (Fig. 2.13).

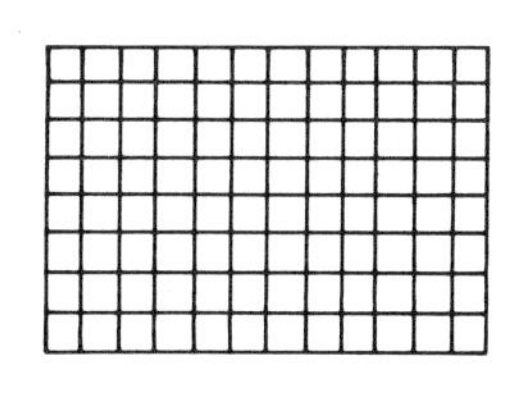

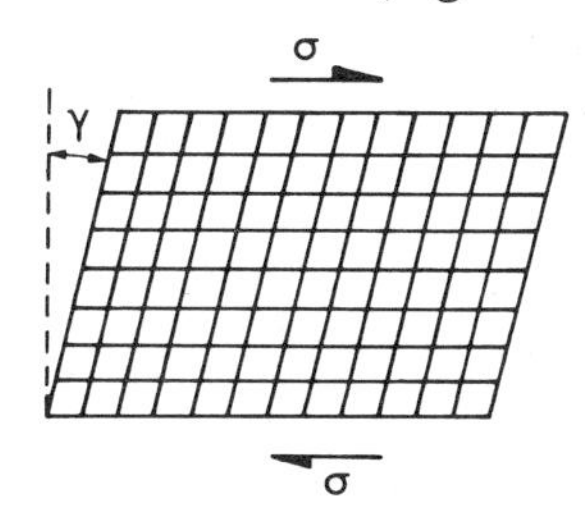

Fig. 2.13. Elastic shear of a crystal (atoms are represented by the nodes of the lattice). σ: shear stress; γ: shear angle. The atomic bonds are distorted

In the same way, the bulk modulus B is a manifestation of the difficulty to displace the atoms from their equilibrium position by shortening the interatomic distance, as a result of the application of an hydrostatic pressure (Fig. 2.14).

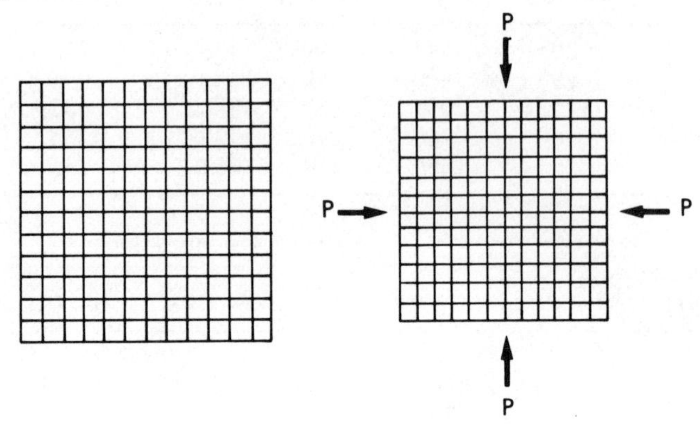

Fig. 2.14. Compression of a crystal (atoms are represented by the nodes of the lattice), P: hydrostatic pressure. The interatomic distances are shortened

It is therefore easy to understand that the atoms return to their equilibrium position as the stress is removed and the crystal recovers its primitive shape and volume.

2.4.2. *Plastic strain*

The plastic strain is the permanent shear strain that is obtained for shear stresses greater than the elastic limit. Plastic deformation is a constant volume deformation, that is to say that it arises only from the deviatoric part of the applied stress. A hydrostatic pressure applied on a crystal can cause only a recoverable elastic deformation and no permanent deformation. This can be understood by considering the physical meaning of plastic strain for a crystal.

The fact that the strain is permanent and does not go to zero as the stress is removed, means that the atoms in the crystal have been moved to new equilibrium positions by the stress.

The examination of the surface of experimentally deformed crystals usually shows microscopic steps or *slip lines* which correspond to an offset along crystallographic planes: the crystal has permanently altered its shape as a result of *slip* or *glide* on slip planes. The atoms on each side of a slip plane have changed neighbours, but the atomic bonds are undistorted and have the same length as in the undeformed crystal (Fig. 2.15). Hydrostatic pressure can lead to no such change in equilibrium positions of the atoms, since it can only bring them together isotropically. Repulsive interatomic forces always oppose this change and the atoms always swing back to their equilibrium position as the stress is removed: There is only an elastic deformation.

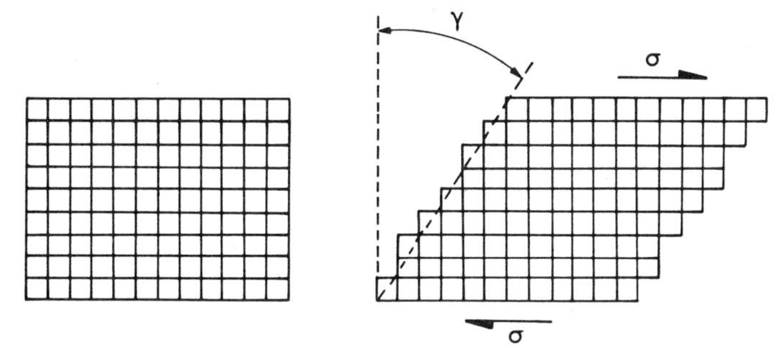

Fig. 2.15. Plastic strain by glide on atomic planes. σ: shear stress; γ: shear angle. The atomic bonds are undistorted

A condition always met for plastic strain is therefore:

$$\frac{\Delta V}{V} = \varepsilon_{11} + \varepsilon_{22} + \varepsilon_{33} = 0 \tag{2.15}$$

(*a*) *Plastic glide of single crystals*

The simplest case of plastic deformation is that of a single crystal subjected to a uniform regime of applied stress. As we have seen, for a shear stress greater than the elastic limit the crystal begins to slip on a family of crystallographic planes (*slip planes*) in a crystallographic direction (*slip direction*).

The *slip system* is determined by the knowledge of the slip plane and the slip direction. As planes and direction are labelled by Miller indices, a given slip system will be indicated by the Miller indices of the plane and the direction: $(hkl)[uvw]$.

The problem of understanding why the crystal glides on certain planes and not on others is one of the most important in physical metallurgy.

(a) In metals, a lattice of positive ions is immersed in a sea of nearly free electrons. The crystal generally glides most easily in the densest direction of the densest crystallographic plane: this can be rationalized by noticing that since the number of ions per cm^3 is fixed, the planes containing the greatest density of ions are the most widely spaced and are thus less strongly bonded (Fig. 2.16); also, it can be expected that slip will be easier along a dense direction, since, the smaller the interatomic distance, the smaller is the work needed to slip by such a distance. For example, metals described by a face-centred cubic cell (Al, Ag, Cu, Au) slip on the 12 $\{111\}$ $\langle 1\bar{1}0 \rangle$ systems* (Fig. 2.16). Most hexagonal close-packed metals (Cd, Zn, Mg) slip on the basal plane (0001) in the dense direction $\langle 11\bar{2}0 \rangle$.

(b) In ionic crystals, where the lattice is formed by ions alternately charged + or −, the easiest slip system will generally be the one that does not bring

* Equivalent planes of the same family are labelled by Miller indices in curly brackets: $\{hkl\}$. Equivalent directions are labelled by Miller indices in carets: $\langle uvw \rangle$.

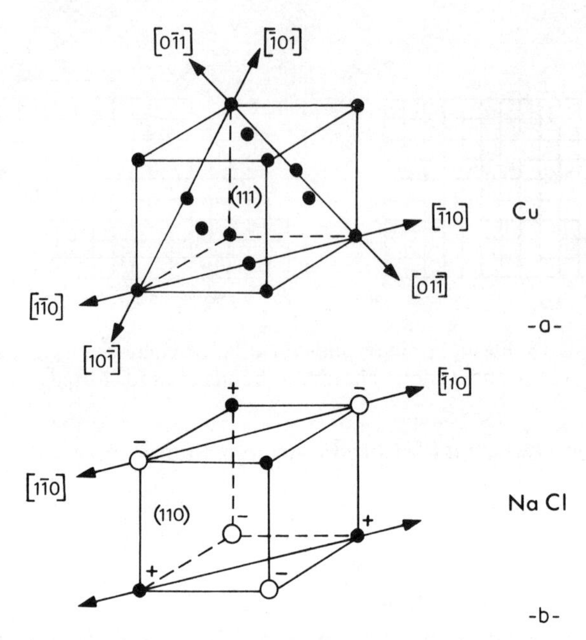

Fig. 2.16. (a) Face-centred cubic metal. Glide occurs in the dense diagonal plane (111) in the dense directions (arrows); (b) ionic crystal, NaCl structure. Glide occurs in plane (110) in the dense direction (arrows)

together ions of like charge during the glide process. Thus alkali halides like NaCl, or simple metallic oxides like MgO, with a NaCl structure, glide on the 6 $\{110\}\langle 1\bar{1}0\rangle$ systems (Fig. 2.16). Although the NaCl structure can be described by a face-centred cubic cell, glide on the densest planes $\{111\}$ would bring together ions of like charges, whereas glide on less dense $\{110\}$ planes does not.

(c) In more complex crystals the easiest glide planes are generally those whose shear breaks the weakest bonds between groups. For instance, in phyllosilicates the easiest slip plane is parallel to the sheets of linked SiO_4 tetrahedra: a shear parallel to this plane breaks only the weak bonds between the metallic cations and the strongly bonded SiO_4 groups.

Of course, the examples given here are relative to simple cases where the choice of the easiest slip plane is obvious. In many instances it is impossible to predict the slip systems by simple inspection, and as we will see later, the slip systems change with temperature.

Pencil glide. Although slip on low index crystallographic planes in compact directions is the general rule, there are a few important cases where macroscopic slip is seen to occur on non-crystallographic planes, although still in a crystallographic compact direction.

This was first reported in the case of iron single crystals by Taylor and Elam (1926) and has since been verified by many experimenters on Fe and other

BCC metals: the slip direction is a compact crystallographic direction ($\langle 111 \rangle$ in this case) but slip occurs on non-crystallographic planes of the $\langle 111 \rangle$ zone, close to the plane with maximum resolved shear stress.

At the time, Taylor and Elam proposed an explanation based on the idea that 'the crystal does not divide itself into sheets when a shearing stress is applied, but into rods or pencils'. The deformation of the crystal would then be similar to that of a bundle of pencils: hence the name *pencil glide*. This model was apparently supported by the fact that the observed slip lines were *wavy* although in a general straight direction.

In view of the fact that pencil glide in Taylor and Elam's sense has been observed repeatedly in olivine at high temperatures on planes in zone with the [100] direction (Raleigh, 1968), it is of some interest to note that a satisfactory explanation of macroscopic glide on non-crystallographic planes has been provided in the case of Fe and the other BCC metals.

This explanation rests on dislocation mechanisms and has entirely superseded Taylor and Elam's pencil glide theory, which had been proposed at a time when the microscopic processes of slip were unknown. We will present the modern views on 'pencil glide' when we discuss the so-called pencil glide in

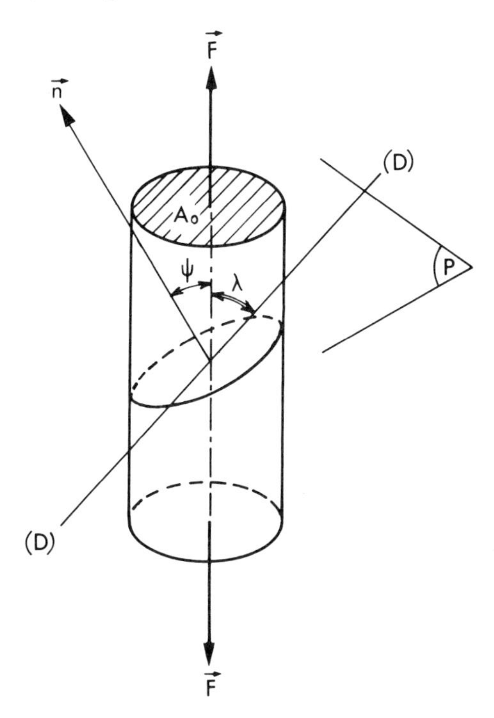

Fig. 2.17. Cylindrical single crystal undergoing plastic deformation under axial load **F**. **n**: normal to slip plane (P); ψ: angle between **n** and load axis: λ: angle between slip direction (D) in plane (P) and load axis

olivine later on (§ 5.2.2). Suffice it to say that the wavy non-crystallographic slip is interpreted as a microscopic composite slip on very small portions of low index crystallographic planes. The amount of slip on each plane differs according to the orientation with respect to the applied shear stress, and the macroscopic 'slip plane' is close to the plane with maximum resolved shear stress.

Critical resolved shear stress, Schmid's law. Let us consider a single crystalline sample in the shape of a cylinder with a cross-section area A_0, subjected to a uniaxial load F (Fig. 2.17). For loads $F < F_e$ the crystal deforms elastically and for a value $F = F_e$ of the load, plastic deformation sets in. The elastic limit is therefore the normal stress $\sigma_e = F_e/A_0$.

For simplicity, let us examine the case where the crystal deforms by slip on a unique slip plane (P) along the direction (D). Let ψ be the angle between the normal to the slip plane and the direction of application of the load, and λ the angle between the slip direction and the direction of application of the load.

Experience has shown that the value of the elastic limit σ_e depends on the orientation of the slip system with respect to the direction of the normal stress, i.e. on ψ and λ.

However, Schmid and Boas have found that for any orientation of the slip system, slip begins for a given value σ_c of the shear stress resolved on the slip plane along the slip direction (*Schmid's Law*) (Schmid and Boas, 1950). The *critical resolved shear stress* (CRSS) is therefore an intrinsic characteristic of the crystal slip system, whereas the elastic limit σ_e, varying with the orientation, is not a good parameter for defining the onset of slip.

The resolved shear stress σ_R can be easily expressed as a function of ψ and λ (Fig. 2.18)

$$\sigma_R = \frac{F}{A_0} \cos\psi \cos\lambda \tag{2.16}$$

We therefore have:

$\sigma_R < \sigma_c$ Elastic deformation

$\sigma_R > \sigma_c$ Plastic deformation by glide over the slip system whose CRSS is equal to σ_c

If there are several slip systems with the same CRSS (for example, the 12 $\{111\}\langle 1\bar{1}0\rangle$ slip systems of the FCC metals), slip begins on the system for which the resolved shear stress has the highest value for a given load F.

This relative ease of slip of otherwise identical systems is expressed by the *Schmid factor S*.

$$S = \cos\psi \cos\lambda \tag{2.17}$$

The slip system with the highest Schmid factor is the first to start.

If the possible slip systems are known, it is easy to determine which will be the active system for a given orientation of the crystal by means of the standard stereographic projection: ψ and λ can be measured and S calculated for all systems (see Hirth and Lothe, 1967) (Fig. 2.19).

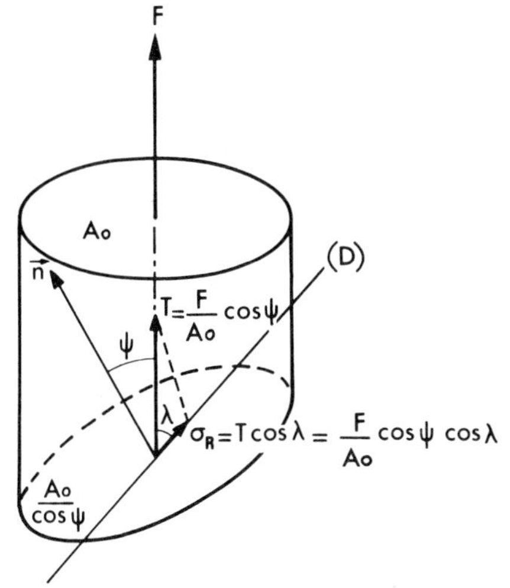

Fig. 2.18. Resolved shear stress. A_0: cross-section area of crystal; T: stress vector on slip plane normal to **n**, of area $A_0/\cos\psi$; σ_R: shear stress on slip plane, resolved in the slip direction (D); $\cos\psi\cos\lambda = S$: Schmid factor

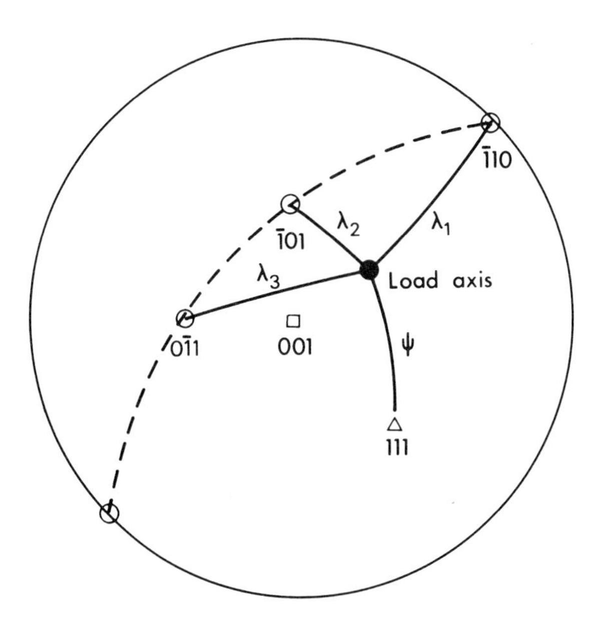

Fig. 2.19. Stereographic projection of a face centred cubic crystal. 111: Pole of the glide plane, at angle ψ from load axis; great circle (broken lines): trace of the glide plane; $\bar{1}10$, $\bar{1}01$, $0\bar{1}1$, possible slip directions at respective angles λ_1, λ_2, λ_3 from load axis

In most crystals there are several families of slip systems, each with its own CRSS. As the CRSS usually varies with temperature, the relative ease of slip of the different families likewise varies and there are domains of temperature where one slip system predominates. The active slip system under a given load for a given temperature and a given orientation is the one with the lowest CRSS and the highest Schmid factor.

(*b*) *Plastic deformation of crystalline aggregates*

Let us first consider the case of a *polycrystal*: aggregate of *grains* of the same crystalline material.

A polycrystalline sample will deform plastically mostly by slip of individual grains (*intragranular slip*) or by the relative motion of the grains along their *grain boundaries* (*grain-boundary sliding*). Intracrystalline slip is the only deformation mode at low temperature, and can be accompanied by grain boundary sliding at high temperature. We will first examine intragranular slip.

If a polycrystal has no *texture*, i.e. its grains are randomly oriented, the grains will clearly not all yield plastically for the same applied stress, for the Schmid factor of the active slip systems will be different in different grains.

Favourably oriented grains will start gliding before others and the *macroscopic elastic limit* will have to be defined conventionally as the stress for which a given detectable plastic strain is obtained. It will clearly be higher than the CRSS for the active slip system (1·3 to 3·0 times higher, in the case of plagioclase for instance).

Now, each grain is surrounded by neighbouring grains which deform differently since, for a given applied stress, all slip systems have not come into play in all grains.

We can imagine the following experiment:

Let us cut up the polycrystal into its separate grains, then let each grain deform by slip on its own active systems under the given stress. After a certain amount of deformation, let us try to glue back every grain and reconstitute a deformed compact polycrystal. This will be impossible, as voids will exist between certain grains and other grains will penetrate into each other (Figs. 6.32 and 6.33). The deformation of a polycrystal is therefore in general *incompatible* on the grain scale. If we wanted to obtain a compact body by gluing together the deformed grains, we could proceed in the following way:

(i) Deform every grain elastically, by applying stresses on it, until it fits with its neighbours.

(ii) Glue the grains together, maintaining the stresses during the operation.

(iii) Release the externally applied stresses. The elastic strains are locked-in.

The compactness is therefore obtained by means of an *internal elastic strain*, the material is self-stressed and the *internal stress* can be expressed in terms of the internal strain and the elastic constants of the material by Hooke's law.

The grains of a plastically deforming polycrystal without a texture are thus constrained by their neighbours and internal stresses arise at the grain boundaries and especially at the triple points.

Now, the internal stresses rise as deformation proceeds and two situations may occur:

(1) Existing slip systems are activated in sufficient number to relieve the internal stresses. The polycrystal remains coherent and we can define at every point the 6 components of strain ε_{ij} as in a continuous solid.

This supposes that every grain can deform on 5 independent slip systems (by independent we mean that no deformation on one slip system can be reproduced by a combination of deformations on the other slip systems).

This results from the fact that the 6 components of the strain are locally imposed by the constraints due to the neighbouring grains. As plastic deformation is a constant volume deformation, there is a relationship between the 6 ε_{ij}:

$$\frac{\Delta V}{V} \simeq \varepsilon_{11} + \varepsilon_{22} + \varepsilon_{33} = 0$$

There are therefore only 5 independent ε_{ij} that can be achieved by slip on 5 independent slip systems.

Thus, a polycrystal can deform coherently to large strains only if the grains can slip on 5 independent slip systems. This is the Von Misès ductility criterion (Paterson, 1969); it is met for example in face-centred cubic metals like copper, gold, aluminium, lead, etc., whose ductility is well known.

(2) There are fewer than 5 slip systems. This is a frequent case for minerals having a low crystalline symmetry (triclinic, orthorhombic, rhombohedric) (Paterson, 1969). Several situations may arise according to the value of the internal stresses (and therefore of the imposed strain).

For small stresses, the incompatibility between grains is taken up (as in the thought experiment) by an elastic non-uniform strain, which can be relaxed by *bending* of the crystallographic planes (Fig. 2.20) (Plate 2(a)). This is the cause of undulatory extinction in thin sections of rocks viewed through crossed polarizers. The bending can also be concentrated into sharp kinks (*kink bands*).

For larger stresses, there comes a point when the strength of the bonds between atomic planes can no longer bear the elastic strain. As no plastic deformation can relieve the stresses entirely, *cleavage cracks* develop along the less strongly bonded crystallographic planes (Plate 1(a) and (b)). In crystals with only one glide system, the cleavage plane is often the only glide plane.

For large shearing strains in simple shear, which is the common case for grains with a unique glide plane in a polycrystal, the glide planes bend, kink band boundaries eventually form, and the splitting of the latter may create a cleavage crack along the glide plane (Fig. 2.20(c)) (Plate 1(b)).

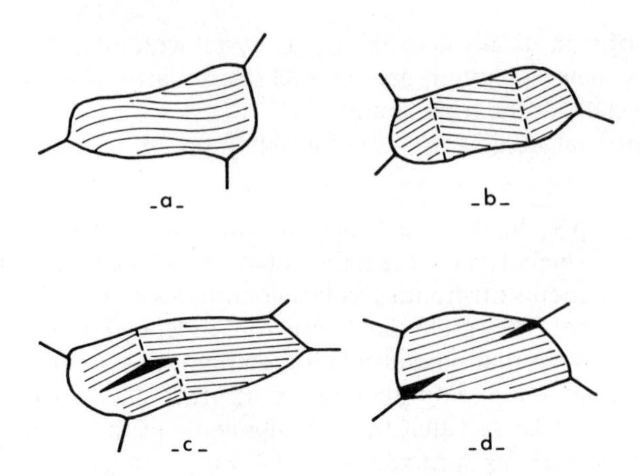

Fig. 2.20. Deformation of polycrystal with 1 slip system. Internal stresses arising from incompatibilities between the deformation of adjacent grains can be relieved by: (a) bending of the glide planes; (b) kinking perpendicular to the glide planes; (c)(d) cleavage cracks along the glide planes

More generally cleavage cracks occur where stress concentrations arise (e.g. at grain boundary triple points, at the tip of a twin lamella). At high temperatures, grain boundary sliding may relieve the internal stresses, but only to a limited extent for two neighbouring crystals are prevented by their neighbours from sliding freely. Where grain boundary sliding is dominant internal stresses also arise and if they cannot be relieved by intracrystalline plastic deformation, *intergranular cracks* occur.

Rocks are usually aggregates of crystals of more than one mineral. In addition to all the problems raised for polycrystals of one species, we have to take into account the fact that all minerals do not have the same number of slip systems and do not deform plastically with equal facility under the same applied stress. The plastic deformation of a rock depends, therefore, on the proportion of hard to ductile minerals and on their orientation with respect to the applied stress. Hard minerals with only one slip system tend to cleave because they cannot relieve all the internal stresses arising from the incompatibility. Besides, hard minerals act as stress concentrating inclusions in the matrix and often promote its cracking.

Fracture of an aggregate is caused by extensive or catastrophic *growth of cracks nucleated at points where there are stress concentrations.* The growth of cracks causes an increase of volume of the solid (*dilatancy*), which can be opposed by an applied hydrostatic pressure. *Hydrostatic pressure therefore opposes the growth of cracks.* Hence, fracture can be delayed or even suppressed if deformation takes place under a high enough hydrostatic pressure, and normally brittle materials can be deformed up to large strain (Bridgman, 1964). Most experiments on deformation of rocks are performed under confining pressures. These are of course the natural conditions in which plastic flow of rocks takes place in the lower crust or upper mantle.

Twinning. Deformation (or mechanical) twinning is another mode of plastic deformation which may be important when there are few slip systems, as in the case of feldspars or calcite, for instance (see Chapter 5). Twinning consists in the formation within the crystal of a volume with a different crystalline orientation specifically related to the orientation of the host lattice (*usually by a rotation about an axis*). In most cases of interest here, the twinned lattice can be deduced from the host lattice by a simple shear S, parallel to a plane K_1 (*composition plane or twinning plane*) along a direction η_1 (*twinning direction*). The shear is constant for a given twin (unlike slip where the shear can be imposed). S is expressed by: $S = \tan \psi' = 2 \tan \psi/2$ (Fig. 2.21). By such a simple shear, a sphere is transformed into an ellipsoid, and all the planes rotate except the planes containing the two circular sections of the ellipsoid: planes K_1 and K_2 (Fig. 2.21). It can be seen that the twinned lattice could have been obtained equivalently by shear parallel to the plane K_2 and along direction η_2 (intersection of K_2 with the plane normal to K_1, containing η_1). Most authors dealing with the crystallographic description of twins define the twinning system by the 4 elements $K_1 K_2 \eta_1 \eta_2$. As we are concerned only with the effective deformation process induced by a known applied shear stress, we will define twinning systems in Chapter 5 by the actual twinning plane K_1, and twinning direction η_1, which, together with the shear S, define completely the system (see Klassen-Neklyudova, 1964).

It is important to note that although the lattice may undergo a large shear in twinning, this is obtained through minor local rearrangements of the atoms. Twinning is a very rapid deformation mode and occurs in bursts (like other

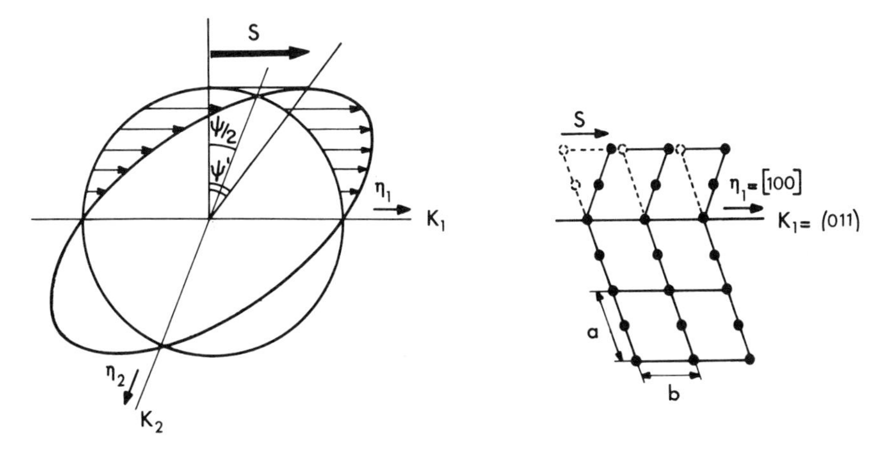

Fig. 2.21. Twinning on the twin plane K_1 along the twin direction η_1. Twinning shear $S = \tan \psi' = 2 \tan \psi/2$

displacive transformations); it differs in this from slip which is a progressive shear.

Twinning proceeds by nucleation and growth (see Reed-Hill et al., 1964): as a large stress is needed to nucleate a twinned volume, the nucleation sites are usually stress concentrators (like inclusions of a second phase, or triple points in grain boundaries). Once a twinned lenticular lamella has been nucleated it can grow very rapidly by concentrating at its sharp edge the stress it needs to propagate, in much the same way as a cleavage crack. As the effective stress for nucleation depends on the degree of stress enhancement, there is no such thing as a critical resolved shear stress characteristic of a twinning system, as there is one for slip systems.

Thickening of the twin lamella results from twin-boundary migration (see § 3.5.3).

Twinning is very seldom observed in ductile materials (like FCC metals) unless at very low temperatures under impact; the reason is that concentrated stresses can be relaxed by operation of the numerous slip systems available, before they reach the high value necessary to nucleate a twin. Conversely, low temperatures and a high imposed strain rate, raise the CRSS for slip and therefore promote twinning.

The same reasons that inhibit twinning in ductile materials make it an important deformation mode in crystals of low symmetry having few slip planes (feldspars, calcite) and can provide some ductility by supplementing the existing systems in crystals where Von Misès criterion is not met.

Deformation twins developing inside grains under stress, must not be confused with *annealing twins*, formed during recrystallization and which are only grains whose lattice is in twin relationship with that of a neighbouring grain.

Kinking. Kinking is a deformation mode commonly found in crystals with only one slip plane (e.g. mica), when no extensive glide on this system is geometrically possible and when bending moments are present. The internal stresses set up in the crystal correspond to a curvature of the lattice which can concentrate in well-defined sharp regions, thus giving a macroscopic strain in response to the stress (Plate 2(a) and 2(b)).

We have already mentioned this phenomenon in the case of the deformation of aggregates (§ 2.3.2) but it can occur also in single crystals stressed in appropriate conditions. A typical and illustrative example is the case of a compression test of a single crystal of hexagonal metals which glide only on the basal plane (Cd, Zn), when the compression axis is almost parallel to the basal plane (Fig. 2.22). The bending of the lattice developed close to the platens concentrates in sharply defined *kink-band boundaries* (KBB) separating regions of the crystal where the lattice is differently oriented but not curved (*kink bands*); this configuration corresponds to an overall shortening of the crystal.

A kink band can be defined as a region of the crystal which has undergone an external rotation with respect to the matrix, the axis of external rotation lying in the KBB. (A twin can often be considered as a special case of kink band.)

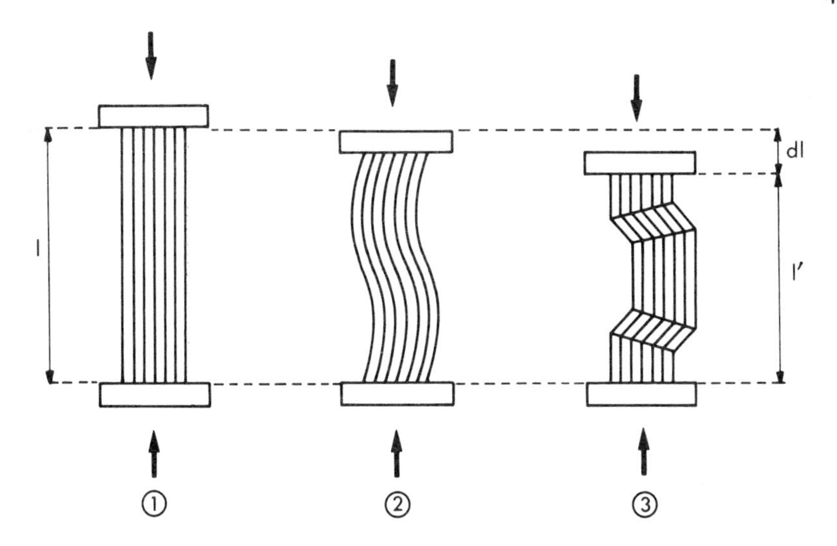

Fig. 2.22. Crystals with one slip system compressed parallel to the slip plane. The slip planes bend and the curvature concentrates into sharp kink band boundaries

Geometrical analyses of kink bands have been made in relation to the crystalline structure (e.g. Starkey, 1968), but it must be noted that the geometrical configuration called kink band can result from many different physical processes. This will be made clear later on, when we analyse KBB's as dislocation walls (Chapter 3). Thus a kink band can be produced by the dynamical deformation process mentioned above, to which the name of *kinking* should be reserved (see § 3.4.6), but it can also result from a quasi-static process by which dislocations gather into walls by slip or by climb (see § 3.5.2); in this case, the production of kink bands is a side effect of deformation and does not constitute a deformation mode by itself. This process can be distinguished from kinking by the name *polygonization* (glide polygonization or climb polygonization) and the kink bands are then more aptly termed *subgrains.*

N.B.: In all cases, Starkey's (1968) macroscopic analysis of the misorientation (or external rotation), corresponds to the microscopic analysis in terms of dislocations expressed by equation 3.71 (equivalent to Starkey's equation 1).

2.5. RELATIONS BETWEEN STRESS AND STRAIN-RATE—FLOW REGIMES

In the previous paragraph we have seen that elastic or plastic strains are produced if a stress is applied to a solid. In the case of elastic deformation there is a simple constitutive relation: Hooke's law, between the stress σ and the elastic strain ε_e. In the case of plastic deformation things are less simple, since the application of stress produces a permanent strain ε_p and changes the state of the solid. The elastic limit of a prestrained solid generally is higher than the elastic limit it had when first strained: this is the phenomenon of *work hardening.*

The *theory of plasticity* is a purely phenomenological theory which analyses the stress–strain relations in solids by making various assumptions. The most important of these is that the response of the solid is insensitive to strain rate. Thus, it is possible to derive incremental relations between stress and strain where the plastic properties of the material are introduced by means of experimental coefficients, but where time does not appear.

However, although the theory of plasticity is quite useful in many quasi-static cases, strain rate does play an important role in the plastic deformation of most materials and it is interesting and fruitful to consider plastic deformation as a flow.

Flow is a process of transport of matter, driven by a stress. The flow of plastic solids, like that of viscous fluids at constant volume, is a shear process. Its velocity under a given stress can be expressed by the strain rate (here rate of shear) $\dot{\varepsilon} = \mathrm{d}\varepsilon/\mathrm{d}t$.

It can be shown (Hart, 1970) that it is generally possible to write an equation of state:

$$f(\sigma, \dot{\varepsilon}, \varepsilon, T) = 0$$

where T is the temperature.

(a) If the strain rate $\dot{\varepsilon}$ is imposed on the solid, as when it is sheared at constant rate, the *flow stress* is therefore a function of $\dot{\varepsilon}$, ε, T:

$$\sigma = \sigma(\dot{\varepsilon}, \varepsilon, T)$$

This equation can, in most cases, be written

$$\sigma = \sigma_0 \dot{\varepsilon}^m \varepsilon^\gamma \qquad \text{for } T = \text{const.} \tag{2.18}$$

$$m = \frac{\partial \ln \sigma}{\partial \ln \dot{\varepsilon}} \tag{2.19}$$

is the *strain-rate sensitivity of stress*

$$\gamma = \frac{\partial \ln \sigma}{\partial \ln \varepsilon} = \frac{\epsilon}{\sigma} \frac{\mathrm{d}\sigma}{\mathrm{d}\varepsilon} \tag{2.20}$$

depends on the *work-hardening coefficient* $\mathrm{d}\sigma/\mathrm{d}\varepsilon$.

The influences of $\dot{\varepsilon}$ and T on the flow stress are strongly linked and this will be made clear when we examine the microscopic aspects of flow. For the present, suffice it to say that when the flow stress is *athermal*, i.e. does not depend on T, it does not depend on $\dot{\varepsilon}$ either; this is usually the case in only a limited domain of T. At low temperatures or at high temperatures (we define high temperatures as temperatures higher than about 0·4 T_M, where T_M is the melting temperature) the flow stress usually decreases as T increases and $\dot{\varepsilon}$

decreases. It is therefore equivalent to increase the temperature and to decrease the strain rate.

(b) If a constant stress σ is imposed on the solid, we may write:

$$\dot{\varepsilon} = \dot{\varepsilon}(\sigma, \varepsilon, T) \qquad \text{This is the case of } \textit{creep}$$

Two cases may be considered:

There is work hardening; the flow rate or *creep rate* decreases as the strain ε increases, since deformation becomes more and more difficult and the stress if constant.

There is no work hardening.

This occurs at intermediate or high temperatures depending on whether stress is low or high. We will see, later on, that work hardening results from a change in the microscopic crystalline structure which was brought about by deformation and which can be *annealed* out by thermal agitation (*recovery*). In this case a *steady state flow* or *steady state creep* is obtained and the flow proceeds at constant strain rate under a constant stress

$$\dot{\varepsilon} = \dot{\varepsilon}(\sigma) \qquad \text{for } T = \text{const.} \tag{2.21}$$

For a given σ, $\dot{\varepsilon}$ always increases with T.

It is always possible to write, for steady state creep:

$$\dot{\varepsilon} = \alpha\sigma \qquad T = \text{const.} \tag{2.22}$$

This expression is identical to that of the rate of flow of a viscous fluid, proportional to the shear stress.

$$\alpha \text{ is called the } \textit{fluidity} \text{ and } \eta = \frac{1}{\alpha} = \frac{\sigma}{\dot{\varepsilon}} \text{ is the } \textit{viscosity} \tag{2.23}$$

Two cases must be considered (Fig. 2.23):

(1) $\alpha = \text{const.}$ independent of σ.

This is the case of Newtonian viscous flow: the creep rate is proportional to the shear stress; this occurs only for low stresses.

(2) $\alpha = f(\sigma)$; the fluidity depends on stress.

It generally increases with stress. The creep rate is then empirically expressed for intermediate stresses, as a power function of σ

$$\dot{\varepsilon} = \dot{\varepsilon}_0\sigma^n \qquad T = \text{const.} \tag{2.24}$$

$$n = \frac{\partial \ln \dot{\varepsilon}}{\partial \ln \sigma} \tag{2.25}$$

is the *stress sensitivity of strain rate* and is constant ($n \simeq 3$–5) for intermediate stresses; by comparison with (2.19) we see that $n = 1/m$.

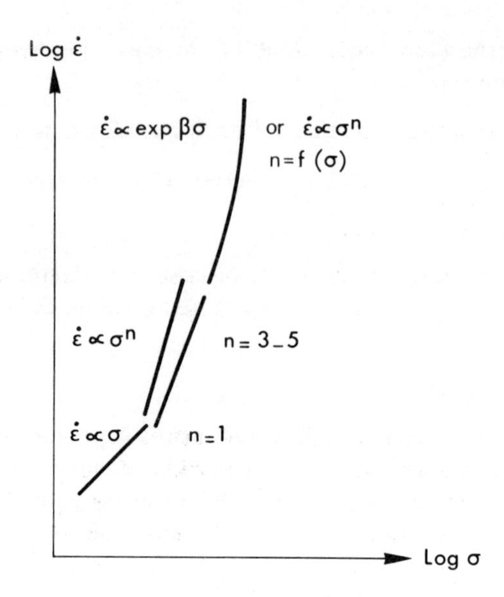

Fig. 2.23. Viscous flow: strain rate ($\dot{\varepsilon}$) versus stress (σ), logarithmic diagram. The slope n is equal to the stress sensitivity of $\dot{\varepsilon}$. $n=1$ Newtonian flow; $n>1$ non-Newtonian flow

For very high stresses, n would depend on the stress. The creep rate can be empirically expressed by an exponential function of σ:

$$\dot{\varepsilon} = \dot{\varepsilon}_0 \exp(\beta\sigma) \tag{2.26}$$

For the Newtonian viscous regime, viscosity is expressed in *poises*: a stress of 1 dyne/cm^2 produces a shear rate of 1 s^{-1} when the viscosity is 1 poise. When the flow is non-Newtonian the viscosity can still be expressed in poises, but for a given stress only.

The fact that $\dot{\varepsilon}$ increases with T can be expressed by saying that viscosity decreases as T increases. From what precedes it can be seen that no qualitative difference is made between the flow of fluids and of solids. In fact the line between fluids and solids is conventionally drawn at a Newtonian viscosity of 10^{15} poises (Cottrell, 1964).

Some orders of magnitude of viscosities are given below:

Water at RT	$\eta = 10^{-2}$ poise
Glycerine at RT	$\eta = 8$
Soda lime glass at glass blowing T	$\eta = 10^7$
Soda lime glass at stress relief annealing T	$\eta = 10^{13}$
Asthenosphere	$\eta \simeq 10^{21}$

Clearly, variation of viscosity with strain rate can be expressed by the strain-rate sensitivity of stress: m, since $\sigma = \eta\ (\dot{\varepsilon})\ \dot{\varepsilon} = \dot{\varepsilon}^m$. A substance with a high strain-rate sensitivity can therefore flow like a fluid if it is slowly strained or develop high stresses leading to fracture if it is rapidly strained (e.g. silicone putty).

CHAPTER 3

Elements of Physical Metallurgy

3.1. GENERALITIES ON LATTICE DEFECTS IN CRYSTALS

Most of the macroscopic properties of crystals and, particularly, the fact that they can plastically deform and flow, can be explained by the existence of lattice defects. Indeed, it may not be an exaggeration to say that the physical metallurgy approach, which will be presented here, depends entirely on knowing the properties of lattice defects. Fortunately, to derive the most fruitful results and understand the physical meaning of the deformation models, it is not necessary to delve in depth into the sometimes intricate theory of crystal defects. This chapter will therefore introduce only the properties of crystal defects which must be known to achieve a good grasp of the physical metallurgy approach.

Definition of lattice defects

A crystal is a three-dimensional periodic array of atoms or groups of atoms. In a perfect crystal every lattice site is indistinguishable from its neighbours. A crystal defect or lattice defect is a site or a group of sites where the periodicity of the lattice is broken. Several kinds of lattice defects can be distinguished:

(i) *Point defects.* The simplest break in periodicity occurs when a lattice site which should be occupied by an atom is empty: this point defect is called a *vacancy*. This defect plays an essential role in transport of matter by solid state *diffusion*. If an atom is inserted between regular lattice sites, the corresponding defect is called a *self-interstitial*. There is also a break in periodicity if an atom different from the atoms of the host crystal is present in the lattice: an *impurity atom* is therefore a point defect; it can be substitutional if it sits on a regular lattice site instead of a matrix atom, or it can be interstitial (Fig. 3.1).

(ii) *Line defects—dislocations.* When the break in periodicity occurs at every point of a line in the crystal, one speaks of a line defect. One could of course think of a line defect produced by the juxtaposition of point defects, but this type of defect would not be stable and is accordingly never found in crystals. The only line defects present in crystals are *dislocation lines*. These defects are extremely important as it is their displacement through the lattice which causes

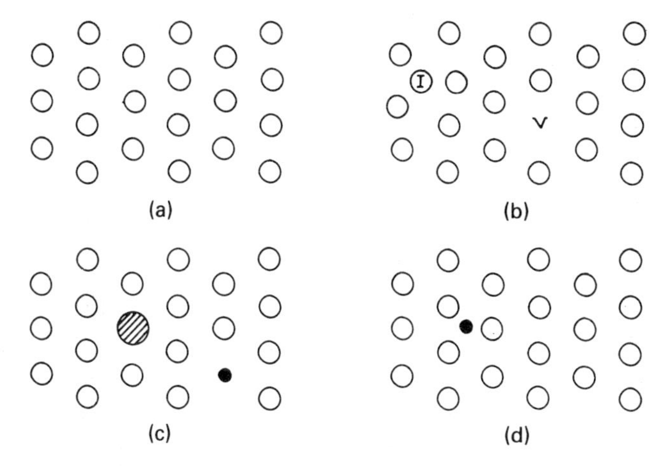

Fig. 3. 1. Point defects: (a) perfect lattice; (b) I: self-interstitial, V: vacancy; (c) substitutional impurity atoms; (d) interstitial impurity atom

plastic deformation; however, the idea of the break in periodicity they introduce cannot be conveyed in a few words, and their description will therefore be postponed for the moment: section 3.2 will be devoted to the characteristic properties of dislocations, which can be used as a definition of the defect.

(iii) *Two-dimensional defects.* Finally, when the break in periodicity occurs over a surface, we speak of a two-dimensional defect. Obviously, the *external surface* of a crystal is such a defect. A polycrystalline material can be described as an aggregate of *grains*, crystals of the same material with various crystallographic orientations, separated by *grain boundaries.* Grain boundaries, of which *twin boundaries* are a particular case, are surface defects. By the same token, the *interface* between two grains of different minerals in a polymineralic rock is a two-dimensional defect.

Two-dimensional arrays of dislocation lines can be considered as two-dimensional defects. We will see later on that such *dislocation walls* separate two slightly misoriented regions of a grain called *subgrains.* Dislocation walls are therefore *subgrain boundaries*, of which *kink band boundaries* are a particular instance. *Stacking faults* are also two-dimensional defects.

3.2. VACANCIES

3.2.1. Formation, migration, sources, sinks

In a crystal in equilibrium with its vapour, the number of atoms is constant: we can therefore describe the formation of a vacancy by means of the following thought experiment:

(1) Take an atom inside the crystal and extract it from its site, leaving a vacancy.

(2) Deposit it on a free surface of the crystal, at a step of the surface, so that the total external area is not changed.

This operation is accompanied by a change of free energy and volume in the crystal.

(i) The change of free energy results from the fact that we had to cut the bonds between the extracted atom and its neighbours. Although we re-establish some bonds by placing the atom on the surface, there is a net number of broken bonds, since an atom at the surface has fewer neighbours than an atom in the bulk. The internal energy of the crystal has increased by ΔE_f corresponding to the work done in breaking the bonds. On the other hand, there has been a change in entropy ΔS_f due to the fact that the frequency of vibration of the atoms has been slightly modified in the vicinity of the vacant site. The free energy of formation of one vacancy is defined as the difference between the free energy of the crystal containing one vacancy and the free energy of the perfect crystal. We have:

$$\Delta F_f = \Delta E_f - T\,\Delta S_f$$

(ii) The change of volume can be analysed in two terms:

An increase by one atomic volume ΔV_a as the extracted atom is deposited on the surface (Fig. 3.2).

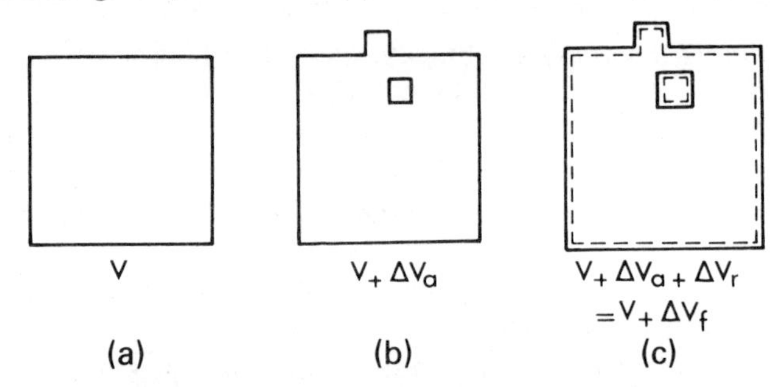

Fig. 3.2. Formation of a vacancy: (a) perfect crystal of volume V; (b) an atom is extracted from its site and deposited at the surface, the volume changes by an atomic volume ΔV_a; (c) the atoms relax around the vacancy. Here the relaxation volume ΔV_r is positive (the atoms relax outward). The total volume change is ΔV_f: formation volume of a vacancy

A variation by a smaller volume ΔV_r due to some relaxation of the atoms around the vacancy. This *relaxation volume* ΔV_r is taken as negative if the atoms relax inward and positive if they relax outward (in ionic crystals ΔV_r is usually positive since the suppression of, say, a positively charged ion is equivalent to the creation of a negative charge which repels the negatively charged neighbours).

We can therefore define the *formation volume* of one vacancy as:

$$\Delta V_f = \Delta V_a + \Delta V_r \tag{3.1}$$

The enthalpy of formation of one vacancy is therefore by definition:

$$\Delta H_f = \Delta E_f + P\,\Delta V_f \tag{3.2}$$

where P is the hydrostatic pressure, which opposes the volume change. The free enthalpy of formation of one vacancy is:

$$\Delta G_f = \Delta H_f - T\,\Delta S_f = \Delta E_f + P\,\Delta V_f - T\,\Delta S_f \tag{3.3}$$

Once a vancancy is formed it can migrate in the crystal by exchange with a neighbouring atom: the atom jumps into the vacancy and leaves a vacancy in its place. To jump into the vacancy the atom has to go through a saddle-point position, forcing apart its neighbours. We have therefore to introduce a *migration energy* ΔE_m, a *migration volume* ΔV_m, and a *free enthalpy of migration* ΔG_m.

When a vacancy migrates to the surface it will exchange with an atom of the surface and disappear. The free surface can therefore act as a *sink* for vacancies by absorbing them and as a *source* of vacancies, as atoms just under the surface jump on to it, leaving vacancies that will migrate throughout the crystal; however, free surfaces (or grain boundaries) are the main vacancy sources and sinks only for very small crystals (grains) when the ratio of surface to volume is high. This is the case for grains whose size does not exceed a few microns. For larger crystals, the contribution of the external surface is minor and the main sources and sinks for vacancies are the dislocation lines as we will see later on.

3.2.2. Equilibrium concentration of vacancies at a given temperature

All real crystals contain vacancies and there exists an equilibrium concentration of vacancies which is an increasing function of temperature.

The equilibrium concentration of vacancies for a given temperature is the number of vacancies per unit volume that the crystal must contain to be in the state of minimum free enthalpy. It may sound surprising that a crystal containing vacancies may have a lower free enthalpy than the perfect crystal, since we have seen that the formation of one vacancy increases the free enthalpy of the crystal by ΔG_f. However, in computing the increase in total free enthalpy of a crystal containing n_v vacancies distributed over n sites, one must not forget to take into account the entropy of configuration S_c resulting from the fact that there are W different ways of distributing n_v vacancies over n sites. The increase in free enthalpy of a crystal containing n_v vacancies is then:

$$\Delta G = n_v\,\Delta G_f - TS_c$$

The existence of a configuration entropy, therefore, lowers G.

By definition:

$$S_c = k \ln W$$

where k is Boltzmann's constant.

There are n possible sites to put the first vacancy, $(n-1)$ to put the second one, etc., and $(n-n_v+1)$ possible sites to put the last of the n_v vacancies.

The total number of different configurations is therefore:

$$W = \frac{n(n-1)(n-2)\ldots(n-n_v+1)}{p}$$

where p is the number of permutations of n_v vacancies

$$p = 1\,.\,2\,.\,3\ldots n_v = n_v!$$

(We have to divide by p, since the vacancies are indistinguishable, and we arrive at the same configuration, independently of the order in which we arrange the n_v vacancies.)

We can write:

$$W = \frac{n(n-1)(n-2)\ldots(n-n_v+1)}{n_v!} = \frac{n!}{(n-n_v)!\,n_v!}$$

The increase in the total free enthalpy of the crystal due to n_v vacancies is:

$$\Delta G = n_v\,\Delta G_f - kT \ln W$$

A first approximation of W can be found by using Stirling's formula

$$\ln(N!) \simeq N \ln N - N$$

hence:

$$\ln W = \ln(n!) - \ln(n_v!) - \ln[(n-n_v)!] \simeq n_v \ln\left(\frac{n}{n_v} - 1\right) - n \ln\left(1 - \frac{n_v}{n}\right)$$

as $n/n_v \gg 1$, we can write:

$$\ln\left(\frac{n}{n_v} - 1\right) \simeq \ln \frac{n}{n_v}$$

and

$$\ln\left(1 - \frac{n_v}{n}\right) \simeq -\frac{n_v}{n}$$

hence:

$$\Delta G \simeq n_v\,\Delta G_f - kT\left(n_v \ln \frac{n}{n_v} + n_v\right)$$

The condition for minimum free enthalpy can now be written as a function of the number of vacancies:

$$\frac{\partial \Delta G}{\partial n_v} = 0$$

$$\Delta G_f - kT \ln \frac{n}{n_v} = 0$$

which gives the equilibrium atomic fraction of vacancies at the temperature T:

$$\boxed{N_v = \frac{n_v}{n} = \exp - \left(\frac{\Delta G_f}{kT}\right)} \tag{3.4}$$

(For metals at the melting point, n_v/n is of the order of 10^{-4} or 10^{-5}) (Table 3.1).

The equilibrium concentration can be calculated in the same way for interstitials, but the free enthalpy of formation of one interstitial is extremely high since an interstitial atom causes a very strong distortion of the atomic bonds. Consequently, the equilibrium concentration of interstitials at all temperatures is so small that it can be completely neglected. We will therefore never consider interstitials (Table 3.1).

Table 3.1. Atomic fraction of defects in copper

T °K	300	800	1300
$\frac{T}{T_M}$	0·22	0·59	0·96
Vacancies	10^{-17}	6×10^{-7}	$1{\cdot}25 \times 10^{-4}$
Interstitials	10^{-67}	10^{-25}	10^{-15}

3.3. SELF-DIFFUSION BY A VACANCY MECHANISM

We have seen that, as temperature increases, there is an increasing atomic fraction of vacant sites in a crystal. These vacancies migrate throughout the crystal by exchange with lattice atoms. From the viewpoint of the lattice atoms, it means that a given atom does not stay on one lattice site but wanders through the lattice by jumping into vacancies. This is the most widespread and most documented mechanism for solid-state diffusion. Diffusion is an extremely important process in crystals at temperatures above about one-third of the melting point (expressed in Kelvins). Knowing the laws of diffusion is indispensable for the physical understanding of phase transformation kinetics, high temperature plastic flow, recrystallization and in general all high-grade metamorphic processes. In what follows, the basic relevant concepts and laws are introduced in the simplest case of pure elements (one type of atom); the case of ionic crystals is then considered briefly. (For a complete treatment, see Adda and Philibert, 1966.)

3.3.1. Concepts and laws

We will consider a crystal containing only one type of atom, say a face-centred-cubic metal like copper, and examine successively the free wandering of a vacancy through the crystal and the transport of matter occurring in the presence of a driving force for diffusion. It is important to point out, even at this stage, that the only driving force for diffusion is a concentration gradient (see § (ii) in this section). It must be clearly recognized that, when a non-hydrostatic stress for instance, causes a transport of matter by diffusion (as in Kamb's theory of piezocrystallization or Nabarro–Herring creep), the stress is not the driving force for diffusion but sets up a vacancy concentration gradient which causes matter transport by self-diffusion. The problem of the influence of stress on the diffusion processes is theoretically very difficult and, experimentally, it does not seem that shear stresses affect diffusion in a noticeable way.

The reader should therefore not be surprised at the absence of a paragraph devoted to diffusion under non-hydrostatic stress. The stress-directed transport of matter will be dealt with when the flow mechanisms are studied.

(i) Let us assume that all vacancies in the crystal can jump independently of each other, that the vacancy has no memory (that is, its successive jumps are independent) and that at every moment it has an equal probability to exchange with one of its nearest neighbour atoms. The motion of the population of vacancies in the crystal is then analogous to the Brownian motion of tiny particles in suspension in a fluid.

If we focus our attention on one vacancy, situated on a given lattice site at a given time, it behaves much as a drunken sailor staggering away from a bar: every one of his steps is taken in a random direction independent of the previous one; however, after a long time, the sailor has gone a certain distance from the bar. This distance can only be known on the average (the average being taken over many trial runs) and determining its value constitutes the answer to the so-called Random Walk Problem. Einstein has shown that the mean square distance of a particle from the origin after N random steps, each of length L, is given by:

$$\langle R_N^2 \rangle = NL^2 \tag{3.5}$$

(the carets mean average value).

If the number of steps is proportional to time, the mean square distance is proportional to time. Thus, in random walk, the average distance of a particle from the origin is proportional to $\sqrt{t}$, whereas the distance would be proportional to t if the particle were drifting away at constant velocity.

A simple demonstration of Einstein's formula can be found in Feynman's Lectures on Physics: if the position of a particle with respect to the origin is given by the vector $\mathbf{R}_{N-1}$ after $N-1$ steps defined by L, after N steps it is given by:

$$\mathbf{R}_N = \mathbf{R}_{N-1} + \mathbf{L} \qquad \text{(see Fig. 3.3)}$$

the mean square distance is given by the scalar product averaged over many trials:

$$\begin{aligned}\langle \mathbf{R}_N \cdot \mathbf{R}_N \rangle &= \langle R_N^2 \rangle = \langle (\mathbf{R}_{N-1} + \mathbf{L})^2 \rangle \\ &= \langle \mathbf{R}_{N-1} \cdot \mathbf{R}_{N-1} \rangle + \langle 2\mathbf{R}_{N-1} \cdot \mathbf{L} \rangle + \langle \mathbf{L} \cdot \mathbf{L} \rangle \\ &= \langle R_{N-1}^2 \rangle + L^2\end{aligned}$$

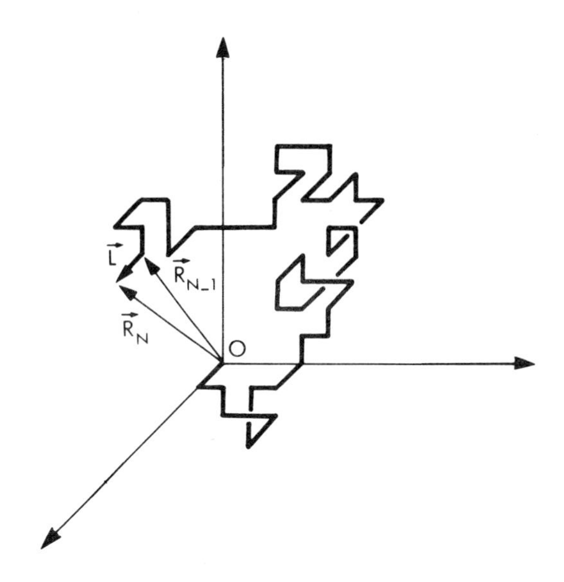

Fig. 3.3. Random walk in a simple cubic lattice. The mean square distance from the origin after N steps of length L is $\langle R_N^2 \rangle = NL^2$

(the average of the scalar product $\mathbf{R}_{N-1} \cdot \mathbf{L}$ is equal to zero, since on the average it will be possible to group the values in pairs of opposite signs). By recurrence, we obtain (3.5)

$$\langle R^2 \rangle = NL^2$$

For a vacancy in a crystal, we can define several jump distances and jump frequencies toward the nearest neighbours which are not in general at the same distance. However, in the case of a cubic crystal we have only one jump distance δl and one jump frequency Γ_j. The total jump frequency Γ, equal to the total number of jumps per second is given by:

$$\Gamma = Z\Gamma_j \tag{3.6}$$

where Z is the number of nearest neighbours.

Einstein's formula (3.5) is directly applicable and the mean square distance covered by a vacancy during time t is:

$$\langle R^2 \rangle = \Gamma t \, \delta l^2$$

The ratio:

$$D_v = \frac{\langle R^2 \rangle}{6t} = \frac{\Gamma}{6} \delta l^2 \tag{3.7}$$

is called the *diffusion coefficient of the vacancy*. (The factor 1/6 is introduced to preserve consistency with another definition of the diffusion coefficient, as we will see later on.)

The diffusion coefficient of the vacancy is therefore a property of the crystal expressing the mobility of vacancies; it depends only on the jump distance and the jump frequency.

In all cubic crystals of lattice parameter a it is easily shown that:

$$D_v = \Gamma_j a^2 \tag{3.8}$$

(in face-centred cubic crystals

$$\delta l = \frac{a}{\sqrt{2}} \qquad \Gamma = 12\Gamma_j$$

in body-centred cubic crystals

$$\delta l = \frac{a\sqrt{3}}{2} \qquad \Gamma = 8\Gamma_j$$

Now, Γ_j is equal to the jump frequency of an atom into a vacancy; it is proportional to the number of times per second ν the atom tries to jump and to the probability it has of succeeding in jumping over the potential barrier.

The attempt frequency ν depends on the lattice vibration frequencies of the atoms and is usually of the order of 10^{13} s^{-1}.

For the atom to make a successful jump, the thermal agitation must provide the free enthalpy of migration ΔG_m, and the probability of this occurring is proportional to: $\exp -\Delta G_m/kT$ (Boltzmann statistics). We can therefore write, for a cubic crystal:

$$\boxed{D_v = a^2 \nu \exp -\left(\frac{\Delta G_m}{kT}\right)} \tag{3.9}$$

We notice immediately that the mobility of vacancies increases considerably as the temperature increases. As we have seen, the wandering of vacancies through the lattice obviously induces a wandering of atoms, the diffusion coefficient of the atoms of the crystal is called *self-diffusion coefficient* D_{Sd}. It expresses their mobility in the crystal; however, it must be clearly recognized that it is not equal to the diffusion coefficient of vacancies D_v, even though the basic step, the exchange of an atom and a vacancy, is the same. The reason is that there are always Z neighbouring atoms around a vacancy, so that the jump of the vacancy only depends on the probability of any of the neighbours jumping into it. But if we now focus our attention on one atom in the crystal, it can move only if there is a vacancy close to it, which is not always the case; the jump frequency of an atom is therefore proportional to the probability of a vacancy being present in a neighbouring site multiplied by the probability of making a successful jump into the vacancy. The probability of a vacancy being present can be taken as equal to the atomic fraction of vacancies present in the crystal in thermal equilibrium, given by (3.4):

$$N_v = \exp -\left(\frac{\Delta G_f}{kT}\right) \tag{3.4}$$

We have therefore:

$$\begin{aligned} D_{Sd} = D_v N_v &= a^2 \nu \exp\left(-\frac{\Delta G_f}{kT}\right) \exp\left(-\frac{\Delta G_m}{kT}\right) \\ &= a^2 \nu \exp -\left(\frac{\Delta G_f + \Delta G_m}{kT}\right) \\ &= a^2 \nu \exp\left(\frac{\Delta S_f + \Delta S_m}{k}\right) \exp -\left(\frac{\Delta H_f + \Delta H_m}{kT}\right) \end{aligned}$$

which we can write:

$$\boxed{D_{Sd} = D_0 \exp -\left(\frac{\Delta H_{Sd}}{kT}\right)} \tag{3.10}$$

with:

$$\Delta H_{Sd} = \Delta H_f + \Delta H_m \tag{3.11}$$

$$D_0 = a^2 \nu \exp\left(\frac{\Delta S_f + \Delta S_m}{k}\right)$$

is the *frequency factor.* ΔH_{Sd} is the *activation enthalpy for self-diffusion* by a vacancy mechanism, and is equal to the sum of the enthalpy of formation and the enthalpy of migration of the vacancies.

It is interesting to make explicit the *effect of hydrostatic pressure* on the self-diffusion coefficient by writing (3.10):

$$D_{Sd} = D_0 \exp -\left(\frac{\Delta E_f + \Delta E_m}{kT}\right) \exp\left(-P\frac{\Delta V_f + \Delta V_m}{kT}\right)$$

or:

$$D_{Sd} = D_0 \exp -\left(\frac{\Delta E_{Sd} + P\Delta V_{Sd}}{kT}\right) \tag{3.12}$$

ΔE_{Sd} is the activation energy for self-diffusion and

$$\Delta V_{Sd} = \Delta V_f + \Delta V_m \tag{3.13}$$

the activation volume for self-diffusion. (The physical meaning of the formation volume ΔV_f and the migration volume ΔV_m of a vacancy has been given in § 3.2.)

In most cases ΔV_{Sd} is positive, and the coefficient of diffusion decreases as the hydrostatic pressure increases:

$$\frac{d(\ln D_{sd})}{dP} = -\frac{\Delta V_{Sd}}{kT} \tag{3.14}$$

(ii) Up to now we have considered diffusion only from the viewpoint of the mobility of vacancies and atoms. It is clear that this mobility will play an important part in the transport of vacancies or matter if there is a driving force for transport.

Two physical kinds of driving forces for transport of particles must be considered: the first one can be looked upon as a force acting on the particle and imparting a velocity to it, as in the case of an electrically charged particle being carried along the lines of force of an electric field. This force results from the existence of a potential energy gradient. The particles drift down the energy gradient and this lowers the total free enthalpy of the system.†

The second kind of driving force is the only one which will concern us here as it is the main cause of the transport of matter by diffusion in crystals. It is still instrumental in bringing about a decrease in the total free enthalpy of the system but this time by increasing the entropy, that is to say, the disorder of the system. This will be made clear by an example: let us consider, at time t_0, a system formed by a thin layer of radioactive isotope of a substance A^* sandwiched between two crystals of the same non-radioactive substance A. The atoms A^* are identical with atoms A but they can be distinguished from A by their radioactivity. The total system is in a highly ordered state since all A^* atoms are lumped together in the thin layer. Now A^* atoms move about by exchange with vacancies exactly as A atoms do. They will therefore wander away from the initial thin layer according to Einstein's law. Even though no external force acts on them they will be carried by diffusion from the layer where their concentration is 1 to the edges of the sandwich where their initial concentration was zero. After an infinite amount of time, A^* will be homogeneously distributed on the whole specimen, and as there are many different ways of distributing A^* atoms among all the lattice sites, the entropy of the system will be higher and its free enthalpy lower.

A concentration gradient is therefore a 'driving force' for diffusion.

Let us now find an expression for the transport by diffusion in these conditions.

It must be pointed out that the diffusion law we are going to derive is extremely general, whatever the diffusion particles or species may be (vacancies, interstitials, radioactive tracer atoms, lattice atoms, impurity atoms, etc.). In fact, the diffusion need not necessarily take place in a crystal.

So let us consider a population of particles able to move about by random walk (with all the attached hypotheses). Let us choose a system of axes Ox Oy Oz and suppose that

† The gravitational potential of the Earth could, in principle, give rise to such a driving force but it is so small as to be entirely negligible even on the geological time scale.

the concentration of particles in a small volume is a function of its position. We will assume that the concentration gradients are very small, for, if this were not the case, the environment of a particle could not be considered as identical at different points, and the hypothesis of the random walk would not be fulfilled. The gradient of concentration gives rise to a flux of particles. Let us first calculate the flux J_x along the axis Ox. J_x is the number of particles which cross per unit time, in the positive direction, the unit area of a plane P perpendicular to Ox; let us consider on each side of plane P two planes P_1P_2 distant from P by a distance $\Delta = \sqrt{(\langle X^2 \rangle)}$, $\langle X^2 \rangle$ being the mean square distance covered by the particles during time t (Fig. 3.4).

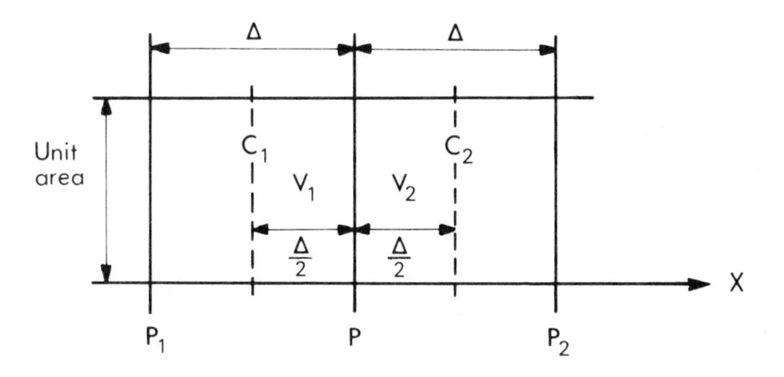

Fig. 3.4. Diffusion under a gradient of concentration: C_1: Concentration of particles left of P; C_2: Concentration of particles right of P; (see text)

The unit area of plane P will be crossed during time t by particles coming from the two volumes bounded by P_1 and P_2 on the left and right of P. But P will be crossed by only half of the particles contained in these volumes since the other half has crossed planes P_1 and P_2. So P is crossed from left to right by the particles contained in volume V_1 and from right to left by the particles contained in volume V_2 (Fig. 3.4). The concentration (number per unit volume) of particles is supposed to be uniform in V_1 and V_2 and respectively equal to C_1 and C_2, the concentrations halfway between P and P_1 and P and P_2. So volume V_1 contains $\frac{1}{2}\Delta \cdot 1 \cdot C_1$ particles and V_2 contains $\frac{1}{2}\Delta \cdot 1 \cdot C_2$ particles.

The net number of particles crossing the unit area of P per unit time, in the positive sense on Ox is therefore:

$$J_x = \frac{1}{2}\frac{\Delta}{t}(C_1 - C_2)$$

As we have assumed that the gradient of concentration was very small we can write:

$$\frac{\mathrm{d}c}{\mathrm{d}x} \simeq \frac{C_2 - C_1}{\Delta}$$

as $\Delta^2 = \langle X^2 \rangle$ we can therefore write

$$J_x = -\frac{1}{2}\frac{\langle X^2 \rangle}{t}\frac{\mathrm{d}c}{\mathrm{d}x}$$

This can be written

$$\boxed{J_x = -D_x \frac{\mathrm{d}c}{\mathrm{d}x}} \tag{3.15}$$

with

$$D_x = \frac{\langle X^2 \rangle}{2t} \tag{3.16}$$

This is *Fick's law*, which states that the flux of diffusing particles is proportional to the gradient of concentration (driving force). The proportionality coefficient is the *diffusion coefficient*, and it is constant; therefore Fick's law is linear only in the limit of validity of the assumption that the concentration variations are very small.

The diffusion coefficient given by (3.16) is identical with the diffusion coefficient defined by (3.7) if the diffusion is isotropic. The mean square distance is:

$$\langle R^2 \rangle = \langle X^2 \rangle + \langle Y^2 \rangle + \langle Z^2 \rangle$$

If the diffusion is isotropic the mean square distances along every one of the axes are equal and $\langle R^2 \rangle = 3\langle X^2 \rangle$, (3.16) can then be written:

$$D = \frac{\langle R^2 \rangle}{6t}$$

which is identical with (3.7).

Diffusion in crystals is isotropic only for cubic crystals; for other crystals, 3 diffusion coefficients D_x, D_y, D_z can be defined in the most general case. However, the anisotropy of diffusion is seldom very important.

3.3.2. Measurement of diffusion coefficients

Fick's law (3.15) is independent of time and valid only for steady-state diffusion. But the diffusion coefficient is usually measured in experiments such as the one described earlier, where the concentration of a radioactive tracer A* is measured as a function of the distance from the original thin layer in a sandwich, after a given time t. Here, we have to deal with a transient regime where concentration depends on time. The law of conservation of matter must then be taken into account and can then be simply expressed in the following way:

Let us consider an infinitely small cylindrical volume element bounded by two planes PP′ distant by dx along the Ox axis (Fig. 3.5), and of unit cross-section. Let $c(x, t)$ be the concentration of particles in the volume. In a time dt, $J(x)$ dt particles enter the volume through P and $J(x + \mathrm{d}x)$ dt leave it

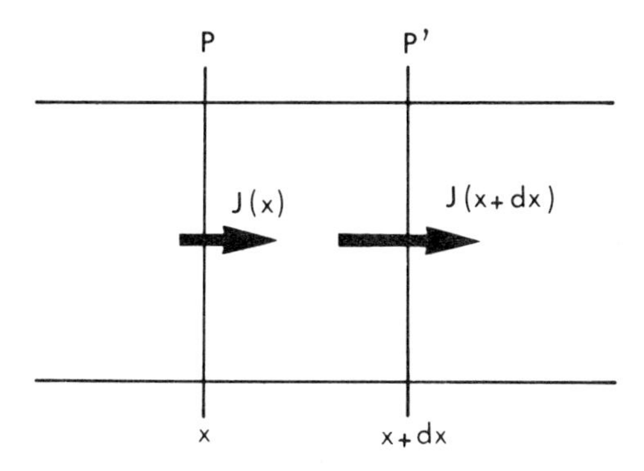

Fig. 3.5. Diffusion under a gradient of concentration: $J(x)$, $J(x+\mathrm{d}x)$: net number of particles crossing planes at x and $x+\mathrm{d}x$ from left to right per unit time (see text)

through P′. The net number of particles which enter the volume is:

$$[J(x)-J(x+\mathrm{d}x)]\,\mathrm{d}t=-\frac{\partial J}{\partial x}\,\mathrm{d}x\,\mathrm{d}t$$

This causes a variation in concentration

$$\mathrm{d}c=\frac{[J(x)-J(x+\mathrm{d}x)]\,\mathrm{d}t}{1\cdot\mathrm{d}x}=\frac{\partial c}{\partial t}\mathrm{d}t$$

The law of conservation of matter can therefore be written

$$\frac{\partial c}{\partial t}=-\frac{\partial J}{\partial x} \tag{3.17}$$

(3.17) and (3.15) give Fick's 2nd law:

$$\boxed{\frac{\partial c}{\partial t}=D\frac{\partial^2 c}{\partial x^2}} \tag{3.18}$$

The solution $c(x, t)$ of this differential equation depends of course on the boundary conditions set by the experiment. For instance, in the tracer sandwich experiment where we start with a thin layer of M atoms of A* per unit area between two 'infinite' crystals of A, the solution can be shown to be:

$$c=\frac{M}{2\sqrt{(\pi Dt)}}\exp\left(-\frac{x^2}{4Dt}\right) \tag{3.19}$$

The curve $c(x)$ for a given time is gaussian (Fig. 3.6).

Thus, the coefficient of self-diffusion of the tracer atoms A* is usually measured by this type of experiment. After a diffusion anneal at a temperature

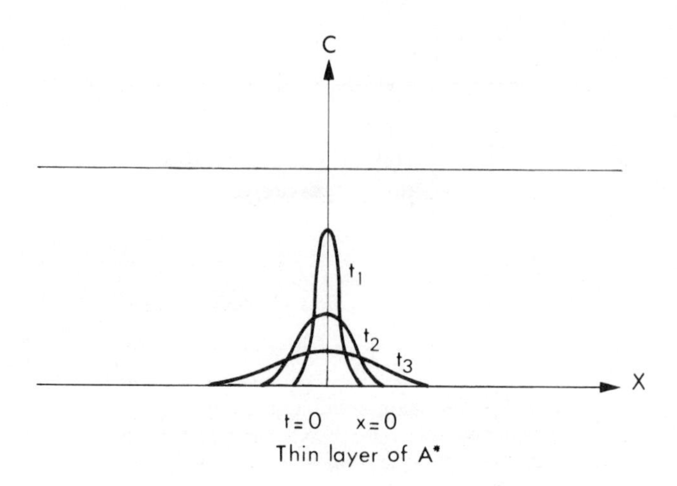

Fig. 3.6. Diffusion of a thin layer of radioactive tracer A^* at $x=0$ for $t=0$. Curves concentration–distance for times $t_1<t_2<t_3$

T during a time t, the concentration of A^* is determined as a function of the penetration x by slicing the specimen and counting the activity of the slices. The tracer self-diffusion coefficient D^* is determined from the curve.

We have mentioned that it is important to know the coefficient of self-diffusion of the atoms of a crystal D_{SD}. Now, the type of experiment we have just considered enables us to determine the self-diffusion coefficient D^* of atoms identical in every respect with the atoms of the crystal except that they can be traced by their radioactive emission; this is precisely why we are able to follow their diffusion and measure the coefficient. There would be no direct way to follow the diffusion of unmarked atoms. It may therefore come as a surprise that the tracer self-diffusion coefficient is *not* equal to the self-diffusion coefficient of the crystal atoms. This stems from the fact that the tracer atom can be traced and keeps its identity after a jump, whereas a crystal atom is indistinguishable from its neighbours. In the case of self-diffusion, each jump of an atom is a new event, independent of the previous jumps and there is truly no memory. Statistically we have the right to apply the random walk theory. On the contrary, after a tracer atom has jumped into a vacancy, it has not an equal number of chances of jumping in any direction; we still know where it is and we know that it has more chances of falling back into the vacancy it has left, unless the vacancy has migrated away. This is called a *correlation* effect and is expressed by a *correlation factor* f which can be calculated for every structure and which makes the tracer diffusion coefficient slightly smaller than the self-diffusion coefficient:

$$D^* = fD_{SD} \tag{3.20}$$

(For diffusion by a vacancy mechanism in a FCC crystal for instance, $f \simeq 0{\cdot}78$.)

The self-diffusion coefficients can therefore be determined from tracer diffusion measurements, if we apply the relation

$$D_{SD} = \frac{D^*}{f} \tag{3.21}$$

3.3.3. *Formulae*

Table 3.2 sums up the principal relations we have just established.

Table 3.2.

$$N_v = N_{0v} \exp -\left(\frac{\Delta E_f + P\Delta V_f}{kT}\right)$$

$$D_v = D_{0v} \exp -\left(\frac{\Delta E_m + P\Delta V_m}{T}\right)$$

$$D_{SD} = D_{0Sd} \exp -\left(\frac{\Delta E_{Sd} + P\Delta V_{Sd}}{kT}\right) = N_v D_v$$

$$\begin{cases} \Delta E_{SD} = \Delta E_f + \Delta E_m \\ \Delta V_{SD} = \Delta V_f + \Delta V_m = -\dfrac{d(\ln D_{SD})}{dP} \\ \Delta H_{SD} = \Delta E_{SD} + P\,\Delta V_{SD} = -k\dfrac{d(\ln D_{SD})}{d(1/T)} \end{cases}$$

$$D^* = fD_{SD} \qquad f \leqslant 1$$

Units

(1) Diffusion coefficients are expressed in $cm^2\ s^{-1}$.

(2) Activation enthalpies or energies can be expressed in two units:

in electron volts per atom if the formula is written

$$D_{SD} = D_{0SD} \exp -\left(\frac{\Delta H_{Sd}}{kT}\right)$$

with $k = 8{\cdot}6 \times 10^{-5}$ eV/°K (Boltzmann's constant).

in calories per mole if the formula is written

$$D_{SD} = D_{0SD} \exp -\left(\frac{\Delta H_{Sd}}{RT}\right)$$

with $R = 1{\cdot}986 \simeq 2$ calories/mole °K (gas constant).

(3) T is in Kelvins.

For most metals $D \simeq 10^{-8}\ cm^2\ s^{-1}$ at melting temperature, and $D \simeq 10^{-16}\ cm^2\ s^{-1}$ at half the absolute melting temperature.

For most oxides the coefficient of self-diffusion of oxygen at melting temperature is between 10^{-11} and $10^{-9}\ cm^2\ s^{-1}$.

Knowing the role of pressure is especially interesting for diffusion in minerals at great depths. Unfortunately the experimental determination of activation volumes is not an easy one; however, some empirical formulae do give a correct order of magnitude. Let us quote for instance the formula:

$$D = D_0 \exp -\left(\frac{gT_m}{T}\right) \qquad g \simeq 18 \text{ (in metals)} \tag{3.22}$$

Weertman (1970) has verified that it gives a reasonable estimate of D for most materials if T_m is set equal to the absolute melting temperature under the given hydrostatic pressure.*

3.3.4. Diffusion in ionic crystals

While the generalities previously given remain valid, the treatment of diffusion is much more complicated in crystals containing several kinds of atoms or groups of atoms in complex lattices, which is the case of practically all minerals. Besides, there are very few experimental data on diffusion coefficients.

In this paragraph we will merely point out the principal particularities of diffusion in the simplest cases of ionic crystals: alkali halides and monometallic oxides MO, with NaCl structure.

An ionic crystal consists of at least two intermeshed independent sub-lattices. In the case of a NaCl type crystal for instance, we have two face-centred cubic lattices, one of negatively charged anions and one of positively charged metallic cations. Vacancies exist on each of the sub-lattices and they are of course electrically charged: cationic vacancies corresponding to the removal of a positive ion are negatively charged, whereas anionic vacancies are positively charged. The fact that the vacancies are charged has several important consequences:

(i) Vacancies are formed in pairs of oppositely charged vacancies (Schottky pairs) which may migrate independently on their respective sub-lattices or remain bound and neutral.

The atomic fractions of cationic and anionic vacancies N_c and N_a in thermal equilibrium are limited by two relations:

$$N_c = N_a \tag{3.23}$$

expressing the electrical neutrality of the crystal, and:

$$N_0^2 = N_a N_c = \exp\left(\frac{\Delta S_f}{k}\right) \exp -\left(\frac{\Delta H_f}{kT}\right) = \exp -\left(\frac{\Delta G_f}{kT}\right) \tag{3.24}$$

where ΔG_f is the free enthalpy of formation of a Schottky pair. This relation is analogous to the expression of a solubility product.

* Formula (3.22) is not necessarily valid for binary systems, where the solidus and the liquidus temperatures may be widely different.

Hence the equilibrium concentration of one type of vacancy is given by:

$$N_a = N_c = \exp\left(\frac{\Delta S_f}{2k}\right) \exp-\left(\frac{\Delta H_f}{2kT}\right) \tag{3.25}$$

The enthalpy of migration, however, may be very different for the anionic vacancy and the cationic vacancy. So the activation enthalpy for the self-diffusion of each type of ion on its own sub-lattice will not be the same.

$$\text{For cations:} \quad \Delta H_{SDc} = \frac{\Delta H_f}{2} + \Delta H_{mc} \tag{3.26}$$

$$\text{For anions:} \quad \Delta H_{SDa} = \frac{\Delta H_f}{2} + \Delta H_{ma} \tag{3.27}$$

In most cases the cations are much more mobile than the anions (the ionic conductivity of alkali halides is mainly due to the diffusion of cations).

(ii) In all cases, electrical neutrality of the crystal must be preserved; the role of foreign aliovalent atoms then becomes very important: let us take as an example the case of NaCl where calcium ions are commonly present in substitutional solid solution.

When a Ca^{++} ion sits on a lattice site where a Na^+ ion should be, there is an excess of one positive charge. This is compensated by the creation of a negatively charged cationic vacancy (extrinsic vacancy). If the atomic fraction of substitutional divalent ions is C, the electrical neutrality condition is written:

$$N_c = C + N_a \tag{3.28}$$

instead of (3.23) for a pure crystal.

As the mass action law (3.24) remains valid, we can carry (3.28) into (3.24) and we get a 2nd degree equation in N_a or N_c whose solutions are:

$$N_c = \frac{C}{2}\left[1 + \left(1 + \frac{4N_0^2}{C^2}\right)^{1/2}\right] \tag{3.29}$$

$$N_a = \frac{C}{2}\left[\left(1 + \frac{4N_0^2}{C^2}\right)^{1/2} - 1\right] \tag{3.30}$$

We can now see that there will be two different regimes of self-diffusion according to the value of temperature and concentration in aliovalent impurities.

If $C \ll N_0$, that is, if the atomic fraction of impurities is negligible before the atomic fraction of thermal vacancies in equilibrium, the self-diffusion enthalpies are given by (3.26) and 3.27). This is the case for very pure crystals (C very small) and/or crystals at high temperature (N_0 very large). This is called the *intrinsic regime.*

If $C \gg N_0$, the concentration of cationic vacancies is now fixed by the concentration of impurities.

$$N_c \simeq C \tag{3.31}$$

As these vacancies did not have to be formed thermally, the activation enthalpy for the cation diffusion is only:

$$\Delta H_{SDc} = \Delta H_{mc} \tag{3.32}$$

On the other hand, the concentration of anionic vacancies is much reduced (3.30), so the self-diffusion of the anion is also much reduced. This is the *extrinsic regime*, which occurs for more impure crystals and/or lower temperatures.

For still lower temperatures and even more impure crystals the extrinsic vacancies may remain bound to the impurity atom and so contribute less to diffusion (*association regime*).

In addition to the effect of impurities on diffusion just considered, we must mention another very important effect resulting from the *imperfect stoichiometry* of many ionic crystals, especially metallic oxides where there is a deficit or an excess of oxygen and accordingly a different number of vacancies in the two sub-lattices. It is sufficient to say here that the diffusion of the ions then depends on the partial pressure of oxygen in the surrounding medium.

(iii) Let us now consider the problem of transport of matter by a vacancy mechanism in ionic crystals. The importance of diffusive transport of matter cannot be over-emphasized, since all physical models for high temperature deformation of solids depend on it directly or indirectly, as will be seen later; unfortunately, due to the existence of several differently charged ions, to the presence of numerous impurities and non-stoichiometry defects, very little is known theoretically or experimentally on diffusion in rock-forming minerals. It is therefore not unusual to extend to complex minerals (e.g. silicates) the essential conclusions of studies made on the much simpler case of pure stoichiometric and purely ionic crystals of NaCl type.

In this case, if transport of matter occurs by vacancy diffusion, the existence of opposite electrical charges on the anionic and cationic vacancies imposes a coupling between the diffusive flux of each species: even though the anionic and cationic vacancies migrate independently with different mobilities, their fluxes are coupled so that the net electrical current is zero in any volume element of the crystal.

In a crystal $A_\alpha B_\beta$ where the electrical charges of A and B ions are respectively q_A and q_B, the condition of zero net electrical current is written:

$$q_A J_A + q_B J_B = 0 \tag{3.33}$$

where J_A and J_B are the fluxes of A and B ions.

In a purely ionic crystal:

$$\frac{q_A}{q_B} = -\frac{\beta}{\alpha}$$

Note that the condition of no net current (3.33) is not the same as the condition of no net charge (3.23); the latter involves only the equilibrium concentration of vacancies, hence their free enthalpy of formation, whereas the former involves also their mobility, hence their free enthalpy of migration.

The coupling of the vacancy fluxes ensures that the composition of transported matter remains constant. It can be shown (Ruoff, 1965) that Fick's equation for each type of ion can be written with a global diffusion coefficient D':

$$J_A = -D'\frac{dC_A}{dx}$$
$$J_B = -D'\frac{dC_B}{dx} \tag{3.34}$$

C_A and C_B are the respective concentrations of A and B vacancies.

$$D' = \frac{D_A D_B}{D_A n_B + D_B n_A} \tag{3.35}$$

where D_A and D_B are the diffusion coefficients of A and B ions and n_A, n_B the atomic fractions of A and B ions:

$$n_A = \frac{\alpha}{\alpha + \beta}$$

$$n_B = \frac{\beta}{\alpha + \beta}$$

For alkali halides or monometallic oxides of NaCl type, $\alpha = \beta$ and:

$$D' = \frac{D_A D_B}{D_A + D_B} \tag{3.36}$$

If the species A diffuses much faster than B and we can write:

$$D_A \gg D_B$$

it follows that

$$D' \simeq D_B$$

The transport of matter is then controlled by the diffusion coefficient of the slower moving species, which is usually the bulkier anion.

Table 2.3 gives experimental diffusion coefficients for anions and cations, in a few typical materials.

Table 3.3.

Mineral	Diffusing ion	T °C	D cm²/s	D_0 cm²/s	ΔH kcal/mole	Ref.
NaCl (Halite)	Na^+	500	$5{\cdot}34\times10^{-12}$	33·16	45·54	1
		750	$7{\cdot}15\times10^{-9}$			
	Cl^-	500	$9{\cdot}09\times10^{-13}$	61	49·22	2
		750	$2{\cdot}18\times10^{-9}$			
MgO (Periclase)	Mg^{++}	1000	$6{\cdot}22\times10^{-15}$	$4{\cdot}19\times10^{-4}$	63·48	3
		2400	$2{\cdot}92\times10^{-9}$			
	O^{--}	1300	$6{\cdot}08\times10^{-15}$	$2{\cdot}5\times10^{-6}$	62·4	4
		1750	$5{\cdot}01\times10^{-13}$			
Al_2O_3 (Sapphire)	O^{--}	1500	$4{\cdot}60\times10^{-16}$	1900	152	5
		1780	$1{\cdot}59\times10^{-13}$			
$MgAl_2O_4$ (Spinel)	O^{--}	1350	$8{\cdot}84\times10^{-16}$	$1{\cdot}5\times10^{3}$	136	6
		1600	$2{\cdot}56\times10^{-13}$			
SiO_2 (Quartz)	O^{--}	1000	$1{\cdot}5\times10^{-18}$	$3{\cdot}7\times10^{-9}$	55	7
		1220	$3{\cdot}7\times10^{-17}$			
		600	$1{\cdot}2\times10^{-17}$			8
Olivine (Fo 90)	O^{--}	400–800		$2{\cdot}5\times10^{-8}$	26	8
Grossular	O^{--}	400–800		$1{\cdot}3\times10^{-12}$	20	8
Anorthite (An 96)	O^{--}	400–800		$1{\cdot}5\times10^{-7}$	27	8
Albite (Ab 98·6)	O^{--}	400–800		$2{\cdot}5\times10^{-8}$	26	8
		800	$2{\cdot}7\times10^{-13}$	4×10^{-13}	37	9

1. F. Bénière, M. Bénière, and M. Chemla, *J. Phys. Chem. Sol.*, **31**, 825 (1970).
2. F. Bénière, and M. Chemla, *C.R. Acad. Sci., Paris,* **267**, 633 (1968).
3. B. J. Wuensch, W. C. Steele, and T. Vasilos, *J. Chem. Phys.*, **58**, 5258 (1973).
4. Y. Oishi, and W. D. Kingery, *J. Chem. Phys.*, **33**, 905 (1960).
5. Y. Oishi, and W. D. Kingery, *J. Chem. Phys.*, **33**, 480 (1960).
6. K. Ando, and Y. Oishi, *Yogyo Kyokai Shi,* **80**, 324 (1972).
7. R. Haul, and G. Dümbgen, *Z. Elektrochemie,* **66**, 636 (1962).
8. B. J. Giletti, M. P. Semet, and R. A. Yund, *Geol. Soc. Amer. Annual Meeting* (1975).
9. H. Mérigoux, *C.R. Acad. Sci., Paris,* **263 D**, 1017 (1966).

3.4. DISLOCATIONS

3.4.1. Generalities

We have seen (§ 2.4) that plastic deformation of a crystal is the permanent shear strain which results from the application of a shear stress greater than the elastic limit. Physically, it can be analysed into translations of parts of the crystal along *slip planes*—usually dense atomic planes.

Let us now consider the elementary unit of plastic deformation: the slip (or glide) by one atomic distance of the part of the crystal above a slip plane, relative to the part situated below the plane. When this unit slip has taken place, all atoms on each side of the slip plane have changed neighbours across the plane, which means that atomic bonds had to be broken and reformed. We can roughly calculate the shear stress necessary to break all atomic bonds across the slip plane; we will call this shear stress the *theoretical elastic limit* or theoretical shear strength.

(*i*) *Estimate of the theoretical elastic limit.* Let us consider (Fig. 3.7) a crystal undergoing slip along dense atomic planes. Let b be the interatomic distance along the slip direction and a the interatomic distance across the planes.

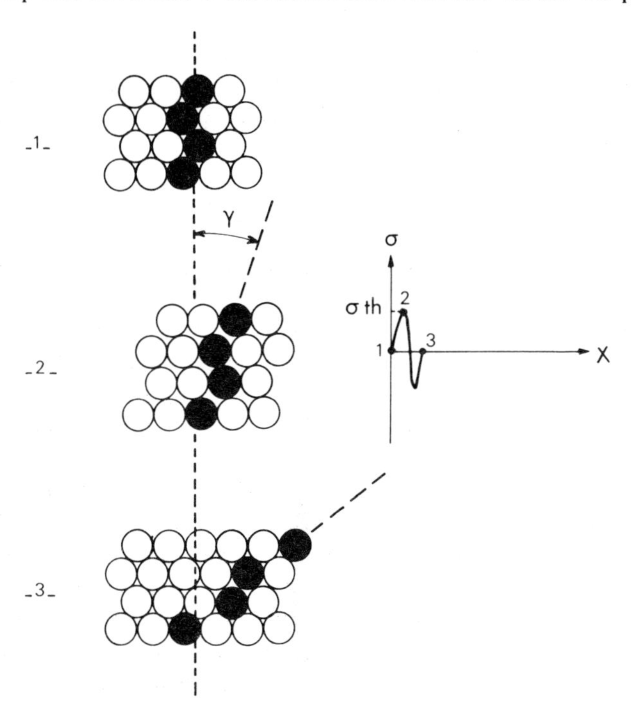

Fig. 3.7. Shear of a crystal along a lattice plane: 1: unsheared crystal; 2: elastic shear by an angle γ for a shear stress $\sigma < \sigma_{th}$; 3: plastic shear (slip) for $\sigma > \sigma_{th}$; σ_{th}: theoretical shear strength. The atoms represented by closed circles have changed neighbours (after Weertman, 1964, reproduced by permission of MacMillan Publishing Co., Inc.)

Let σ be the shear stress needed to shear the crystal through an angle γ (two neighbouring planes are then shifted by x). The elastic state of the crystal is the same after two neighbouring planes have been shifted by b since the atoms have fallen back into new equivalent equilibrium positions; it is therefore reasonable to assume that σ is periodic with a period b and we can approximate its variation by a sinusoidal function of x, the distance by which two neighbouring planes have been shifted ($x < b$). Now, for small values of x, the atoms are only slightly displaced from their equilibrium positions and the shear strain $\varepsilon = x/a$ is elastic; σ must tend towards the value given by Hooke's law $\sigma_0 = \mu(x/a)$ (μ = shear modulus). As for small values of x we have

$$\sin\frac{2\pi x}{b} \simeq \frac{2\pi x}{b}$$

we can therefore write:

$$\sigma = \frac{\mu b}{2\pi a} \sin \frac{2\pi x}{b} \tag{3.37}$$

The maximum value of σ corresponds to the stress necessary to break *simultaneously* all the atomic bonds and take all the atoms from one equilibrium position to the next at a distance b. It is therefore equal to the theoretical shear strength:

$$\sigma_{\text{th}} \simeq \frac{\mu b}{2\pi a} \simeq \frac{\mu}{10}$$

The theoretical shear strength is therefore of the order of magnitude of a tenth of the shear modulus. Thus, a copper wire of 1 mm^2 cross-section should be able to withstand, without even starting to stretch permanently, a load of almost 400 kilograms; everyday experience clearly shows that metals deform plastically under much lower stresses: the *actual elastic limit* for all crystalline materials is commonly of the order of 10^{-3} μ or 10^{-4} μ and can be even lower for well-annealed metallic single crystals.

This important discrepancy of several orders of magnitude between the theoretical and actual elastic limit was explained in 1934 by G. I. Taylor, E. Orowan and M. Polanyi who independently postulated that plastic deformation was caused by the movement along the slip planes of linear lattice defects which could move under stresses much lower than the theoretical elastic limit: the dislocations. It was only in 1956 that Hirsch, Horne and Whelan at the Cavendish Laboratory in Cambridge could see dislocations moving, while observing thin metal foils by Transmission Electron Microscopy. In order to understand what kind of defect can explain such an important lowering of the elastic limit, we must remember one assumption which underlies the calculation of the theoretical elastic limit: it is that all atoms on each side of a slip plane *simultaneously* change neighbours, i.e. all atomic bonds across the plane must be broken at the same time. This clearly involves an important expenditure of energy. If, on the other hand, we consider slip as being progressive, or propagated by the spreading of a slipped area into unslipped regions of the plane, this process obviously requires less force since, at a given time, the only atoms which change neighbours lie along the border line between the slipped and unslipped areas of the slip plane. This border line constitutes a defect in the crystalline lattice: the *dislocation*.

It is equivalent to say that *slip occurs by the movement of dislocations along the slip plane or that slip is propagated by the outward spreading of the boundary of a slipped area of the plane.*

This can be visualized by considering the progression of a caterpillar (Fig. 3.8). A caterpillar does not move forward by sliding its whole body over the ground or moving all its feet simultaneously (for a crystal this would be the equivalent of simultaneously breaking all the atomic bonds across the slip plane).

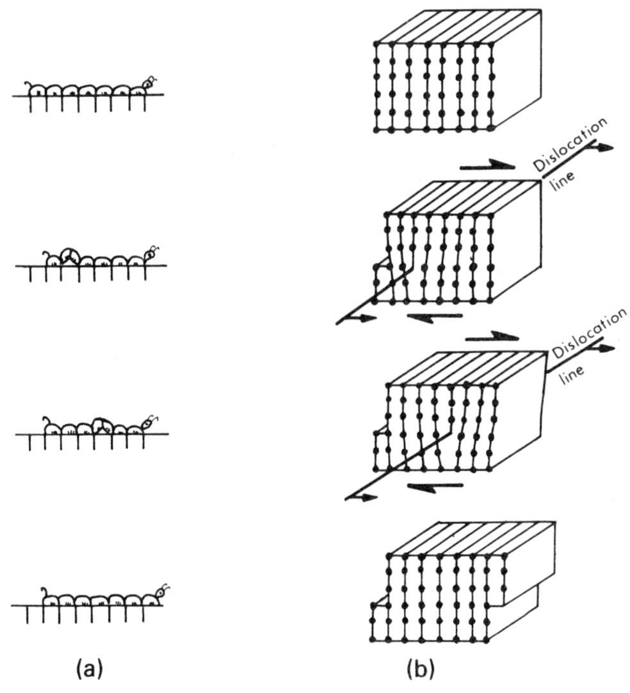

Fig. 3.8. Slip by propagation of an edge dislocation. (a) progression of a caterpillar; (b) slip of a crystal by movement of an edge dislocation. The extra-half plane corresponds to the ruck in the caterpillar (see text)

Instead, the creature nucleates a ruck in its body by drawing in its tail and moves forward the ruck, i.e. the boundary of the slipped area. For a crystal the equivalent of the ruck is an extra half-plane of atoms. The edge of the extra half-plane, constituting the boundary between the slipped and unslipped areas, can propagate by flipping of bonds (Fig. 3.9).

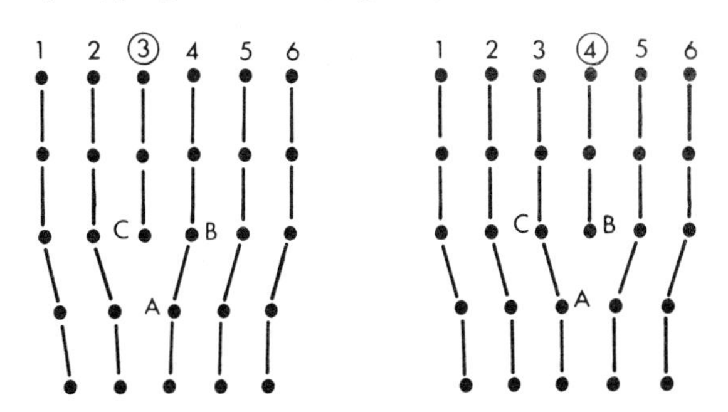

Fig. 3.9. Movement of an edge dislocation. The dislocation originally in (3) moves by one interatomic distance to (4). The bonds have flipped over from AB to AC

The *dislocation line*, thus defined by the edge of an extra half-plane (Fig. 3.8) is perpendicular to the shear direction. It is called an *edge dislocation*. Another type of dislocation can also be visualized by tearing a sheet of paper (Fig. 3.10). The dislocation line running through the sheet at the propagating tip of the tear is parallel to the shear direction. In a crystal no extra half-plane can be associated with this type of dislocation, but the atomic planes are distorted helicoidally: this is a *screw dislocation*. As seen in Fig. 3.10, the dislocation line is parallel to the shear direction, but shearing is obtained by its displacement normal to the shear direction.

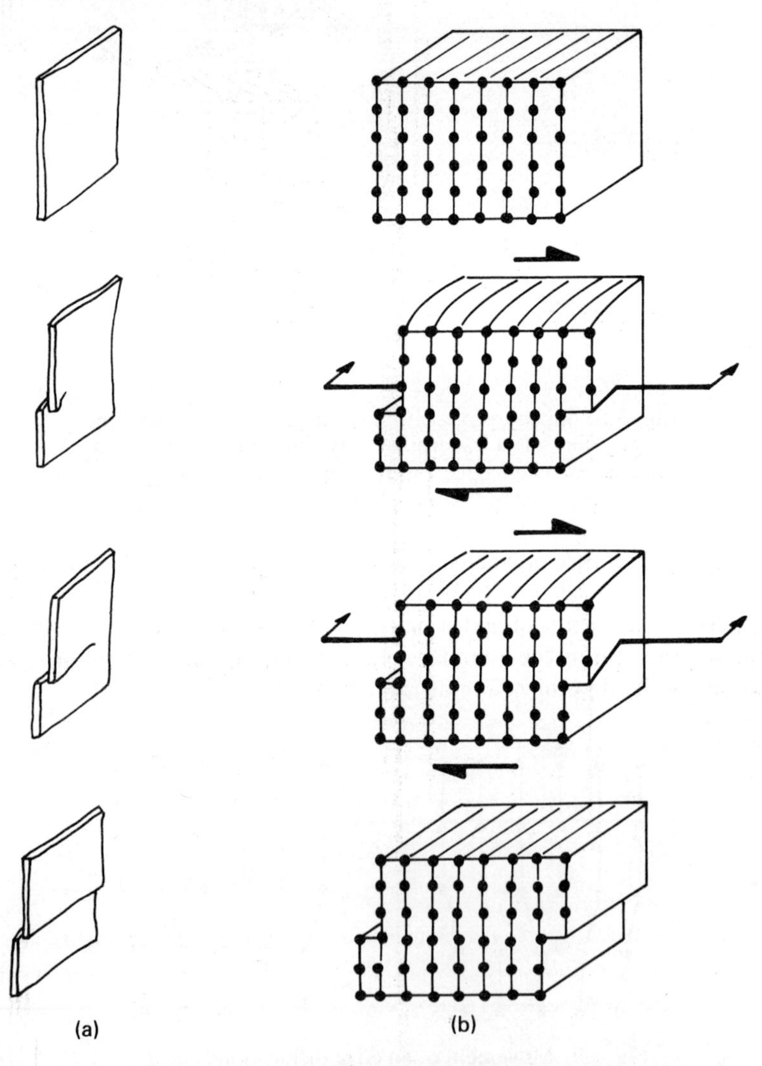

Fig. 3.10. Slip by propagation of a screw dislocation: (a) tearing of a sheet of paper. There is a screw dislocation at the tip of the tear (see text); (b) slip of a crystal by movement of a screw dislocation

We have only considered, in these examples, straight dislocations running through the crystal from one surface to another. More generally, the slipped area is completely enclosed by a *dislocation loop*. This loop consists of screw portions where the tangent to the line is parallel to the shear and edge portions where it is perpendicular. All the other parts are of *mixed* character (Fig. 3.11). Although the shear has a definite direction, the loop expands outwards in all directions perpendicular to itself (Plate 6(b)). When the loop goes out of the crystal, a shear by one atomic distance has taken place.

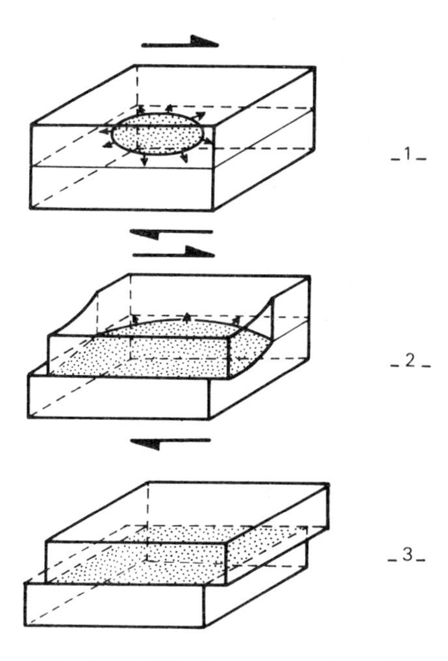

Fig. 3.11. Slip by propagation of a dislocation loop. Under the action of the shear stress (arrows), the slipped area (stippled) expands.

3.4.2. Burgers vector

We have defined a dislocation line or loop as the boundary between a slipped area and an unslipped area on a slip plane. The dislocation can therefore be characterized by a vector expressing the amount and direction of slip caused by its propagation on the slip plane. This vector is called the *Burgers vector* and is of course identical for any portion, edge or screw of the same dislocation loop.

The glide plane of a dislocation contains the dislocation line and the Burgers vector: for a screw dislocation any crystallographic plane in zone with the dislocation line is potentially a glide plane.

A dislocation line is also a linear lattice defect: the periodicity of the lattice is broken at a dislocation. Therefore the Burgers vector can also be defined as a characteristic quantity of a linear lattice defect in the following way:

Let us consider a crystal containing a straight dislocation line and a section by an atomic plane normal to the dislocation line, let us also consider a reference perfect crystal containing no dislocation. Starting at any point in the perfect crystal we can describe a closed circuit; for instance, in the simple cubic lattice in Fig. 3.12(1) we can take one step up, two steps right, two steps down, two

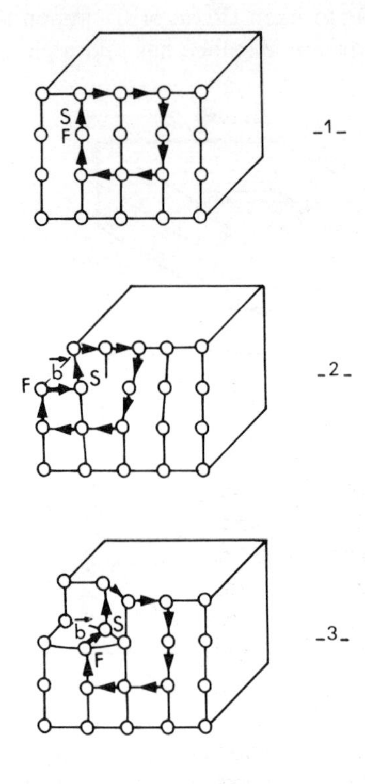

Fig. 3.12. Burgers circuit (arrows): S: Start, F: Finish; 1. closed Burgers circuit in perfect crystal; 2. Burgers circuit around an edge dislocation; 3. Burgers circuit around a screw dislocation; b: closure failure = Burgers vector (see text)

steps left and one step up (each step being equal to an interatomic distance) and the point where we start the circuit S is the same as the point where we finish it, F. If we now take the same circuit, or Burgers circuit, in the crystal containing a dislocation, two cases can be considered:

The circuit does not encircle the dislocation; it is still closed, although distorted.

The circuit encircles the dislocation; if from point S we take one step up, two steps right, two steps down, two steps left and one step up we arrive at F which is distinct from S. To close the circuit we have to go one further step FS (Fig. 3.12 (2) (3)). *The Burgers vector* **b** *is the closure failure of the Burgers circuit.*

The modulus of the Burgers vector is independent of the Burgers circuit around the dislocation, but its sign can be chosen arbitrarily, and depends on the orientation of the dislocation line. The Burgers vector is usually taken as equal to FS for a clockwise (right-hand) circuit, when one sights down the arbitrary positive sense of the dislocation line.*

Fig. 3.12 (2) shows the Burgers circuit around an edge dislocation and Fig. 3.12 (3) the Burgers circuit around a screw dislocation. Fig. 3.13 shows a

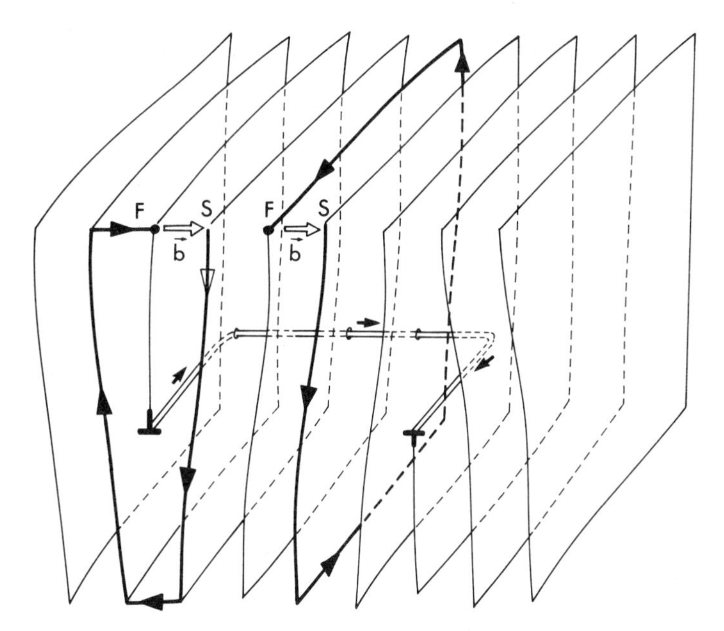

Fig. 3.13. Dislocation half-loop. The crystal is represented by the lattice planes parallel to the extra-half plane of the edge portions. Burgers circuits are represented around the edge and the screw portions. The upper extra-half plane on the left is continuously connected to the lower extra-half plane on the right by lattice planes distorted, in the way of a winding staircase, by the screw portion which threads them

* To define **b** independently of the width of the Burgers circuit it is usual to close the circuit in the crystal containing the dislocation. The closure failure then occurs in the perfect crystal. The Burgers vector is taken as equal to FS in the perfect crystal for a right-hand circuit closed in the imperfect crystal, sighting down the positive sense of the dislocation (FS/RH convention). This is a sign convention opposite to the one used in Figs. 3.12 and 3.13.

rectangular dislocation loop with Burgers circuits around the edge and screw portions. As can be seen, the Burgers vector is the same for the edge and screw portions of the same loop, which is to be expected since its modulus is the interatomic distance in the slip direction.

Seeing that a dislocation line is defined as the boundary of an area, it cannot terminate within the crystal. It must form a loop or end at another crystalline defect (surface, interface) or meet two other dislocation lines at a triple point. In this case the total Burgers vector must be conserved. If $\mathbf{b}_1$ is the Burgers vector of a dislocation branching into two dislocations of vector $\mathbf{b}_2$ and $\mathbf{b}_3$, we must have if the dislocations are oriented in the same sense:

$$\mathbf{b}_1 = \mathbf{b}_2 + \mathbf{b}_3 \tag{3.38}$$

(or $\mathbf{b}_1 + \mathbf{b}_2 + \mathbf{b}_3 = 0$ if the dislocations are all oriented toward or outward from the node).

The conservation of the Burgers vector is visualized in Fig. 3.14: as can be seen, the closure failure of a Burgers circuit encircling two dislocations is the sum of the closure defects of their individual circuits; the common Burgers circuit can be obtained by displacing the Burgers circuit of the third dislocation.

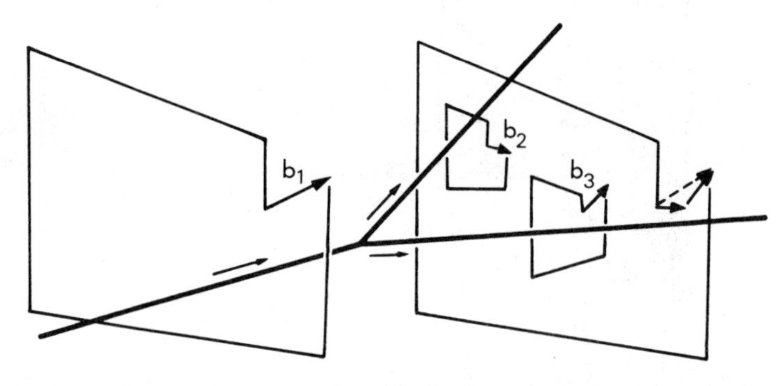

Fig. 3.14. Conservation of the Burgers vector. The Burgers circuits around the two right-hand branches of the Y can be deformed continuously and made to merge into a larger Burgers circuit enclosing both branches, identical with the one around the left-hand side stem of the Y

3.4.3. Stress field of a dislocation

The atomic planes are distorted around a dislocation, and the latter therefore causes an internal elastic strain in the crystal; to this strain corresponds an internal stress expressed with the help of the theory of elasticity: a crystal containing a dislocation is self-stressed. Knowing the stresses induced by a dislocation is important, for, as we will see later, dislocations interact with one another through their stress fields. To calculate the stress by the theory of elasticity we need not use the fact that the dislocation is in a crystal.* In fact, we

* Indeed, the concept of a dislocation as the boundary between a slipped and unslipped area has been extended to earthquake shear-faults (Steketee, 1958); Reid's elastic rebound theory (Benioff, 1964) can be viewed in terms of dislocation slip.

can define a dislocation in an elastic continuum (like a block of plastic) and construct it by the following procedure:

Take a surface (S) bounded by a line (L) inside the solid.

Make a cut along (S).

Shift one side of the cut relative to the other by a vector **b** by applying an external stress.

While maintaining the applied stress, glue together the two sides of the cut in the shifted position, so that the surface (S) physically disappears.

Release the applied stress.

The solid is then self-stressed (the internal stress being equal to the stress applied to shift the sides of the cut). The line (L) is a dislocation of Burgers vector **b**. We can see that defining the dislocation in a crystal as the boundary of a plane slipped area is a particular case of the above definition. The atoms within one or two atomic distances of the dislocation line are displaced so far from their equilibrium positions in a perfect lattice that an elastic strain cannot be defined. This region of bad crystal is the '*core*' of the dislocation. In the elastic continuum model there is a singularity in the stress field at the dislocation line, so it is hollowed out, for calculation purposes. Fig. 3.15 shows how

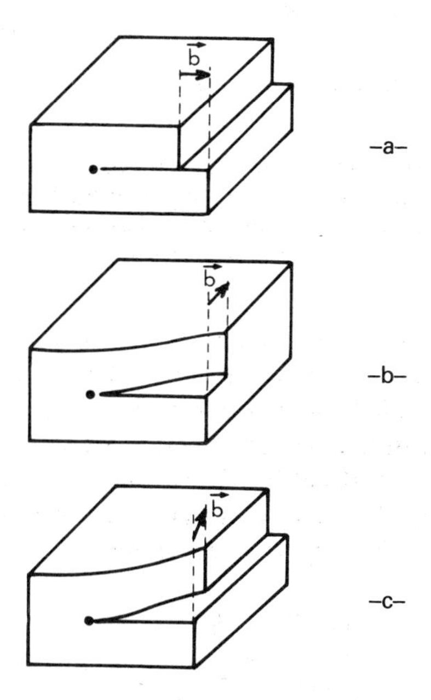

Fig. 3.15. 'Preparation' of a dislocation: (a) edge dislocation: displacement of the upper surface of the cut perpendicular to the dislocation line; (b) screw dislocation: displacement of the upper surface of the cut parallel to the dislocation line; (c) mixed dislocation

straight edge, screw, and mixed dislocations can be 'prepared' in this way. It must be noted that the cut surface can be arbitrarily chosen as long as the sides are displaced parallel to **b**; this can correspond to a slip by **b** along the cut or to the insertion of a layer of extra matter of thickness **b** (equivalent to the atomic extra half-plane in a crystal) (Fig. 3.16).

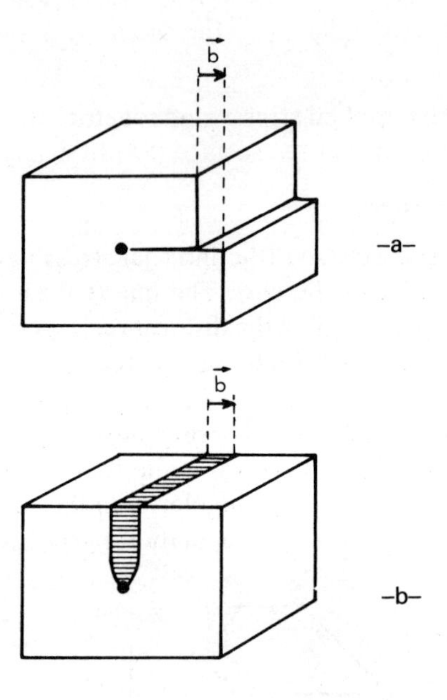

Fig. 3.16. 'Preparation' of the same edge dislocation: (a) by shearing a cut surface containing the Burgers vector; (b) by inserting a layer of extra matter in a cut perpendicular to the Burgers vector

As an example of stress field calculation we shall take the case of a straight screw dislocation along the axis of an infinite cylinder of matter. So that we may apply the theory of elasticity, we will consider a cylinder hollowed out along its axis (where the dislocation lies). Let us introduce the dislocation by making a cut along a radial plane and shifting the two sides by **b** parallel to the cylinder axis (Ox_3) (Fig. 3.17). The elastic displacement **u** at any point in the solid (x_1, x_2, x_3 or r, θ, z) is given by its components:

$$\begin{cases} u_1 = u_2 = 0 \\ u_3 = \dfrac{b\theta}{2\pi} = \dfrac{b}{2\pi}\tan^{-1}\dfrac{x_2}{x_1} \end{cases} \tag{3.39}$$

The only non-zero component is u_3 and it is periodic in θ, since for every turn we take around the dislocation, we move up by one distance b.

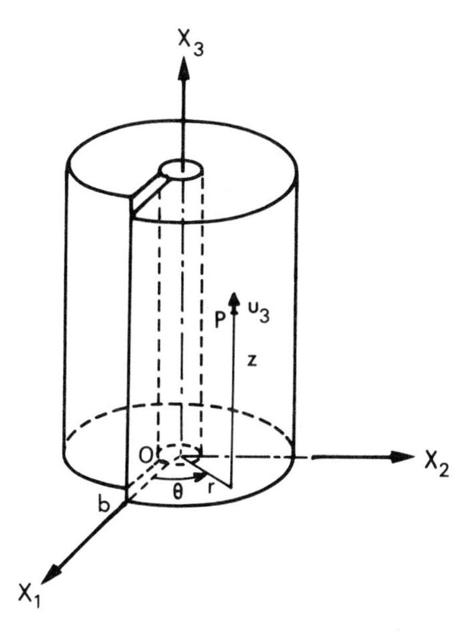

Fig. 3.17. Screw dislocation along the axis of a cylinder of matter

Using the definition of the elastic deformation:

$$\varepsilon_{ij} = \frac{1}{2}\left(\frac{\partial u_i}{\partial x_j} + \frac{\partial u_j}{\partial x_i}\right)$$

We obtain the only non-zero deformations:

$$\begin{cases} \varepsilon_{13} = -\dfrac{1}{2}\dfrac{b}{2\pi}\dfrac{x_2}{x_1^2+x_2^2} = -\dfrac{b}{4\pi}\dfrac{\sin\theta}{r} \\ \varepsilon_{23} = +\dfrac{b}{4\pi}\dfrac{\cos\theta}{r} \end{cases} \tag{3.40}$$

and using the relationship between σ_{ij} and ε_{ij} for isotropic elasticity (equation 2.10), for a screw dislocation along Ox_3 we get:

$$\begin{cases} \sigma_{13} = 2\mu\varepsilon_{13} = -\dfrac{b\mu}{2\pi}\dfrac{\sin\theta}{r} \\ \sigma_{23} = 2\mu\varepsilon_{23} = +\dfrac{b\mu}{2\pi}\dfrac{\cos\theta}{r} \end{cases} \tag{3.41}$$

or in cylindrical polar coordinates

$$\begin{cases} \sigma_{rr} = \sigma_{\theta\theta} = \sigma_{zz} = \sigma_{r\theta} = \sigma_{rz} = 0 \\ \sigma_{\theta z} = \sigma_{z\theta} = -\sin\theta \cdot \sigma_{13} + \cos\theta \cdot \sigma_{23} = +\dfrac{b\mu}{2\pi r} \end{cases} \tag{3.42}$$

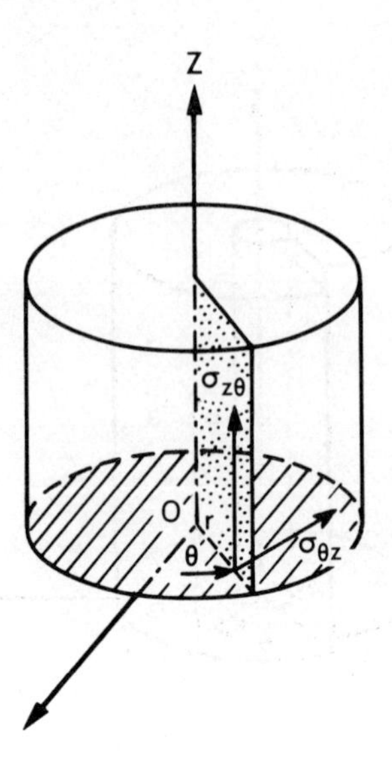

Fig. 3.18. Shear stresses introduced by a screw dislocation along the axis of a cylinder of matter: $\sigma_{z\theta}$: shear stress on radial plane (stippled); $\sigma_{\theta z}=\sigma_{z\theta}$: shear stress on plane normal to Oz (hatched)

A screw dislocation introduces only shear stresses but no hydrostatic pressure (Fig. 3.18); the values of the stresses decrease slowly away from the dislocation as $1/r$. The calculations are more complicated in the case of an edge dislocation, due to the lower symmetry of the problem. The stress field for an edge dislocation along Ox_3 with its Burgers vector along Ox_1 is given by the following expressions (Weertman and Weertman, 1964):

$$\left\{\begin{aligned}
\sigma_{11} &= -\frac{\mu b}{2\pi(1-\nu)}\,\frac{x_2(3x_1^2+x_2^2)}{(x_1^2+x_2^2)^2}\\
\sigma_{22} &= \frac{\mu b}{2\pi(1-\nu)}\,\frac{x_2(x_1^2-x_2^2)}{(x_1^2+x_2^2)^2}\\
\sigma_{33} &= \nu(\sigma_{11}+\sigma_{22})\\
\sigma_{12} &= \frac{\mu b}{2\pi(1-\nu)}\,\frac{x_1(x_1^2-x_2^2)}{(x_1^2+x_2^2)^2}\\
\sigma_{13} &= \sigma_{23}=0
\end{aligned}\right.
\quad \text{or} \quad
\left\{\begin{aligned}
\sigma_{rr} &= \sigma_{\theta\theta} = \frac{\mu b}{2\pi(1-\nu)}\,\frac{\sin\theta}{r}\\
\sigma_{zz} &= \frac{\mu b}{\pi(1-\nu)}\,\frac{\sin\theta}{r}\\
\sigma_{r\theta} &= -\frac{\mu b}{2\pi(1-\nu)}\,\frac{\cos\theta}{r}
\end{aligned}\right. \tag{3.43}$$

$\nu=\lambda/[2(\lambda+\mu)]$ is Poisson's ratio. For most materials $\nu\simeq 0{\cdot}3$.

The stress field of an edge dislocation has a hydrostatic pressure term. The stresses decrease as $1/r$, as for screw dislocations.

We have assumed that the solid was elastically isotropic, which is of course unjustified in the case of crystals, even cubic. However, the expression of the stresses, in anisotropic elasticity, although much more complicated does not differ widely from the ones we have given (Hirth and Lothe, 1967). For most cases it is sufficient to use (3.42) and (3.43).

3.4.4. Self-energy of a dislocation. Line tension

The increment of energy $\mathrm{d}W$ required to impose an increment of strain $\mathrm{d}\varepsilon_{ij}$ on a volume element of a solid subjected to stresses σ_{ij} is:

$$\mathrm{d}W = \sum_{ij} \sigma_{ij}\, \mathrm{d}\varepsilon_{ij} \tag{3.44}$$

If the solid is elastic and if the strain results from the application of the stress (or the other way round), σ_{ij} and ε_{ij} are related by Hooke's law:

$$\sigma_{ij} = \mu \varepsilon_{ij}$$

where μ is the appropriate elastic modulus.

Thus the elastic energy per unit volume expended to stress a solid to a uniform level of stress $(\sigma_{ij})_{\mathrm{Max}}$ is:

$$W = \frac{1}{\mu} \int_0^{(\sigma_{ij})_{\mathrm{Max}}} \sum_{ij} \sigma_{ij}\, \mathrm{d}\sigma_{ij} = \sum_{ij} \frac{(\sigma_{ij})^2_{\mathrm{Max}}}{2\mu} = \frac{1}{2} \sum_{ij} (\sigma_{ij})_{\mathrm{Max}} (\varepsilon_{ij})_{\mathrm{Max}} \tag{3.45}$$

This elastic energy is stored in the stressed (and elastically strained) solid.

A solid containing a dislocation is elastically strained and the corresponding stresses have been calculated in the previous paragraph. It is therefore clear that a solid of a given volume containing a dislocation has a greater internal (elastic) energy than a perfect volume. The energy stored in a solid by the existence of a unit length of dislocation line is called the *self-energy of the dislocation*, but it must not be forgotten that, in reality, this is not the energy of the dislocation but that of the volume of solid containing it; this energy is spread throughout the whole volume and depends on the dimensions of the solid.

It is now possible to calculate the self-energy of a screw dislocation from (3.40), (3.41) and (3.45). Let us consider a cylinder of matter of unit length containing a screw dislocation along its axis. Let R be the outside radius of the cylinder and b_0 the radius of the dislocation core, within which the elastic theory does not apply. As the stress field of a dislocation varies with the distance r to the dislocation line, we can calculate the energy $\mathrm{d}W(r)$ stored in a cylindrical shell with a thickness $\mathrm{d}r$ (Fig. 3.19) and integrate from $r = b_0$ to $r = R$ to obtain the self-energy $W(R)$ of the unit length of dislocation.

From (3.45)

$$\mathrm{d}W(r) = \frac{1}{2\mu}(\sigma_{13}^2 + \sigma_{23}^2) \cdot 2\pi r\, \mathrm{d}r$$

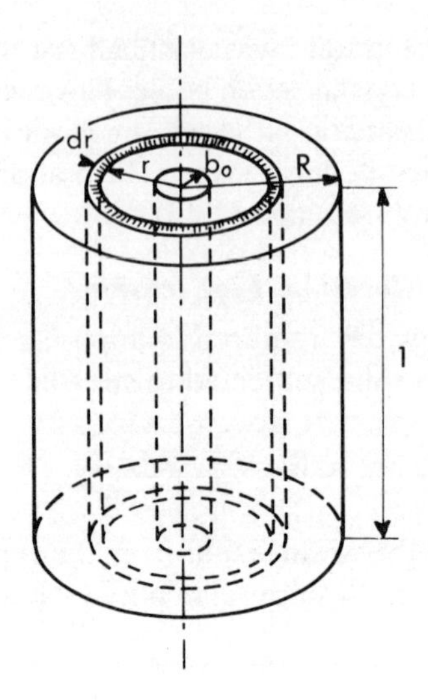

Fig. 3.19. Energy stored by an axial dislocation in a thin cylindrical shell (see text)

and from (3.41)

$$\sigma_{13}^2+\sigma_{23}^2=\left(\frac{\mu b}{2\pi}\right)^2\cdot\frac{1}{r^2}$$

Hence

$$\mathrm{d}W(r)=\frac{1}{2\mu}\left(\frac{\mu b}{2\pi}\right)^2\frac{2\pi r\,\mathrm{d}r}{r^2}$$

$$W(R)=\int_{b_0}^{R}\mathrm{d}W(r)=\frac{1}{2\mu}\left(\frac{\mu b}{2\pi}\right)^2\int_{b_0}^{R}\frac{2\pi r\,\mathrm{d}r}{r^2}=\frac{\mu b^2}{4\pi}\int_{b_0}^{R}\frac{\mathrm{d}r}{r}$$

$$W(R)=\frac{\mu b^2}{4\pi}\ln\frac{R}{b_0}\quad\text{per unit length}\tag{3.46}$$

The core radius is chosen arbitrarily equal to a few b; in a crystal, R can be taken as equal to the distance at which the stress field of the dislocation is comparable with the stress field of a neighbouring dislocation, hence it is of the order of the average distance between dislocations.

For an edge dislocation, the self-energy per unit length is given by:

$$W(R)=\frac{\mu b^2}{4\pi(1-\nu)}\ln\frac{R}{b_0}\tag{3.47}$$

It is interesting to note that the internal energy of a crystal increases with the length of dislocation it contains. In a crystal containing a given length of dislocation line, this length can be arranged according to a finite number of configurations; as for point defects (see 3.2.2), a configuration entropy term therefore exists, but it can be shown that this term is always smaller than the energy term arising from the self-energy of the dislocations. Hence the free enthalpy of a crystal always increases as the length of dislocation line increases and—contrasting with the case of point defects—there is no minimum value of the free enthalpy corresponding to an equilibrium concentration of dislocations. This is important, for it means that a crystal containing a high *dislocdtion density* (length of dislocation line per unit volume) can always reach a lower free enthalpy state by getting rid of dislocations: the elastic energy of dislocations is one of the main driving forces for such important processes as recovery or recrystallization by which the crystals partially or totally clean themselves of dislocations.

As an example we can evaluate by equation (3.46) or (3.47) the strain energy stored in a crystal of olivine containing a dislocation density $\rho = 10^8$ cm of dislocation line per cm^3 (which is a reasonable value in many cases).

Let us take $b[100] \simeq 4{\cdot}8 \times 10^{-8}$ cm and an average value of the shear modulus $\mu \simeq 8 \times 10^{11}$ dynes/cm^2.

Let us assume that the dislocations are of edge character, thus (3.47) applies and that $\nu \simeq 0{\cdot}3$.

If $\rho = 10^8$ cm^{-2}, the average distance between dislocations is $R \simeq 1/\sqrt{\rho} \simeq 10^{-4}$ cm.

Let us take the core radius $b_0 \simeq 2b \simeq 10^{-7}$ cm. For 1 cm of dislocation line, we therefore have:

$$W = 0{\cdot}8\mu b^2 \simeq 1{\cdot}5 \times 10^{-3} \text{ erg/cm}$$

For 10^8 cm of line we have the energy stored in 1 cm^3 of olivine:

$$W' = \rho W = 1{\cdot}5 \times 10^5 \text{ erg/cm}^3 = 1{\cdot}5 \times 10^{-2} \text{ J/cm}^3 \simeq 4 \times 10^{-3} \text{ cal/cm}^3$$
$$\simeq 0{\cdot}2 \text{ cal/mole}$$

For $\rho = 10^{11}$ cm^{-2} (very high density) we would have:

$$W = 0{\cdot}4\mu b^2 \simeq 7{\cdot}5 \times 10^{-4} \text{ erg/cm}$$

and

$$W' = 7{\cdot}5 \times 10^7 \text{ erg/cm}^3 \simeq 2 \text{ cal/cm}^3 \simeq 88 \text{ cal/mole}$$

Gross (1967) deduced from X-ray line broadening measurements that the stored energy of dislocations in deformed calcite could be as high as 3 to 5 cal/cm^3 or 110 to 180 cal/mole.

A wavy dislocation line tends to minimize its energy by reducing its length to the extent compatible with the forces exerted on it (see § 3.4.5). If no force is exerted, the undulations tend to straighten out. By analogy with the case of an elastic string, one can define a *line tension* τ: If an elastic string of length l is stretched by dl under a tension τ, the increase dW in its energy is equal to the work done by τ during the stretching:

$$dW = \tau \, dl \tag{3.48}$$

Similarly, we can define the line tension of a dislocation as the energy per unit length:

$$\tau = \frac{W' - W}{dl} \tag{3.49}$$

where:

W is the total energy of a straight dislocation of length l
W' is the total energy of the same line with a bulge (or undulation), whose length has increased by dl.

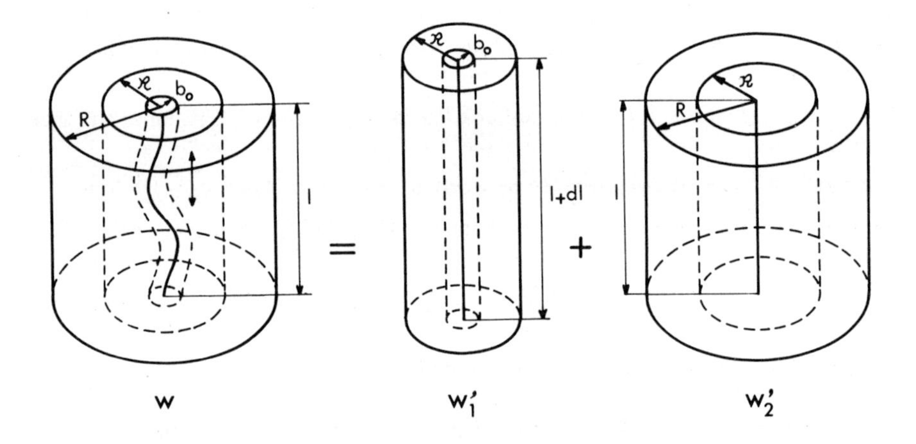

Fig. 3.20. Energy stored by a wavy dislocation in a cylinder of matter (see text)

For such a line the energy of a cylinder of outer radius R can be decomposed into two terms (Fig. 3.20) (Friedel, 1964):

$$W' = W'_1 + W'_2$$

Within a radius $\mathcal{R}$, of the order of the length of the bulge (or the wavelength of the undulations), the stresses at a given point are about the same as those created by a straight dislocation along the tangent to the line at its closest point. In other words, at short range, the bulge is not seen and the energy of a cylinder of outer radius $\mathcal{R}$ and inner radius b_0 is the same as if the cylinder contained a straight dislocation of length $l + \mathrm{d}l$

$$W_1' = \left[\frac{\mu b^2}{4\pi K} \ln \frac{\mathcal{R}}{b_0}\right](l+\mathrm{d}l) \tag{3.50}$$

$K = 1$ for a screw dislocation and $K = 1 - \nu$ for an edge dislocation.

For distances between $\mathcal{R}$ and R (the outer radius of the cylinder of matter considered), the perturbations in the stresses caused by the bulge have died out and the energy is the same as would be caused by a straight dislocation of length l

$$W_2' = \left[\frac{\mu b^2}{4\pi K} \ln \frac{R}{\mathcal{R}}\right] l \tag{3.51}$$

The initial energy W is written from (3.46) or (3.47):

$$W = \left[\frac{\mu b^2}{4\pi K} \ln \frac{R}{b_0}\right] l \tag{3.52}$$

(3.49) can then be written:

$$\tau = \frac{\mu b^2}{4\pi K} \cdot \frac{1}{\mathrm{d}l}\left[\left(\ln \frac{\mathcal{R}}{b_0}\right)(l+\mathrm{d}l) + \left(\ln \frac{R}{\mathcal{R}}\right) \cdot l - \left(\ln \frac{R}{b_0}\right) \cdot l\right]$$

or

$$\tau = \frac{\mu b^2}{4\pi K} \ln \frac{\mathcal{R}}{b_0} \tag{3.53}$$

As $\mathcal{R}$ and b_0 are somewhat arbitrarily chosen within reasonable orders of magnitude, it is justifiable in most cases to assume that $\log \mathcal{R}/b_0 \simeq 4\pi$ and take for practical purposes:

$$\tau \simeq \frac{\mu b^2}{K} \tag{3.54}$$

The line tension can be viewed as a force tending to straighten the dislocation.

Let us consider a curved dislocation arc AB (Fig. 3.21) of length $\mathrm{d}s$ and radius of curvature R_c. As for an elastic string, the line tension τ will appear as

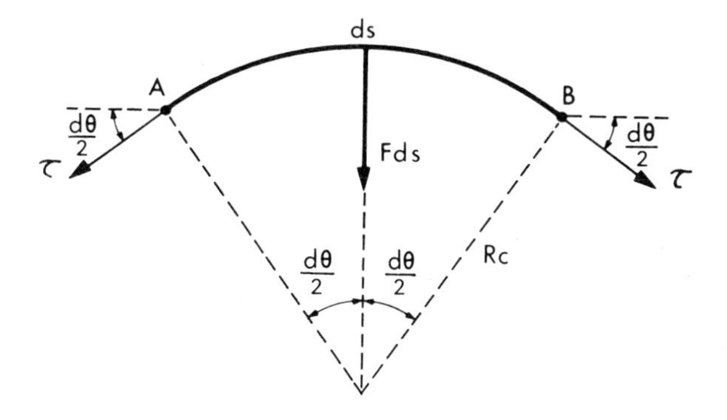

Fig. 3.21. Line tension of a curved dislocation arc (see text)

two forces τ in opposite senses along the tangents at A and B. Their resultant is a force directed toward the centre of curvature:

$$2\tau \sin\frac{d\theta}{2} \simeq \tau\, d\theta = \tau\frac{ds}{R_c}$$

The inward force per unit length is therefore

$$F = \frac{\tau}{R_c} \simeq \frac{\mu b^2}{KR_c} \tag{3.55}$$

A dislocation arc can keep an equilibrium curvature only if an outward force is applied to it (see § 3.4.5). Thus the existence of the line tension will cause a mobile dislocation loop of small radius to shrink and disappear if no force is exerted on it.

3.4.5. Forces exerted on a dislocation by stresses

When we defined dislocations as the defect whose propagation under an applied shear stress causes plastic shear along planes of crystals, we implicitly admitted that shear stresses exert forces on dislocations, making them move. In this section we are going to show that this is really so and give the expression of the force F per unit length of dislocation of Burgers vector $\mathbf{b}$ in a stress field given by its components σ_{ij}.

It must be pointed out that the force is exerted on the *configuration* of atoms that constitute the dislocation (even remote atoms in the crystal participate in the configuration, since the associated strain goes to the limit of the crystal), and that it must not be taken as a force acting on the atoms themselves.

Let us consider the simple case of an edge dislocation (Fig. 3.22(a)) in a

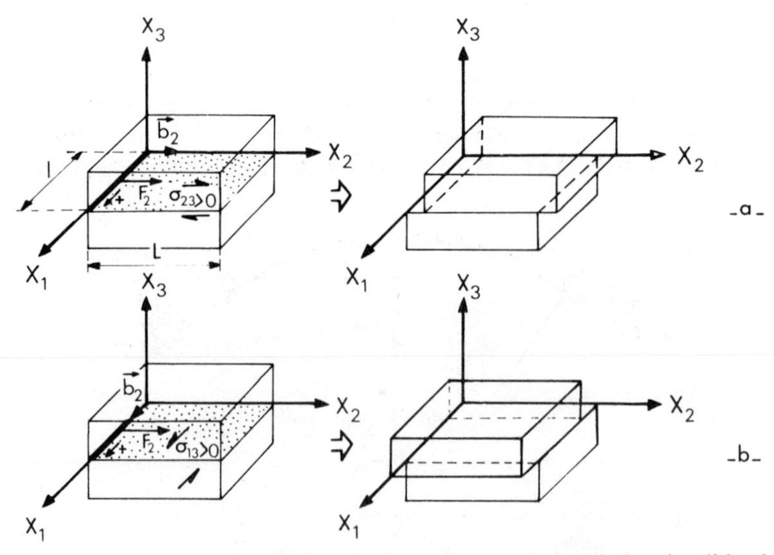

Fig. 3.22. Force exerted on a dislocation by a shear stress applied on its glide plane. (a) Edge dislocation; (b) Screw dislocation

crystal referred to a system of axes $x_1x_2x_3$. Let us take the dislocation as parallel to axis Ox_1, positively oriented in the Ox_1 positive sense, the extra half-plane being in the Ox_1x_3 plane on the positive side of Ox_3. With the FS/RH convention (§ 3.4.3) the Burgers vector $\mathbf{b}_2$ is directed toward the positive side of Ox_2, and the slip plane is Ox_1x_2. If a positive shear stress σ_{23} is applied on plane Ox_1x_2 (perpendicular to Ox_3) in the direction of Ox_2, according to the convention in Chapter 2.1, it will tend to displace the part of the crystal on the positive side of Ox_3 (upper part) in the positive Ox_2 sense, with respect to the lower part. Let us now suppose that the dislocation moves to the right under the action of the (as yet unknown) force F_2 (per unit length) obviously parallel to the Ox_2 axis and directed towards the positive sense of Ox_2, in agreement with the prescribed sign σ_{23}. When the dislocation of length l has travelled the length L of the crystal parallel to Ox_2, the upper part of the crystal has been displaced by a distance $+b_2$, positive, along Ox_2 with respect to the lower part, and the force F_2l on the dislocation has done the following work:

$$W = F_2 l \cdot L$$

Now, the externally applied shear stress σ_{23} corresponds to a force $\sigma_{23}lL$ applied to the slip plane of area lL. Since the upper part of the crystal has slipped by a distance $+b_2$, the work done by the externally applied force is:

$$W' = +\sigma_{23} lL \cdot b_2$$

The force F_2 on the dislocation is defined by the fact that the work it has done is equal to the work done by the externally applied force. Hence:

$$W = W'$$

and

$$F_2 = +\sigma_{23} b_2 \tag{3.56}$$

The case of the screw dislocation can be worked out in the same way (Fig. 3.22(b)).

In both cases the force on the dislocation is perpendicular to the dislocation line, and it can easily be seen that a shear stress acting on the slip plane of a dislocation loop creates forces normal to the loop at every part, thus tending to expand the loop (Fig. 3.23) (Plate 6(b)).

The force on a mixed dislocation exerted by a general stress field σ_{ij} can be calculated (Friedel, 1964).

Let u_i ($i = 1, 2, 3$) be the components along the axes Ox_1Ox_2Ox_3 of the unit vector tangent to the dislocation line at the point where we want to know the force.

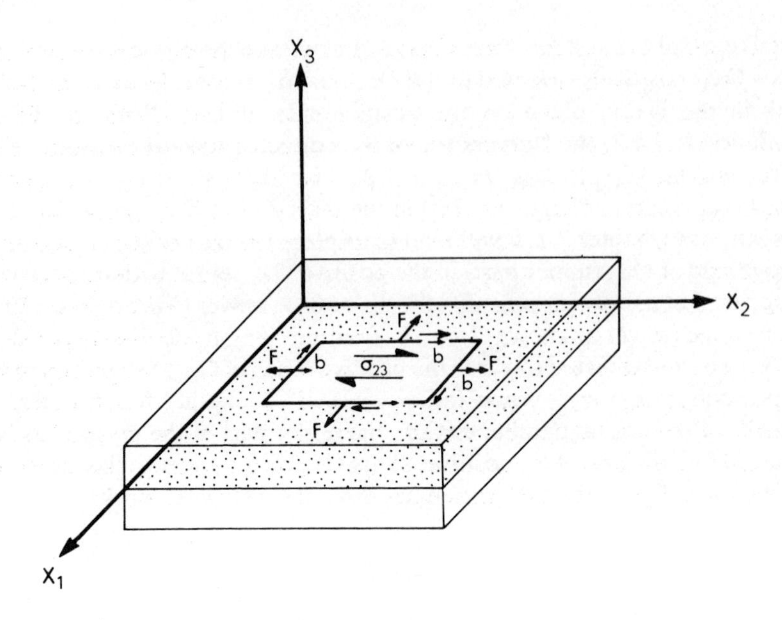

Fig. 3.23. Expansion of a square dislocation loop under the action of a shear stress applied on its glide plane

Let b_i be the components of the Burgers vector and σ_{ij} the components of the stress tensor.

The components F_i of the force per unit length on the dislocations are given by the Peach and Koehler formula which can be written:

$$F_i = \sum_{jk} \varepsilon_{ijk} p_j u_k$$

with (3.57)

$$p_j = \sum_l b_l \sigma_{jl}$$

$\varepsilon_{ijk} = 0$ if any two of the indices are equal
$= +1$ if the indices are in a direct circular permutation order: 123, 231, 312
$= -1$ if the indices are in an inverse circular permutation order: 132, 213, 321

Example

What are the components of the force exerted on an edge dislocation of olivine, with a Burgers vector **b** directed along [100] and mobile in the plane (010) under a general stress field σ_{ij}?

We have:

$$\begin{cases} b_1 = b \\ b_2 = b_3 = 0 \end{cases} \qquad \begin{cases} u_1 = u_2 = 0 \\ u_3 = 1 \end{cases}$$

$$\sigma_{11}, \sigma_{12}, \sigma_{13}, \sigma_{22}, \sigma_{23}, \sigma_{33}$$

$$\begin{cases} p_1 = b_1\sigma_{11} + b_2\sigma_{12} + b_3\sigma_{13} = b\sigma_{11} \\ p_2 = b_1\sigma_{21} + b_2\sigma_{22} + b_3\sigma_{23} = b\sigma_{12} \\ p_3 = b_1\sigma_{31} + b_2\sigma_{32} + b_3\sigma_{33} = b\sigma_{13} \end{cases}$$

$$\begin{cases} F_1 = \varepsilon_{123}p_2u_3 + \varepsilon_{132}p_3u_2 = b\sigma_{12} \\ F_2 = \varepsilon_{231}p_3u_1 + \varepsilon_{213}p_1u_3 = -b\sigma_{11} \\ F_3 = \varepsilon_{312}p_1u_2 + \varepsilon_{321}p_2u_1 = 0 \end{cases}$$

The force is perpendicular to the dislocation as it lies in the plane (001). The resolved component in the slip direction is $F_1 = b\sigma_{12}$ (σ_{12} is the shear stress in the slip plane, along the slip direction).

Equilibrium radius of curvature of a dislocation under a shear stress σ

Let us consider a dislocation segment of length l pinned at points A and B and free to glide in its slip plane between the pinning points. Under a shear stress σ, the segment will bow out in the direction of the applied force $F = -b\sigma$. But in doing so, it increases its length and is pulled back by its line tension. It will eventually stop and remain in equilibrium under the opposite forces. As the applied force is always normal to the dislocation line, the equilibrium shape will be an arc of a circle, whose radius will be given by equating the two forces (3.55) and (3.56):

$$b\sigma = \frac{\tau}{R_c} \simeq \frac{\mu b^2}{KR_c}$$

whence:

$$R_c = \frac{\tau}{b\sigma} \simeq \frac{1}{K}\frac{\mu b}{\sigma} \tag{3.58}$$

3.4.6. Origin of dislocations

In the preceding paragraphs, we have reviewed the most important properties of dislocations, but we have not yet answered one important question: How are dislocations physically created?*

(i) First of all, dislocations are already present in the crystal before straining, and were built-in during the growth of the crystal. This can be easily understood if one considers the fact that a crystal growing from the melt is built progressively by the accretion of individual atoms to the surface. A slight

* The 'recipe' given in 3.4.2 to 'prepare' a dislocation has evidently nothing to do with the way they are created in reality. It was only a list of successive operations devised to form a defect with the same stress field as a dislocation.

'error' in the position of some atoms can produce a defect which will be perpetuated during further growth.

An example of this can be found in the two-dimensional crystals formed by kernels on a corn cob: a slightly misplaced kernel frequently gives rise to an extra-half line of kernels (edge dislocation) (Fig. 3.24). As a matter of fact,

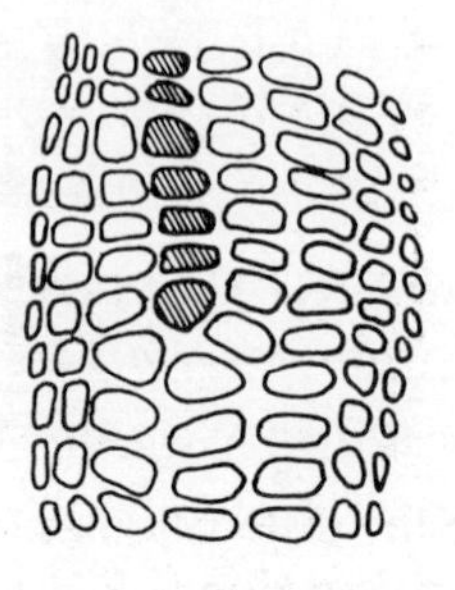

Fig. 3.24. Extra-half row of kernels on a corn cob

dislocations are even necessary to the growth of crystals: for instance a screw dislocation piercing a growing face creates helicoidal steps to which the incoming atoms will more readily attach themselves. This situation is self-perpetuating and the dislocation line, providing growth sites, is grown-in as the crystal grows (Fig. 3.25).

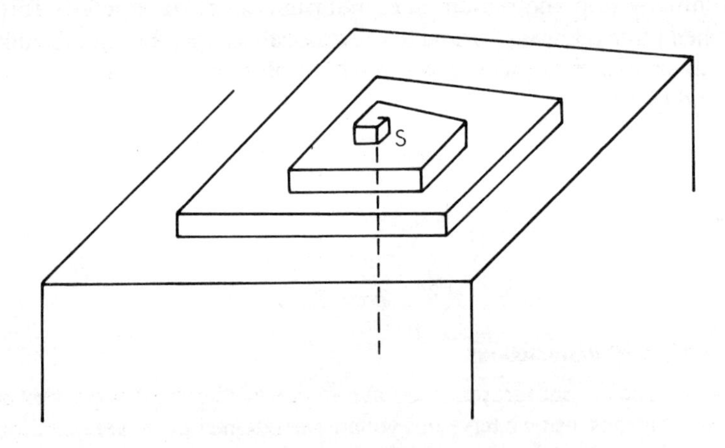

Fig. 3.25. Growth spiral on the surface of a crystal pierced by a screw dislocation

(ii) Dislocations can also be nucleated at stress concentration centres (Plate 6(b)). The process is then analogous to the one we described in §3.4.1: in an initially dislocation free region, if the elastic stress is raised to high values, of the order of $\mu/10$, it can be relieved by nucleating dislocation loops which thus create a local plastic strain. For instance, if a crystal contains second phase inclusions with a different thermal expansion coefficient, thermal stresses will develop during heating or cooling. If the inclusion expands more than the matrix, atomic planes of the matrix will be pushed in and positive edge dislocation loops will be punched out. In externally stressed crystals, stress concentrations commonly occur at heterogeneities such as second phase particles and grain boundary triple points.

(iii) In a homogeneous crystal without stress concentrations, extensive plastic deformation could never take place if we had only to rely on the mobile grown-in dislocations, for after gliding through the crystal they would leave it and be lost for further deformation. Now, not only extensive plastic deformation is possible but also the dislocation density is known to increase with plastic strain. A process responsible for the *multiplication* of dislocations must necessarily exist and this will of course be the most important one in plastic deformation.

The dislocations responsible for plastic deformation are practically all turned out by mills operating on the following principle, and called *Frank–Read sources.*

Let us consider a loose 3-D network of grown-in dislocations and focus our attention on a segment AB (Fig. 3.26) mobile by slip in a glide plane (*glissile*)

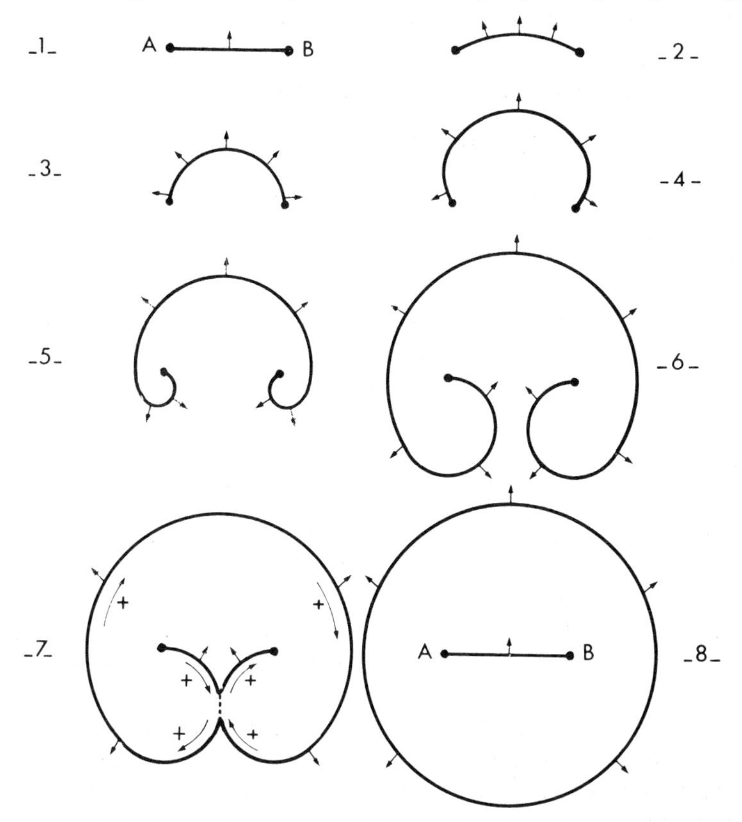

Fig. 3.26. Successive stages in the formation of a dislocation loop by operation of a Frank Read source in the plane of the sheet

and pinned at each end at junctions with dislocations threading the plane. Under an applied stress σ the segment AB bows out and takes the equilibrium shape of an arc of a circle (§ 3.4.5) with a radius R_c given by (3.58).

For a critical stress σ_{crit}, the segment can take the shape of a semi-circle; from then on the dislocation is not stable and it expands outwards without any further increase of stress. It sweeps around the junctions at A and B and portions with opposite signs meet and annihilate. We are left with a loop which expands outwards and glides away and a segment of dislocation anchored at A and B which swings back and starts the process again. As long as $\sigma > \sigma_{crit}$ the source functions and emits loops of dislocation.

The order of magnitude of σ_{crit} is easily found from equation (3.58) by taking $R_c = l/2$, where l is the length of the segment AB. We have:

$$\sigma_{crit} \simeq \frac{\mu b}{2Kl} \tag{3.59}$$

For l of the order of 1 μm, which is the usual case, we have $\sigma_{crit} \simeq \mu/10^4$. The stress for dislocation multiplication is therefore often comparable with the experimental shear strength.

3.4.7. *Relation between lattice curvature and stationary dislocations*

Let us consider a long crystal (Fig. 3.27 (1)) and let us bend it plastically by applying opposite sign torques at both ends. Several cases may arise.

(*i*) *There are active slip planes perpendicular to the axis of the single crystalline bar*

For a shear stress higher than the critical resolved shear stress, slip will occur on these planes and relieve the elastic state of bending. Dislocations will sweep through the planes, produce a plastic shear strain and move out of the crystal.

After we release the applied torques, the crystal will have assumed a permanently bent shape but supposing that all glide dislocations have left the crystal, the lattice will remain perfect and there will be no curvature of the lattice plane (Fig. 3.27 (2)).

(*ii*) *The only available slip planes are parallel to the axis of the single crystalline bar*

For the crystal to assume a permanently bent shape, the planes would have to be sheared in opposite directions at the opposite ends of the bar (Fig. 3.27 (3)) (as when one bends a telephone directory).

The lack of register between the atomic planes perpendicular to the axis can be taken up by edge dislocations, mobile in the slip planes. The dislocations have to have the same sign and must remain in the solid if the bend is to be permanent after the torque is released (Fig. 3.27 (4)). A curvature of the lattice planes, due to an excess of locked-in dislocations of the same sign corresponds to the macroscopically bent shape of the crystal. These dislocations are termed '*geometrically necessary*' (Cottrell, 1964).

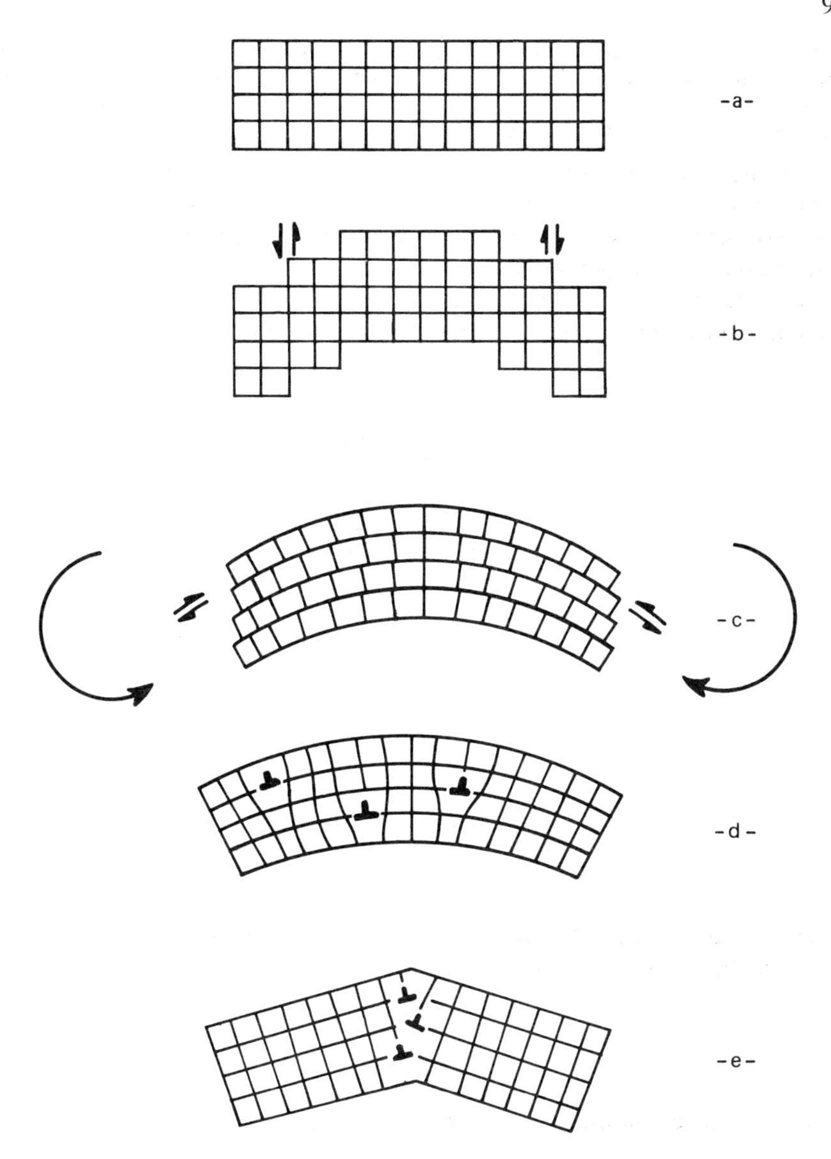

Fig. 3.27. Bending of a crystal: (a) undeformed crystal; (b) plastic bending by slip (shear folding); (c) elastic bending by curvature of the lattice plane (slip flexural folding); (d) lattice curvature due to geometrically necessary dislocations (flow flexural folding); (e) kinking (kink folding)

N.B. (1) In the case of a permanent shear strain caused by slip (as in Fig. 3.27 (2)) the mobile dislocations may not all move out of the crystal. The stored dislocation would obviously cause some local curvature of the lattice planes, but this is purely accessory and not necessary to achieve the imposed shape. Besides, there would be a statistically equal number of dislocations of both signs, hence no long-range curvature of the planes.

N.B. (2) A parallel may be drawn between the various ways in which bending of the crystal is achieved (Fig. 3.27) and the mechanisms of folding in rocks: Fig. 3.27 (2) = shear folding—Fig. 3.27 (3) = slip flexural folding—Fig. 3.27 (4) = flow flexural folding—Fig. 3.27 (5) = kink folding.

The lattice planes curvature may be simply related to the excess dislocation density (Nye, 1953): let us consider a region of the crystal where the density of geometrically necessary dislocations, of Burgers vector **b**, is ρ (number of dislocations threading a unit area).

Let us define a unit area Burgers circuit in the shape of a segment of a circular annulus centred on the centre of curvature of the lattice planes (Fig. 3.28).

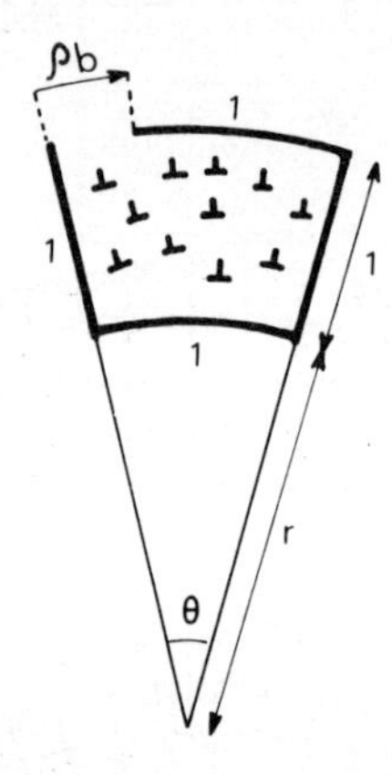

Fig. 3.28. Burgers circuit around a region containing geometrically necessary dislocations (see text)

The Burgers circuit encloses ρ dislocations and its closure failure is ρb. The angle θ (Fig. 3.28) is:

$$\theta = \frac{1}{r} = \frac{1+\rho b}{1+r}$$

where r is the curvature radius of the planes. We therefore have:

$$\frac{1}{r}(1+r) = 1 + \rho b$$

and:

$$\frac{1}{r} = \rho b \tag{3.60}$$

As a rule, in all cases where there is a gradient of deformation (non-uniform deformation), geometrically necessary dislocations must be stored in the crystal, i.e. lattice planes are curved (Ashby, 1970).

This is particularly the case for the deformation of polycrystalline aggregates. When the von Misès criterion (§ 2.4.2) is not met, that is when there are fewer than five available slip systems, the deformation of a grain on, say, one slip system is impeded by the neighbouring grains, or in rocks for instance by a grain of a mineral with different properties. Internal stresses and bending moments will arise and for coherency to be maintained (at least within certain limits) the lattice planes must acquire a curvature (Plate 2(a)).

In birefringent minerals examined between crossed polarizers the lattice curvature is the cause of the undulatory extinction observed in minerals of deformed rocks.

Deformation energy is stored in the crystals by the excess dislocations. The energy of the crystal may be lowered if the dislocations are rearranged by glide into dislocation walls (or kink band boundaries). Fig. 3.27 (5) shows such a wall of edge dislocations (tilt wall): the curvature is sharply concentrated around the wall and there is no long range stress, the planes are not curved far from the wall.

3.4.8. Glide of dislocations in a lattice

We have seen that an applied stress exerts a force on a dislocation. If there is a glide plane available, that is if the dislocation line and its Burgers vector are situated in a low index (compact) crystallographic plane, the dislocation will be able to move and thus propagate a shear strain on the slip plane. It is, however, clear that there must be some sort of resistance to the motion of the dislocation or else it would be accelerated to very high velocities. Indeed, the dislocation, which is an elastic perturbation in the crystal, could be accelerated until it reached the limiting speed for such perturbations: the velocity of sound. Experience shows that this is not the case and that on the contrary the velocity of dislocations can sometimes be very low. In the next paragraph we will relate the plastic strain rate to the dislocation velocity. It is therefore important to know what the causes of the resistance to the dislocation motion are, since they control the strain rate, together with the applied stress.

(a) The resistance to motion arises from two main sources:

(i) Interaction with other defects: point defects, impurity atoms, and chiefly other dislocations and grain boundaries. These defects act as obstacles which momentarily stop the dislocation and have to be overcome with the help of the applied stress for deformation to continue (*hardening*). In most cases this is the main source of resistance to long-range dislocation motion and the details of the hindering interactions and the way they are overcome, condition most of the plastic deformation mechanisms (see Chapter 4).

(ii) Between defects, that is to say in the perfect lattice, the crystalline lattice itself exerts a continuous friction force on the dislocation so that it acquires a constant velocity (which depends of course on the stress).

This interaction between crystalline lattice and dislocation is designated by the name of 'Peierls force'. The physical cause of the friction can be found in the fact that atomic bonds must be broken for the dislocation to move one atomic distance. In other words, a dislocation parallel to an atomic row lies in a valley of potential and has to go over the top of an energy hill before it can fall in the next valley, one interatomic distance ahead. The energy provided to overcome the hill is dissipated in the form of heat when the dislocation falls into the valley.

The Peierls force is understandably greater in solids with highly directional bonds, which are more difficult to break. This is the case in particular for covalent crystals like silicon. The dislocations lie straight along dense atomic rows (Plate 6(b)) and move only with great difficulty, hence Frank–Read sources are difficult to activate and the dislocation density is very low. These two facts account for the considerable hardness of such crystals.

In metals, where the ions are held together by a 'glue' of nearly free electrons, the interionic bonds have a low directionality and are rather easy to break. Consequently, the dislocations are generally quite mobile in the perfect lattice and the sources are readily activated. In most cases, the deformation is controlled more by the overcoming of obstacles than by the lattice friction. Silicates have electronic structures which are often intermediate between covalent and ionic ones and the Peierls force may vary considerably according to the silicate or even to the slip plane.

(b) The propagation of a dislocation in a geometrically possible slip plane (containing the line and its Burgers vector) can be greatly helped by the phenomenon of *splitting* or *dissociation* into two (or more) parallel *partial dislocations*, in that plane.

The geometrical sum of the Burgers vectors $\mathbf{b}_1$, $\mathbf{b}_2$ of the partial dislocations is equal to the Burgers vector of the total perfect dislocation:

$$\mathbf{b} = \mathbf{b}_1 + \mathbf{b}_2$$

This means that the dislocation moves one interatomic distance b, so to speak in two zig-zagging steps: The leading partial dislocation moves along the direction $\mathbf{b}_1$ by a distance $|\mathbf{b}_1|$ which does not correspond to a lattice vector (there is no atom at the end of $\mathbf{b}_1$); it is followed by a planar defect across which the atoms do not match as in the perfect crystal and which is called a *stacking fault* (see §3.5) (Plate 9). Finally there comes the trailing partial dislocation which wipes out the fault by moving the atoms in the direction $\mathbf{b}_2$ so that $\mathbf{b}_1 + \mathbf{b}_2 = \mathbf{b}$ (Fig. 3.29).

This splitting presents two advantages which greatly facilitate glide:

(i) The Burgers vectors $\mathbf{b}_1$, $\mathbf{b}_2$ are shorter than b if the angle between $\mathbf{b}_1$ and $\mathbf{b}_2$ is acute, and

$$|\mathbf{b}_1|^2 + |\mathbf{b}_2|^2 \leqslant |\mathbf{b}|^2$$

As the self-energy of a dislocation line is proportional to the square of its Burgers vector modulus, the total elastic energy of two partial dislocations is

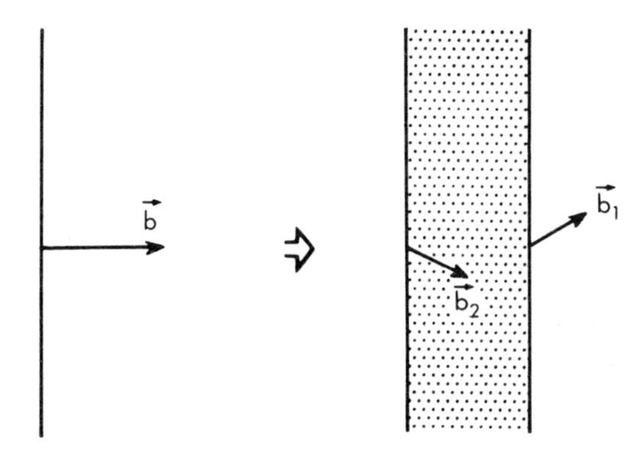

Fig. 3.29. Splitting of a dislocation into two partial dislocations separated by a ribbon of stacking fault (stippled)

lower than the energy of the total dislocation. One must also take into account the stacking fault energy (see § 3.5) but even so the energy balance is favourable to the split dislocation.

As a dislocation always seeks to lower its energy, it will tend to stay in planes where it can split, so there will be a longer length of dislocation available for slip. In addition the fault stabilizes the dislocation in the plane.

(ii) The decomposition of the slip vector **b** into two components $\mathbf{b}_1$ and $\mathbf{b}_2$ corresponds to circumventing the maximum energy barrier and adopting a devious path of lower energy.

This point can be made clear in considering the simple case of a face-centred cubic metal—say copper—whose crystalline structure can be described by stacked compact {111} planes of hard spheres (Fig. 3.30). If the first layer of spheres in contact is called A, the spheres of the second layer fit into holes (position B) of the first layer and the third layer spheres are situated over the other holes (position C) in layer A. We therefore have a sequence ABCABC...

Let us now consider a dislocation of Burgers vector $\mathbf{b} = a/2\,[110]$ in the glide plane $(1\bar{1}1)$ (Fig. 3.30). Let B_1 be one atom of the dislocation line. For the dislocation to slip by **b**, B_1 must come into position B_2 so that $\mathbf{B}_1\mathbf{B}_2 = \mathbf{b}$. The atom originally in B_1 must, in the process, go through a saddle point position S over sphere A_2 and we can roughly estimate the energy involved as being proportional (in the hard sphere model) to the height Δh to which the centre of the sphere B_1 must rise to go through the saddle point.

It is clear on Fig. 3.30, that the same result can be achieved by going round sphere A_2: the sphere B_1 goes first to the hole C_1 (unoccupied in the perfect structure) by going through the saddle point S′, and then to the final position B_2 through an equivalent saddle point. We have $\mathbf{B}_1\mathbf{C}_1 = \mathbf{b}_1 = a/6\,[121]$ and $\mathbf{C}_1\mathbf{B}_2 = \mathbf{b}_2 = a/6\,[21\bar{1}]$.

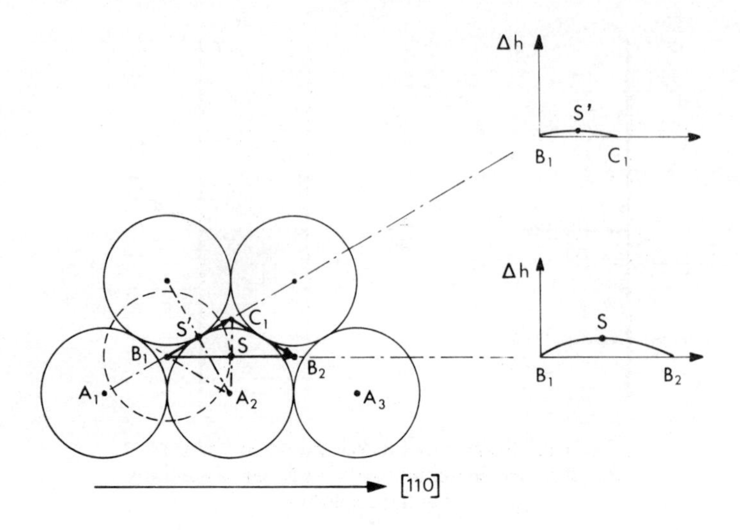

Fig. 3.30. Splitting of a dislocation of the face-centred cubic lattice (Burgers vector B_1B_2) into two partial dislocations of Burgers vector $B_1C_1 = \mathbf{b}_1$ and $C_1B_2 = \mathbf{b}_2$. In the hard sphere model, an atom of the upper layer (dotted circle) has to go through a saddle point S along the path B_1B_2 or a lower saddle point S′ along B_1C_1

Δh corresponding to saddle point S′ is much lower than Δh corresponding to saddle point S. Thus, in face-centred cubic metals glide dislocations split according to the reaction:

$$\frac{a}{2}[110] \rightarrow \frac{a}{6}[121] + \frac{a}{6}[21\bar{1}]$$

Splitting is a very important factor of ease of glide in all crystals. In most cases the easy glide planes are planes where a dissociation is possible.

In layered structures, such as phyllosilicates, where weakly bonded planes slip very easily, dislocations can split into more than two partials separated by wide ribbons of stacking faults.

3.4.9. Climb of dislocations

The glide of a dislocation on a slip plane under the action of a shear stress involves no transport of matter by diffusion: atomic bonds are shifted but there is no need for atoms to be brought in or evacuated. This is not the case for the movement of an edge dislocation (or a dislocation with an edge component) perpendicular to its glide plane. For the dislocation line to *climb* one interatomic distance up or down, a line of atoms along the edge of the extra-half plane has to be removed or added according to the position of the extra-half plane (Fig. 3.31).

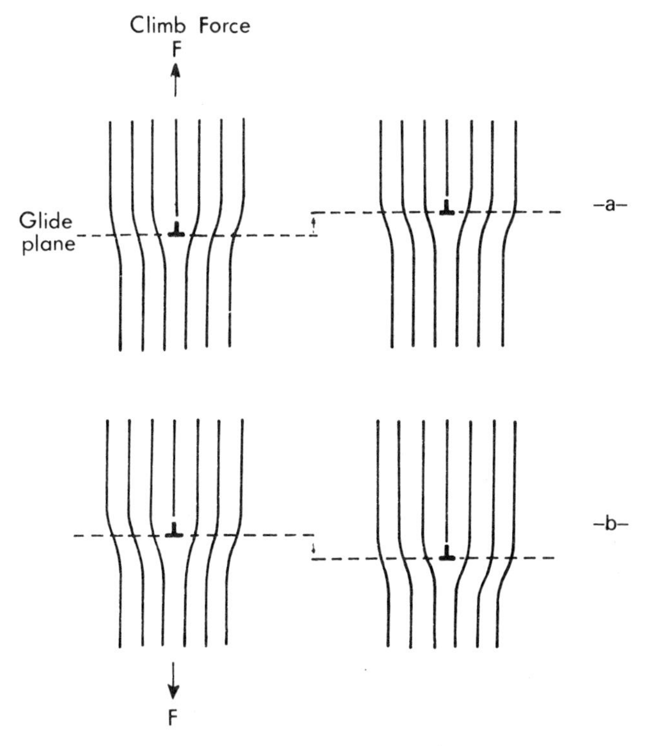

Fig. 3.31. Climb of an edge dislocation: (a) a line of atoms is removed at the edge of the extra-half plane; (b) a line of atoms is added at the edge of the extra-half plane

The atoms are brought or removed by *diffusion*. In other words, in the frequent cases where diffusion occurs by a vacancy mechanism, the atoms exchange with vacancies. A climbing dislocation will act therefore as a source or a sink of vacancies—in fact dislocations are the principal sources and sinks of vacancies.

Let us now take a closer look at the mechanism of climb.

First of all, it must be realized that the dislocation does not climb one interatomic distance, at the same time, along its whole length, as this would necessitate the instantaneous addition or removal of a whole row of atoms.

In reality, the edge of the extra-half plane is always jogged (Fig. 3.32) (the configuration entropy term introduced by the possible positions of the *jogs* along the line, lowers its free enthalpy). It is easier to add or remove an atom on the extra-half plane at a jog than on a straight portion of the line since it means forming or breaking fewer atomic bonds. Hence, if the force on the dislocation tends to make it climb by reducing the area of the extra-half plane, climb will occur piecewise, as the jogs migrate along the line absorbing vacancies, nibbling away the extra-half plane in the process. The same process would take place, but with emission of vacancies, for climb in the reverse direction.

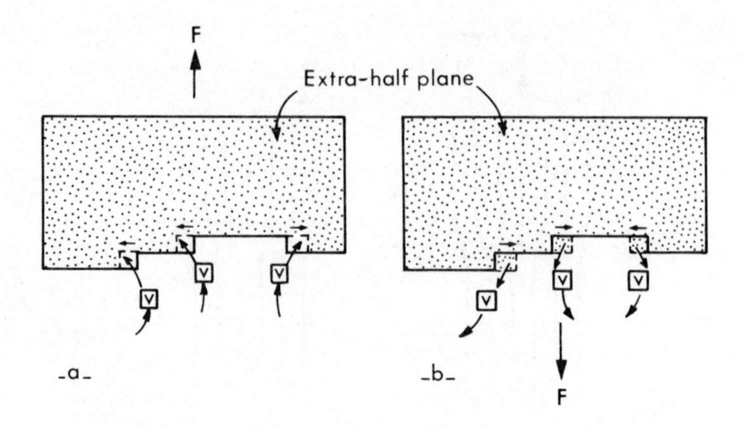

Fig. 3.32. Climb of an edge dislocation by migration of jogs: (a) vacancy-absorbing jogs. The extra half-plane decreases in area; (b) vacancy-emitting jogs. The extra half-plane increases in area

According to the sense of the force on the dislocation the jogs will tend to migrate in one direction or the other, emitting or absorbing vacancies, in order to allow the dislocation to yield to the climb force.

The climb process can be decomposed into two stages in series:

A vacancy is brought close to the jog by diffusion. The atom at the jog jumps into the vacancy, annihilating it, hence the jog migrates one interatomic distance (Fig. 3.22(a)).

or conversely:

A vacancy is created at a jog (that is, an atom of the lattice jumps out of its site, leaves a vacancy and sticks to the jog), hence the jog migrates one atomic distance (Fig. 3.32(b)). The vacancy created is carried away by lattice diffusion. This is necessary or else the reverse process could take place at the jog.

Climb is therefore controlled by the slower of the two processes in series:

If diffusion is very rapid and there are few jogs on the dislocation, climb is absorption or emission controlled.

If diffusion is slow and the jog concentration high, there will always be enough jogs to absorb incoming vacancies and the rate of migration of vacancy-absorbing jogs, for instance, will be diffusion controlled (the same is of course true for vacancy-producing jogs).

It is practically always assumed that there are enough jogs on dislocations and the *climb is diffusion controlled*; obviously, then, climb is a much slower process than glide. With the assumption of diffusion control, we can now proceed to evaluate the *climb velocity* of a dislocation as a function of the vacancy flux. To obtain an expression of the climb velocity as a function of the applied stress we must first relate the vacancy flux to the applied stress.

We will consider only the case of a vacancy-absorbing jog (the reasoning is identical for vacancy-producing jogs).

For a length l of dislocation to climb one interatomic distance b during time t, a jog has to travel along the length of the line during that time. We have:

$$t = \frac{l}{v_j}$$

where v_j is the jog velocity. The climb velocity of the dislocation is

$$v_c = \frac{b}{t} = \frac{b}{l} v_j$$

for n_j jogs of length b on the length l, we have clearly:

$$v_c = \frac{bn_j}{l} v_j$$

or

$$v_c = v_j C_j \tag{3.61}$$

where C_j is the concentration of jogs (number per unit length) along the line.

We will assume that the concentration of jogs is very high and take $C_j \simeq 1$, hence $v_c = v_j$.

Let us evaluate v_j if a jog absorbs the flux Φ of vacancies arriving by diffusion (number of vacancies arriving per unit time and unit length of dislocation). The absorption of one vacancy causes the jog to move one interatomic distance b. The length of the jog being b, it receives in one second Φb vacancies, and moves by Φb^2. We have therefore, for diffusion-controlled climb:

$$v_c = v_j = \Phi b^2 \tag{3.62}$$

The problem now is understanding why the application of a stress, giving a force acting on the dislocation perpendicularly to the glide plane, can induce a vacancy flow which will allow the dislocation to yield to the force by climbing. The reason can be understood by considering the simple case of an edge dislocation in a cubic crystal subjected to a compressive stress σ normal to the climb plane (Fig. 3.33).

We can verify by Peach and Koehler formula (3.57) that the force F on the dislocation favours climb by reduction of the extra-half plane (intuitively, one feels that the compressive stress tends to squeeze the extra-half plane out of the crystal). Absorbing (or creating) a vacancy at the dislocation is roughly equivalent to absorbing (or creating) a vacancy at the surface of the crystal parallel to the extra-half plane. In both cases the compressive stress works by

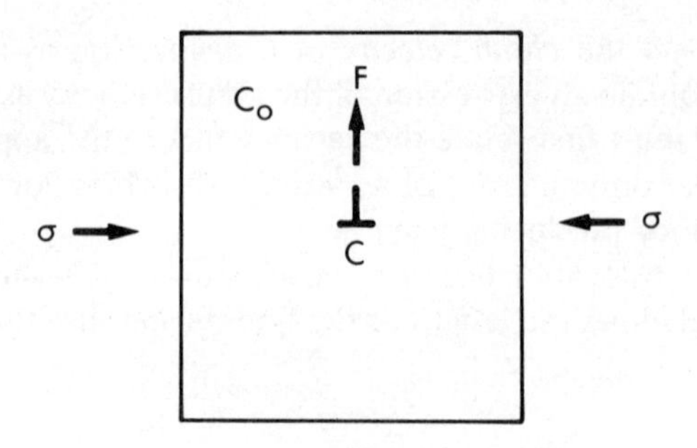

Fig. 3.33. Climb force on an edge dislocation in a crystal subjected to a compressive stress σ normal to the extra-half plane

$\sigma\Omega$ for the absorption and against the creation of the vacancy ($\Omega \simeq b^3$ is the atomic volume). This is equivalent to saying that the formation energy of a vacancy at the dislocation under stress is increased by $\sigma\Omega$; the equilibrium vacancy concentration (number of vacancies per unit volume) close to the dislocation is therefore:

$$C = \frac{N_v}{\Omega} \simeq \frac{N_v}{b^3} = \frac{1}{b^3} \exp -\left(\frac{\Delta G_f + \sigma\Omega}{kT}\right) = C_0 \exp -\left(\frac{\sigma\Omega}{kT}\right) \tag{3.63}$$

where N_v is the equilibrium atomic fraction of vacancies given by (3.4), and C_0 the equilibrium vacancy concentration far from the dislocation (C_0 is unaffected by σ). Because C is smaller than C_0, there will be a flow of vacancies toward the dislocation down the concentration gradient. As we are only interested here in a general expression of the climb velocity, we may take as a rough approximation for the concentration gradient

$$\frac{dC}{dx} \simeq \frac{C - C_0}{L}$$

L being here the average distance from the dislocation at which the vacancy concentration is C_0.

For $\sigma\Omega \ll kT$ (3.63) gives

$$\frac{dC}{dx} \simeq \frac{C_0}{L}\left[\exp -\left(\frac{\sigma\Omega}{kT}\right) - 1\right] \simeq -C_0 \frac{\sigma\Omega}{kTL}$$

The number of vacancies Φ crossing a cylinder of radius b around the unit length of dislocation per second is $\Phi = 2\pi bJ$, when J is given by Fick's equation (3.15), hence:

$$\Phi \propto bD_v C_0 \frac{\sigma\Omega}{LkT}$$

D_v is the vacancy diffusion coefficient and:

$$D_v C_0 = \frac{D_v N_{v0}}{b^3} = \frac{D_{SD}}{b^3}$$

where D_{SD} is the self-diffusion coefficient.

As $v_c \simeq \Phi b^2$, we have:

$$\boxed{v_c \propto \frac{D_{SD}}{L} \cdot \frac{\sigma\Omega}{kT}} \tag{3.64}$$

which we will use later on as a general expression—although a very approximate one—of the climb velocity.

3.4.10. Relations between plastic strain and mobile dislocations

In this paragraph we are going to derive the main relationship for the physics of deformation: Orowan's equation, which expresses the strain rate caused by the motion (glide or climb) of dislocations of Burgers vector **b**, with a velocity v, the density of mobile dislocations being ρ_m.

We will examine successively the cases of glide and climb in the extremely simple situation where straight dislocations thread a parallelepipedic crystal.

(*a*) *Glide*

Let us consider (Fig. 3.34) a block of matter in the shape of a parallelepiped of length L, width l and height h and introduce one straight edge dislocation of Burgers vector **b** near one end, by pushing in an extra-half plane; the glide plane is chosen parallel to the face of length L and width l. When the dislocation has swept through the whole glide plane by glide, i.e. moved the distance L, it will have left a step b on the exit face and the corresponding shear strain of the crystal will be equal, to first order, to:

$$\varepsilon = \frac{b}{h}$$

If the dislocation only moves by ΔL (or sweeps through an area ΔA) we can admit that the average shear strain will be:

$$\varepsilon = \frac{b}{h}\frac{\Delta L}{L} = \frac{b}{S}\Delta L$$

or, equivalently

$$\varepsilon = \frac{b}{h}\frac{\Delta A}{A} = \frac{b}{V}\Delta A$$

where $A = Ll$, $S = Lh$ and $V = Llh$.

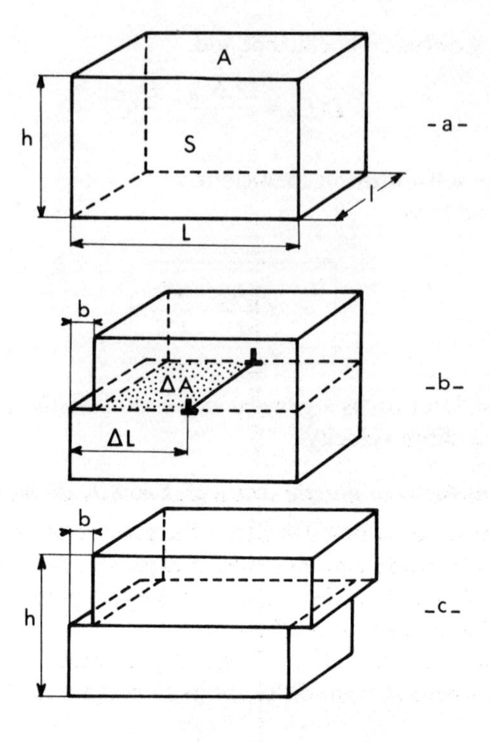

Fig. 3.34. Shear strain caused by the glide of an edge dislocation. Slipped area stippled

If we now consider N_m straight mobile and parallel dislocations threading the block we have:

$$\varepsilon = \frac{N_m}{S} b \,\Delta L$$

or

$$\varepsilon = \frac{N_m}{V} b \,\Delta A$$

Let us introduce the *dislocation density*

$$\rho_m = \frac{N_m}{S} = \frac{N_m}{V} \cdot l;$$

this is the length of dislocation line per unit volume or the number of emergence points of dislocations per unit area. $\mathcal{N}_m = N_m / V$ is the number of dislocation lines per unit volume. We therefore have:

$$\boxed{\varepsilon = \rho_m b \,\Delta L} \tag{3.65}$$

or

$$\boxed{\varepsilon = \mathcal{N}_m b\, \Delta A} \tag{3.66}$$

We assume that these formulae remain valid on the average for the real distribution of curved dislocations in crystals, although it is much more complicated, and that we thus have:

$$\varepsilon = \bar{\rho}_m b\, \overline{\Delta L} \tag{3.67}$$

where $\bar{\rho}_m$ is the average dislocation density, and $\overline{\Delta L}$ the average distance slipped by dislocations.

The strain rate is therefore:

$$\dot{\varepsilon} = \frac{\mathrm{d}\varepsilon}{\mathrm{d}t} = b\frac{\mathrm{d}}{\mathrm{d}t}(\bar{\rho}_m \overline{\Delta L}) = b\left[\Delta L \frac{\mathrm{d}\bar{\rho}_m}{\mathrm{d}t} + \bar{\rho}_m \frac{\mathrm{d}\,\overline{\Delta L}}{\mathrm{d}t}\right] \tag{3.68}$$

$\mathrm{d}\bar{\rho}_m/\mathrm{d}t$ is the rate of variation of the dislocation density by production or annihilation, and $\mathrm{d}\,\overline{\Delta L}/\mathrm{d}t = \bar{v}$ is the average velocity of the dislocations.

Except in cases of sudden and heterogenous deformations where the dislocation density varies very rapidly, it is generally justified at steady state to consider a population of dislocations of constant density. We therefore have $\mathrm{d}\bar{\rho}_m/\mathrm{d}t = 0$ and (3.68) can be written:

$$\boxed{\dot{\varepsilon} = \bar{\rho}_m b \bar{v}} \tag{3.69}$$

This is *Orowan's equation*

(*b*) *Climb*

Let us consider, in the same geometry as Fig. 3.34 a dislocation climbing by absorbing vacancies (Fig. 3.35). When the dislocation has moved through the

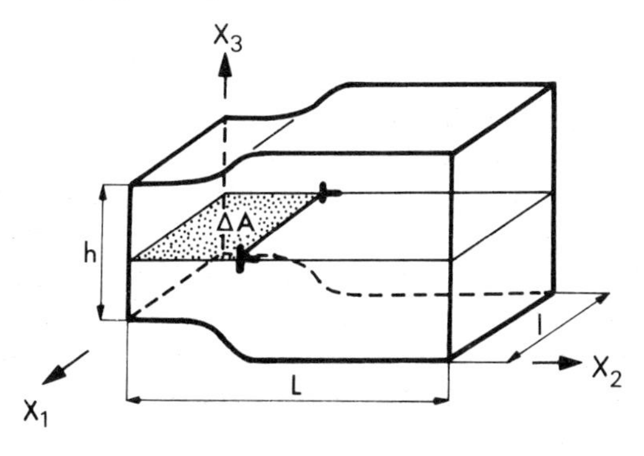

Fig. 3.35. Strain caused by the climb of an edge dislocation. Slipped area stippled

whole block, an atomic plane will have disappeared and the block will have undergone a compression strain (negative stretch):

$$\varepsilon_3 = -\frac{b}{h}$$

for a swept area ΔA we therefore have:

$$\varepsilon_3 = -\frac{b}{h}\frac{\Delta A}{Ll} = -b\frac{\Delta A}{V}$$

As matter has to be conserved, we must introduce a second dislocation (Fig. 3.36) which will provide the vacancies necessary for the climb of the first one. In

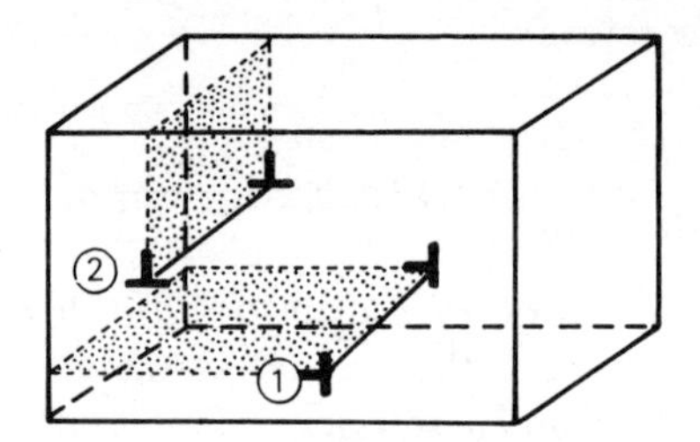

Fig. 3.36. Two edge dislocations climbing in opposite sense. A shear strain results. Extra half-planes stippled

so doing it will itself climb with the same velocity. After sweeping through ΔA in the same time as the first dislocation, it will give a tension strain:

$$\varepsilon_2 = \frac{b}{L}\frac{\Delta A}{hl} = b\frac{\Delta A}{V}$$

The total dilatation is therefore

$$\Delta V = \varepsilon_1 + \varepsilon_2 + \varepsilon_3 = \varepsilon_2 + \varepsilon_3 = 0$$

and we have a total deformation by pure shear.

In any case it is clear that if one dislocation sweeps an area ΔA by climb there is a resulting strain given by $\varepsilon = b(\Delta A/V)$. The same line of reasoning as for glide appears and Orowan's equation $\dot{\varepsilon} = \bar{\rho}_m b\bar{v}$ is valid for climb as well as for glide ($\bar{v}$ being taken as climb velocity or glide velocity).

A point worth noting is that the strain rate is related to the density of *mobile* dislocations. This may be quite different from the total dislocation density obtained by transmission electron microscopy measurements on thin foils. This point has to be kept in mind when trying to inject measured values of dislocation density into the strain rate expression obtained from a theoretical model.

3.5. TWO-DIMENSIONAL DEFECTS

When the break of periodicity in the lattice occurs on a surface within the crystal, we have a two-dimensional defect. We will briefly consider four types of defects.

The stacking faults which were introduced in the previous paragraph; they are closely linked to dislocation glide.

The dislocation walls, or *subgrain boundaries* (or even *sub-boundaries*), which play an important role in the processes of kinking and recovery.

The grain boundaries, which although they play no active role comparable with that of dislocations, are nevertheless very important at low and high temperatures in the deformation of polycrystals and in the process of recrystallization at high temperatures.

The interfaces between crystals of different nature. Although their role is certainly important in the deformation at low and high temperatures of polymineralic rocks, very little is, as yet, known of their properties and influence.

3.5.1. Stacking faults

We have seen (§ 3.4.8) how stacking faults were introduced by the splitting of dislocations and we will only illustrate here the resulting break of periodicity by taking once again the simple case of the face-centred cubic structure. It must be noted that the crystallographic description of stacking faults, which is beyond the scope of this book, is generally complex and especially so in the case of minerals.

Let us consider the ABC . . . stacking of {111} planes (Fig. 3.37); as we have seen in § 3.4.8, the stacking fault is introduced by shifting atoms B to position C above a certain plane, the other atoms being accordingly shifted C→A, A→B. . . . This stacking fault is therefore bounded by two partial dislocations and the stacking order across it is:

A B C A ⋮ C A B C . . .   (Fig. 3.38)

We can see that the operation by which we introduced the fault is equivalent to removing one plane B.

That kind of fault (*intrinsic fault*) can be actually created in this way by the collapse of a disk of vacancies. This situation arises when a crystal is quenched from a high temperature and the high temperature concentration of vacancies is frozen-in at low temperatures. The vacancies in supersaturation which have clustered during the quench can collapse and give intrinsic stacking faults, surrounded in this case by one loop of partial dislocation, with Burgers vector $\mathbf{b}=\frac{1}{3}\langle 111\rangle$, which is therefore not glissile (Fig. 3.39). Other types of stacking

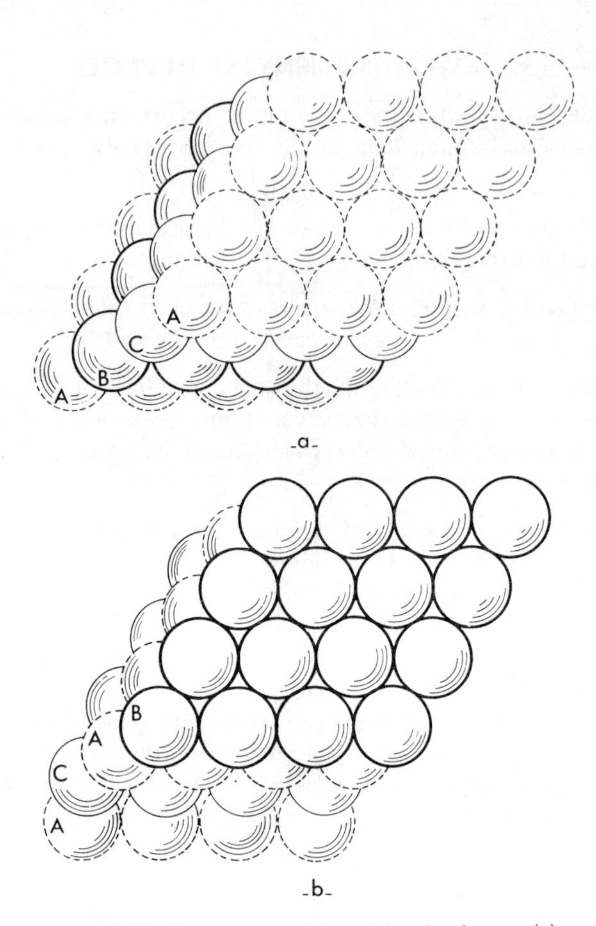

Fig. 3.37. Compact stacking of hard spheres: (a) face-centred cubic stacking: ABCA . . . , (b) stacking fault: ACAB . . .

faults may be considered, for instance *extrinsic* stacking faults corresponding to the insertion of one plane:

A B C A C B C A B C . . .
↑

In all cases the stacking fault corresponds to violations of the proper FCC stacking sequence for distant atoms, although not for first neighbours.

For instance, in the FCC intrinsic stacking fault there are two second nearest neighbour violations (two A atoms for instance should not be second neighbours).

A B C A C A B C . . .

The fault therefore introduces atomic distortions in the crystal.

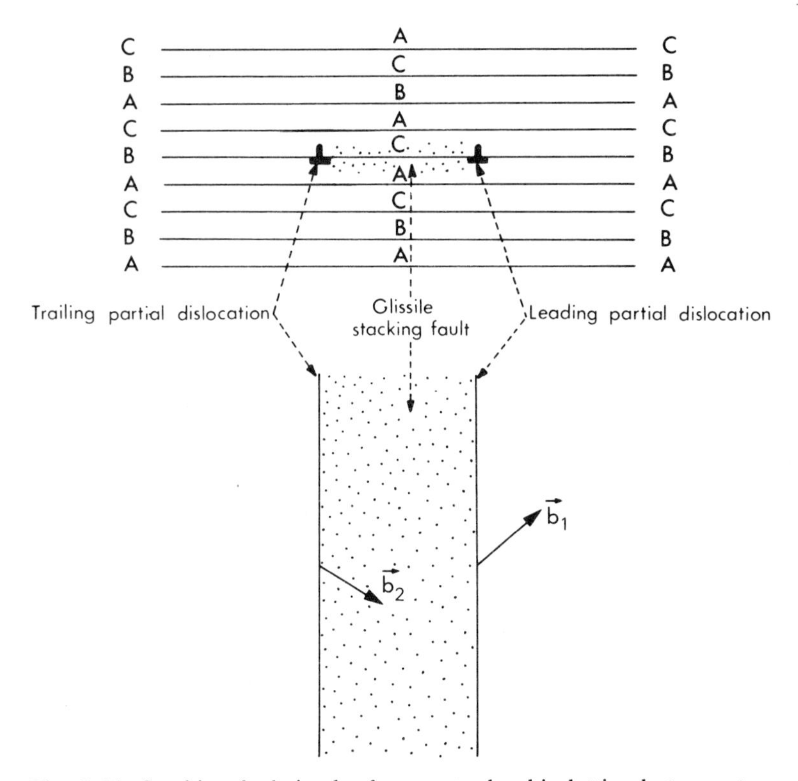

Fig. 3.38. Stacking fault in the face-centred cubic lattice between two partial dislocations (⊥): *Top*: section perpendicular to the (111) planes stacked in ABC . . . sequence outside the ribbon of fault, in ACAB . . . sequence inside the ribbon of fault. *Bottom*: split dislocation with ribbon of fault (stippled) in a (111) plane

The energy of the defect per unit area can be, in principle, calculated if one knows the number of violations and the distortion energy corresponding to each. The stacking fault energy (usually expressed in ergs/cm^2) γ_F is an important physical quantity for on it depends the width of splitting of the dislocations: the two partial dislocations, in each other's stress field repel with a force varying as $\mu b/r$ for a unit length of line; for a distance r between the partials a surface $l \times r$ of stacking fault is created, with a total energy $W = \gamma_F r$. A force $F = \partial W/\partial r = \gamma_F$ tends, therefore, to decrease the area of fault and hence reduce the distance between partials. The equilibrium width of the stacking fault r_e is obtained when the attractive and repulsive forces balance each other. We therefore have:

$$r_e \propto \frac{\mu b}{\gamma_F} \qquad (3.70)$$

Stacking faults may exist in any structure, whenever it can be described by a periodic stacking of planes. However, they may be very complicated and the

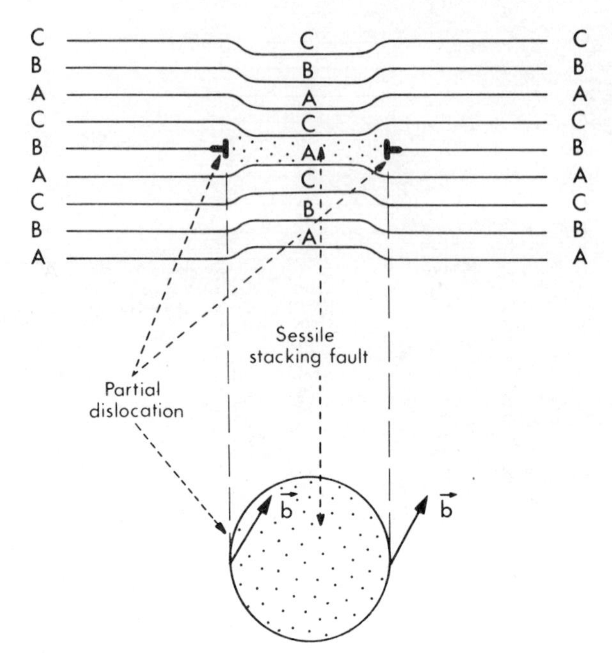

Fig. 3.39. Stacking fault surrounded by a loop of partial dislocation, resulting from the collapse of a disk of vacancies

equilibrium distance of dissociation so small that it is better in these cases to speak of a slightly flattened core of a perfect dislocation rather than of two partials separated by a planar ribbon of fault.

3.5.2. Dislocation walls

There is, in general, an interaction between dislocations in a crystal: as each dislocation is immersed in the stress field of the others it is subjected to a force. Thus, all the dislocations in a population tend to move by glide or climb or a combination of both in order to achieve a spatial arrangement by which the energy of the crystal is lowered. During this process, some dislocations of opposite signs are annihilated and others enter stable two-dimensional configurations known as dislocation walls.

These walls, or *sub-boundaries*, introduce a slight misorientation between the two adjacent regions of the crystal or *subgrains* (Plates 11–14). The dislocation composition of the walls may be more or less complex (Plates 5(a), 8, 9 and 10(b)); we will consider here the two simplest fundamental cases.

(*a*) *Symmetrical tilt walls*

These walls are formed by parallel straight edge dislocations of the same signs, stacked at equal intervals on top of each other, in parallel slip planes (Fig. 3.40) (Plates 8 and 9). The two adjacent subgrains are misoriented by an angle θ which depends on the Burgers vector **b** of the dislocations and their distance d.

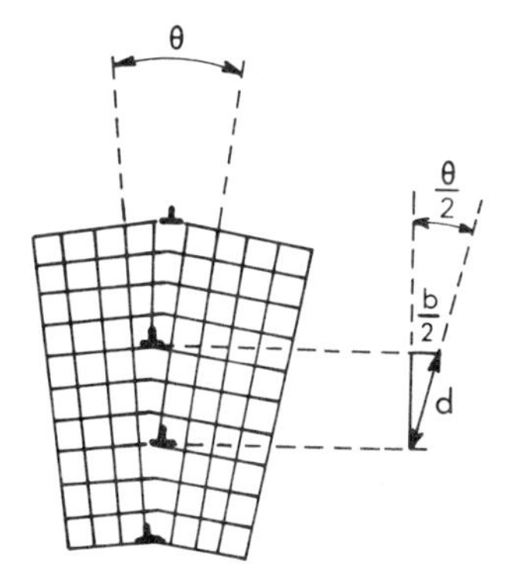

Fig. 3.40. Symmetrical edge dislocations tilt wall separating two subgrains misoriented by an angle θ

The axis of rotation is in the plane of the boundary. It can be seen in Fig. 3.40 that:

$$d = \frac{b}{2 \sin \theta/2} \simeq \frac{b}{\theta} \tag{3.71}$$

This is of course only valid for small values of $\theta (\theta \lesssim 15°)$. For larger values of θ, the sub-boundary becomes a grain boundary.

(*b*) *Pure twist walls*

These walls are formed by two families of screw dislocations as in Fig. 3.41, constituting a planar network.

They separate two subgrains deduced from each other by a rotation about an axis normal to the wall (twist). The pattern of the network depends on the dislocations involved and on the crystalline system: in FCC crystals for instance a twist boundary in {111} planes consists of a hexagonal network of screw dislocations.

(*c*) *Stress field and energy of dislocation walls*

The stress fields of the dislocations constituting the wall cancel out at long distances. It can be shown that stresses decrease exponentially as the distance from the wall increases. At a distance d of the order of the distance between dislocations in the wall, the stresses practically vanish. A dislocation wall therefore introduces no long-range lattice curvature. The energy of the wall can be taken as approximately equal to the sum of the energies of the constituent dislocations.

For a tilt wall composed of edge dislocations distant of d, the energy of the wall per unit area E_W is equal to the energy per unit length of dislocation given by (3.47) multiplied by the number $1/d$ of dislocations per unit length

$$E_W \simeq \frac{\mu b^2}{4\pi(1-\nu)} \left(\ln \frac{d}{b_0} \right) \cdot \frac{1}{d}$$

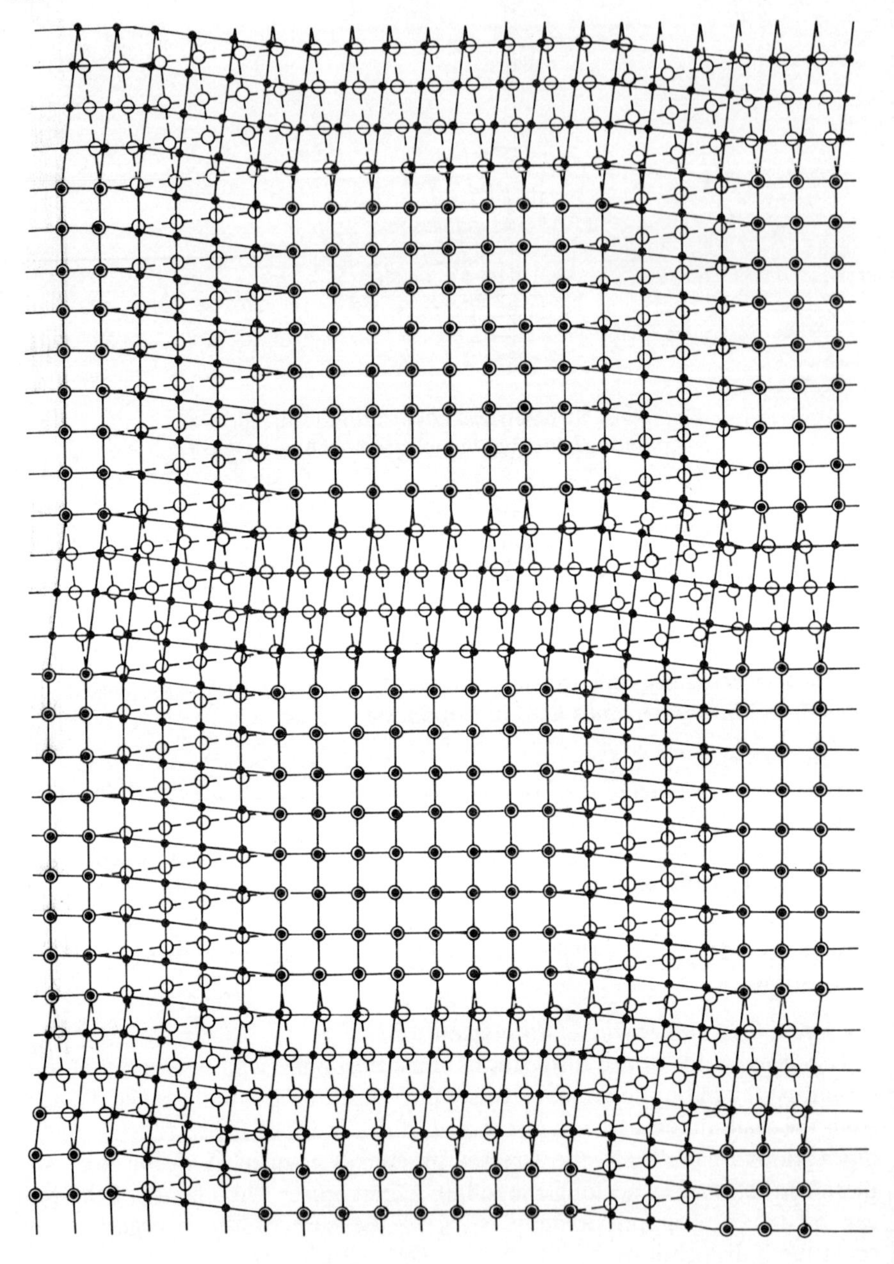

Fig. 3.41. Two slightly rotated cubic lattices superimposed give rise to a cross-grid network of screw dislocations (after Read). The pure twist boundary thus formed is parallel to the plane of the figure. Open circles = atoms just above the boundary. Solid circles = atoms just below. (Reproduced by permission of McGraw-Hill Company)

(we have taken $R = d$ in (3.47)). Using (3.71) we obtain:

$$E_W \simeq \frac{\mu b}{4\pi(1-\nu)}\theta\left[\ln\frac{b}{b_0} - \ln\theta\right] \tag{3.72}$$

We can see that the energy per unit area of wall increases with the angle of misorientation θ (Fig. 3.42). This approximation is valid for angles $\theta \lesssim 15°$. For larger angles of misorientation the wall must be treated as a grain boundary.

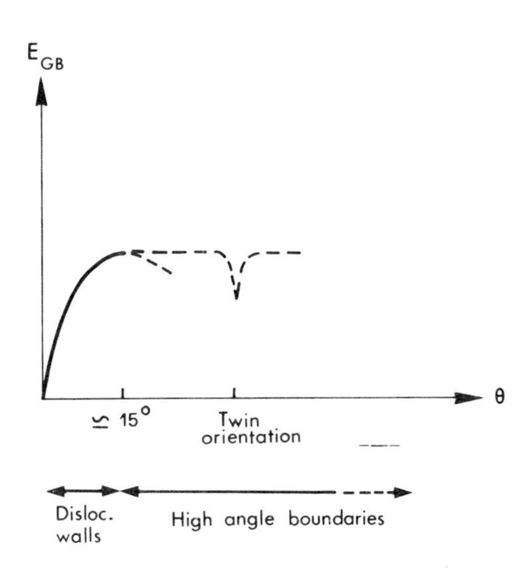

Fig. 3.42. Energy of a grain boundary as a function of misorientation angle θ

3.5.3. *Grain boundaries*

A grain boundary can be defined as the surface of junction between two crystals of the same material but of different orientation. The grain boundary can therefore be characterized by:

The angle of rotation θ, by which the crystalline lattice of one grain can be deduced from the lattice of the other grain (3 degrees of freedom).

The orientation of the boundary in one of the lattices (2 degrees of freedom).

Although we have dealt with them separately, the dislocation walls or low-angle grain boundaries belong, of course, to the broader category of grain boundaries. (The transition from a low-angle to a high-angle grain boundary is naturally ill defined). Conversely, for all grain boundaries, the rotation θ can always be analysed into a tilt component and a twist component which can, in principle, be accounted for by the existence of one family of edge dislocations and two families of screw dislocations. However, this analysis of a grain boundary in terms of dislocations is physically meaningful, only if the dislocation cores do not touch or overlap, that is for $\theta \lesssim 20°$ to $30°$.

Different models exist for the structure of large-angle grain boundaries (see Bollmann, 1970) but it would be beyond the scope of this book to deal with them here. We will limit ourselves to the following observations. For certain specified values of θ in a given crystalline structure it is possible to define a common superlattice or *coincidence lattice*, to which a certain proportion of the atoms in each grain belongs (Fig. 3.43(a)). Such adjacent crystals are said to be

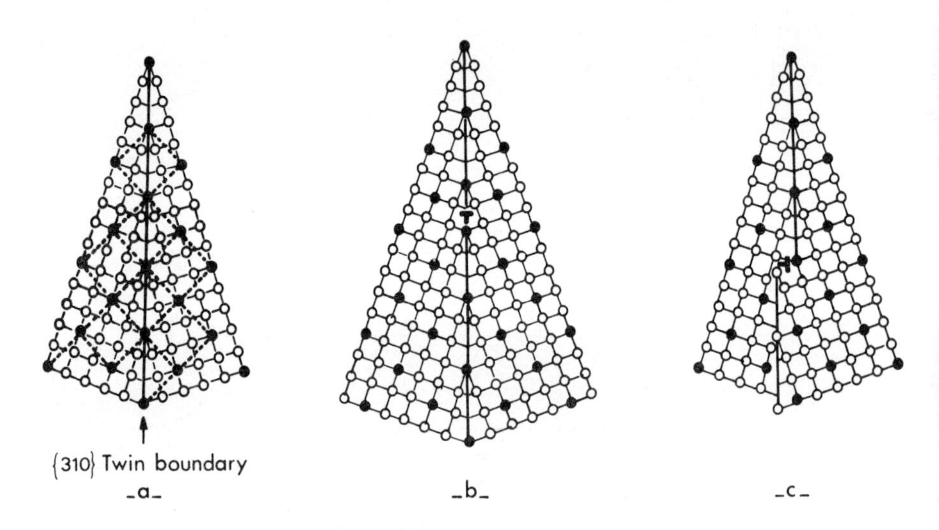

Fig. 3.43. Grain boundaries and coincidence lattice (closed circles): (a) coincidence boundary between grains in twin orientation; (b) coincidence boundary with grain boundary dislocation between grain deviating slightly from twin orientation; (c) boundary with ledge, deviating slightly from a coincidence boundary between grains in coincidence orientation. (After Hirth, 1972, reproduced by permission of the American Society for Metals)

in *twin orientation.* Note that such a relationship between the lattices of the adjacent grains implies nothing about the orientation of the grain boundary.

If the grain boundary is planar and coincides with an atomic plane of the coincidence lattice, it is called a *coincidence boundary* (Fig. 3.43(a)); (coherent *twin boundaries*, separating annealing or mechanical twins from the matrix, belong to this type). If the crystals deviate slightly from a coincidence orientation, coincidence boundaries may still exist, the elastic distortions that would result from the jointing of the two crystals along a plane which deviates slightly from a low index plane are taken up by grain boundary dislocations (Fig. 3.43(b)).

If the grain boundary between two crystals in coincidence orientation deviates slightly from a coincidence lattice plane, it can still be analysed as a coincidence boundary with *ledges* (which sometimes correspond to grain boundary dislocation) (Fig. 3.43(c)). Non-planar grain boundaries can often be analysed locally as coincidence boundaries with ledges.

The energy per unit area of coincidence boundaries is lower than the energy of a grain boundary between two randomly oriented crystals, which does not generally depend on the orientation (Fig. 3.42). Kretz (1966) has noticed that the boundary between two grains of anisotropic minerals like hornblende or biotite tends to be oriented parallel to a low index plane in one of the grains.

Grain boundary migration

Grain boundaries may migrate, which means that one grain expands at the expense of the other. The migration therefore involves a short-range transport of matter between the two grains, which can be analysed as a movement of ledges or a climb of the grain boundary dislocations. The grain boundary migration is thus effective only at high temperatures.

The driving forces for grain boundary migration (see §4.4) are:

The decrease of free energy brought about when a strain-free grain consumes a grain where the dislocation density is high.

The decrease of grain boundary energy produced by a reduction in the total grain boundary area.

Growth of mechanical twins

An important case of grain boundary migration is the rapid displacement of a mechanical twin boundary leading to a thickening of the twin lamella. In contrast with the case of grain growth where the migration of the boundary involves the climb of grain boundary dislocations, here the displacements of the twin boundary involve the glide of twinning dislocations in the boundary.

Fig. 3.44 shows how a twin boundary can be displaced normal to itself by the propagation of a twinning dislocation parallel to the twinning plane. The

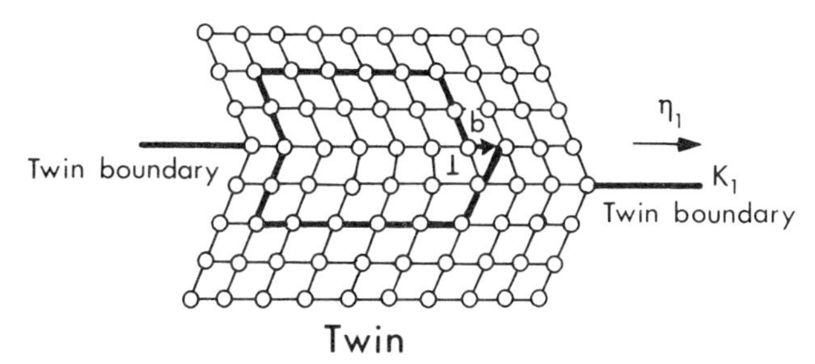

Fig. 3.44. Growth of a mechanical twin in a direction perpendicular to the twin plane by propagation of a twinning dislocation in the twin boundary. The boundary advances one interatomic distance up when the dislocation has swept across the boundary from left to right

twinning dislocation has a Burgers vector equal to the twinning shear which can be defined by making a Burgers circuit around a ledge in the twin boundary. As the twinning shear does not correspond to a lattice period, the twinning dislocation is imperfect.

Twinning dislocations can be seen by electron microscopy or X-ray topography (Plate 4(a)).

Grain-boundary diffusion

Diffusion is easier along grain boundaries than in the bulk of the crystal. This can be qualitatively understood if one considers the grain boundary as a perturbed zone in the lattice.

Experiments yield the quantity $D_{GB}\delta$, where D_{GB} is the grain boundary diffusion coefficient and δ a length which in theoretical treatments is taken as the width of the zone in which diffusion takes place ('width of the boundary'). The activation energy for self-diffusion in the grain boundary is generally about half the activation energy for bulk self-diffusion. It is therefore clear that at temperatures too low for bulk diffusion to be significant, transport of matter can still be active along grain boundaries.

Impurity segregation

A high-angle grain boundary is a perturbed zone and impurity atoms, larger or smaller than the crystal atoms, can be attracted to the grain boundary where they distort the crystal less. This impurity segregation often contributes to the preferential etching or decoration of grain boundaries.

3.5.4. Interfaces

We will use here the word interface in the restrictive meaning of: surface of junction between two crystals of different materials (or the same material with a different crystal structure).

Interfaces between the various minerals of a rock constitute obviously important defects in rocks. However, due to their complexity, very little is known about them.

In some cases there may be a simple orientation relationship between two crystals across a planar interface. This happens when the interface is a lattice low index plane for both crystals and when the atoms in this plane are common to both lattices. We then have a *coherent interface* (Fig. 3.45).

If the parameters of the planar lattices of both crystals in the coherent interface are slightly different, elastic strains arise thus allowing the crystals to match across the interface. For larger (but still very small) differences, the elastic energy involved would be too large and it is energetically favourable to take up the mismatch by *interface or misfit dislocations.* We thus have a *semi-coherent interface.* An example of this type of interface can be found in exsolution lamellae of clinopyroxene in an orthopyroxene matrix (Champness and Lorimer, 1973).

The interfacial energy (per unit area of interface) is probably smaller for an interface lying parallel to a low index plane in one of the adjacent crystals.

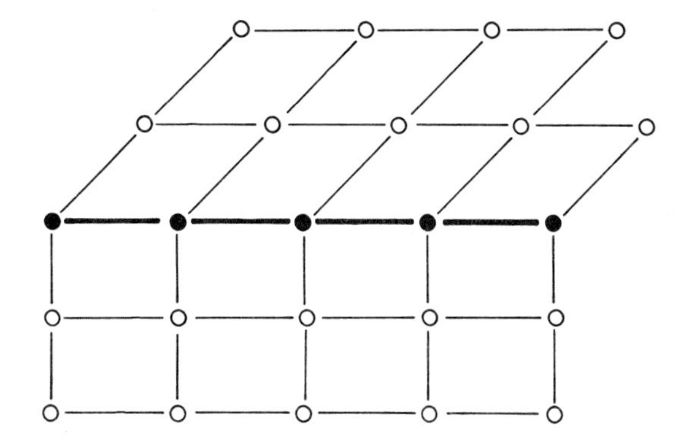

Fig. 3.45. Coherent interface between different lattices

Kretz (1966) observed in a specimen of amphibolite that the interface between grains of biotite and hornblende may lie parallel to the (110) plane of hornblende or to the (001) plane of biotite depending on the relative orientation of the two grains (Fig. 3.46). Although not enough is known as yet of the

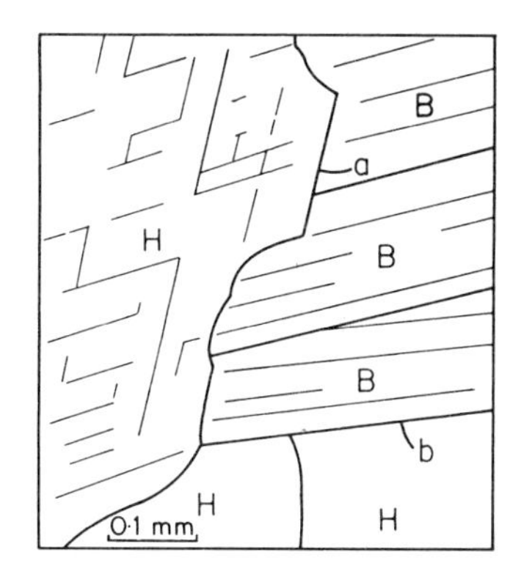

Fig. 3.46. Interfaces between grains of hornblende and biotite parallel to low index planes of hornblende or biotite (after Kretz, 1966, reproduced by permission of Oxford University Press, Oxford)

interface energies, interesting conclusions can be drawn concerning the shape of some mineral grains in metamorphic rocks, which may often result from a minimization of interfacial energy (Kretz, 1966).

CHAPTER 4

Flow and Annealing Processes in Crystals

4.1. GENERALITIES: WORK HARDENING, RECOVERY

Having introduced the principal crystalline defects, we are now in a position to understand, at least qualitatively, how the dynamic and kinematic quantities describing plastic flow depend on the production, interaction and annihilation of these defects.

Let us first review the various mechanical and physical parameters which we will try to relate, with the help of laboratory experiments and models.

(1) Plastic flow is described by an applied stress σ and a strain rate $\dot{\varepsilon}$. Only one of these mechanical parameters can be chosen arbitrarily and imposed on a sample in laboratory experiments:

In tensile (or compression) tests the sample is forced to deform at a given constant strain rate $\dot{\varepsilon}$, the stress σ necessary to achieve this can be measured as a function of time t (hence, of strain ε).

In creep tests, a given constant stress σ is applied to the sample, and the resulting strain rate $\dot{\varepsilon}$ can be measured as a function of time t.

In both of these tests, the physical parameters (temperature T and pressure P) are held constant. If, in a creep test, at constant σ, T, P, the resulting strain rate $\dot{\varepsilon}$ is constant, the important case of *steady-state flow* arises. In a tensile (or compression) test at imposed $\dot{\varepsilon}$, steady-state flow occurs when the applied stress remains constant in time.

(2) The physical parameters related to the crystalline defects can be divided into two groups:

(i) Microstructural parameters which can be described qualitatively or measured experimentally on samples deformed in known conditions of σ, $\dot{\varepsilon}$, T, P. Laboratory experiments yield two kinds of results with respect to these parameters:

Characteristic microstructural features can be qualitatively associated to certain domains of $\dot{\varepsilon}$ or T (fabric, shape of grains, aspect of grain boundaries, presence of subgrains, aspect and distribution of dislocations, etc.).

Quantitative microstructural parameters can be related by empirical laws to the values of the physical or mechanical parameters ($\dot{\varepsilon}$, σ, P, T) prevailing during the deformation: dislocation density $\rho = f(\sigma, T, \varepsilon \ldots)$, subgrain size $d = f(\sigma)$, for example.

(ii) Parameters whose relationship to σ, T, $P \ldots$ are known only from physical models. Such are, for instance, the dislocation climb or glide velocity, or the vacancy concentration and mobility.

The empirical relationship derived from laboratory experiments and the theoretical relationships can be injected into equations expressing microscopic deformation models (most of them being derived from Orowan's master equation $\dot{\varepsilon} = \rho_m b\bar{v}$). The microstructural parameters are thus eliminated and one obtains constitutive relations between macroscopic parameters $f(\dot{\varepsilon}, \sigma, T, P) = 0$ valid in certain domains of σ, $\dot{\varepsilon}$, T, P.

These relationships obtained by the conjunction of laboratory experiments and microscopic models are quite important since they can be used, obviously with a number of precautions, to figure out what the flow rate of rocks may be (or may have been) in certain assumed conditions.

The qualitative microstructural characteristics of flow in various conditions of $\dot{\varepsilon}$ and T can also be quite useful in allowing, by simple microscopic inspection of naturally deformed rocks, a reasonable estimation to be made of the strain rate or temperature conditions prevailing during (and/or after) deformation of the rock. We will critically examine the validity of this microstructural evidence in § 4.2.

In order to be able to understand the conclusions that can be drawn from the laboratory experiments on minerals and rocks (reported in Chapters 5 and 6), we will proceed to examine the background of the physical models of flow in crystals. Various simple models of high temperature flow will be discussed in § 4.3.

Most of the high temperature flow models rely on the competition between two elementary processes:

Work hardening, which expresses the fact that the more a crystal is deformed the more it is difficult to deform; it can be discussed in terms of internal stresses.

Recovery which expresses the fact that the internal stresses leading to work hardening can be relieved by thermally activated processes, thus allowing deformation to go on.

We will successively examine both processes and show that steady-state deformation results from a balance between them.

4.1.1. Work hardening

The process of work hardening can be best understood by referring to a deformation test at an imposed strain rate. At a given temperature and pressure, the sample is deformed at a constant rate $\dot{\varepsilon}$ and the applied flow stress σ is recorded as a function of time t. As $\dot{\varepsilon}$ is constant, the strain ε is proportional to t and the resulting graph is usually presented as a σ–ε curve (Fig. 4.1).

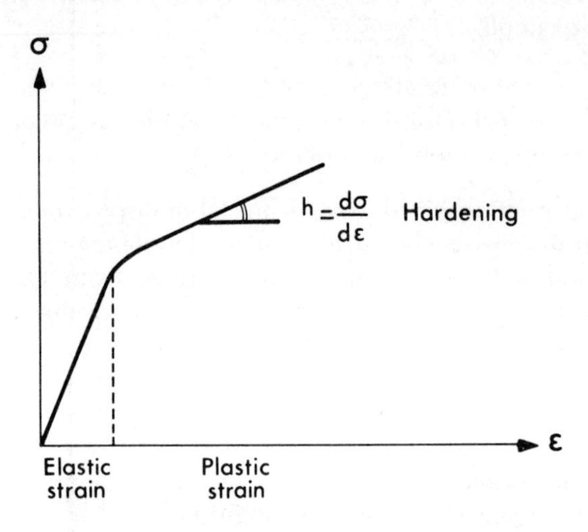

Fig. 4.1. Stress–strain curve at constant strain rate

For a given strain (or stress), work hardening is defined by the *work hardening coefficient*:

$$h = \frac{d\sigma}{d\varepsilon} \tag{4.1}$$

h represents the slope of the σ–ε curve at point (σ, ε).

A positive work hardening coefficient means that deformation at a given strain rate can go on only if the applied stress increases with time at an instantaneous rate:

$$\left(\frac{d\sigma}{dt}\right)_h = \frac{d\sigma}{d\varepsilon} \cdot \frac{d\varepsilon}{dt} = h\dot{\varepsilon} \tag{4.2}$$

In general, h depends on ε; if h is constant, one speaks of linear hardening since σ varies linearly with ε.

We know that plastic flow is caused by the motion of dislocations, and that Orowan's equation is generally valid:

$$\dot{\varepsilon} = \rho_m b \bar{v} \tag{3.69}$$

ρ_m is the mobile dislocation density and obviously depends on the number of sources which can be activated under a given shear stress. $\bar{v}$ is the average velocity of dislocations and depends principally on the number of obstacles and their resistance to motion.

Let us examine separately how the two terms ρ_m and $\bar{v}$ can depend on stress and strain.

(i) We have seen (§ 3.4.6) that the critical stress for the activation of a source is given by equation (3.59)

$$\sigma_{\text{crit}} \simeq \frac{\mu b}{2Kl}$$

where l is the length of the dislocation segment acting as a source. Dislocations in a crystal generally interact and form a 3-dimensional network, the lengths l of the dislocation segments being distributed around a mean value $\bar{l}$.

The source is activated when the local shear stress becomes equal to σ_{crit}. Now, the local shear stress can be taken as equal to: $\sigma - \sigma_i$, where σ is the applied stress and σ_i the internal stress due to other dislocations counteracting σ. Although in some places σ_i can be additive to σ, only substractive σ_i's control the multiplication of dislocations, for they give rise to more restrictive conditions.

Let us now consider a source of length l active under an applied stress σ at a given moment during deformation at an imposed strain rate. For the sake of argument, we can reasonably assume that the source segment is plane and emits dislocations in a slip plane. As dislocations are emitted, they propagate through the crystal but, eventually, the first dislocation of the train is blocked on some obstacle (usually other dislocations interacting with it); its stress field then stops the second dislocation and so on, until the internal back stress σ_i of all the piled up dislocations reaches a value at which the effective stress on the source $\sigma - \sigma_i$ becomes too low to activate the source:

$$\sigma - \sigma_i < \sigma_{\text{crit}}$$

For the deformation to continue at the imposed strain rate, the applied stress must increase, thus allowing either the obstacle to be overcome, or the source to emit other dislocations, and sources of slightly shorter length to become active. As deformation goes on, more dislocations are trapped in the pre-existing network and this gives rise to another contribution to work hardening by decreasing the mean mesh length l of the network, thus making the sources activable only under larger applied stresses.

(ii) The increase in the total density of dislocation ρ_t (not to be confused with the mobile dislocation density ρ_m) brought about by the continuation of deformation, has also as a consequence to decrease the mean velocity $\bar{v}$ of dislocations by increasing the number and the strength of obstacles to motion.

Dislocations can act as an obstacle to motion of other dislocations in two ways:

(a) By their long-range stress field when they are situated in planes parallel to the glide plane.

(b) By short-range interaction when they cut the glide plane. Dislocations cutting glide planes are called *trees* and they form a *forest* impeding motion on all glide planes cutting the forest. Trees impede the motion of dislocations in two different ways according to the nature of their interaction with the incoming dislocation:

If the stress field of the tree exerts a repulsive force on the dislocation (*repulsive tree*), the latter has to be pushed against the tree by the applied stress and can eventually cross it by acquiring a jog (Fig. 4.2). This process is usually important only at low temperatures.

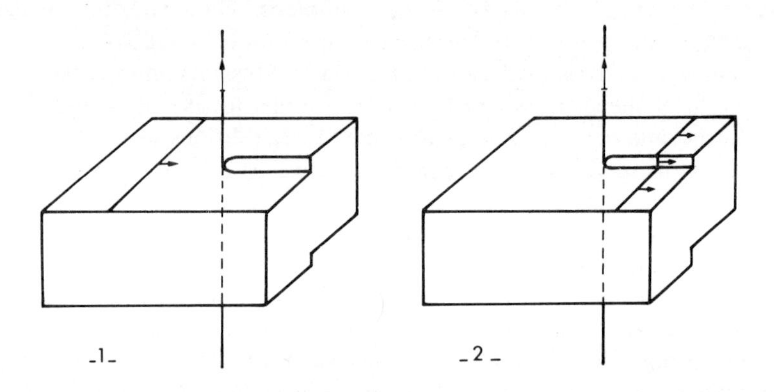

Fig. 4.2. Edge dislocation in a horizontal glide plane threaded by a screw dislocation (tree). The edge dislocation acquires a jog when it cuts the tree

If the stress field of the tree exerts an attractive force on the dislocation (*attractive tree*), the dislocation and the tree can usually react to form a dislocation segment between two triple nodes (Fig. 4.3). This segment is very stable (*attractive junction*), the dislocation is effectively trapped and can pursue its slip only if the applied stress is high enough to destroy the junction.

Hardening by the forest due to attractive junctions is a very effective process in metals at moderate temperatures.

When the total dislocation density increases up to a certain value ($\rho_t \simeq 10^9$ to 10^{10} disloc./cm^2 in metals) the repartition of dislocation ceases to be uniform: dislocations interlock in *tangles*. As strain continues to increase, a *cellular structure* appears: *cell walls* of numerous entangled dislocations separate *three-dimensional cells*, where dislocation density is low (Plate 10(a)). Tangles and cell walls are obviously very effective obstacles to the motion of dislocations and at the same time the dislocations emitted by sources inside the cells can pile up against them and block the sources. They are therefore an important source of work hardening.

Tangles are also very frequently initiated by the interaction of dislocations on one slip system with dislocations on secondary slip systems crossing the

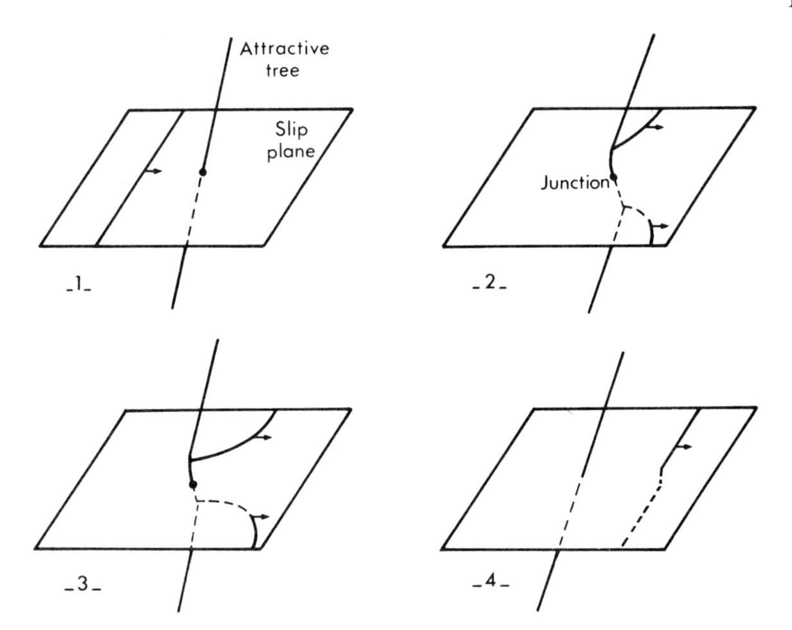

Fig. 4.3. Attractive junction: a dislocation gliding in its glide plane (1) cuts an attractive tree. A dislocation segment (junction) is formed (solid lines above the slip plane, dotted lines below the slip plane) (2). The dislocation bows out under the stress (3) and the junction is eventually destroyed leaving a jog on the dislocation (4)

primary slip planes. Dislocations of the two systems may lock together in immobile configuration (or *barriers*), against which dislocations of both systems pile up and interact, giving rise to tangles and a cellular structure. The work hardening coefficient is usually much higher when several slip systems are simultaneously active than when only one is active.

From what precedes, it must now be clear that intragranular work hardening (in a crystal containing no second phase inclusions) is essentially due to the increase in the total dislocation density with strain. To know the exact dependence of h on ε or σ, it would be necessary to make a model and also to know how the dislocation density varies. If the deformation is homogeneous and the dislocations which remain in the crystal are more or less in equilibrium in the stress fields of their neighbours at an average distance L from one another, we will have:

$$\rho_t \simeq \frac{1}{L^2}$$

The mean internal stress field will be

$$\bar{\sigma}_i \propto \frac{\mu b}{L}$$

so that we will have

$$\rho_t \propto \bar{\sigma}_i^2$$

If it is further assumed that $\bar{\sigma}_i \propto \sigma$ (σ applied stress) then:

$$\rho_t \propto \sigma^2 \tag{4.3}$$

This relationship is often found experimentally, but its theoretical foundation rests on so many assumptions that it is always questionable to derive the value of the applied stress from counting the dislocations in a previously (or naturally) deformed crystal. We will return to this point later (§ 4.2).

Grain-boundary hardening

Although the increase in dislocation density inside the grains is a very important source of work hardening, we must not forget that grain boundaries are also quite effective obstacles to slip and therefore represent also a source of work hardening. Grain-boundary hardening is predominant at low deformations when dislocation density inside the grains is still low and grain boundaries constitute the major obstacles to slip; its relative importance decreases as it becomes harder and harder for dislocations to cross the forest or tangles inside the grains, i.e. for large values of the strain. In what follows we will briefly examine how the presence of grain boundaries increases the stress necessary to deform a sample and how this stress (for a given strain) can be related to the grain size.

It is, first, necessary to say a few words about *pile-ups* of dislocations against obstacles (in this instance grain boundaries).

When the leading dislocation of a train of dislocation on a slip plane is held up at an obstacle too strong to be overcome under the applied stress σ, all the following dislocations pushed by the applied stress σ pile up behind it (Fig. 4.4).

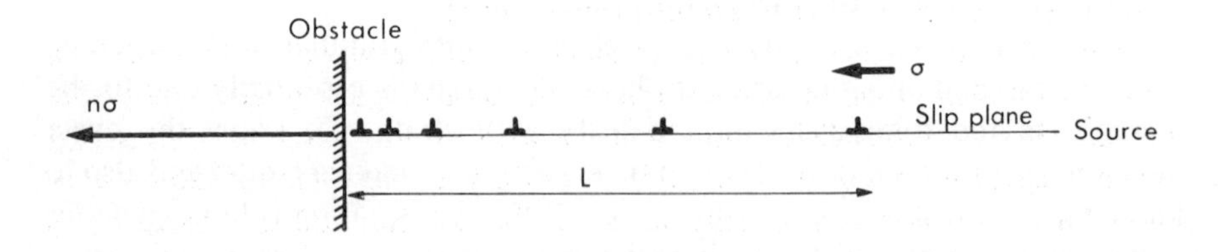

Fig. 4.4. Edge dislocations piled up behind an obstacle under a stress σ. The stress ahead of the obstacle is multiplied by n (number of piled-up dislocations)

Let us calculate the actual shear stress σ^* at the head of the pile up: if the head of the pile up is allowed to move a distance $\mathrm{d}x$, all n dislocations will move by $\mathrm{d}x$. They are all subjected to a force per unit length $F \simeq \sigma b$ (equation 3.56), so that the total work done is: $nF\,\mathrm{d}x = n\sigma b\,\mathrm{d}x$. This must be equated to the work done by the force on the first dislocation resulting from the back stress σ^*.

We therefore have:

$$n\sigma b \, dx = \sigma^* b \, dx$$

hence:

$$\sigma^* = n\sigma \tag{4.4}$$

This shows that a dislocation pile-up enhances the applied stress and acts as a *stress concentrator*, in the same way as the tip of a crack does.

We can now evaluate the hardening arising from the presence of grain boundaries. Let us assume that the intrinsic resistance of a grain boundary to shear is given by σ_{GB}, the shear stress that it must undergo for slip to go through. Under the applied stress σ, dislocations will pile up until:

$$\sigma_{GB} = n\sigma$$

Now, the number of dislocations in a pile up of length L can be shown to be (Hirth and Lothe, 1967):

$$n \propto L^{1/2}$$

The length of the pile up can be assumed to be proportional to the grain size d, so that the stress σ necessary to overcome the grain boundary will be:

$$\sigma = \frac{\sigma_{GB}}{n} \propto \sigma_{GB} L^{-1/2} \propto \sigma_{GB} d^{-1/2} \tag{4.5}$$

It can thus be seen that a *fine grained polycrystal is harder to deform than a coarse grained one.* The excess hardening over a single crystal is proportional to $d^{-1/2}$. This is known in metallurgy as *Hall–Petch law* and is generally well verified experimentally.

4.1.2. Recovery

We have seen that intragranular hardening can be ascribed to the increase in the total dislocation density with strain, due to the fact that glide dislocations remain trapped in the crystal, thus increasing the internal stress and acting as obstacles to slip. (Note that these dislocations are not geometrically necessary.)

The increase in total dislocation density ρ_t, of course, increases the energy stored in the crystal (see § 3.4.4), and a driving force arises tending to bring back the crystal to a lower energy state by reducing the dislocation density.

At low temperatures, dislocations can move only in their slip planes and as they form networks and tangles, there is practically no way for the crystal to get rid of them. The situation is different at high temperatures ($T > 0{\cdot}3T_m$, T_m being the melting temperature) for diffusion processes become important and dislocations can climb (§ 3.4.9) and are thus granted another degree of freedom for moving in the crystal. The consequences are the following:

(i) Dislocations with Burgers vectors of opposite sign in parallel glide planes can climb toward each other and annihilate, thus reducing ρ_t and eliminating obstacles (Fig. 4.5).

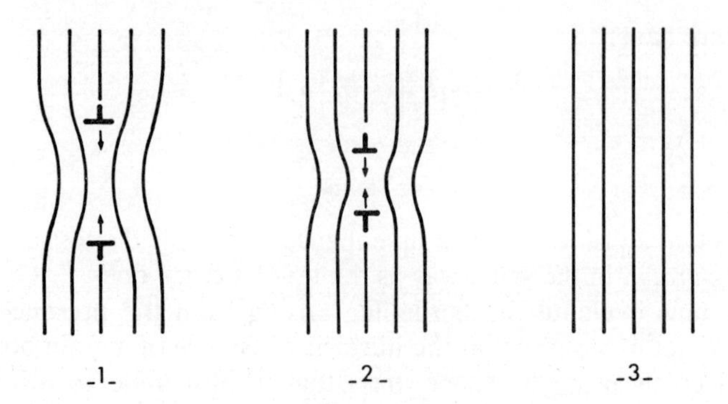

Fig. 4.5. Annihilation of two edge dislocations of opposite sign climbing toward each other

(ii) The dislocation network can coarsen its mesh (Friedel, 1964), thus reducing the internal stress and providing sources of greater length activable under lower stresses.

(iii) Dislocations piled up against obstacles in their glide planes can climb past the obstacles, thus relieving the back stress, and allowing other dislocations to be emitted (Fig. 4.6).

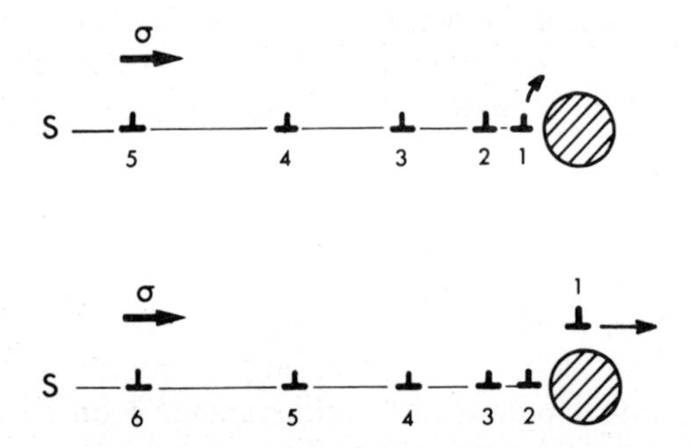

Fig. 4.6. Climb of the leading dislocation of a pile-up over an obstacle

(iv) Excess dislocations with the same Burgers vector can rearrange themselves into dislocation walls with a short-range stress field, separating slightly misoriented blocks (subgrains) containing fewer dislocations. This phenomenon is known as *polygonization by climb* and is a typical feature of recovery during high temperature deformation (or recovery during a post-deformation anneal) (Fig. 4.7) (see § 4.4).

Another phenomenon, active only at high temperatures (or high stresses), may act in conjunction with climb to further recovery in certain conditions: this

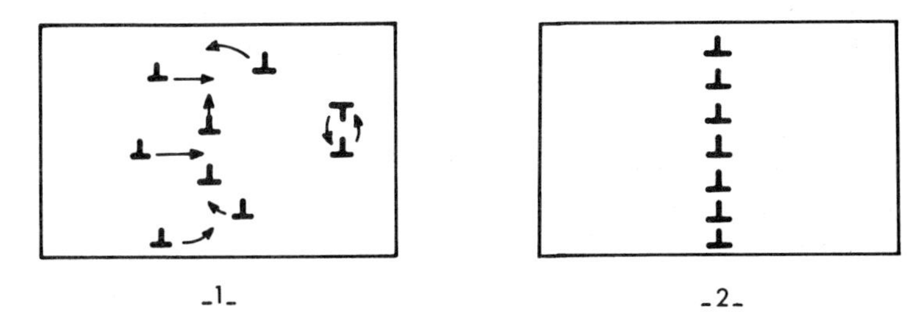

Fig. 4.7. Rearrangement of edge dislocations by glide and climb leading to the formation of a tilt wall (polygonization by climb)

is *cross slip* or deviation of screw dislocations from one glide plane to another (Plate 3(a)). When cross slip is possible, opposite sign screw dislocations can cross-slip toward each other and annihilate (Fig. 4.8(a)) or screw dislocations

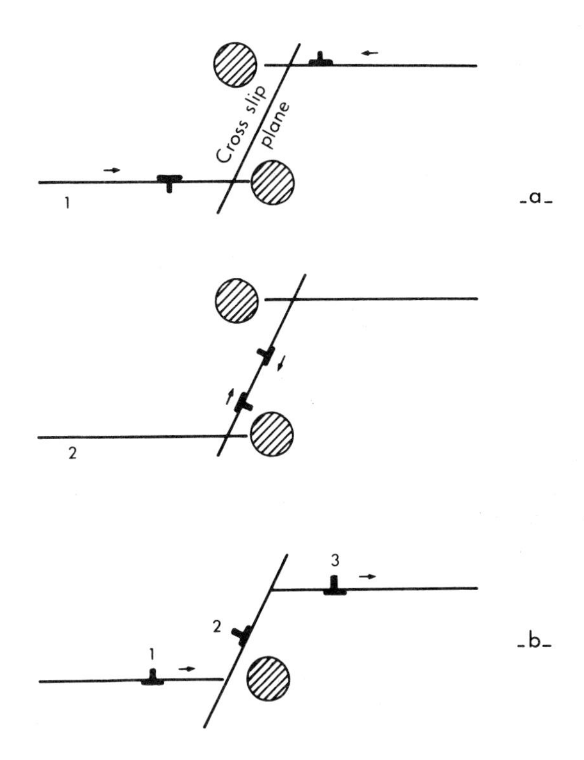

Fig. 4.8. Cross slip of screw dislocations in front of an obstacle: (a) two screw dislocations of opposite sign annihilate by cross slip. (b) a screw dislocation cross-slips to avoid the obstacle and resumes gliding in a plane parallel to the former glide plane

can circumvent an obstacle by cross-slipping in another plane (Fig. 4.8(b)). (edge dislocations have only one glide plane, hence they cannot cross-slip.)

Of course, this is possible only if there exist several glide planes containing the screw dislocation Burgers vector; cross slip is therefore much less general than climb. However, in view of the fact that it may happen in certain minerals, it is of some interest to give a rough description of the mechanism of cross-slip.

Let us consider a screw dislocation split in a lattice plane, into two partial dislocations separated by a ribbon of stacking fault (see § 3.4.8) (Fig. 4.9).

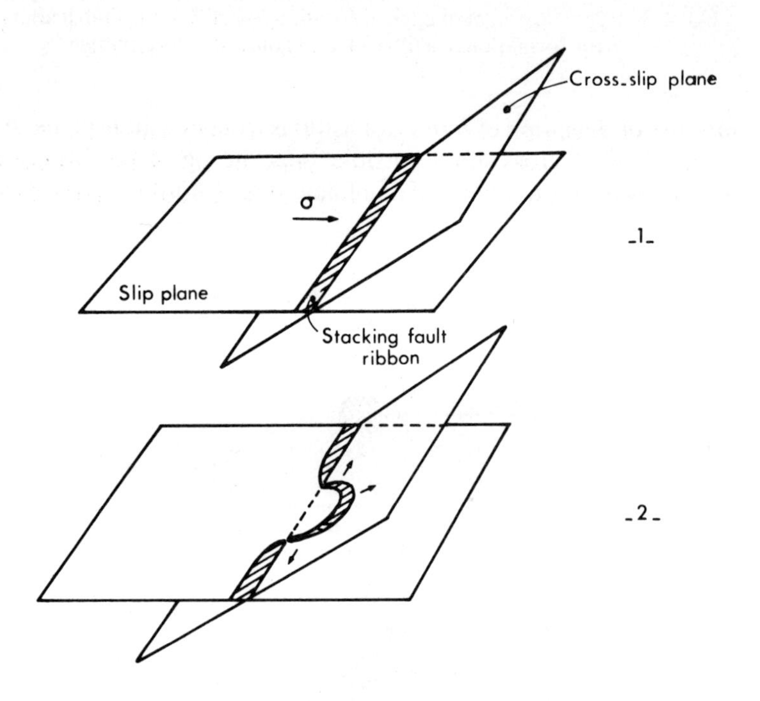

Fig. 4.9. Cross slip of a split screw dislocation: (1) a dislocation split into two partials separated by a stacking fault ribbon is pushed against an obstacle; (2) a length of dislocation has recombined and has cross-slipped, it splits in the cross-slip plane where it can bow out and expand under the applied stress (see text)

Let us assume that the crystal structure and the dislocation line orientation are such that one or several other low index lattice planes are in zone with the dislocation line. Although it would be possible for the unsplit total dislocation to glide in these planes, it is, however, impossible for the split dislocation to do so, since the stacking fault lies in another plane. Cross slip in planes in zone with the screw dislocation is nevertheless possible if the split dislocation recombines, that is if the stacking fault is pinched. Pinching of the stacking fault can occur locally if the dislocation is pushed in its glide plane against some obstacle by the applied stress but this process is thermally activated and becomes important only above a certain temperature (for a given applied stress). A length of dislocation can then recombine and the resolved shear stress on the secondary cross-slip plane will pull the dislocation on to that plane, increasing at the same time the length that can cross slip. The cross-slipping dislocation arc may split in the cross-slip plane if a stable stacking fault exists in it; if not, its screw portions splits into a plane parallel to the first glide plane and the process has to start again.

In any case, whether cross slip of such dislocations is present or not, diffusion controlled climb of edge dislocations can always occur and this is sufficient for recovery to take place.

Recovery undoes the effects of work hardening, partially or totally, by decreasing the internal stress and making the continuation of slip possible at a lower stress than if there was only work hardening.

We can define a recovery rate:

$$r = \left(\frac{\mathrm{d}\sigma}{\mathrm{d}t}\right)_r \tag{4.6}$$

as the rate at which the flow stress would decrease with time for zero strain rate. We can reasonably assume that it is proportional to the rate of decrease in the total dislocation density, which, in its turn, is proportional to the climb velocity:

$$r = \left(\frac{\mathrm{d}\sigma}{\mathrm{d}t}\right)_r \propto \left(\frac{\mathrm{d}\rho_t}{\mathrm{d}t}\right)_r \propto v_c$$

hence

$$r = \left(\frac{\mathrm{d}\sigma}{\mathrm{d}t}\right)_r \propto v_c \propto D_{SD} \tag{4.7}$$

after equation (3.64).

D_{SD} is the coefficient of self-diffusion, itself a function of temperature and pressure: *the recovery rate increases with temperature and decreases as hydrostatic pressure increases.*

Recovery by diffusion-controlled polygonization and annihilation of dislocations can be superseded by *recrystallization* which is a much more drastic and effective way of lowering the strain energy stored in deformed crystalline bodies. For high enough stored energy, the driving force to decrease the dislocation density may become very important; if the temperature is high enough to allow grain boundary migration, grains which happen to be relatively dislocation free will consume heavily deformed grains. This will result in a new undeformed state, although with a generally different grain size.

More will be said in § 4.4 about the mechanisms of recrystallization, which can occur, as does recovery, during deformation (*dynamic or syntectonic recrystallization*) or during a post-deformation anneal (*static recrystallization*). In any case, it is a process which decreases hardening.

4.1.3. Competition between work hardening and recovery

Although we have introduced the processes of work hardening and recovery separately, they do occur simultaneously in crystalline solids. The relative importance of work hardening and recovery rates may serve as a criterion for defining several domains of temperature and strain rate, where deformation has distinct characteristic features.

We will define the following regimes with reference to the simple laboratory experiments already mentioned:

Uniaxial tensile or compression tests at imposed constant strain rate.

Uniaxial creep tests at imposed constant applied shear stress.

We must recall here that the only rule of the confining hydrostatic pressure in the considered mechanisms is to decrease the diffusion coefficient, hence the recovery rate, or equivalently to decrease the relative temperature T/T_M, since the melting temperature T_M usually increases with pressure.

The overall effect of work hardening and recovery can be deduced from the sign of the expression:

$$h\dot{\varepsilon} - r$$

From equations (4.2) and (4.6) we have:

$$h\dot{\varepsilon} - r = \left(\frac{\mathrm{d}\sigma}{\mathrm{d}t}\right)_h - \left(\frac{\mathrm{d}\sigma}{\mathrm{d}t}\right)_r = \frac{\mathrm{d}\sigma}{\mathrm{d}t} \tag{4.8}$$

Four main cases can be distinguished (Table 4.1):

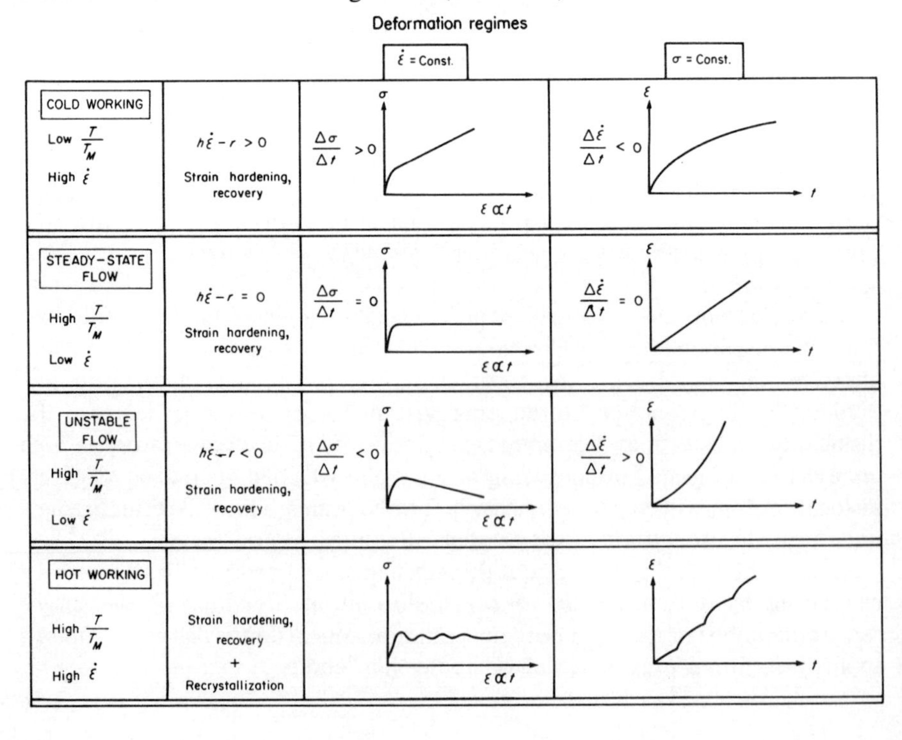

(*i*) $h\dot{\varepsilon} - r > 0$

In constant strain rate conditions, this means that the applied stress has to increase by $\Delta\sigma > 0$ during a time interval Δt, to maintain a constant $\dot{\varepsilon}$.

Conversely, in constant applied stress conditions, $d\sigma/dt = 0$, the strain rate $\dot{\varepsilon}$ will necessarily decrease, if $h = \text{const.}$

This regime occurs at *low relative temperatures* T/T_M (say $T/T_M < 0{\cdot}3$), since the recovery rate is low (small diffusion coefficients), but its occurrence depends also on the value of $\dot{\varepsilon}$. Even for a fairly high relative temperature, the expression $h\dot{\varepsilon} - r$ can be positive if the strain rate is high enough. It is a question of whether hardening is fast enough compared with recovery. The boundary of the ($\dot{\varepsilon}$, T/T_M) domain where this regime exists depends on the material properties of the crystals (number of slip systems, grain size, activation energy for diffusion etc.); for simplicity of exposition we will, however, refer to this domain as a *low temperature–high strain rate domain.* In metallurgical language this is the domain of *cold working.*

The typical structures of this regime will be studied in § 4.2; we will only note here that dislocation density and internal stresses increase with strain; for high values of strain, intragranular fracture and cataclastic flow eventually occur. Most superficial crustal deformation belongs to this type.

(*ii*) $h\dot{\varepsilon} - r = 0$

This is the condition for *steady-state flow*: In constant strain rate conditions, the applied stress remains constant ($d\sigma/dt = 0$) and in constant applied stress conditions, the strain rate remains constant (*steady-state creep*):

$$\dot{\varepsilon} = \frac{r}{h} = \text{const.} \tag{4.9}$$

This regime corresponds to the *balance between strain hardening and recovery.* Recovery is usually achieved through climb and polygonization and the strain energy does not usually reach high enough values for recrystallization to play an important role (except possibly at grain boundaries).

This is typically a *high temperature–low strain rate* regime which can occur in the deep crust or in the upper mantle.

(*iii*) $h\dot{\varepsilon} - r < 0$

The recovery rate exceeds the strain hardening rate, so that the material offers less and less resistance to deformation. At constant strain rate the applied stress necessary to drive the flow decreases with time, and at constant applied stress, strain rate increases. This is an unstable regime which must obviously give way either to catastrophic failure or to a new stage of hardening due to the increase in $\dot{\varepsilon}$.

(*iv*) For *high temperatures and high strain rates,* we can reach a domain where, despite recovery, the strain rates are high enough to increase the strain energy to a point where recrystallization takes over. The new grains are rapidly hardened and the process starts again. Periods of strain hardening therefore alternate with dynamic recrystallization, and this can be observed as stress oscillations in constant $\dot{\varepsilon}$ tests and strain-rate oscillations in creep tests. However, on the average, we may have steady-state flow up to extremely large strains (Jonas, Sellars, Tegart, 1969). This regime occurs during *hot working* of metals and we will designate it thus.

4.2. CHARACTERISTIC DISLOCATION STRUCTURES

In what follows, we will describe the more characteristic microstructural features which are observable optically or by transmission electron microscopy, and which can serve as markers of the $\dot{\varepsilon}$, T conditions during the deformation (the techniques will be described in Chapter 10). We will also briefly review the applicability of empirical or theoretical relationships which can sometimes give an estimate of the applied stress.

4.2.1. Dislocation structures of the cold-worked state

The cold-worked state is generally characterized by rather high dislocation densities in grains and little evidence of climb or recovery. More specifically, we can distinguish two cases:

(*a*) *Deformation by slip on a unique slip plane.* A slight deformation is characterized by the presence of long *straight or smooth dislocation lines lying in their slip plane.* They interact elastically with other dislocations in parallel slip planes and form typical structures known as *debris*: their main features are *dipoles* (self-locking pairs of opposite sign edge dislocations on parallel slip planes). Dipoles can pinch off and give *elongated loops* and *cusps.* (Plates 7(a) and 16). As deformation goes on, dipoles and loops stop other dislocations and *dislocation tangles* are formed.

For higher values of strain the dislocation tangles grow more and more dense and eventually become *cell walls* with a high dislocation density, separating cells relatively free of dislocations (Plate 10(a)).

Dislocations stay in their slip plane and may *pile up* against grain boundaries. Due to the presence of only one slip plane, the incompatibility arising from the neighbouring grains is taken up by geometrically necessary dislocations. Hence planes are bent and give rise to *undulatory extinction* between crossed polars; *glide polygonization* and *kink band* formation may occur. For certain orientations where slip is not easy, deformation may occur only by *kinking* (compression kinks) or *twinning* if twinning is possible.

(*b*) *Deformation by slip on several systems.* For small strains, the repartition of dislocations is homogeneous: dislocations react and form a three-dimensional network and tangles. The dislocation density increases with strain and for high strains a *cellular structure is formed.* The cell walls are *tangled* and contain a high density of dislocations.

4.2.2. Dislocation structures of the high temperature flow

The most typical feature is the existence of a *polygonized substructure* consisting of slightly misoriented blocks (*subgrains*), relatively dislocation free, and separated by *dislocation walls.* The dislocation walls are formed by climb and are generally sharp in contrast with the diffuse tangled cell walls observed in the cold worked state.

When there is only one slip system, the dislocation walls are usually perpendicular to the slip plane and the tilt walls are geometrically similar to kink-band boundaries observed after low temperature deformation. As the excess dislocations within the blocks have climbed into the walls there are usually no bent planes and no undulatory extinction. However, a succession of very narrow subgrains can give undulatory extinction (White, 1973). When there are several glide systems the blocks tend to become more equiaxed.

Inside the blocks, the dislocations do not usually stay in planes. Debris may be observed consisting of contorted dislocation lines and mostly *dislocation loops* formed during climb of non-planar dislocation lines. The polygonized blocks are clearly visible between crossed polars. As a rule their misorientation is greater than 1° and may be as large as 10° or 15°, their size is of the order of 100 μ or more and depends on the stress (see below). Dislocations often rearrange themselves by climbing inside the blocks producing dislocation walls with a much lower angle (of the order of a few minutes of arc) which can be seen by TEM or etch-pits.

At grain boundaries, pile-ups are never visible since they are usually relaxed by climb. The incompatibility of deformation between neighbouring grains may be relieved by diffusion accommodated *grain-boundary sliding.* A finer polygonization is often seen at the vicinity of the grain boundaries and these are often corrugated due to local *grain-boundary migration.*

Recrystallization may occur at the grain boundaries where strain energy is high. Recrystallization is a typical feature of the hot-worked state, but it is difficult to distinguish between recrystallization during deformation at high temperature (syntectonic) and static recrystallization during a post-deformation anneal (cf. § 4.5.4). The qualitative characteristics of the cold-worked state and high-temperature deformation are summed up in Table 4.2.

4.2.3. Empirical relations between dislocation microstructure and applied stress

Even a rough estimate of the value of the stress prevalent during the natural deformation of a rock would be of the utmost importance to the structural geologist or the geophysicist.

Now, certain quantitative microstructural parameters have been empirically related to the applied stress during *steady-state flow* in a number of crystals; it is therefore interesting to try to assess the validity of such relations in making retrodictions about the stress applied to a rock during its natural deformation.

Table 4.2. Typical microstructures

	Low T/T_M High $\dot{\varepsilon}$	High T/T_M Low $\dot{\varepsilon}$
Unique slip system	Straight or smooth dislocations	Contorted dislocations
Low strains	Dipoles	
	Elongated loops	Rounded loops
	Cusps	
	Dislocation pile-ups	
	KBB normal to slip plane (glide polygonization)	*Climb polygonization*:
	Bent planes: undulatory extinction	Tilt walls
	(Twinning)	Twist walls
Several slip systems and/or high strains	Tangles Cellular structure Thick tangled cell walls	*Climb polygonization*: Misoriented equant subgrains Thin, well-defined sub-boundaries Tilt and twist dislocation walls Corrugated grain boundaries Fine subgrains close to grain boundaries Grain-boundary migration *Recrystallization*

(*a*) *Relation between applied stress and subgrain size.* In all materials deformed at high temperature, where climb is important, the most noticeable feature whether in steady-state or hot-working conditions, is the presence of misoriented subgrains.

It is a well-established fact in physical metallurgy (Weertman, 1968) that these subgrains form at the beginning of the deformation and that their size remains constant once they are formed. Their size does not practically depend on temperature but only on the applied stress, according to the empirical law, best written under a dimensionally correct form:

$$d = K\frac{\mu b}{\sigma} \tag{4.10}$$

where d is the average size of the subgrains, μ is the average shear modulus, b is the Burgers vector of the active dislocations, K is a dimensionless constant of proportionality.

Table 4.3 presents some values of K for different materials. The observation techniques are also mentioned (see Chapter 10 for their description).

Table 4.3

Material	$K = d\sigma/\mu b$	Technique of observation	Reference
Metals (compilation)	100	Mostly TEM	Weertman, 1968
Au	70	Electron diffraction orientation contrast	Dorizzi, 1974
NaCl	50	Etch-pits + Berg Barrett	Poirier, 1972
LiF	60	Etch-pits	Cropper, Pask, 1973
AgCl	44	Decoration + Electron diffraction orientation contrast	Pontikis, Poirier, 1974
MgO	50	Etch-pits	Hurm, Escaig, 1973
Olivine	800	Polarized light microscopy	Raleigh, Kirby, 1970

To make a valid comparison between the results of Table 4.3, it must first be borne in mind that what the authors call 'subgrains' fall into two distinct categories; in other words, one must recognize the existence of two orders of polygonization substructure (Streb and Reppich, 1973; Pontikis and Poirier, 1975; Green and Radcliffe, 1972; White, 1973).

(i) *Subgrains* proper, resulting from the fragmentation of the coarse grains into blocks misoriented by an angle $\theta \lesssim 1°$. These subgrains are formed during the first stages of the deformation at high temperature, probably by rearrangement by climb of the dislocations lying in the glide bands (Poirier, 1972). Their size is larger than 50 or 100 μm and can be of the order of 1 mm according to the stress. Consequently, they can exist only in grains whose size is at least of the order of 1 mm or in single crystals. They can be observed at low magnifications either by optical microscopy in polarized light, or by etch-pits or decoration techniques which reveal the subgrain boundaries or by orientation contrast in X-ray or electron diffraction (Plates 11–14). They are invisible by transmission electron microscopy, since their dimensions are too large.

They are typical of the high-temperature deformation process and in addition to the fact that their size depends on the applied stress and not on temperature, they are very stable during a post-deformation anneal. As a consequence, if *one can determine experimentally the constant K on the same material (or a similar material), it is possible to estimate an applied stress by optical measurement of the subgrain size of naturally deformed crystals and application of the formula*

$$\sigma = K\frac{\mu b}{d}$$

It is likely that the value obtained will be a reasonable estimate (within less than one order of magnitude) of the maximum shear stress applied during deformation.

In Table 4.3 it is seen that for a single crystalline metal (Au) and single crystalline ionic crystals (NaCl, LiF, AgCl), the values of K are very close. This is probably due to the fact that all these materials have many slip systems and are rather easy to deform under very comparable stresses; on the other hand, olivine deforms much less easily and the stresses needed to obtain significant deformation in steady state are much higher. This may perhaps explain why K is one order of magnitude larger in olivine than in less competent ionic crystals.

(ii) *Cells* limited by very low-angle dislocation walls ($\theta \simeq$ a few minutes of arc) inside the subgrains. Their size is generally of the order of a few microns or at most a few tens of microns. Due to their small size and very low misorientation, these cells are usually invisible by optical and orientation contrast techniques, but they are seen by transmission electron microscopy (Plate 10(b)).

Although the cells form during high-temperature deformation (Staker and Holt, 1972) they may also, and often do, result from a rearrangement by climb during post-deformation anneals of the dislocations present in subgrains. Consequently, they may not be typical of the deformation process. The empirical relation (4.10) is also generally verified for the cells, with a smaller K, but the stresses that can be derived from it do not necessarily represent the stresses applied during deformation.

Green and Radcliffe (1972) and White (1973) have drawn attention to the fact that a relation between σ and d established for optically visible subgrains should not be extended indiscriminately to cells visible by TEM. It seems advisable to be extremely circumspect in trying to evaluate the applied stress from the size of cells visible by TEM.

(*b*) *Relation between applied stress and dislocation density within cells.* We have seen in § 4.1 that if a homogeneous population of elastically interacting dislocations is in equilibrium in a closed volume, the square of the internal stress is necessarily proportional to the total dislocation density:

$$\sigma_i = K'\mu b\sqrt{\rho_t} \qquad (4.11)$$

This of course will hold, whatever the detailed mechanisms leading to the obtention of cells enclosing dislocations may be (interaction of gliding dislocations leading to tangles, fluctuations in a dense uniform population of dislocations or climb polygonization); only the constant K' would be different.

Many experimenters have counted dislocations by etch-pits or TEM after deformation of crystals and have generally reported a good correlation between the dislocation density and the square of the applied stress, although they not infrequently propose relations of the type:

$$\sigma = \sigma_0 + K\mu b\sqrt{\rho} \qquad \text{(Challenger and Moteff, 1973) (Stainless steel)}$$

or

$$\sigma^{1\cdot 4} \propto \rho \qquad \text{(Hüther and Reppich, 1973) (MgO).}$$

This may be attributed to the unavoidably wide scatter in the experimental dislocation density and need not be in grave contradiction with the commonly accepted empirical relationship:

$$\sigma = K''\mu b\sqrt{\rho_t} \tag{4.12}$$

This relation has been verified recently in the case of high temperature deformation of olivine by Kohlstedt and Goetze (1974) who found $K'' \simeq 5$.

It is, of course, tempting to use such a simple relation to evaluate an unknown applied stress from a measured dislocation density. However, this method is beset by serious difficulties and the values obtained for the stress applied to a naturally deformed specimen are certainly questionable.

Here are the principal objections:

(i) The dislocation population is often heterogeneous and it is not always justifiable to take a mean value from observations which are necessarily local and small in number. Besides, the experimental values of ρ are extremely inaccurate (see Chapter 10).

(ii) In experiments, precautions are usually taken to preserve the dislocation density present during deformation by rapidly cooling the sample under load. Thus, the constant K'' in (4.12) is determined. But it is unjustifiable to use this value of K'' to find an unknown stress by counting dislocations in a naturally deformed specimen, since there may have been a long static anneal at unknown temperatures after the deformation. Dislocations may have been annihilated or rearranged by climb and, unless it is otherwise warranted, *there is no reason to believe that the dislocation density is the same as it was during deformation.*

In short, although it would be theoretically justifiable to use (4.11) to find the internal stress, if we knew K' we would know only the internal stress existent at the present moment, not during deformation. Using the empirical formula (4.12) to find σ, could be theoretically justified only if we assumed that

(a) $\sigma = \sigma_i$, i.e. $K' = K''$.

(b) σ_i has not changed since deformation occurred.

Both these assumptions are probably unwarranted in the absence of complementary information.

(*c*) *Relation between applied stress and curvature of dislocation segments.* Formula (3.58) $R = \mu b/K\sigma$ relates the radius of curvature of a segment of dislocation to the applied stress that maintains it in a bowed-out configuration. As bowed-out dislocation segments are often seen in TEM, one could think of using (3.58) to find the applied stress. This, however, would be highly questionable for the following reasons:

To measure the real value of R on a TEM micrograph, one should know the plane in which the segment is curved, and that plane is often impossible to determine for an isolated dislocation segment.

Equation (3.58) is theoretically sound and experimentally valid, but it will give only the effective stress acting on the dislocation, determined in part by the applied stress but mostly by its interaction with neighbouring dislocations; this interaction is very dependent on post-deformation anneals. Therefore, equation 3.58 can only give local values of the internal stress at the time of the observation. This should not be related to the applied stress during previous deformation.

4.3. MECHANISMS OF HIGH-TEMPERATURE PLASTIC FLOW

4.3.1. Generalities

For simplicity of exposition, we will distinguish three cases.

(i) *Strain is caused by diffusional transport of matter between surfaces of crystals differently oriented with respect to stress.* This includes *Nabarro–Herring creep* where vacancies diffuse through the grain between two areas of its boundary and *Coble creep* where diffusion takes place along the grain boundary. Certain models of piezocrystallization can also give a strain. In view of the importance of the topic and the confusions that often arise, we will discuss the problem of piezocrystallization separately in § 4.4.

(ii) *Strain is caused by diffusional transport of matter between dislocations subjected to different climb forces.* This has been proposed by Friedel (1964) as an extension of the Nabarro–Herring model, and developed by Nabarro (1967).

(iii) *Strain is caused by dislocation slip, but steady state can be attained only if recovery takes place* (§ 4.1). Thus, strain rate is controlled by dislocation climb, hence by diffusion. This is the most common mechanism for high temperature creep and it is commonly termed *dislocation creep* or *Weertman creep,* for Weertman was the first to propose a detailed dislocation mechanism, in 1955.

In all three cases the strain rate is obviously proportional to some diffusion coefficient D and is a function of stress. If the stress is not too high, it is empirically possible to fit the dependence of $\dot{\varepsilon}$ to a power law of σ. We will therefore have generally:

$$\dot{\varepsilon} = \dot{\varepsilon}_o \sigma^n D \tag{4.13}$$

with $n > 1$.

In the following paragraphs we will briefly derive the particular form of this equation for each mechanism.

4.3.2. Stress-induced diffusional flow of matter between surfaces

(*a*) *Lattice diffusion: Nabarro–Herring creep.* This deformation mechanism was first proposed by Nabarro in 1948 and improved by Herring (1950).

We will derive the fundamental equation in the simple case of one grain of a polycrystal, with the initial shape of a cube of edge length *d*, subjected to a pure shear (Fig. 4.10). The shear is obtained by applying a normal tensile stress in

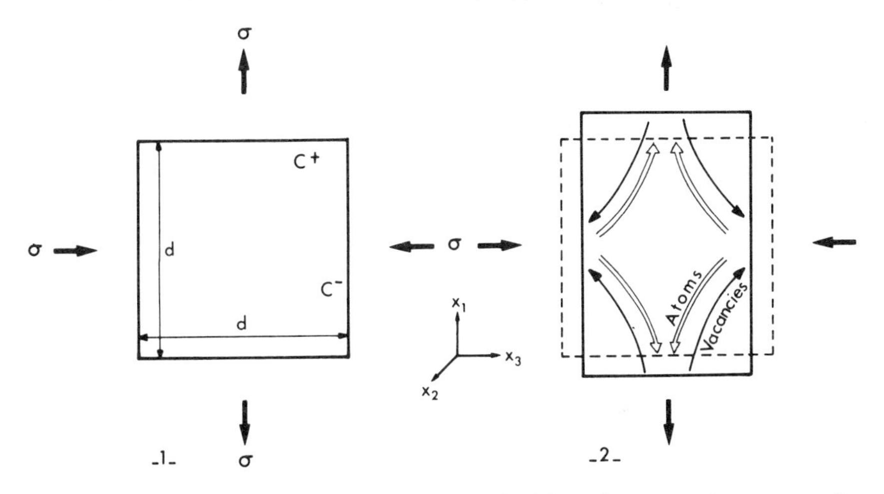

Fig. 4.10. Nabarro-Herring creep. Cubic crystal subjected to a pure shear; vacancies flow from the faces in tension where their concentration C^+ is higher, to the faces in compression where their concentration C^- is lower. Matter flows in the opposite sense (see text)

the direction x_1, and a normal compressive stress in the perpendicular direction x_3, with:

$$\sigma = \sigma_3 = -\sigma_1$$

Let us assume that vacancies can only be formed or annihilated at the surfaces of the grain, that is to say at the grain boundaries. At a given temperature T and hydrostatic pressure P, the thermal equilibrium concentration of vacancies would be:

$$C_0 = \frac{N_v}{b^3} \exp - \left(\frac{\Delta G_f}{kT} \right)$$

(see § 3.2.2 and § 3.4.9).

However, the formation of a vacancy can be schematized as extracting an atom from the bulk and pushing it out through the surface. The stress applied on the surface will do the work $\pm\sigma b^3$ for or against the extraction of an atom (of atomic volume $\Omega \simeq b^3$). Following the same reasoning as in § 3.4.9 we see that the equilibrium concentration of vacancies at the surface in tension, C^+, will be higher than C_0 and that the equilibrium concentration of vacancies at the surfaces in compression, C^-, will be lower than C_0:

$$C^+ = C_0 \exp\left(\frac{+\sigma b^3}{kT}\right)$$
$$C^- = C_0 \exp\left(\frac{-\sigma b^3}{kT}\right) \tag{4.14}$$

The equilibrium concentration of vacancies is therefore not the same at all surfaces under non-hydrostatic stress. The crystal is not in *global equilibrium* with respect to vacancies (although there is a local equilibrium at the surfaces). There will therefore be a flux of vacancies from the surfaces in tension toward the surfaces in compression given by Fick's equation:

$$J = -D_v \text{ grad } C \tag{4.15}$$

(D_v = lattice diffusion coefficient of vacancies).

We know nothing about the local concentrations of vacancies along the path between source and sink, hence grad C is not known locally. It is, however, assumed that it is constant and equal to:

$$\text{grad } C \propto \frac{C^+ - C^-}{d} \tag{4.16}$$

The number ϕ of vacancies transported through a face of section d^2 per second is:

$$\phi = Jd^2 \tag{4.17}$$

The formation of a vacancy on a face corresponds to one atomic volume popping out. On the average we will consider that the volume ϕb^3 corresponding to Φ atoms per second, is spread over the whole surface d^2. This corresponds to a layer of thickness $\phi b^3/d^2$, and the corresponding strain in extension of the crystal of initial length d is

$$\delta\varepsilon = \phi\frac{b^3}{d^2}\cdot\frac{1}{d} \quad \text{per second,}$$

the strain rate is therefore:

$$\dot{\varepsilon} = \phi\frac{b^3}{d^3} \tag{4.18}$$

Now, from (4.15) (4.16) (4.17) and (4.14) we have:

$$\phi = -D_v d^2 \text{ grad } C \propto \mathrm{d} D_v C_0 \left[\exp\left(\frac{\sigma b^3}{kT}\right) - \exp\left(\frac{-\sigma b^3}{kT}\right)\right]$$

for $\sigma b^3 \ll kT$ we may write:

$$\phi = \frac{\mathrm{d} D_v C_0}{\alpha} \sinh \frac{\sigma b^3}{kT} \simeq \frac{\mathrm{d} D_v C_0}{\alpha} \cdot \frac{\sigma b^3}{kT}$$

where α is a numerical coefficient depending on the grain shape and the boundary conditions for σ.

$$D_v C_0 = \frac{D_{SD}}{b^3}$$

where D_{SD} is the coefficient of self-diffusion (see § 3.4.9). Hence, writing $b^3 \simeq \Omega$:

$$\boxed{\dot{\varepsilon} \simeq \frac{D_{SD} \sigma \Omega}{\alpha d^2 kT}} \tag{4.19}$$

The following remarks must be made:

(1) The activation energy of creep is equal to the activation energy for self-diffusion.

(2) The stress dependence of the strain rate is linear: $n = 1$ in (4.13). A linear relationship between $\dot{\varepsilon}$ and σ suggests a *Newtonian viscosity* but in the expression of the viscosity $\eta = \sigma_S / \dot{\varepsilon}$, it is a shear stress σ_S which must enter and we have calculated here an *extension* strain rate; it can be shown that the equivalent viscosity may be expressed by

$$\eta \simeq \frac{1}{3} \frac{\sigma_N}{\dot{\varepsilon}} \tag{4.20}$$

where σ_N is a normal stress (in our example $\sigma = \sigma_N$).

The factor $\frac{1}{3}$ appears in the literature or is omitted, apparently in a haphazard fashion, due to the fact that for various authors σ stands for a normal applied stress or a shear stress.

(3) This type of creep can only appear if the sole sources and sinks of vacancies are the grain boundaries, hence there must be few dislocations inside the grains. The stress must be small so that no dislocations can be nucleated (which justifies the approximation $\sinh \sigma \simeq \sigma$). The size of the grain d must be

very small, so that dislocation segments cannot bow out and multiply under low stresses; besides if dislocations could multiply, the diffusion creep strain would be negligible compared with the much higher dislocation glide strain.

(4) The strain rate depends on the grain size as $\dot{\varepsilon} \propto 1/d^2$; this is another reason for this type of creep to reach noticeable strain rates only for very small grain sizes.

All these arguments show that Nabarro-Herring creep can take place in a significant fashion only in polycrystals with a very fine grain size at high temperatures and small stresses. To obtain values of creep rates measurable in usual experimental conditions, d must be typically smaller than a few hundred microns. For a fine-grained polycrystal with a uniform grain size, it is reasonable to assume that the strain of a macroscopic sample is equal to the strain of one grain and can therefore be given by equation (4.19).

Although theoretically predicted as early as 1948, Nabarro–Herring creep was not discovered experimentally until much later: Squires, Weiner and Phillips (1963) examined a fine-grained alloy of Mg 0·5 % Zr containing fine precipitates of zirconium hydride after creep at 500 °C. They found that there were denuded zones without precipitates along grain boundaries normal to the principal tensile stress. This could be explained only by a transport by diffusion of Mg atoms toward the zones where the vacancy concentration was higher. Harris and Jones (1963) conclusively showed that this was the case and found a very good agreement between the strain rate for one grain calculated from the width of the denuded zone appearing in a given time, the macroscopic strain rate and the theoretical strain rate. Since then, Nabarro–Herring creep has been recognized in many materials, especially ceramic ones where the dislocation density is usually low. Elliot (1973) found convincing evidence for diffusion creep in quartzites deformed under greenschist facies at about 352 °C. The dust rings enclosing the detrital grains served as inert markers: at grain-boundaries under tension, new crystal is formed outside the dust ring, whereas dust accumulates at grain boundaries under compression.

The numerical coefficient α depends on the shape of the grain and on the boundary conditions for the stress at the grain boundary. It was determined by Herring (1950) for a spherical grain in the cases where the tangential stress is relaxed and unrelaxed at the grain boundary—i.e. for grain-boundary sliding possible and impossible. The values of α are given in Table 4.4.

The theoretical equation (4.19) corresponds to steady-state creep, giving a constant creep rate under constant applied stress. However, Green (1970) remarked that for large strains the grains become very elongated and the diffusion path between sources and sinks must become longer, this corresponds to a larger α. Accordingly, the creep rate for large strains must decrease as strain increases.

(*b*) *Grain-boundary diffusion*: *Coble creep.* Coble (1963) proposed a model for the diffusional creep of a fine-grained polycrystal in which the creep rate is controlled not by diffusion through the lattice but by diffusion along the grain boundary. For a spherical grain he found:

$$\dot{\varepsilon} = \frac{D_{GB}\delta\sigma\Omega}{\alpha d^3 kT} \tag{4.21}$$

D_{GB} is the grain boundary self-diffusion coefficient, δ the thickness of the grain boundary, α a numerical coefficient as in (4.19) whose values for different boundary conditions are given in Table 4.4.

Table 4.4

Conditions	α (Herring, 1950)†	α (Coble, 1963)†
Spherical grain, unrelaxed tangential stress	$\frac{1}{16}$	$\frac{1}{7{\cdot}1}$
Spherical grain, relaxed tangential stress	$\frac{1}{40}$	$\frac{1}{141}$

† The values of α given here apply to the case where σ in (4.19) and (4.21) stands for a shear stress. They can be directly used to express an equivalent viscosity. If σ stands for a uniaxial tensile stress ($\sigma = \sigma_1 - \sigma_3$), α should be multiplied by 3.

For Coble creep the activation energy must be the same as for grain boundary diffusion. It has been measured in very few cases, but it is generally assumed to be about half the activation energy for lattice diffusion. Grain-boundary diffusion is predominant over lattice diffusion in the lower part of the temperature range where diffusion (of any kind) becomes important.

The other difference with Nabarro–Herring creep is the dependence $\dot{\varepsilon} \propto 1/d^3$ instead of $\dot{\varepsilon} \propto 1/d^2$. Jones (1973) gave a good review of the experiments in which Nabarro–Herring and Coble creep have been found to operate.

4.3.3. Stress-induced diffusional transport of matter between dislocations

This mechanism was proposed by Friedel (1964) as a generalization of Nabarro–Herring creep, where the sources and sinks of vacancies would be climbing dislocations instead of the grain boundaries.

Nabarro (1967) treated the problem of steady-state diffusional creep between dislocations as resulting from the competition between an increase in internal stress (hardening) by the multiplication of dislocations and a decrease in internal stress (recovery) by climb (see § 4.1.3). Multiplication of dislocations is due to the operation of Bardeen–Herring sources (Bardeen–Herring sources are climb sources, they function exactly like Frank–Read sources but instead of bowing out in their glide plane they bow out in their climb plane under the action of a climb force).

Nabarro considers a regular network of dislocations whose mesh can increase and decrease as a consequence of recovery and work hardening. He calculates the equilibrium size of the dislocation arcs and uses it to find an expression for the flux of vacancies between source and sink dislocations oriented differently with respect to the stress. The creep rate is found to be:

$$\dot{\varepsilon} \propto D_{SD}\sigma^3 \tag{4.22}$$

Note that here the strain is due to the climb motion of the dislocations. As we saw in § 3.4.10, the creep rate could be calculated using Orowan's formula $\dot{\varepsilon} = \rho_m b v_c$, where v_c would be the climb velocity of dislocations and ρ_m the density of climbing dislocations.

4.3.4. Creep by climb-controlled glide of dislocations (Weertman creep)

This is the most common creep mechanism when slip systems are available at high temperature. In this case *the strain is due to slip of dislocations* but *the strain rate is controlled by climb* (hence by diffusion). The most popular dislocation model for this type of creep was first proposed by Weertman in 1955 and recently improved (Weertman, 1968, 1972).

The strain is due to the glide of dislocations, issued from Frank–Read sources in neighbouring parallel planes. The edge portions of the dislocations can interact and become mutually locked in arrays of opposite sign dislocations, called multipoles, at a mean distance L from the sources (Fig. 4.11). These

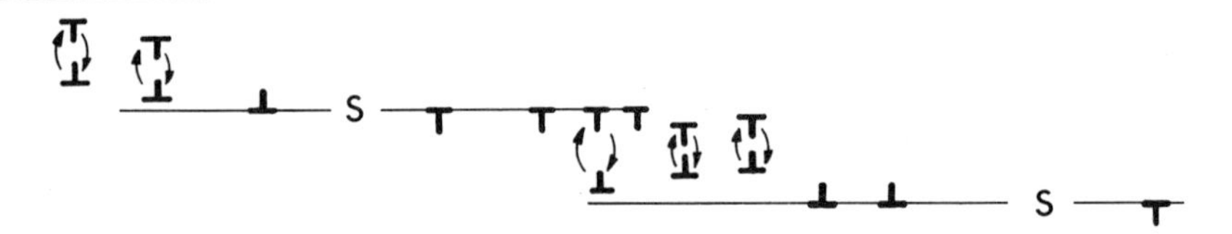

Fig. 4.11. Dislocations issued from sources (S) in parallel slip planes and locked in multipoles. Dislocations of opposite sign climb toward each other and annihilate (arrows). (Reproduced by permission of the American Society for Metals)

configurations exert a back stress on the source. This stress is relieved only if the leading dislocations in neighbouring planes, distant in average of d, climb toward each other and are annihilated. The source can then emit a new dislocation that slips over L before being blocked in its turn. The average velocity of the slipping dislocation is $\bar{v} = L/t$, where t is the time taken by the total process, which can be decomposed in the following way: the leading dislocation climbs a distance d in a time t_c allowing the emission of a dislocation, which will slip in a time t_s, thus:

$$t = t_c + t_s$$

As slip is considered to be unimpeded between obstacles, it is by far the fastest step of the process and the strain rate is controlled by climb: $t_s \ll t_c$.

Orowan's equation can be written:

$$\dot{\varepsilon} = \rho_m b \bar{v} = \rho_m b \frac{L}{t_c}$$

By introducing the climb velocity $v_c = d/t_c$, we have

$$\dot{\varepsilon} = \rho_m b \frac{L}{d} v_c \tag{4.23}$$

The terms ρ_m, L, d can be expressed as function of the density of sources and of the applied stress. v_c is proportional to σ and D_{SD}.

One gets finally the expression:

$$\boxed{\dot{\varepsilon} = \beta \frac{D_{SD}}{b^2}\left(\frac{\sigma}{\mu}\right)^3 \frac{\mu b^3}{kT}} \tag{4.24}$$

(4.24) is valid if the density of sources is supposed to be proportional to the stress. (β is a numerical factor depending on the geometrical assumptions.) If the density of source is supposed to be independent of the stress, or if stress concentration factors (due to pile-ups) are introduced, the stress exponent n is found to be equal to $n = 4{\cdot}5$.

Although $n = 4{\cdot}5$ is currently found experimentally, the value $n = 3$ is at least as common if not more so (Stocker and Ashby, 1973). Weertman (1972) considers it is the most 'natural' stress exponent.

This can be seen by very simple reasoning:

Whatever the geometrical assumptions, we generally have for a homogeneous population of dislocations in equilibrium under the action of their stress fields:

$$\rho \propto \frac{1}{l^2}$$

where l is a characteristic length of the repartition (mean distance between dislocations, mesh length, etc.).

As the stress field of a dislocation varies inversely with the distance, we have $\sigma \propto 1/l$, hence $\rho \propto \sigma^2$. Now, in (4.23), L and d are characteristic lengths so that L/d is usually independent of σ. Finally, $v_c \propto D_{SD}$

It follows that in general we must have

$$\dot{\varepsilon} \propto \sigma^3 D_{SD}$$

Other models do exist for steady-state creep due to climb-controlled glide of dislocations; as a rule, unless they make specific assumptions, they come up with a stress exponent $n = 3$. In particular, there are models where dislocations are supposed to glide freely in the polygonized subgrains and are blocked in front of the sub-boundaries into which they must climb to allow further glide (see Weertman (1972) for a review of recent models).

Dislocation creep by climb-controlled glide can therefore be represented by:

$$\dot{\varepsilon} = A\left(\frac{\sigma}{\mu}\right)^n \frac{D_{SD}\mu b}{kT} \tag{4.25}$$

or

$$\dot{\varepsilon} = \dot{\varepsilon}_0\left(\frac{\sigma}{\mu}\right)^n \exp-\left(\frac{\Delta H_{SD}}{kT}\right) \tag{4.26}$$

(Weertman's equation or Dorn's equation).

Usually with $n \simeq 3$ to 5.

The activation energy of creep is equal to the activation energy of self diffusion.

It must be clear that the fact of experimentally finding a relation of the type (4.26) supports a mechanism of dislocation creep. However, the coincidence between the experimental value of n and a value predicted by a dislocation model is no proof at all that this particular model is applicable. On the whole it is reasonable to consider that dislocation creep by Weertman's model, or some other related model, takes place if the following characteristics are found together.

(i) Activation energy = activation energy for diffusion of the slowest species.

(ii) Stress exponent of the strain rate $n \simeq 3$ to 5.

(iii) Existence of a substructure of polygonized blocks, whose size, proportional to σ^{-1}, does not change during steady state.

(iv) Little or no dependence of $\dot{\varepsilon}$ on grain size.

N.B. Weertman creep belongs to the category of steady-state creep resulting from the balance between strain hardening and recovery.

It can be shown (Poirier, 1976) that equations $\dot{\varepsilon} = r/h$ and $\dot{\varepsilon} = \rho_m b(L/d)v_c$ are equivalent.

4.3.5. Superplastic deformation

(*a*) *Generalities.* It is now a well-known fact in metallurgy that polyphased fine-grained alloys can be deformed in tension under certain temperature and strain-rate conditions, up to strains of more than 1000 % without necking or fracture. This phenomenon has been called superplasticity for obvious reasons and it has been well analysed from a mechanical point of view.

Superplastic deformation is a stable deformation; in other words, minor local variations of cross-section, or incipient necking, do not grow catastrophically with time. It can be shown, by purely mechanical reasoning, that for a local strain gradient not to increase catastrophically, the strain-rate sensitivity of the material must be high.

The strain-rate sensitivity is defined by equation (2.19)

$$m = \frac{\partial \ln \sigma}{\partial \ln \dot{\varepsilon}}$$

The deformation is superplastic if $m \gtrsim 0{\cdot}3$, its limiting value being 1.

If we can write:

$$\dot{\varepsilon} \propto \sigma^n$$

as we usually do for high temperature steady-state deformation, we must have:

$$n = \frac{1}{m} < 3$$

($n = m = 1$ would correspond to Newtonian viscous flow, obviously stable).

Much research has been done recently on the conditions of apparition of superplasticity and on the physical causes of a high strain-rate sensitivity (see, for instance, recent reviews by Daves et al., 1970 and Dingley, 1973). In view of the fact that these conditions may sometimes be realized in mylonites formed at high temperature (Boullier and Nicolas, 1974; Gueguen and Boullier, 1975; Boullier and Gueguen, 1975), it is of some interest to give a concise description of superplastic deformation and its mechanisms.

Typical materials with a high strain-rate sensitivity are amorphous materials like pitch or glass: they deform under low stresses in a viscous fashion when slowly strained ($\dot{\varepsilon}$ low) but a moderate increase in $\dot{\varepsilon}$ produces a considerable increase in the stress required to achieve deformation and it can be high enough to cause fracture.

Most crystalline materials do not have a constant strain-rate sensitivity for all values of the strain rate and temperature: there is a domain of superplasticity, however small or transient for most materials. If a material is strained at a given value of $\dot{\varepsilon}$ in the non-superplastic range and if it enters the superplastic range (for instance by a rise in temperature, or recrystallization, as we will see later), this is shown by a weakening of the material which can deform under a stress lower by $\Delta\sigma$ than the stress at which it would be deforming if the same non-superplastic mechanism continued to operate.

(*b*) *Conditions for superplastic deformations.* The two indispensable conditions for a polycrystalline material to be superplastic are the following:

(i) The deformation must occur at high temperatures ($T \gtrsim 0{\cdot}5\, T_m$).

(ii) The grain size must be very small, of the order of a few microns or tens of microns. The smaller the grain size, the higher the strain rate below which superplasticity begins. The grain size must be stable; that is to say, no grain growth must take place during deformation at high temperature (in this case, superplasticity would soon disappear). The presence of two phases in the solid prevents the grains from growing, this is why most superplastic alloys are two-phased alloys (usually eutectics) with a very fine grain size (*microduplex* structure).

The superplastic domain is clearly seen on a plot ln σ vs ln $\dot{\varepsilon}$; at high $\dot{\varepsilon}$ the deformation occurs by Weertman creep with a stress exponent $n > 3$ ($m < 0{\cdot}3$), and for lower $\dot{\varepsilon}$ this domain gives way to the superplastic one where the slope of the curve is higher ($m > 0{\cdot}3$). The curve is shifted toward higher strain rates when the grain size decreases: Fig. 4.12 shows typical curves for an Mg 34 % Al eutectic (Lee, 1969).

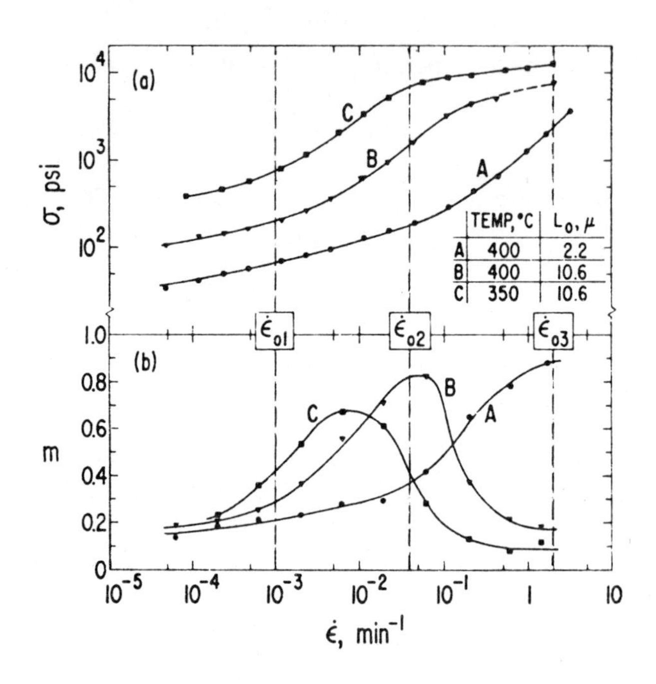

Fig. 4.12. Logarithmic curves σ–$\dot{\varepsilon}$ for various temperatures and grain sizes (MgAl eutectic). The values of the slope m are given below. In the superplastic region, $m > 0{\cdot}3$ (Reprinted with permission from Lee, *Acta Met.*, **17**, 1057 (1969), Pergamon Press)

(*c*) *Experimental observations.* Clearly, the possibility of reaching considerable elongations in tensile tests is only a consequence of the microstructural mechanisms that take place during superplastic deformation. This type of deformation can and does occur in other types of solicitation (compression, creep, etc.), but the microstructural characteristics have been best recognized in highly deformed tensile samples.

In contrast with dislocation steady-state creep, superplastic deformation is characterized by the following features (Fig. 4.13):

(i) Although the deformation of the polycrystal may be considerable, the grains themselves are not deformed: they remain equant, instead of being elongate as after dislocation creep.

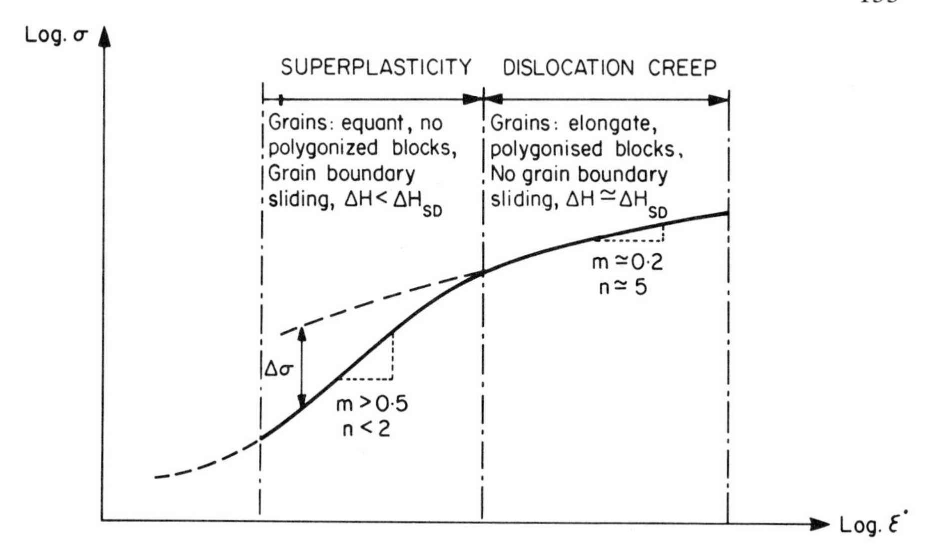

Fig. 4.13. Typical stress–strain rate logarithmic curve for a superplastic material ($\Delta\sigma$; weakening with respect to non-superplastic behaviour)

(ii) The polygonized substructure of misoriented blocks separated by dislocation sub-boundaries, typical of dislocation creep, is conspicuously absent in the superplastic domain.

(iii) Whereas grain-boundary sliding is almost absent in dislocation creep, it contributes as much as 60 to 90 % of the total strain in the superplastic domain.

(iv) Finally, the activation energy is smaller than the activation energy for bulk self-diffusion and is often equal to about $\frac{1}{2}\,\Delta H_{SD}$, of the order of magnitude of the activation energy for self-diffusion along grain boundaries.

(*d*) *Mechanism.* The before-mentioned characteristics strongly suggest that global deformation occurs by sliding at the grain boundary and rotation of the grains rather than by intragranular slip as in dislocation creep.

However, a grain in a polycrystal is surrounded by neighbours and cannot rotate independently and freely: grain-boundary sliding necessarily creates incompatibilities of deformation which must be accommodated by some other process. The absence of a dislocation substructure in grains, in most cases, rules out dislocation glide and the best candidate as a process for accommodation is diffusion creep. The low value of the activation energy points to Coble creep or transport of matter by diffusion along the grain boundaries. The most likely process is therefore *diffusion-accommodated grain-boundary sliding.*

Ashby and Verrall (1973) have recently proposed a model based on this idea. In this model the grains glide past one another and shift neighbours without significant elongation. The strain is due to the relative translation of grains so that the number of grains increases along one direction and decreases along the perpendicular direction. The basic neighbour-shifting event (Fig. 4.14) involves an intermediary step which necessitates accommodation strains,

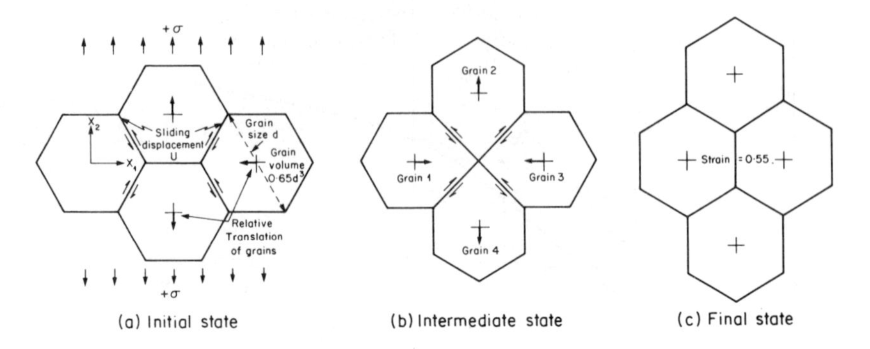

Fig. 4.14. Elementary neighbour shifting event involving four grains. The grains translate past each other by diffusion-accommodated boundary sliding. In the final state the grains are undeformed but the group of four grains has suffered a strain $\varepsilon = 55\ \%$ (Reprinted with permission from Ashby and Verrall, *Acta Met.*, **21**, 149 (1973), Pergamon Press)

provided here by Herring–Nabarro and Coble creep. This model predicts rather well the shape of the ln σ vs ln $\dot{\varepsilon}$ curve (with a contribution of dislocation creep at higher $\dot{\varepsilon}$) and the main features of superplastic flow; moreover it is in good quantitative agreement with the experiments.

4.3.6. Transformational superplasticity

It is unfortunate (and due to historical reasons) that the name superplasticity applies to two entirely different deformation mechanisms: 'structural' superplasticity, reviewed in the preceding paragraph, and 'transformational' superplasticity which is the subject of the present one.

Both these mechanisms have in common the fact that they can provide very large elongations without necking but they are otherwise physically quite different.

Transformational superplasticity is a discontinuous phenomenon which may occur when a crystal is repeatedly cycled in temperature under stress above and below a phase transition. Every time the phase transition point is crossed, an increment of deformation takes place in the sense of the applied stress, irrespective of the sense in which the phase transformation occurs. After many cyclings the increments of deformation add up to large deformations. This phenomenon has been clearly evidenced in iron or carbon steel cycled under stress around the α-γ transformation point (ferrite–austenite) (de Jong and Rathenau, 1959, 1961; Oëlschlagel and Weiss, 1966). The transformation under stress produces a deformation ε_1 in the sense $\alpha \rightarrow \gamma$ and a deformation ε_2 in the sense $\gamma \rightarrow \alpha$. These deformations add up and the total deformation per cycle $\varepsilon = \varepsilon_1 + \varepsilon_2$ is a linear function of the applied stress. This has been shown to correspond to an average deformation rate proportional to the stress (Newtonian effective viscosity) and the material behaves as though it had a vanishingly small elastic limit. The elastic limit lowering is proportional to the relative difference in specific volume of the two allotropic forms.

Greenwood and Johnson (1965) proposed the following mechanism: during the phase transition in a polycrystal, the variation in specific volume between the grains which have transformed and those which not yet transformed sets up internal stresses. If these internal stresses reach the value of the elastic limit they are relaxed by plastic deformation, but as the transformation goes on they build up again. The material reacts to an applied stress as though its elastic limit was very small and it yields as a Newtonian fluid during the time it takes to transform entirely. The same phenomenon occurs when materials possessing a large thermal expansion anisotropy are cycled in temperature; the internal stresses are set up by the differences in thermal expansion between differently oriented grains. This is well known in uranium (Cottrell, 1964).

It has been proposed that this type of flow could occur in the upper mantle at the olivine-spinel phase transition boundary at 400 km (Sammis and Dein, 1974).

4.4. STRESS-INDUCED RECRYSTALLIZATION (PIEZOCRYSTALLIZATION)

4.4.1. Generalities—chemical potential

The mechanism of stress-induced recrystallization (or piezocrystallization) has been invoked time and again to explain the presence of a preferred orientation of minerals in non-hydrostatically stressed metamorphic rocks (Kamb, 1959, 1961; Hartmann and den Tex, 1964; Goguel, 1965, 1967; de Vore, 1969; den Tex, 1970; Schwerdtner, 1964). It has also been presented as a mechanism for flow at high temperatures (Avé Lallemant and Carter, 1970). A more detailed discussion of its significance and validity will be given in § 4.4.3; in order to present the principal models of piezocrystallization we will only define it roughly here as a process by which crystals or grains, whose lattices are in a favourable orientation relative to a non-hydrostatic stress field, grow at the expense of non-favourably oriented crystals. This results from transport of matter by diffusion along grain boundaries or in an intercrystalline fluid phase, from the unfavourably oriented crystals to the favourably oriented ones. We stated in § 3.3 that the driving force for diffusion was a concentration gradient of the diffusing species and, accordingly, it is natural to think that such a gradient is caused by the non-hydrostatic stress. As a matter of fact, Nabarro–Herring creep or Coble creep (§ 4.3.2) are such stress-induced diffusion processes, where atoms are transported from the faces in compression to the faces in tension of a crystal through the bulk of the crystal or along the grain boundary. In the simple treatment given in § 4.3.2 we made use of the fact that diffusion occurred by a vacancy mechanism and calculated the equilibrium concentration of vacancies at surfaces, differently oriented relative to the stress field, to find the concentration gradient responsible for diffusion. However, although this approach is good enough to find the direction in which the matter flows in the very simple case considered, it is not thermodynamically rigorous and could not be used in the more general case of piezocrystallization. This is due to the fact that, strictly speaking, the driving force for diffusion is not a concentration gradient but a chemical potential gradient; indeed, Herring and Coble did use chemical potentials in their treatments. However, in most usual cases the concentration gradient is simply related to the chemical potential gradient. Before presenting the models of piezocrystallization, we will therefore briefly recall the definition of a chemical potential and state the thermodynamical relations between the chemical potential of atoms at the surface of a crystal and the non-hydrostatic stress field.

It must be noted here that we will be concerned only with finding the driving force for diffusion and predicting the evolution of the system toward a final state (preferred orientation of crystals). We will not consider the kinetics of the evolution since this has not been treated and anyway would involve many unwarranted assumptions.

Chemical potential and non-hydrostatic stresses

(i) The chemical potential μ_i of a species i in a composite system whose composition may vary can be defined by:

$$\mu_i = \frac{\partial G}{\partial N_i} \qquad T, P, N_j \neq i \tag{4.27}$$

(Guggenheim, 1967; Kern and Weisbrod, 1964). G is the free enthalpy of the system and N_i the number of moles of the component i.

When two open systems which can exchange the component i are put in communication, the closed system formed by their reunion is at equilibrium with respect to component i, if μ_i is constant over the whole system, the free enthalpy G is then minimum. If μ_i does not have the same value in two regions of the system, or in other words, if a gradient of μ_i exists, a diffusion flux is set up between the two regions, thus tending to equalize μ_i over the system and establish equilibrium.

(ii) Let us consider the simplest case of a crystal containing only one component, the chemical potential will then refer to the atoms of the crystal. Let us suppose that the crystal is in contact along different portions of its surface with a fluid in which it is soluble. If different parts of the crystal surface are in contact with fluids at different pressures, the crystal is then in a state of non-hydrostatic stress. Gibbs (1876) has shown that in this case there is no global equilibrium possible for the crystal. However, local equilibria can be defined at the surface. At a given portion of the surface in contact with the fluid under pressure p_k, it is possible to define a chemical potential μ_k of the atoms of the solid dissolved in the fluid:

$$\mu_k = F + p_k V \tag{4.28}$$

V is the specific volume of the stressed solid and F is the Helmholtz free energy per mole of the solid immediately adjacent to the surface (Kamb, 1961).

Here the solid was non-hydrostatically stressed by different pressures p_k at different portions of its surface; it is more convenient to consider that the solid is stressed by a permeable loading frame and immersed in the fluid at pressure p (Fig. 4.15).

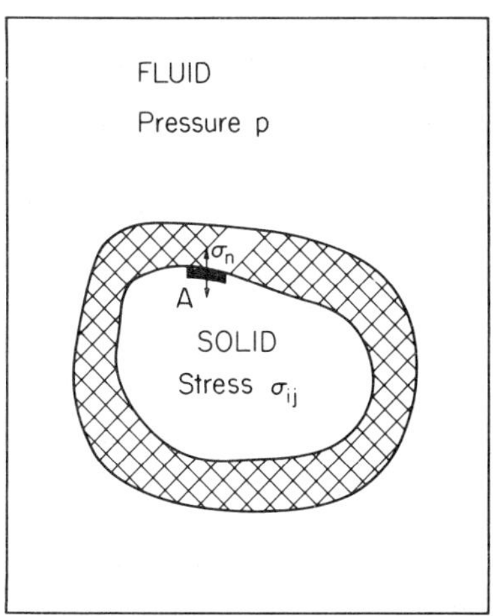

Fig. 4.15. Closed system containing a solid under non-hydrostatic stress σ_{ij} in contact with a fluid at pressure p where it can dissolve. The solid is stressed by a permeable loading frame (cross-hatched) (Reproduced with permission from Paterson, *Rev. Geophys. Space Phys.*, **11**, 355 (1973), copyright by American Geophysical Union)

It can be shown (Paterson, 1970) that when the stressed solid is in local equilibrium with its solution in the vicinity of a given site A, the chemical potential of the atoms in solution close to the surface is:

$$\mu = F + \sigma_n V \tag{4.29}$$

σ_n is the normal stress at the surface at site A.

For a linear elastic solid under infinitesimal strain, the free energy F is equal to the free energy of the unstressed solid plus the elastic strain energy

$$\tfrac{1}{2}\sum_{ij} \sigma_{ij}\varepsilon_{ij} V_0 \tag{4.30}$$

(V_0 = specific volume in the stress-free state) (Kamb, 1961). We therefore have:

$$\mu = \mu_0 + \sigma_n V + \frac{1}{2}\sum_{ij} \sigma_{ij}\varepsilon_{ij} V_0 \tag{4.31}$$

The situation then stands as follows:

It is impossible to define a chemical potential inside a non-hydrostatically stressed solid (Kamb, 1959).

It is possible to define locally the chemical potential at the surface of the stressed solid, in local equilibrium with its solution in a fluid.

The chemical potential can take different values at different portions of the surface. The principal cause for variation of μ is the fact that the normal stress σ_n may have different values at differently oriented parts of the surface.

This has important consequences which have been clearly reviewed by Paterson (1973). In particular, the chemical potential of the atoms of the solid in solution in the fluid will be higher close to a face subjected to a compressive normal stress than it would be if the solid were hydrostatically stressed; hence the solubility is significantly increased. Conversely, the solubility at a face in tension is lower relative to that of a hydrostatically stressed solid. Besides, a non-hydrostatically stressed solid is unstable with respect to an identical unstressed solid if they are both immersed in the same fluid and if the diffusion of the solid atoms is fast enough in the fluid.

4.4.2. *Kamb's models for stress-induced recrystallization*

The most important models for preferred orientation by stress-induced recrystallization were proposed by Kamb (1959, 1961). They were exposed in a very clear fashion by Paterson (1973), who analysed the underlying assumptions and critically assessed the conclusions. Most of what follows in this paragraph has been borrowed from Paterson's review paper.

Kamb (1959) deals with the problem of stress-induced recrystallization in two cases: (Model 1 and Model 2)

(a) *Model 1*: The mineral, whose preferred orientation is to be considered, does not form most of the polycrystalline agregate or there is an intercrystalline fluid phase. Kamb refers specifically, as typical examples, to the case of biotite crystals in a gneiss or calcite crystals in a marble with intercrystalline water.

In this model, preferred orientation results from the growth of favourably oriented crystals at the expense of unfavourably oriented ones. The transport of matter occurs through the matrix, between crystals, but no diffusion is allowed to take place in the bulk of one crystal between its faces.

For local equilibrium between the crystal and matrix at a given site on the boundary of the crystal, the chemical potential of the crystal atoms in the matrix is given by equation (4.29), where σ_n is the normal stress across the boundary at the given site:

$$\mu = F + \sigma_n V \tag{4.29}$$

If F_0 and V_0 are the values of the free energy and the specific volume in the unstressed state, we may write:

$$F = F_0 + \Delta F$$

$$V = V_0 + \Delta V$$

and

$$\mu = F + \sigma_n V = (F_0 + \sigma_n V_0) + (\Delta F + \sigma_n \, \Delta V) \tag{4.32}$$

F_0 does not depend on orientation or stress, the term $\Delta F + \sigma_n \, \Delta V$ depends on the orientation of the crystal in the stress field (the elastic strain energy and the specific volume depend on orientation for an elastically anisotropic crystal), it is a second-order term compared with $\sigma_n V_0$.

Let us consider a crystal of the kind whose preferred orientation is considered. For the crystal to grow the matrix must be saturated on all faces (or else it would dissolve away). The fact that the crystal grows or dissolves away is thus determined by the chemical potential at the faces where the solubility is greatest, i.e. faces normal to the maximum compressive stress. Kamb reaches the conclusion that the preferred (stable) orientation is the one for which the chemical potential for equilibrium at a face normal to the maximum compressive stress is a minimum.

In other words the preferred orientation minimizes the chemical potential at the faces where the solubility is greatest. The criterion for this can then be found by putting $\sigma_n = \sigma_1$ (σ_1 = maximum compressive stress) in (4.32) and minimizing μ. (This is where the second-order term intervenes.) We have:

$$\Delta F = F - F_0 = \tfrac{1}{2} \sum_{ij} \sigma_{ij} \varepsilon_{ij} V_0$$

$$\frac{\Delta V}{V_0} = -\sum_i \varepsilon_{ii} \quad \text{(using the sign convention that a compressive stress is positive)}$$

Hence Kamb's criterion for Model 1 is that the *preferred orientation is the one for which*:

$$\tfrac{1}{2} \sum_{ij} \sigma_{ij} \varepsilon_{ij} - \sigma_1 \sum_i \varepsilon_{ii} \tag{4.33}$$

is a minimum.

Paterson (1973) has remarked that Kamb's Model 1 can be subdivided into two models relative to physically different cases.

(i) *Model 1(a):* The distance between crystals is large compared with their dimensions, hence the shortest diffusion path is from a dissolving face to a growing face of the same crystal. Kamb's criterion applies and the preferred orientation is the one that minimizes (4.33). However, because of the $\sigma_n V_0$ term in (4.32), there is a strong tendency for crystals of *all* orientations to grow at the faces normal to the minimum compressive (or maximum tensile) stress at the expense of faces normal to the maximum compressive stress. This causes a flattening of the grains normal to the maximum compressive stress. To maintain the stress in grains during the transfer of material the aggregate must undergo an overall shortening, parallel to the maximum compressive stress direction (and of course, since the volume is constant, a stretching in a perpendicular direction). This is truly, as Paterson remarks, a *pressure-solution mechanism of deformation.* Coble creep or deformation by stress-induced transport of matter along grain boundaries belongs to this class of deformation mechanisms. Green (1970) described it by using Kamb's equations for the chemical potentials.

(ii) *Model 1(b):* The distance between crystals is small compared with their dimensions (for example, marble with intercrystalline water). Here the shortest diffusion path is between neighbouring faces of adjacent grains. As these faces are parallel, $\sigma_n V_0$ is the same for each and it is the value of $\Delta F + \sigma_n \Delta V$ on each face that decides which crystal grows and which dissolves. The preferred orientation is that of the crystals which have grown most on *all* their faces, hence for roughly equant crystals, *the preferred orientation is the one for which*:

$$\tfrac{1}{2}\sum_{ij} \sigma_{ij}\varepsilon_{ij} - \bar{\sigma} \sum_{i} \varepsilon_{ii} \tag{4.34}$$

is a minimum. ($\bar{\sigma}$ is the mean normal stress.)

(b) *Model 2:* This model considers monomineralic polycrystals which are well below their melting point and in which intergranular fluid is absent so that no bulk diffusional transport of matter is possible (i.e. no solution redeposition). However, atoms belonging to a grain can leave its lattice and stick to an adjacent grain if the chemical potential is lower there. This requires only minor rearrangements of atoms in the boundary and results in grain-boundary migration.

For two elastically anisotropic crystals of different orientations in the same stress field, the elastic strain energies and the specific volumes are different. Thus, μ is unequal for two contiguous crystals of different orientations. The difference in chemical potential between a crystal and its neighbour across a given point on their boundary is given by the difference between the values of $\Delta F + \sigma_n \Delta V$ for the two crystals (σ_n is the same for both). Preferred orientation results from the growth of favourably oriented grains at the expense of unfavourably oriented neighbours through grain-boundary migration.

Kamb assumes that the grains are initially equant and the rate of grain boundary migration at a given point is proportional to the difference between the chemical potential of crystals, averaged over all orientations and the chemical potential of the crystal growing in the preferred orientation. In averaging over all orientations of the grain boundary, the maximum growth rate is found to occur for the orientation which minimizes $\Delta F + \bar{\sigma}\,\Delta V$ ($\bar{\sigma}$ = mean normal stress). *The preferred orientation is the one for which*

$$\tfrac{1}{2}\sum_{ij} \sigma_{ij}\varepsilon_{ij} - \bar{\sigma}\sum_{i} \varepsilon_{ii} \tag{4.35}$$

is a minimum (as for Model 1(b)).

The predicted preferred orientations for a few minerals in rocks under uniaxial compressive stress are given in Table 4.5.

Table 4.5. Predicted Preferred Orientations for Particular Minerals in Rocks under Axisymmetric Compressive Stress

Mineral	Kamb Model 1*a*	Kamb Model 2 (also 1*b*)
α quartz (Kamb, 1959)	Small circle girdle of *c* axes at about 60° to σ_1	Small circle girdle of *c* axes at about 40° to σ_1
Calcite (Kamb, 1959)	*c* axis maximum parallel to σ_1	Great circle girdle of *c* axes normal to σ_1
Olivine (Hartmann and den Tex, 1964; den Tex, 1969, 1970)	Optical α-axis, [010], maximum parallel to σ_1	Optical γ axis, [100], maximum parallel to σ_1
Mica (Schwerdtner, 1964)	Pole of (001) maximum parallel to σ_1	Great circle girdle of (001) poles normal to σ_1*

Predictions are also listed by Kamb (1959) for several metals and other minerals, by Schwerdtner (1964) for hornblende, and by Schwerdtner (1970) for anhydrite.

* This result is not given by Schwerdtner but it follows from substituting the elastic compliances of Alexandrov and Ryzhova (1961) (as given by Clark (1966)) into formulae 39 and 40 of Kamb (1959); however, the preference for the girdle over an (001) pole maximum parallel to σ_1 is small. These results apply to biotite, muscovite, and phlogopite. (Paterson 1973.)

4.4.3. Discussion

(*a*) *Stress-induced vs strain-induced recrystallization.* Kamb's models rest on the indispensable assumption that the internal state of crystals is the same before and after recrystallization except in regard to elastic strain. *Piezocrystallization described by these models is therefore incompatible with dislocation plastic flow and annealing recrystallization* in a metallurgical sense such as will be described in § 4.5.

Kamb (1959) specifically excluded 'plastic flow that would lead to changes in the internal condition of the crystals'.

The fundamental difference between 'stress-induced' recrystallization (or piezocrystallization) and 'strain-induced' recrystallization (or annealing recrystallization) can be stated as follows: in both processes the driving force is the difference in free enthalpy between the recrystallized and unrecrystallized states.

In stress-induced recrystallization, the difference in free enthalpy is attributed only to the difference in *elastic-strain energy* between elastically anisotropic crystals in a favourable orientation and those in an unfavourable orientation in a *non-hydrostatic stress field.*

In the strain-induced recrystallization, the difference in free enthalpy is attributed to the difference in *stored plastic-strain energy* due to the presence of dislocations generated during plastic flow and remained trapped in the crystals; grain growth is due to the difference in *total grain-boundary energy* between a fine-grained and a coarse-grained polycrystal.

Other characteristic features of piezocrystallization models not shared by strain-induced recrystallization are the following:

The possibility of *nucleation* of new grains with a preferred orientation is not considered.

The possible difference in growth rate of faces or in grain-boundary migration rate for certain crystallographic orientations is not taken into account. In fact these differences may be important and often stem from kinetic factors unrelated to the applied stress (influence of the segregation of impurities on the grain boundary mobility, mobility of a coincidence boundary, etc.).

These points certainly lead to questioning the applicability of piezocrystallization models in many cases. However, the main criticism which can be directed to this theory is inherent in its basic tenets: it neglects the role of stored energy due to previous or concomitant plastic flow, and privileges the role of the variation in elastic strain energy with orientation. Paterson (1973) rightly remarks that the magnitude of grain boundary energy or stored energy from plastic deformation can be at least comparable with the variation in elastic strain energy even under high stresses (Paterson, 1959; Gross, 1967; Nicolas, 1974). In the case of rather plastic minerals with several slip systems, one can even predict that the plastic-strain energy will be much higher than the variation in elastic-strain energy, especially since the elastic stresses will be partly relaxed in a plastically deforming material. This is actually what happens in all metals where strain-induced recrystallization is the only documented mechanism. Table 4.6 sums up the principal characteristics of both mechanisms.

Table 4.6.

	Stress-induced recrystallization	Strain-induced recrystallization
Energy lowered by recrystallization	Elastic-strain energy	Stored plastic-strain energy grain-boundary energy
Defects required	None	Dislocations
Stress	Non-hydrostatic	Unspecified or zero
Strain	Elastic, infinitesimal	Plastic, large
Preferred orientation	Depends on the model	Host controlled: various models

(*b*) *Stress-induced recrystallization vs syntectonic recrystallization.* Stress-induced recrystallization models have been devised to account for mineral-preferred orientations observed in rocks, and, as we have seen, they exclude plastic flow by dislocation. Therefore, whereas strain-induced recrystallization can be, and often is, syntectonic (dynamic recrystallization) it is difficult to understand how stress-induced recrystallization can be syntectonic unless it brings about a deformation.

This raises the question of whether stress-induced recrystallization can be termed syntectonic and considered as a mode of flow as suggested for olivine by Avé Lallemant and Carter (1970).

It seems clear that *none of Kamb's models can account for a deformation, except Model 1(a)* insofar as grain-boundary diffusion creep (Coble) dominates, as pointed out by Paterson (1973), *and in this case it should be more aptly termed Coble creep* (Green, 1970).

Avé Lallemant and Carter (1970) reject the possibility of plastic flow and base their contention that piezocrystallization is the operative 'mode of flow' on the fact that 'the fabric observed in the experiment and most olivine tectonites is that predicted by Kamb's (1959) thermodynamical theory for solids in the presence of an intercrystalline fluid as applied to olivine'. It is of some interest to remark that Kamb (1959) himself had noted that 'the type of preferred orientation expected in recrystallization under stress is in most cases the same as would be expected in mechanical deformation of the same materials under the same uniaxial stresses'; it therefore seems difficult to base the rejection of plastic flow on the coincidence between the observed fabric and the fabric predicted by Kamb.

4.5. STRAIN-INDUCED RECRYSTALLIZATION

4.5.1. Generalities

Strain-induced recrystallization is the process by which new, dislocation-free grains develop in a deformed polycrystal at the expense of strain-hardened grains whose dislocation density is high. Strain-induced recrystallization—or simply 'recrystallization' in metallurgical language—is a very important softening process which has been much investigated in metals where it is very well documented and used. As a matter of fact, it is the only known process of recrystallization in metals, as no evidence has ever been found of stress-induced recrystallization.

Two main cases are usefully distinguished:

(*a*) *Dynamic recrystallization*, which occurs during hot-working and is the softening process that competes with strain hardening during deformation at high temperatures and strain rates. The characteristic features of dynamic recrystallization will be examined in § 4.5.4.

(*b*) *Static recrystallization*, which occurs during stress-free high-temperature anneal of cold-worked materials when the strain has exceeded a critical value.

Most of what will subsequently be said about recrystallization has been primarily deduced from studies on static recrystallization. Static recrystallization is a process of nucleation and growth. The final grain size depends on the relative importance of the stages of nucleation and growth. A high annealing temperature and a strong deformation promote a high density of nucleation sites, hence the grain size is small. Conversely, a low density of nucleation sites, together with a high grain-boundary migration rate, favours a large recrystallized grain size. Fig. 4.16 shows a typical variation in grain size with annealing temperature and strain in the case of magnesium.

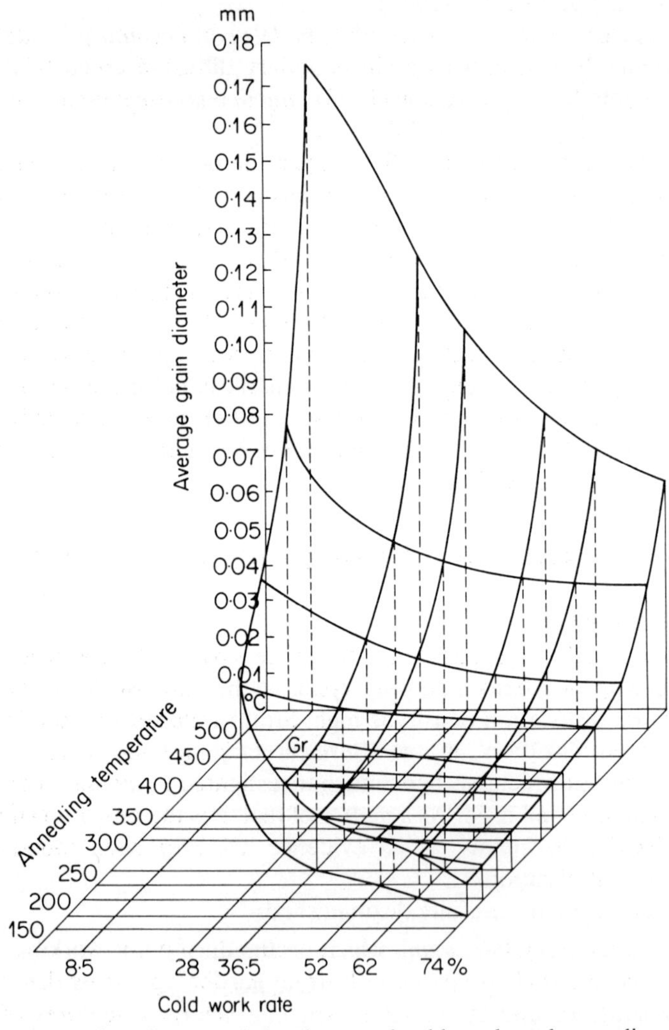

Fig. 4.16. Influence of the degree of cold work and annealing temperature on the recrystallized grain size measured at 20° C (after Guillet, 1957, reproduced by permission of Masson & Cie, Paris)

4.5.2. Nucleation

For all processes involving nucleation and growth in materials science, nucleation is always the least understood stage of the process. Recrystallization is no exception to that rule. Several mechanisms for nucleation of strain-free nuclei which can grow by boundary migration have been proposed:

(*a*) *Homogeneous nucleation.* This model is based on the classical theory of fluctuations. The observation that the grain size decreases with increasing temperature (hence the number of nuclei increases) has led some authors to regard nucleation as a thermally activated homogeneous process. However, this has been shown to be a very unlikely process in recrystallization.

(*b*) *The nuclei are preformed cells or subgrains.* A cellular structure is formed in heavily cold-worked regions and can transform into subgrains in the early stages of annealing. To be an effective nucleus, a dislocation-free cell or a subgrain must be large enough to be able to grow at the expense of its neighbours and also its misorientation must be great enough for its boundaries to migrate easily. (A low angle boundary, which should migrate by cooperative movement of its dislocations, is usually not very mobile.) Nucleation is therefore expected to be more important in heavily deformed regions of the crystal where the lattice distortions are strong, for instance in deformation bands and in the vicinity of grain boundaries.

This theory of nucleation from preformed nuclei has received much experimental support. In particular, recent experiments have shown that the orientation of new recrystallized grains in heavily deformed aluminium (measured by X-ray diffraction techniques) lies within the orientation spread of the various parts of the deformed host material (Bellier, quoted by Doherty, 1974). It has also been frequently observed (Doherty, 1974) that subgrains of the fine polygonization network adjacent to grain boundaries can grow into the neighbouring grain by migration of the high angle part of their boundary (the driving force being the reduction in strain energy). Most of the grain-boundary recrystallization is thought to be due to this mechanism, which can also operate in the transition band between parts of a grain strongly misoriented during heterogeneous deformation.

(*c*) *Nucleation assisted by chemical fluctuations.* In multi-component systems like minerals small fluctuations in chemical composition may exist. Etheridge and Hobbs (1974) have suggested, in the case of mica, that strain-induced nucleation may be assisted by lowering the free energy resulting from a small change in chemical composition between a new grain and the host.

4.5.3. Grain-boundary migration

(*a*) *Driving force.* Let us consider a simple system formed by two grains 1 and 2 of respective volumes V_1 and V_2, separated by a grain boundary of area A_{GB} whose position is defined by a coordinate x (Fig. 4.17). The total free

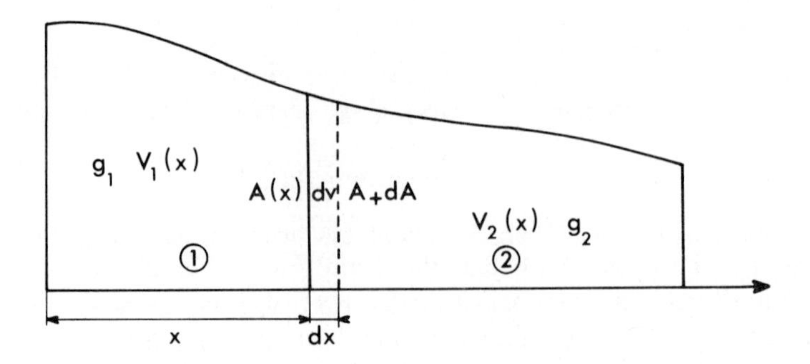

Fig. 4.17. Migration of the grain boundary in a bicrystal (see text)

enthalpy of the system is:

$$G = V_1(x)g_1 + V_2(x)g_2 + A_{GB}(x)g_{GB} \tag{4.36}$$

where g_1 and g_2 are the free enthalpies per unit volume in grains 1 and 2 and g_{GB} the free enthalpy per unit area of the grain-boundary.

If the grain boundary migrates by dx at constant T and P, the free enthalpy of the system varies by:

$$\mathrm{d}G = \left(g_1 \frac{\mathrm{d}V_1(x)}{\mathrm{d}x} + g_2 \frac{\mathrm{d}V_2(x)}{\mathrm{d}x} + g_{GB} \frac{\mathrm{d}A_{GB}(x)}{\mathrm{d}x}\right) \mathrm{d}x \tag{4.37}$$

As the total volume of the system is constant $\mathrm{d}V_1 = -\mathrm{d}V_2$.

The driving force F can be defined, as usual, by:

$$F = -\frac{\mathrm{d}G}{\mathrm{d}x}$$

which can thus be written:

$$F = (g_2 - g_1)\frac{\mathrm{d}V_1}{\mathrm{d}x} - g_{GB}\frac{\mathrm{d}A_{GB}}{\mathrm{d}x} \tag{4.38}$$

A positive driving force corresponds to a spontaneous displacement of the boundary in the sense that minimizes the total free enthalpy of the system.

The expression (4.38) of F comprises two terms with different physical meanings.

$$F_v = (g_2 - g_1)\frac{\mathrm{d}V_1}{\mathrm{d}x} \tag{4.39}$$

and

$$F_A = g_{GB}\frac{\mathrm{d}A_{GB}}{\mathrm{d}x} \tag{4.40}$$

(*i*) *Volume driving force F_v*. If the grain-boundary area remains constant during migration, the only force is the volume one F_v.

If $g_2 > g_1$, F_v is positive, that is, grain 1 grows at the expense of grain 2. The difference in specific free enthalpy between the two grains in the system under consideration can be attributed only to a difference in internal energy. The stored energy of dislocations contributes principally to the internal energy of a deformed crystal. Hence the *volume driving force for grain growth is the decrease in strain energy* corresponding to the consumption of a grain with a high dislocation density by a new dislocation-free grain. This process takes place during *primary recrystallization*: the dislocation-free nuclei grow at the expense of their heavily deformed neighbours. The volume driving force is preponderant, and it is energetically favourable for the polycrystal to get rid of its deformed grains, even if this means increasing the grain boundary area a little.

In a polycrystal, the migration of the boundaries is limited by their being pinned at triple points and during primary recrystallization the boundaries tend to assume a curved shape, with their *convexity in the sense of the migration* (Fig. 4.18(a)).

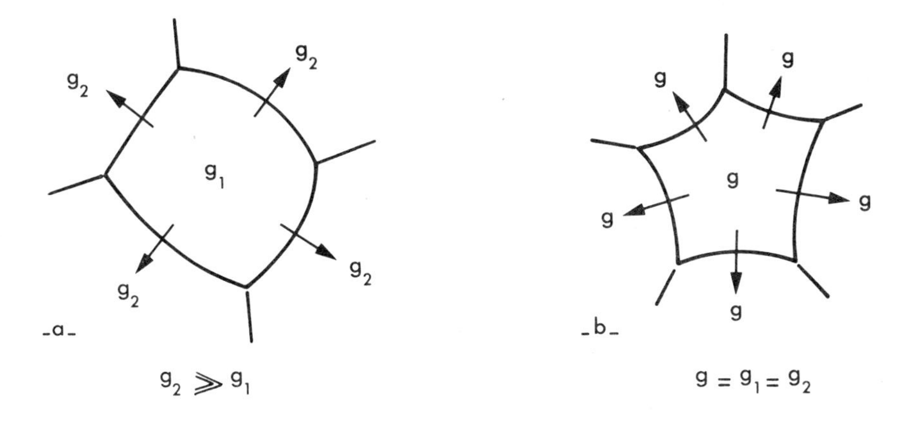

Fig. 4.18. (a) Grain growth under volume driving force (strain energy). The boundaries are convex outward; (b) Grain growth under surface driving force (boundary area). The boundaries are convex inward

(*ii*) *Surface driving force F_A*. If $g_2 \simeq g_1$, the only driving force is the surface one F_A. The system tends to lower its free enthalpy by decreasing the total grain-boundary free enthalpy; hence the *surface driving force for grain growth is the decrease in grain-boundary energy* corresponding to a decrease of the total boundary area, inherent in grain growth. This process takes place during the stage of grain growth after completion of the primary recrystallization. All grains are now dislocation free and the grain boundaries tend to straighten out and reach equilibrium at triple points (with angles of 120 °). In the process some grains disappear and on the whole the *grain boundaries tend to move toward their concavity* (Fig. 4.18(b)).

(*b*) *Migration kinetics.* The grain-boundary migration rate v_{GB} is usually expressed as a function of the driving force F and the grain-boundary mobility M by a relation:

$$v_{GB} = MF^n \tag{4.41}$$

Whatever the detailed mechanism of the migration, it involves transport of atoms across the grain-boundary and one expects the process to be thermally activated. This is confirmed experimentally and the mobility can be expressed by a relation:

$$M = M_0 \exp - \left(\frac{\Delta H_{GBM}}{kT} \right) \tag{4.42}$$

where ΔH_{GBM} is the activation energy for grain-boundary migration.

Experiments have been done in a few simple cases where the driving force can be relatively well known. We will only mention here two typical experiments:

Rutter and Aust (1965) measured v_{GB} in lead bicrystals, in the case where $F = F_v$ (GB area = const.); they found that for high-angle tilt boundaries, at 200 °C, the migration rate of coincidence boundaries was higher and their activation energy lower (Fig. 4.19).

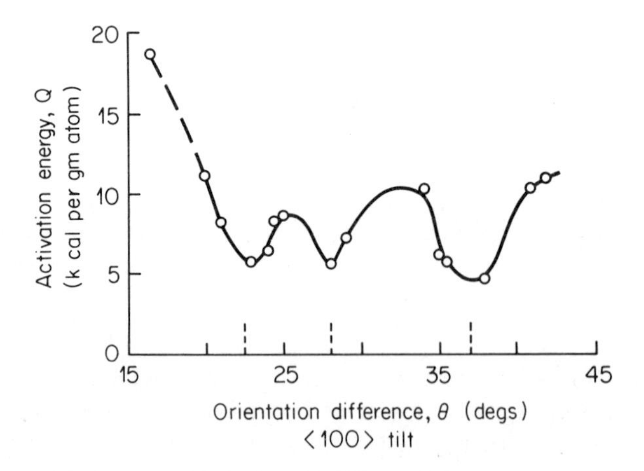

Fig. 4.19. Measured activation energies (Q) for migration and rate of boundary migration at 200 °C for ⟨100⟩ tilt boundaries in zone-refined lead (Reprinted with permission from Rutter and Aust, *Acta Met.*, **13**, 181 (1965), Pergamon Press)

Sun and Bauer (1970) measured v_{GB} in NaCl bicrystals, in the case where $F = F_A$ (annealed grains, wedge-shaped bicrystals). They found a linear relationship between v_{GB} and F above a certain value of $F\,(n = 1)$ in equation 4.41).

Despite the scarcity of reliable experiments and the fact that they are sometimes contradictory, the following facts seem to be rather well documented:

(1) The migration rate varies with the misorientation of the grain-boundary. It is higher for coincidence boundaries.

(2) Impurity atom segregation at grain boundaries has a strong effect. At lower temperatures and for a small driving force, the migration rate is controlled by the diffusion of the cloud of impurity atoms elastically interacting with the boundary; at higher temperatures and for a large driving force, the boundary frees itself from the cloud and migrates as in the pure crystal.

4.5.4. Dynamic recrystallization

Experimental observations on dynamic recrystallization are difficult since the experiment must be stopped to allow a static observation of the sample. Even though the precaution is taken of rapidly quenching the specimen to a low temperature at the same time as the deformation is stopped, it is difficult to be sure that no static recrystallization has quickly taken place while the specimen is being cooled down.

However it is often possible to distinguish between dynamically and statically recrystallized grains. Statically recrystallized grains are equant, whereas dynamically recrystallized grains are usually flattened or elongated due to subsequent deformation (Green et al., 1970; Avé Lallemant and Carter, 1970). Glover and Sellars (1973) observed in the case of iron that *dynamically recrystallized grains have ragged grain-boundaries and have a polygonized substructure whereas statically recrystallized grains generally have straight boundaries and very few dislocations inside.* This is an expected consequence of dynamic recrystallization taking place by grain boundary migration *during* deformation (see also Hobbs, 1968). Luton and Sellars (1969) also found, by TEM, a dislocation substructure in dynamically recrystallized grains of nickel and nickel–iron alloys.

An important point is *that the size of dynamically recrystallized grains does not significantly depend on temperature but depends only on the flow stress* (Luton and Sellars, 1969; Glover and Sellars, 1973). Luton and Sellars (1969) found the following relationship:

$$\sigma \simeq k d^{-0.75} \tag{4.43}$$

This observation is in good agreement with the idea that polygonization subgrains are the nuclei of recrystallization (Honeycombe and Pethen, 1972).

In high temperature deformation their size (hence their number) does not depend on T but only on stress according to a similar relationship (within experimental scatter) to (4.43).

Although polygonized subgrain formation by recovery seems quite well linked to recrystallization, there may be a competition between both processes. If polygonization takes place very rapidly and easily it is quite effective in lowering the strain energy of the grains, thus suppressing the driving force for recrystallization. Hardwick, Sellars and Tegart (1961) did indeed notice that metals which polygonize very easily do not usually recrystallize during creep (e.g. aluminium), whereas metals which do not readily polygonize do in fact recrystallize during creep (e.g. copper or lead).

However, easy polygonization can sometimes lead to another kind of dynamic recrystallization without grain-boundary migration. Poirier (1974) proposed that in steady-state flow the misorientation between adjacent subgrains must increase with strain. Beyond a certain degree of misorientation (10 to 15 °) the subgrains must be considered as new dislocation free grains. This *in situ* recrystallization without grain-boundary migration results in a smaller grain size equal to the subgrain size. Recrystallization of quartz single crystals during experimental deformation had suggested this interpretation to Hobbs (1968) who also considered its influence on the orientations of the new grains relative to their host (see § 6.3.2). Both the increase in misorientation with strain and the formation of new grains with a size equal to the subgrain size were also verified by Poirier and Nicolas (*J. Geology*, in press) in olivine crystals of peridotites from a suite of naturally deformed nodules brought up from the upper mantle by basalts. Boullier and Nicolas (1973) found in peridotite nodules from kimberlites, domains where the orientation relation between grains is quite compatible with the fact that they were former subgrains in large deformed crystals of olivine. The phenomenon of *in situ* dynamic recrystallization was clearly recognized in quartzites by White (1973a), who also attributed it to the rotation of subgrains until they become individual grains.

CHAPTER 5

Plastic Deformation of Rock-forming Minerals

5.1. INTRODUCTION

The properties of rock-forming minerals described here are only those related to plastic flow, that is essentially the glide systems, considered in their respective P, T, $\dot{\varepsilon}$ domains of activity, the recovery and the specific creep laws. Unfortunately, the amount of data, though rapidly growing, is at present very limited. We will, therefore, restrict ourselves to describing only the few minerals for which a reasonable standard of information is available: olivine, ortho and clinopyroxenes, micas, plagioclase, quartz and carbonates. Incidental information is also presented for a few other silicates (kyanite, amphiboles). Spry (1969) (p. 60) enumerates the principal glide systems for some other rock-forming minerals. The question of preferred mineral orientations in polycrystalline aggregates which is well documented for olivine, quartz, and calcite is treated in Chapter 6 and olivine is more extensively studied in Chapter 11.

The glide systems were experimentally evidenced in the 1960's on crystals deformed in hot creep apparatus, using methods of optical observation exposed in Chapter 9. More recently introduced in the domain of geological materials, transmission electron microscopy (TEM) and X-ray techniques, described in Chapter 9, provide the means of direct observation of dislocations and carry the analysis down to the elementary mechanisms.

Although they are not connected to plastic flow, we present the shape ratios of minerals found as cumulus phases in magmatic cumulates, because of their importance in investigating the origin of preferred orientations in rocks where these minerals predominate.

5.2. OLIVINE

5.2.1. Crystallography

Olivine $(Mg,Fe)_2(SiO_4)$ is an orthosilicate of full orthorhombic symmetry. The atomic structure of forsterite, the magnesian end-member to which natural olivines are close, is represented in Fig. 5.1, projected on the

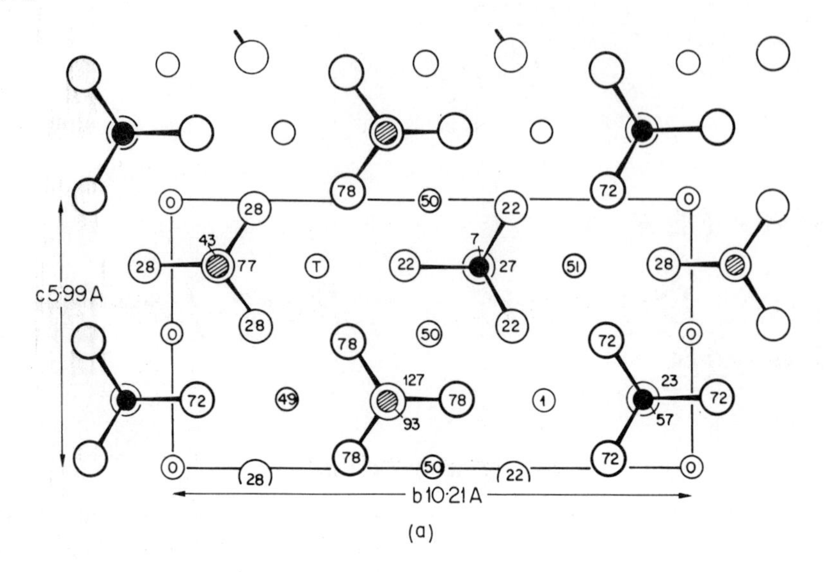

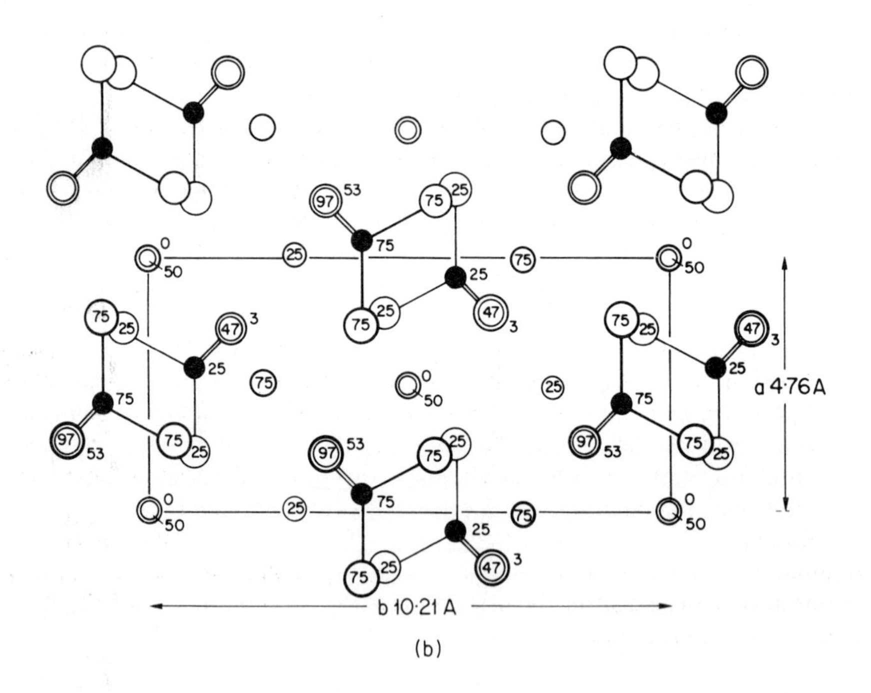

(a) Projection on (100) plane
(b) Projection on (001) plane
(after Bragg, Claringbull and Taylor, 1965, reproduced by permission of G. Bell & Sons Ltd)

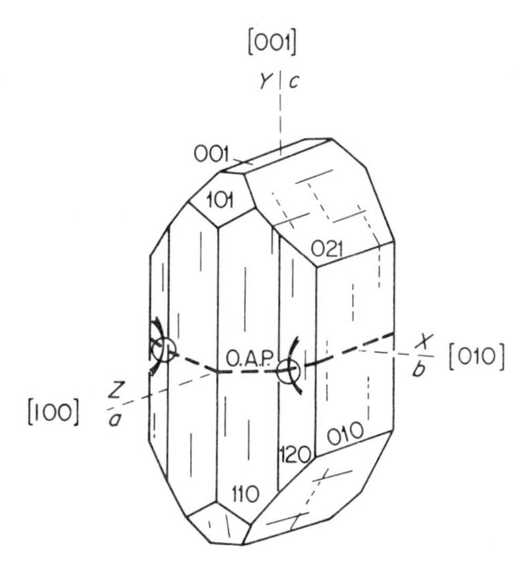

Fig. 5.2. Crystallographic habit and optical direction of forsterite (after Tröger, reproduced by permission of E. Schweizerbart'sche, Stuttgart)

(100) and (001) planes. Crystallographic and optical properties are shown in Fig. 5.2. The optical conventions accepted here are Z = the highest index, Y = the mean one and X the lowest one. In olivine the optical axes plane is (001) with Z parallel to the [100] axis. Olivine has two imperfect cleavages (010) and (100); idiomorphic crystals display a pronounced development mainly of the {010} face and to a lesser extent of the {110} and {021} faces; they are elongated along the c or the a axes. In phenocrysts from basalts Brothers (1959) describes a flattened habit parallel to (010) with a minor development of {0kl} and (001) in the [100] zone and of {110} in the [001] zone. Exceptionally the {110} faces are extraordinarily large with a corresponding reduction in (010) development; flattening is then normal to [100]. Turner (1942) describes phenocrysts in lavas with prominent {021} faces resulting in $a : b : c$ ratios of 2 : 1 : 1. In cumulates from the Stillwater complex, the morphological development $a : b : c$ is 0·77 : 0·63 : 1 (Jackson, 1961).

5.2.2. Slip Systems

(a) Crystallographic considerations

Since the energy of a dislocation is proportional to b^2 (b = length of the Burgers vector), slip generally occurs in directions corresponding to the shortest Burgers vectors. In the case of olivine the lattice parameters of the unit cell are:

$$a\,[100] = 4{\cdot}76\ \text{Å}$$
$$b\,[010] = 10{\cdot}21\ \text{Å}$$
$$c\,[001] = 5{\cdot}99\ \text{Å}$$

Dislocations with the Burgers vector [010] would have an energy about four times as high as dislocations with [100] or [001] Burgers vector. It is therefore to be expected that slip will occur in the [100] or [001] directions.

Silicates can usually be considered as almost close packed crystals of oxygen (the bulkiest ion) with the silicon atoms or metallic cations occupying various interstices of the structure. In order to investigate the possible slip planes geometrically, it is interesting to consider olivine as a hexagonal close packed crystal of oxygen, whose basal plane (0001) corresponds to the (100) plane of the orthorhombic lattice (Poirier, 1975). The densest planes (possible slip planes) are:

(100), corresponding to (0001) of the HCP lattice and (011), (010), $(0\bar{1}1)$, corresponding to the first-order prism planes of the HCP lattice.

However, since olivine is composed of isolated SiO_4 tetrahedra, all the dense planes of the HCP oxygen lattice are not equivalent from the point of view of the ease of slip, since Si–O bonds have to be broken during slip on all planes except (010). On these grounds one would expect slip on (010) to be the easiest, but it is known that hydrolysis of the Si–O bonds may considerably facilitate slip (Blacic, 1971) and the experimental evidence for slip systems (see below) shows that things are not so simple, even though most of the predicted slip systems exist in some range of temperature, strain rate, pressure and possibly water content.

(*b*) *Experimental evidence*

Turner (1942) and Chudoba and Frechen (1950) suggested that the principal glide system in olivine was (010) [100]. Apart from geometrical considerations, they relied on observations of 'undulose banding' (i.e. KBB) in the plane (100) which was thought to be perpendicular to the glide direction (see Chapter 9).

More recently Raleigh (1965, 1968) performed compression experiments on olivine rocks under confining pressure at high temperatures; he also determined the operating slip systems by optical observation of the KBB planes, as well as by direct observation of slip traces on the specimen surfaces and photoelastic effects associated with deformation lamellae. Carter and Avé Lallemant (1970) carried the investigated domain to higher temperatures (800 °C $< T <$ 1200 °C) and used a decoration technique of deformation lamellae to evidence the glide planes. Raleigh and Kirby (1970) optically determined the slip systems in the same range, in specimens deformed by hot creep.

The results of Raleigh, Carter and Avé Lallemant and Raleigh and Kirby are gathered in Fig. 5.3 as a function of T and $\dot{\varepsilon}$. The systems (100) [001], {110} [001] and exceptionally (100) [010], observed at lower temperatures and higher strain rates give way to {0kl} [100] systems at higher temperatures and lower strain rates; at still higher T and lower $\dot{\varepsilon}$, Carter and Avé Lallemant find a predominance of the (010) [100] system.

Raleigh found that slip on {0kl} [100] systems occurred on all planes, not necessarily crystallographic, in zone with [100] and called it 'pencil glide' using the term introduced by Taylor and Elam (1926) for non-crystallographic slip

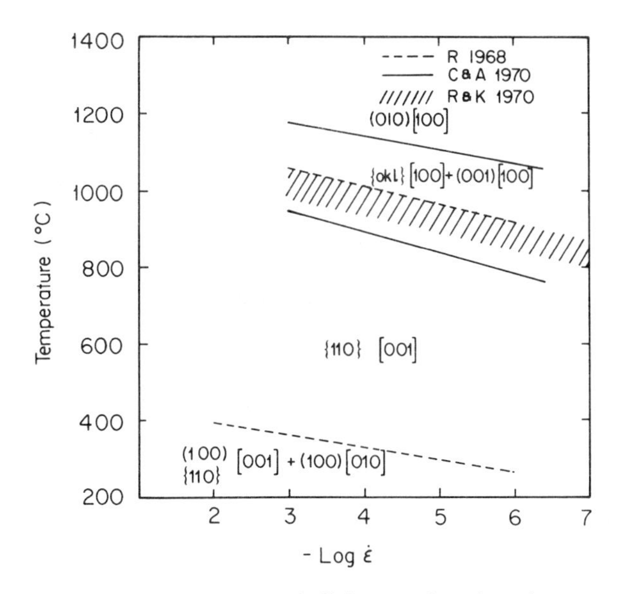

Fig. 5.3. Glide systems of olivine as a function of temperature and deformation rate

on planes in zone with [111], observed in iron. Carter and Avé Lallemant found that although certain definite crystallographic planes are preferred, they are sufficient in number to justify the term 'pencil glide'. This problem will be discussed in the next paragraph. Slip systems in olivine from many peridotites deformed in various natural conditions have been optically studied by the KBB technique (see Chapter 9) (Raleigh, 1968; Avé Lallemant and Carter, 1970; Raleigh and Kirby, 1970; Nicolas et al., 1971; Loney et al., 1971; Mercier and Nicolas, 1975; Boullier, 1975). Some studies have also been carried out on olivine from meteorites (Carter et al., 1968; Klosterman and Buseck, 1973). The {0kl} [100] systems have always been found to be the most common, without any distinct tendency to shift to the (010) [100] system at the higher temperatures.

Finally Carter and Avé Lallemant (1970) also determined the role of increasing confining pressure on the transition from low T, high $\dot{\varepsilon}$ systems to high T low $\dot{\varepsilon}$ ones: an increase in pressure of 1 kB lowers the transition temperature by about 7 °C (Fig. 5.4). The pressure effect is less pronounced than the strain rate effect, at least in the investigated domain, where it is found that a decrease in strain rate of one decade lowers the transition temperature by about 50 °C (Fig. 5.3).

In recent years, various techniques for direct observation of dislocations and especially Transmission Electron Microscopy (TEM) (see Chapter 9) have allowed experimenters to obtain direct microscopic evidence of the existence of optically observed slip systems. Young (1969) experimentally deformed olivine single crystals at pressures ranging from 1 bar to 10 kB and at temperatures up to 1400 °C, and studied the dislocations by etch pit and decoration

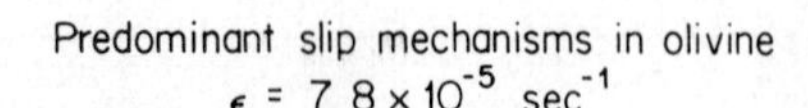

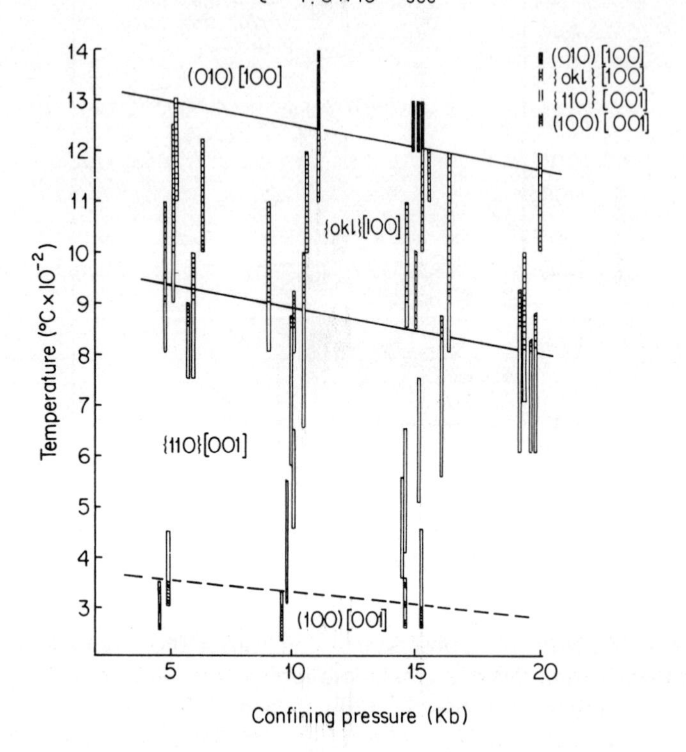

Fig. 5.4. Glide systems of olivine as a function of temperature and confining pressure (Carter and Avé Lallemant, 1970). Vertical bars show results from single specimens in a given temperature range. (Reproduced by permission of The Geological Society of America)

techniques. His results largely confirm the existence of the slip systems found by Raleigh and Carter and Avé Lallemant in the investigated (T, P, $\dot{\varepsilon}$) fields. Young also obtained direct evidence of slip on the (001) [100] system, the [100] dislocations being apparently more mobile above 1200 °C on the (001) plane than on (010). Wegner and Christie (1974) developed the etch pit technique to correlate the optically visible features of slip in olivine with the dislocation substructure (Plate 14).

Deformed olivine crystals were studied in TEM by Boland et al. (1971), Blacic (1971), Phakey et al. (1972), Green and Radcliffe (1972a,b,c), Blacic and Christie (1973), Goetze and Kohlstedt (1973) and Olsen and Birkeland (1973).

Phakey et al. (1972) accurately determined a few active glide systems at different temperatures (Table 5.1) which confirm the systems already found by other authors. However they found no evidence of {0kl} [100] at the microscopic level, except for the well known (010) [100] system. They showed that at

Table 5.1.

Temperature, °C	Types of dislocations	Slip systems
600	Screw, $\mathbf{b}=c[001]$ (edges very short)	(?) [001]
800	Screw and edge, $\mathbf{b}=c[001]$	(100) [001], {110} [001], and (010) [001]
800	Screw and edge, $\mathbf{b}=b[010]$	(100) [010]
800	Screw and edge, $\mathbf{b}=a[100]$	(010) [100]
1000	Screw and edge, $\mathbf{b}=c[001]$, and loops	{110} [001] and probably (100) [001]
1000	Screw and edge, $\mathbf{b}=a[100]$, and some loops	(010) [100]
1250	Helices, loops, and networks	Not determined

(After Phakey et al., 1972)

1000 °C the stresses required to form or move dislocations with $\mathbf{b}=[100]$ and $\mathbf{b}=[001]$ are nearly equal; this is also approximately the boundary temperature between the higher T domain where [100] dominates and the lower T domain where [001] dominates, as found by all experimenters. Green and Radcliffe (1972c) examined the dislocation structures in olivine from a peridotite experimentally deformed 30 % in axial compression at 900 °C under 15 kB confining pressure at a strain rate of $10^{-6}\,s^{-1}$. They found planar arrays of parallel edge dislocations with Burgers vectors [100] on the planes (010) and the plane normal to $[01\bar{1}]$ (second-order pyramidal plane $(\bar{1}2\bar{1}0)$ in the HCP oxygen lattice). These arrays have been identified as slip bands. Green and Radcliffe (1972b), and Goetze and Kohlstedt (1973), deduced the existence of the {0kl} [100] system from the analysis of the orientation of dislocations in tilt walls (see next paragraph).

The experimental evidence on slip systems is summed up in Table 5.2.

Table 5.2. Summary of active slip systems in olivine determined by optical, decoration and TEM methods.
+: common −: uncommon
(For specific P, T, $\dot{\varepsilon}$ domains refer to Figs. 5.3 and 5.4)

Low T High $\dot{\varepsilon}$	Medium T Medium $\dot{\varepsilon}$	High T Low $\dot{\varepsilon}$
(100) $[001]^+$	(100) [001]	
{110} $[001]^+$	{110} $[001]^+$	{110} [001]
(100) $[010]^-$	(100) $[010]^-$	
		(010) $[100]^+$
		{0kl} $[100]^+$
		(001) [100]
		{101} $[010]^-$

(c) Discussion of 'Pencil glide'

The term 'pencil glide' applied to {0kl} [100] glide by Raleigh (1968) was first introduced by Taylor and Elam (1926) to describe figuratively the wavy non-crystallographic slip in the ⟨111⟩ direction observed in crystals of iron. Nye (1949) also used it in the case of similar observations in face-centred-cubic AgCl. Such macroscopically non-crystallographic slip has since been observed in all body-centred-cubic metals and in face-centred-cubic metals of high stacking fault energy like aluminium. In the latter case, it was rapidly recognized that the wavy non-crystallographic slip lines were composed of microscopically visible portions of slip lines on dense {111} planes. This was explained by the cross-slip of ⟨110⟩ screw dislocations from one dense plane to another, the variable amount of slip on each plane giving a macroscopically non-crystallographic orientation. However, it took longer to analyse pencil glide in BCC metals along the same lines. Until recently it was thought that slip could occur on any plane in zone with the slip direction ⟨111⟩ supporting the maximum resolved shear stress. Some experimenters, however, believed that slip could take place on numerous but discrete crystallographic planes. The situation is obviously the same nowadays for olivine: while Raleigh (1968) and Raleigh and Kirby (1970) find that any plane in the zone [100], supporting the highest shear stress, will be active, Carter and Avé Lallemant (1970) suggest that there may be no 'true pencil glide' but slip on numerous discrete crystallographic planes of the zone.

'Pencil glide' is now an almost obsolete term in physical metallurgy, which does not reflect the physical reality of the process. It is known that slip cannot occur on any plane in zone with a direction, but that screw dislocations dissociated in one plane can constrict their stacking fault with the help of the applied stress and thermal agitation, and are thus able to cross-slip in another dense crystallographic plane to avoid a local obstacle. The process can repeat itself leading to composite slip, on a submicroscopical scale. The amount of slip on each of the planes depends on the orientation of the crystal with respect to the applied stress (angle χ), the orientation of the macroscopic 'slip plane' with respect to the crystal (angle ψ) depends on χ. (Šesták, 1972) (Fig. 5.5).

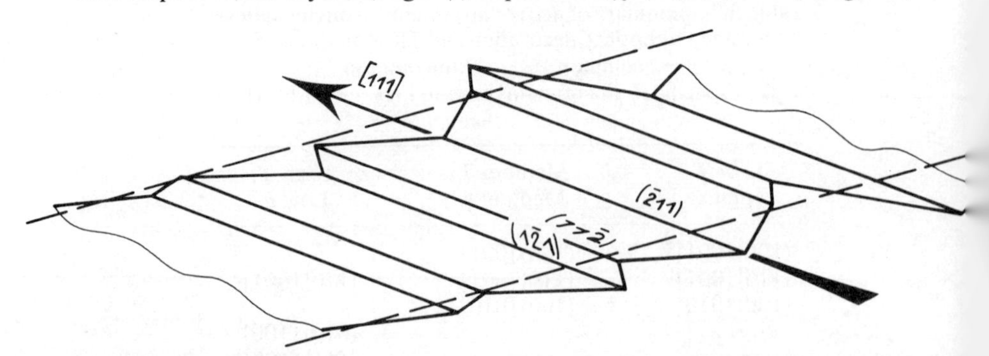

Fig. 5.5. Macroscopically non-crystallographic slip plane (dashed lines) in body-centred-cubic metals resulting from cross-slip of [111] screw dislocations on microscopic portions of various {112} planes in zone with [111]. Arrow indicates the slip direction

It is likely that the so-called 'pencil glide' in olivine can also be explained as resulting from cross-slip. Phakey et al. (1972) first suggested this possibility, as they found evidence of cross-slip in TEM but no evidence of slip on non-crystallographic {0kl} planes. As we will see presently, it seems that no direct unimpeachable evidence of slip on non-crystallographic planes has ever been given in the literature.

The published evidence of {0kl} [100] slip rests, as we have seen, on optical methods and TEM. The most frequently used optical method consists in determining the KBB planes and assuming that the slip plane is perpendicular to them (Chapter 9). This would undoubtedly be correct if the KBBs were symmetrical tilt walls formed by glide polygonization by a single type of edge dislocations, as is the case at low temperatures; on the contrary, at high temperature KBBs are mainly due to climb polygonization (Goetze and Kohlstedt, 1973) and may be constituted by different types of dislocations (screws and edges belonging to different systems). The rotations estimated on both sides of such sub-boundaries are not necessarily representative of the active glide system.

Other methods use deformation lamellae and slip traces. Unfortunately, the interpretation of lamellae as slip bands is not unequivocal (see § 5.9.2) and slip traces are less distinct at high temperatures than at low temperatures, especially if the magnification is low.

As for TEM evidence, it is also inconclusive. True, many experimenters have determined the Burgers vectors of dislocations, which gives only the possible slip direction, but direct evidence of slip bands (Green and Radcliffe, 1972c) confirms only the existence of crystallographic slip on dense planes. The determination of slip planes by measuring the orientation of dislocations in KBBs is objectionable for the aforementioned reason that climb polygonization is prevalent (Goetze and Kohlstedt, 1973).

In conclusion, we think it is most probable that the so-called pencil glide {0kl} [100], *occurring at high temperatures is in fact due to composite cross-slip on a limited number of dense crystallographic planes*, possibly some of the prism planes of the HCP oxygen lattice: (011) (010) $(0\bar{1}1)$ (first-order prism planes) or (001) (031) $(0\bar{3}1)$ (second-order prism planes).

5.2.3. Recovery and recrystallization

(i) Climb of dislocations into sub-boundaries becomes important above about 1000 °C for strain rates of 10^{-3} s^{-1}. After 1 hour of annealing at 1290 °C, Goetze and Kohlstedt (1973) report that the dislocations extensively climb into regular arrays; after 1 hour at 1380 °C, over 95 % of the originally isolated dislocations are organized into sub-boundaries. A distinct substructure is formed with dominantly (100) tilt walls associated with the [100] Burgers vector. Along with the (100) tilt walls, Nicolas and Boudier (1975) and Kirby and Wegner (1973), have optically evidenced (001) tilt walls associated with the [001] slip direction in naturally deformed peridotites. When both (100) and (001) walls are present, the olivine crystals display a rectangular substructure in the (010) plane.

(ii) Avé Lallemant and Carter (1970) report that above 1050 °C for $\dot{\varepsilon} = 10^{-3}\ \mathrm{s}^{-1}$, recrystallization occurs at grain-boundaries, followed by intragranular recrystallization at somewhat higher temperatures. The optical orientations of the new grains relative to their hosts are discussed in § 6.2.2 and in Chapter 10. Recrystallization reduces the dislocation density by a few orders of magnitude: for instance, from 10^7 disl. cm^{-2} to 10^5 disl. cm^{-2} in a peridotite nodule from a kimberlite in which the recrystallization is thought to be static (Y. Gueguen and H. W. Green, pers. comm.).

5.2.4. High temperature creep laws

The creep laws were first determined by Carter and Avé Lallemant (1970) in hot creep experiments performed on 'wet' and 'dry' dunites and lherzolites at confining pressures from 5 to 20 kB, temperatures from 300 °C to 1400 °C and strain rates from $10^{-3}\ \mathrm{s}^{-1}$ to $10^{-8}\ \mathrm{s}^{-1}$. The steady-state regime is best fitted in both types of rocks by a power creep equation; they find for:

$$\begin{cases} \text{Wet Lherzolite} & \dot{\varepsilon} = 3{\cdot}2 \times 10^7 \sigma^{2\cdot3} \exp -\left(\dfrac{80^{\mathrm{kCal/mole}}}{RT}\right) & (5.1) \\ \text{Wet Dunite} & \dot{\varepsilon} = 6{\cdot}2 \times 10^6 \sigma^{2\cdot4} \exp -\left(\dfrac{80^{\mathrm{kCal/mole}}}{RT}\right) & (5.2) \\ \text{Dry Dunite} & \dot{\varepsilon} = 1{\cdot}2 \times 10^{10} \sigma^{4\cdot8} \exp -\left(\dfrac{120^{\mathrm{kCal/mole}}}{RT}\right) & (5.3) \end{cases}$$

Raleigh and Kirby (1970) performed further experiments on dry dunite and proposed a creep law incorporating their results with those of Carter and Avé Lallemant:

$$\dot{\varepsilon} = 10^8 \sigma^5 \exp -\left(\frac{106^{\mathrm{kCal/mole}}}{RT}\right) \tag{5.4}$$

which they also express as:

$$\dot{\varepsilon} = 10^8 \sigma^5 \exp -\left(26{\cdot}4 \frac{T_m}{T}\right) \tag{5.5}$$

Post and Griggs (1973) obtained new experimental results on creep of wet and dry dunite and put forward a model for the Fennoscandian post-glacial uplift which requires a non-Newtonian flow with a stress exponent of $n = 3{\cdot}2$; they propose a creep law for steady state:

$$\dot{\varepsilon} = 1{\cdot}7 \times 10^9 \sigma^{3\cdot18} \exp -\left(31{\cdot}8 \frac{T_m}{T}\right) \tag{5.6}$$

The activation energy for creep of wet and dry dunite was found to be respectively equal to 93 and 130 kCal/mole.

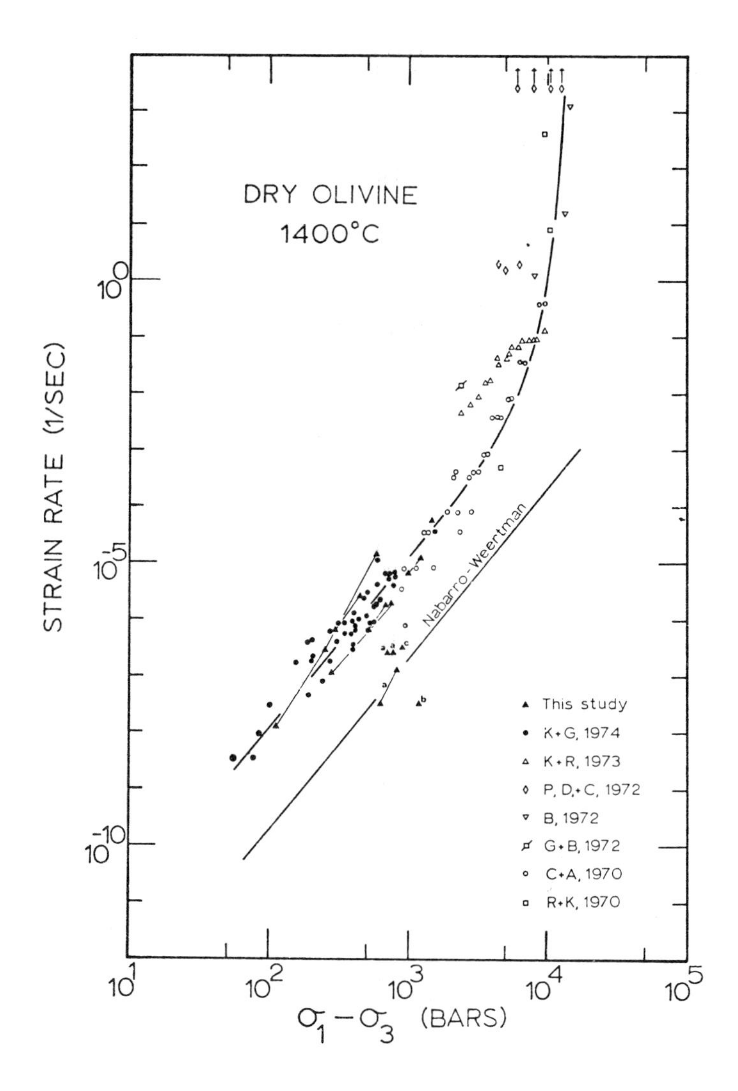

Fig. 5.6. High temperature creep of olivine. Temperature compensated strain rate at 1400 °C versus differential stress. The experimental strain rate is temperature-compensated by multiplying it by $\exp\left(\frac{Q}{RT}-\frac{Q}{RT_0}\right)$, where $Q = 125$ kCal/mole and $T_0 = 1400$ °C. The solid line follows the trend of the data. (Reproduced with permission from Kohlstedt and Goetze, *Jour. Geophys. Res.*, **79**, 2045 (1974), copyright by American Geophysical Union) (corrected by D. Kohlstedt)

Kohlstedt and Goetze (1974) carried out constant stress creep experiments at atmospheric pressure on dry single crystalline peridot, at temperatures between 1400 and 1700 °C and stresses between 50 and 1500 bars. They found a steady state with a creep rate independent of the crystal orientation and an activation energy of 126 kCal/mole.

Fig. 5.6 recapitulates their results, together with those of former studies (Carter and Avé Lallemant, 1970; Raleigh and Kirby, 1970; Goetze and Brace, 1972; Kirby and Raleigh, 1973).

The experimental values of the activation energy for creep of dry olivine range between 106 and 130 kCal/mole (Fig. 5.7) and agree well with the activation energy for climb of dislocations: 135 kCal/mole determined by the kinetics of shrinkage by climb of [001] dislocation loops in (001) planes observed in TEM (Goetze and Kohlstedt, 1973).

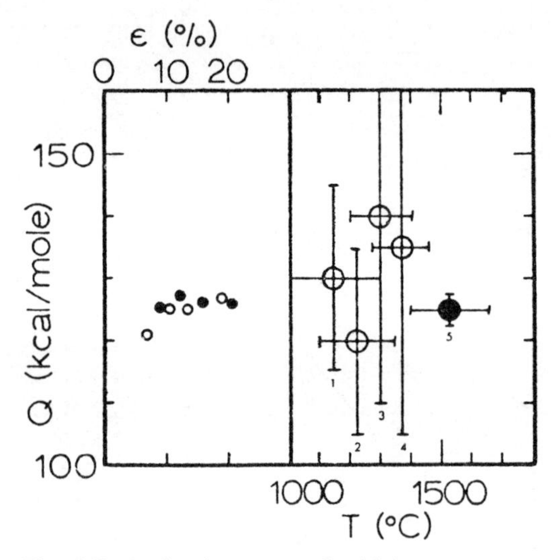

Fig. 5.7. Activation energy for high temperature creep of olivine. (Left) Activation energy as a function of strain from Kohlstedt and Goetze (1974), 1400 °C < T < 1700 °C. (Right) Activation energy from different authors: 1: Post and Griggs, 1973; 2: Carter and Avé Lallemant, 1970; 3 and 4: Goetze and Kohlstedt, 1973; 5: Kohlstedt and Goetze, 1974. (Reproduced with permission from Kohlstedt and Goetze, *Jour. Geophys. Res.*, **79**, 2045 (1974), copyright by American Geophysical Union)

These values are quite comparable with the activation energy for self-diffusion of oxygen in oxides like Al_2O_3 or silicates like spinel (see Table 3.1). They are compatible with the idea that dislocation climb is controlled by the diffusion of the oxygen ion although experimental values are lacking for olivine, as are values for the activation volume of diffusion (estimated between 11 and 37 cm^3/mole by Stocker and Ashby, 1973).

On the other hand, Blacic (1971) performed experiments on dunites and single crystalline olivine, wet and dry at 900 and 1000 °C. He found that specimens increase in strength by a factor 2 when dried in vacuum at 880 °C. By analogy with the water weakening occurring in quartz (§ 5.9) by hydrolysis of the Si–O bonds, he proposes that the rate-controlling factor for the deformation of olivine may be the diffusion rate of water to dislocations, although the diffusing species (H_2O, H^+, OH^-, O^{--} or H_3O^+) is unknown.

Clearly, diffusion experiments at high temperatures under pressure are needed to understand the creep controlling mechanisms better.

5.3. KYANITE

Kyanite is a triclinic polymorph of the aluminium silicate Al_2SiO_5 containing independent SiO_4 tetrahedra and oxygen atoms which are not linked to silicon. The atomic structure is represented on Fig. 5.8.

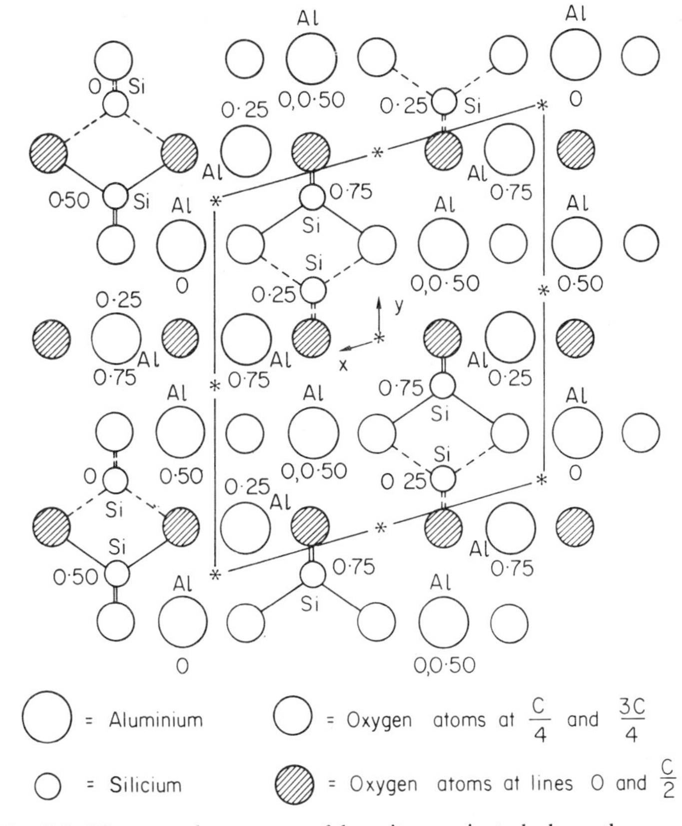

Fig. 5.8. The crystal structure of kyanite, projected along the c-axis (after Naray-Szabo et al., 1929, reproduced by permission of Akademische Verlagsgesellschaft, Frankfurt)

From crystallographic considerations (Raleigh, 1965), it can be predicted that the (100) plane should be an easy slip plane, since it involves no breaking of Si–O bonds; (100) is also a perfect cleavage plane. A probable slip direction is

[001], parallel to close-packed rows of oxygen atoms and for which the Burgers vector is the shortest ($b = 5{\cdot}57$ Å). Mügge (1883) and G. Friedel (1964) indeed found that kyanite slips on the (100) plane. Raleigh (1965) deformed kyanite 5 to 10 % at temperatures between 700 and 850 °C and 5 to 7 kB confining pressure; he observed slip traces corresponding to (100) and well-defined kink band boundaries nearly normal to slip lines with an external rotation about [010]. He concluded that the slip direction was actually [001]. No twinning was observed.

5.4. ENSTATITE AND CLINOENSTATITE

5.4.1. Crystallography

Enstatite and bronzite are near the magnesian pole of the pyroxene series $(Mg,Fe)_2Si_2O_6$, and do not differ chemically from clinoenstatite. They have a

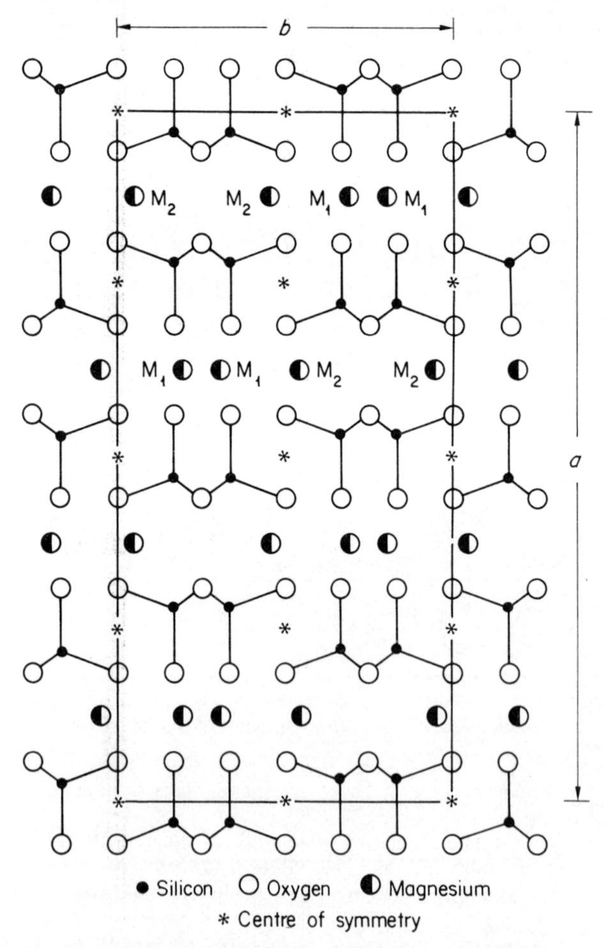

Fig. 5.9. The crystal structure of enstatite, projected on (001) (after Deer, Howie and Zussman, 1963, reproduced by permission of the Longman Group Ltd.)

chain structure with orthorhombic symmetry for enstatite (Figs. 5.9 and 5.10(a)) and monoclinic symmetry for clinoenstatite (Fig. 5.10(b)).

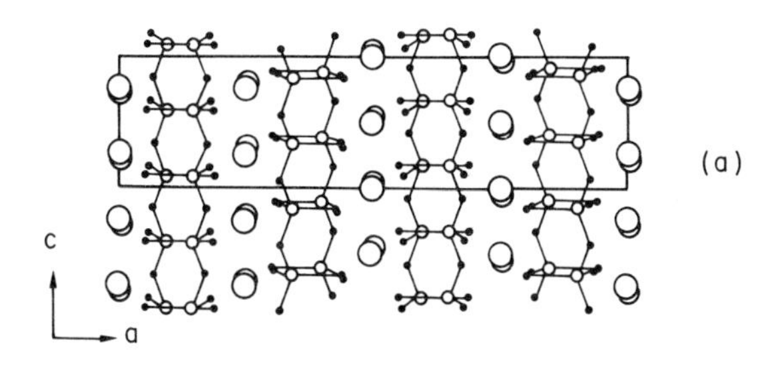

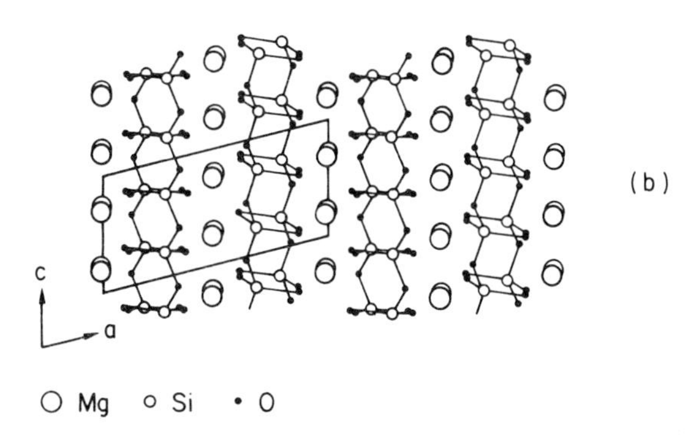

Fig. 5.10. The structure of (a) orthoenstatite, (b) clinoenstatite projected on (010) (after Deer, Howie and Zussman, 1963, reproduced by permission of the Longman Group Ltd.)

Enstatite (100) cleavage is better than the (210) and $(2\bar{1}0)$ ones. In freely growing conditions, its crystallographic habit is characterized by the development of the (100), {210}, (010) and {101} faces (Fig. 5.11); the (100) plane is commonly underlined by Ca-rich clinopyroxene exsolutions. In cumulates from the Stillwater complex, Jackson (1961) reports that the dimensions of bronzite crystals in the $a : b : c$ directions are in the ratios 0·79 : 0·63 : 1.

5.4.2. Slip systems

(a) Crystallographic considerations

The structure of enstatite (Fig. 5.10(a)) consists of chains of SiO_4 tetrahedra sharing an oxygen ion, aligned in the [001] direction. The chains are packed in (100) layers alternating with layers of (Mg, Fe) ions. Slip on the (100) planes

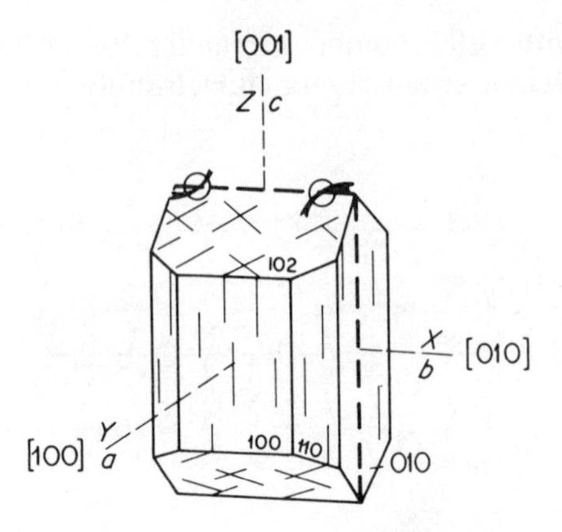

Fig. 5.11. Crystallographic habit and optical directions of enstatite (after Tröger, reproduced by permission of E. Schweizerbart'sche, Stuttgart)

clearly does not involve breaking of Si–O bonds and should be easy especially in the direction of the chains [001]. This slip direction is also favoured energetically, as it corresponds to the shortest Burgers vector ($b = 5{\cdot}2$ Å). The next shortest Burgers vector in the (100) planes has the direction [010] ($b = 8{\cdot}9$ Å).

(*b*) *Experimental evidence*

Mügge (1898) observed a bending of the (100) plane about the [010] axis in natural orthopyroxenes and concluded to the existence of the (100) [001] glide system. These observations were confirmed by Turner et al. (1960) and Griggs et al. (1960) in experimentally deformed enstatite rocks. Raleigh (1965) experimentally deformed peridotite nodules between 700 and 850 °C and 5 to 7 kB confining pressure and, on the surface of enstatite crystals, he directly observed a single set of slip lines corresponding to the (100) plane. The existence of the (100) [001] slip system has been further corroborated by the observation in TEM of dislocations with a Burgers vector parallel to [001], dissociated in partials separated by a wide stacking fault ribbon lying in the (100) plane (Plate 9) (Green and Radcliffe, 1972b; Coe and Muller, 1973; Kohlstedt and Van der Sande, 1973).

Dislocations with a Burgers vector parallel to [010] have been exceptionally observed (Kohlstedt and Van der Sande, 1973).

5.4.3. Ortho-clinoenstatite inversion

Although this transformation is extremely rare in natural conditions (Trommsdorff and Wenk, 1968), it has been theoretically studied (Brown et al., 1961; Coe, 1970) and is well documented experimentally (Griggs et al., 1960; Turner et al., 1960; Borg and Handin, 1966; Riecker and Rooney, 1967;

Raleigh, 1967; Raleigh et al., 1971; Coe and Muller, 1973; Kirby and Coe, 1974). These studies have shown that over a wide range of temperature (27–1200 °C) and pressure (5–40 kB), the orthoenstatite in deformed zones transforms to clinoenstatite. The transformation is sensitive to shearing stresses and requires a shear through 13·3° parallel to [001] in the (100) plane. The inversion is nearly displacive with little or no volume change and involves breaking only half of the Mg–O bonds and none of the Si–O bonds (Coe, 1970).

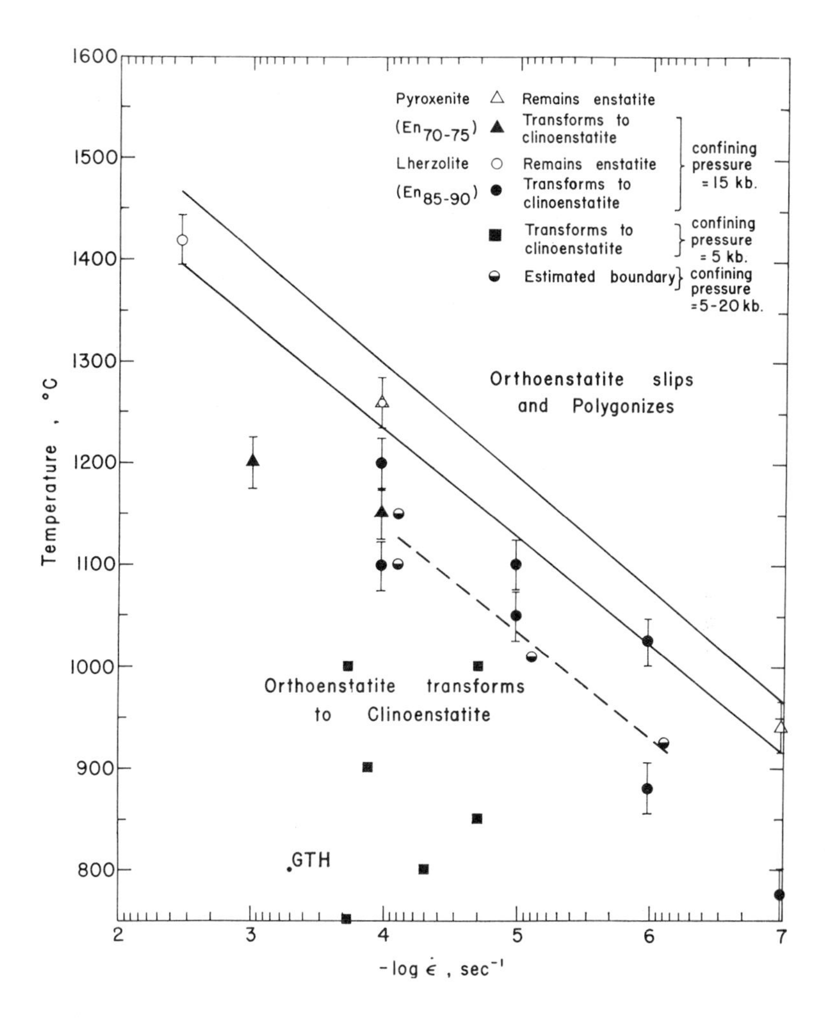

Fig. 5.12. Critical temperature for orthoenstatite to clinoenstatite inversion as a function of strain rate. GTH represents the experimental point of Griggs et al. (1960). (Reproduced with permission from Raleigh et al., *Jour. Geophys. Res.*, **76**, 4011 (1971), copyright by American Geophysical Union)

At a given strain rate there is a critical temperature above which the enstatite slips instead of transforming: the critical temperature is strongly dependent on strain rate (Fig. 5.12) (Raleigh et al., 1971). This behaviour may perhaps be explained by considering that the inversion is a diffusionless transformation of the same type as mechanical twinning or the formation of martensite, and as such depends little on temperature and much more on the stress concentrations necessary to nucleate it. It is likely that at low T and high $\dot{\varepsilon}$, the stress concentrations (at grain boundaries for instance) are not easily relaxed and the crystal can only deform by inversion; conversely at high T and low $\dot{\varepsilon}$ more processes (including diffusion) are available to relax the stresses and prevent the nucleation and propagation of transformed layers. (The same explanation accounts for the well-known absence of twinning at moderate and high temperatures in face-centred-cubic metals, whereas it can be observed for very low T and high $\dot{\varepsilon}$.) Studies of the inversion by transmission electron microscopy (Kirby, 1975) showed clearly the role of partial dislocations in the transformation (Plate 5(b)).

5.4.4. Recovery and recrystallization

Annealing orthopyroxene from peridotites, between 1000° and 1500 °C does not result in any modification of the defect population visible in TEM; on the other hand, at 1380 °C, practically all isolated dislocations in olivine have become organized into sub-boundaries (Kohlstedt and Van der Sande, 1973); these authors estimate that the climb velocity of dislocations in their experiments on enstatite was less than 0·1 % of the climb velocity of dislocations in olivine; they attribute this to the existence of dissociated dislocations with wide stacking faults. Such behaviour should result in a high creep resistance of orthopyroxene compared with olivine in the high temperature domain where creep is climb-controlled. Green and Radcliffe (1972b) also note that climb effects are much less important in enstatite than in olivine, in agreement with the optical evidence of a lesser tendency to polygonize and recrystallize.

5.4.5. Exsolution lamellae

Orthopyroxene is commonly streaked with (100) lamellae approximately 0·5 μ in width and 5 to 50 μ, apart formed by a Ca-rich clinopyroxene (Champness and Lorimer, 1973; Kohlstedt and Van der Sande, 1973). The lamellae are bounded by rows of tightly spaced dislocations accommodating the misfit between the lattices. Kohlstedt and Van der Sande (1973) propose the following formation mechanism for lamellae: To lower the strain energy, the large Ca atoms bind with edge dislocations producing a Ca atmosphere in the zone in tension around the dislocations. The atmospheres of dislocations arranged into boundaries could combine, thus forming lamellae and stabilizing the dislocation structure. Kirby (1973) suggests a more convincing mechanism by which lamellae of clinopyroxene could be produced by shear-induced inversion of orthopyroxene, subsequent or simultaneous diffusion of Ca to clinopyroxene completing the exsolution process.

5.5. DIOPSIDE

5.5.1. Crystallography

Diopside Ca (Mg, Fe) Si_2O_6 is close to the magnesian pole of the clinopyroxene series. It has a chain structure of SiO_4 tetrahedra sharing an oxygen ion and its symmetry is fully monoclinic (Fig. 5.13).

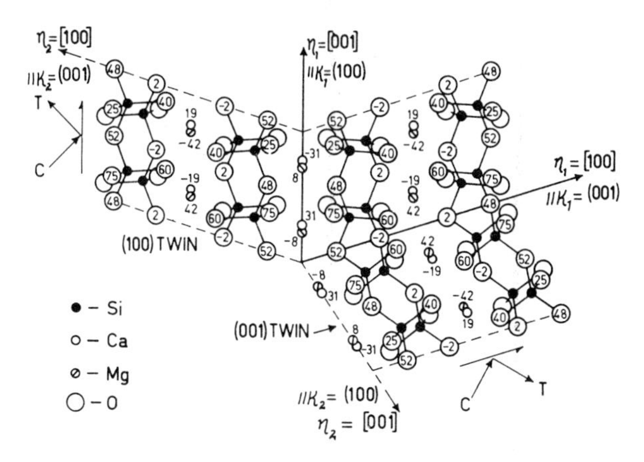

Fig. 5.13. The crystal structure of diopside projected on (010) (after Warren and Bragg. 1928). (100) or (001) twins are produced by compression or tension directed along *C* or *T*. (Reproduced by permission of Akademische Verlagsgesellschaft, Frankfurt)

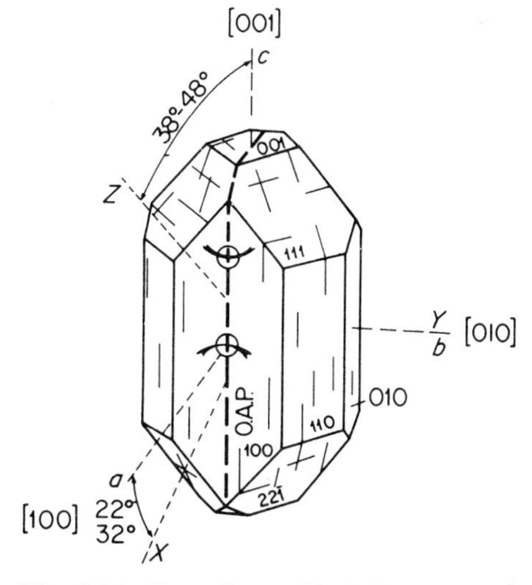

Fig. 5.14. Crystallographic habit and optical directions of diopside (after Tröger, reproduced by permission of E. Schweizerbart'sche, Stuttgart)

Diopside has (100), (110) and $(1\bar{1}0)$ cleavages. Its crystallographic habit and related optical properties are represented in Fig. 5.14; it does not commonly present an idiomorphic shape.

5.5.2. Slip systems

As in the case of orthopyroxene, it can be predicted by simple inspection of the structure that the easy slip system should be (100) [001], along the plane parallel to the chain layers and in the direction of the chains, which also corresponds to the shortest Burgers vector ($b = 5{\cdot}25$ Å).

The first experimental evidence of (100) [001] slip was reported by Griggs et al. (1960), and Raleigh and Talbot (1967) definitely identified it in experimentally deformed diopside. Slip becomes important above 1000 °C for $\dot{\varepsilon} = 5 \times 10^{-5}$ s^{-1}. No other slip system has been reported.

5.5.3. Mechanical twinning

Griggs et al. (1960) studied mechanical twinning in diopside crystals of a pyroxenite experimentally deformed at 500 °C under 5 kB confining pressure. They identified two systems (Table 5.3) which were later reinvestigated by Raleigh (1965) at 700 to 850 °C and Raleigh and Talbot (1967) at temperatures up to 1000 °C. Slip on (100) [001] and twinning on (100) [001] prevail over (001) [100] twinning at high temperatures and low strain rates. When (100) [001] slip and (100) [001] twinning are simultaneously present, slip is active in the sense in which twinning is impossible (anti-twinning sense), (Raleigh and Talbot, 1967) (Fig. 5.13).

Table 5.3

Twin plane	Twinning direction	Shear
$K_1 = (100)$	$\eta_1 = [001]$	$S = 0{\cdot}57$
$K_1 = (001)$	$\eta_1 = [100]$	$S = 0{\cdot}57$

In both naturally and experimentally deformed diopside the (100) twins are narrower and more closely spaced than the (001) twins which also tend to be discontinuous. Mechanical twins of the (100) system could be confused with growth twins or orthopyroxene exsolution lamellae. They can, however, be distinguished by the following criteria: Mechanical twin boundaries are straight and lie exactly in the (100) planes whereas growth twins have stepped boundaries and tend to be broader; mechanical twins extend to the crystal boundaries and have an oblique extinction contrast between crossed polarizers, whereas exsolution lamellae pinch out close to the crystal boundaries and exhibit a parallel extinction contrast (Plate 2(a)).

5.6. AMPHIBOLES

5.6.1. Crystallography

The structure of all amphiboles of various chemical compositions is characterized by the existence of double chains of linked SiO_4 tetrahedra (Fig. 5.15).

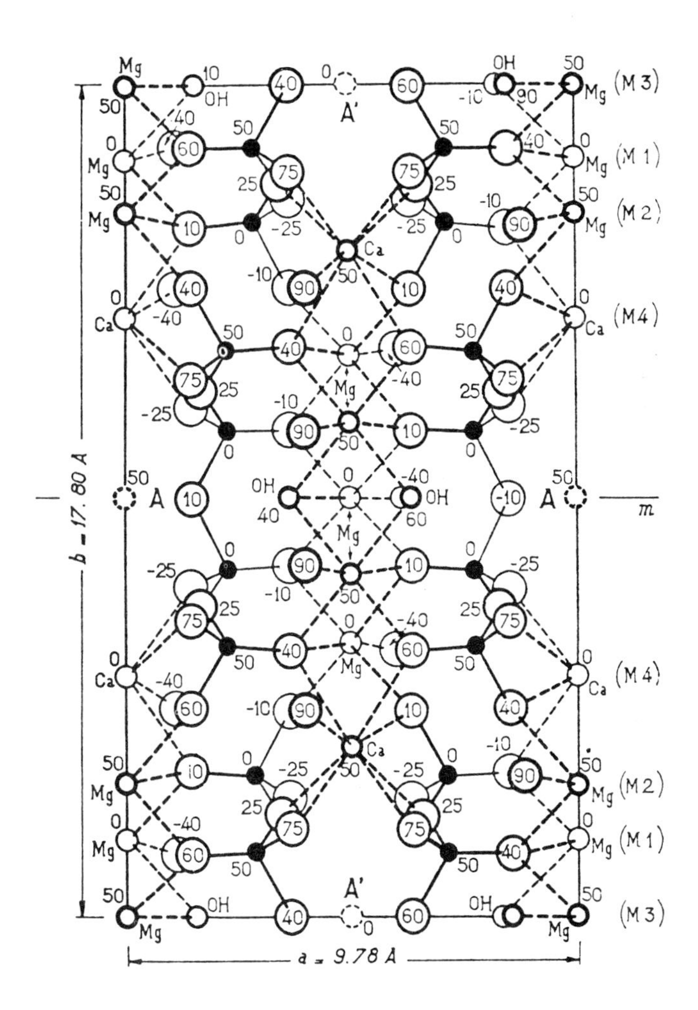

Fig. 5.15. The crystal structure of tremolite projected on (001) (after Warren, reproduced by permission of Akademische Verlagsgesellschaft, Frankfurt)

Amphiboles are thus closely related to pyroxenes which possess a simple chain structure. Hornblende: $(Si_3AlO_{11})_2\ Ca_2(Mg,Fe^{++})_3\ (Al,Fe^{+++},Ti)_2(OH)_2$, is the only amphibole whose plastic properties have been investigated; it is monoclinic and its crystallographic axes can be readily determined by optical measurements on the universal stage: X is close to $a = [100]$, Y coincides with $b = [010]$; the last axis $c = [001]$ is parallel to the length of prismatic crystals; in other cases it is determined as the intersection of the two main {110} cleavages. Fig. 5.16 shows the crystallographic habit of hornblende.

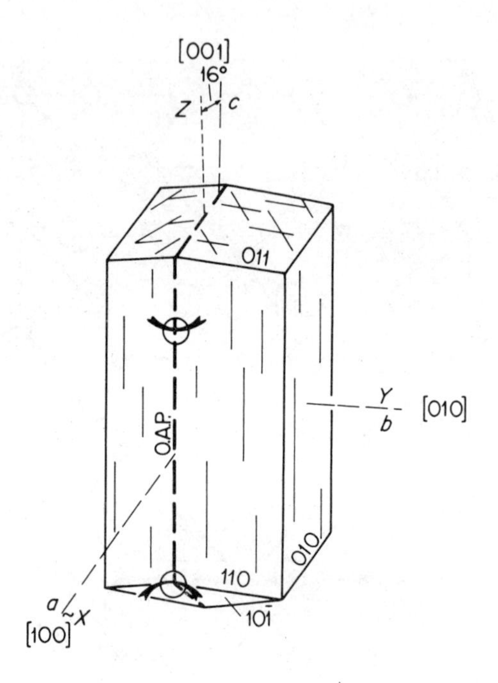

Fig. 5.16. Crystallographic habit and optical directions of hornblende

5.6.2. Slip systems

As for pyroxenes, simple inspection of the structure (Fig. 5.15) suggests that slip should occur preferentially on the (100) [001] system, which does not break any Si–O bonds.

Rooney and Riecker (1969; 1973) performed compression experiments on hornblende at room temperature, $\dot{\varepsilon} = 10^{-1}\,s^{-1}$ and 50 kB confining pressure and at 800 °C, $\dot{\varepsilon} = 10^{-5}\,s^{-1}$ and 10 kB confining pressure. Using Turner's analysis of KBB, they deduced that slip had occurred on the (100) [001] system. Buck (1970) did not succeed in activating this slip system in experiments conducted on single crystals at room temperature, $\dot{\varepsilon} = 10^{-4}\,s^{-1}$ and pressures below 20 kB, even though the resolved shear stress was very high. Dollinger and Blacic (1974), from evidence of external rotation axes, suggest that at T close to the breakdown of hornblende glide may occur on (100) in directions other than [001].

5.6.3. *Mechanical twinning*

The dominant mode of plastic deformation in amphibole is $(\bar{1}01)[\bar{1}0\bar{1}]$ mechanical twinning with a shear $S = 0{\cdot}53$ (Rooney and Riecker, 1969, 1973; Rooney et al., 1970; Buck, 1970). This twinning mode is operative at temperatures below 800 °C but requires confining pressures in excess of 10 kB at room temperature; it is accompanied by extensive parting on $(\bar{1}01)$ planes and by the presence of thin lamellae on (100) planes, which are interpreted as mechanical twins of a second system (100) [001].

Creep experiments on amphibolite (Borg and Handin, 1966; Rooney and Riecker, 1969, 1973) and on amphibole single crystals (Rooney and Riecker, 1969, 1973; Buck and Paulitsch, 1969; Buck, 1970) show that these are among some of the strongest silicate rocks and minerals: the main plastic deformation mechanism, $(\bar{1}01)[\bar{1}0\bar{1}]$ twinning, is not activated unless the resolved stress exceeds 2 to 4 kB at 400°–600 °C, $\dot{\varepsilon} = 10^{-5}\ s^{-1}$ and 5–15 kB confining pressure; if the crystals are not properly oriented so as to deform by twinning, they fail in a brittle fashion.

Above 800 °C a distinct weakening is observed, and the strength drops from more than 7 kB at 700 °C to almost zero near 1000 °C (Riecker and Rooney, 1969; Rooney and Riecker, 1973); this is attributed to pore pressure build-up accompanying the dehydration of amphibole which occurs between 925 and 1125 °C according to the chemical composition. Weakening is also reported in single crystals at low temperatures and confining pressures above 15 kB but no interpretation has yet been found for it.

5.7. MICAS

5.7.1. *Crystallography*

The two principal rock-forming micas are biotite: $K_2(Mg, Fe)_6[Si_6Al_2O_{20}]\cdot(OH)_4$ and muscovite: $K_2Al_4[Si_6Al_2O_{20}](OH)_4$. They are characterized by a layered structure with a monoclinic symmetry (Fig. 5.17). The lattice can be considered as pseudo-hexagonal, having as its 'basal' plane the plane of the layers (001) which corresponds to a perfect cleavage. The crystallographic habit exhibits a strong shape anisotropy with the shortest development normal to (001) (Fig. 5.18).

5.7.2. *Slip systems and kinking*

(*a*) *Crystallographic considerations*

The pseudo-hexagonal structure is formed by double layers of SiO_4 tetrahedra separated by layers of potassium ions parallel to (001). It is therefore to be expected that slip can easily occur on (001) planes, since only ionic bonds would be broken. The possible slip directions are the dense directions of the *K* layer: $[100]\langle 110\rangle$, corresponding to the $\langle 11\bar{2}0\rangle$ directions of the slightly distorted hexagonal lattice.

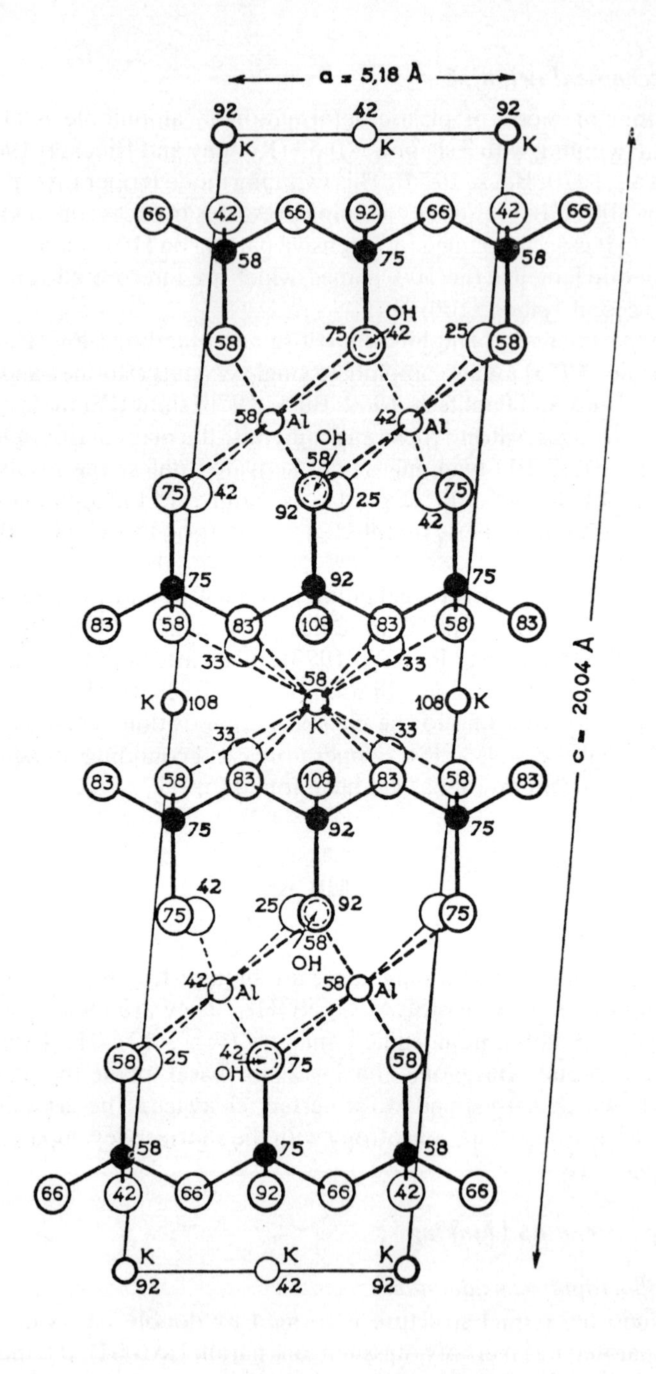

Fig. 5.17. The crystal structure of muscovite projected on (010) (after Jackson and West, 1930, reproduced by permission of Akademische Verlagsgesellschaft, Frankfurt)

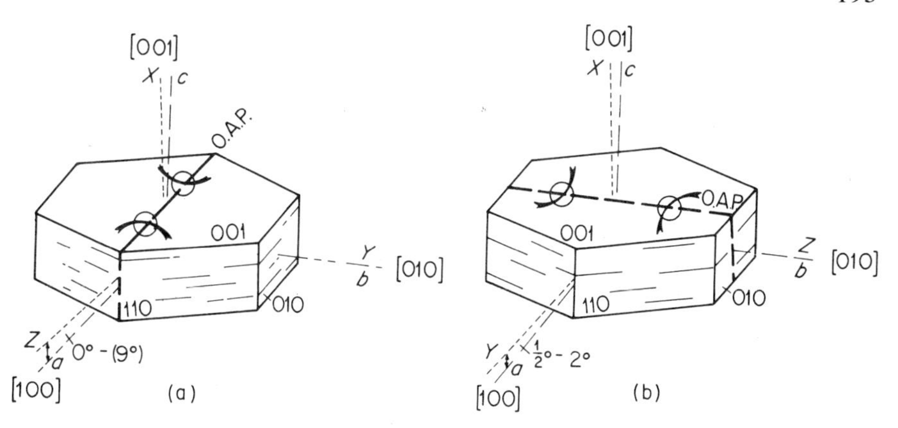

Fig. 5.18. Crystallographic habit and optical directions of micas:
(a) Biotite
(b) Muscovite
(after Tröger, reproduced by permission of E. Schweizerbart'sche, Stuttgart)

(b) Experimental evidence

Mügge (1898) observed 'bend', 'pressure' and 'percussion' figures in mica and concluded that the slip systems were (001) [100] and (001) ⟨110⟩ in the directions normal to the observed axes of bending. No slip plane other than (001) was reported in later investigations* (Borg and Handin, 1966; Hörz and Arhens, 1969; Hörz, 1970; Schneider, 1972; Etheridge et al., 1973). Slip directions, determined by the analysis of external rotation in kink bands, were found to be [100] (Borg and Handin, 1966) and [100] and ⟨110⟩ (Etheridge et al., 1973). Non-crystallographic slip reported by Hörz and Arhens and Schneider can be interpreted as composite slip along several close-packed directions.

Dislocations were directly observed in muscovite by TEM (Silk and Barnes, 1961; Amelinckx and Delavignette, 1962; Demny, 1963) and by X-ray topography (Caslavski and Vedam, 1970). They are not visibly dissociated into partials (Plate 5(a)). The Burgers vectors of dislocations lying in the basal plane (001) were found to be $\frac{1}{2}[110]$, $\frac{1}{2}[\bar{1}10]$ and [100].

Experimental deformation of biotite (Griggs et al., 1960) takes place mostly by bending and kinking with subordinate slip on (001). Slip alone has not been recorded. This may be due to the fact that the compressive stress was, as far as we know, always parallel to the basal plane of the compressed single crystals. However, in the deformation of biotite gneiss compressed with the foliation at an angle of about 45° from the compression axis, kinking is still predominant over slip (Borg and Handin, 1966).

The geometry of kinking in biotite and the way it can accommodate large strains is described at length by the above-mentioned authors (see also Starkey,

* The slip planes (135), (205), (104) at high angles to (001) inferred by Dana (1911) and still listed by Spry (1969) probably do not exist: kink band boundaries were, in all likelihood, mistaken for slip traces (Borg and Handin, 1966).

1968). The angles of rotation associated with kink band boundaries are quite large in biotite and range from 25° to 120° (Etheridge et al., 1973), this precludes any interpretion of the KBB as simple tilt walls formed by edge dislocations of the same sign.

5.7.3. Recrystallization

Etheridge and Hobbs (1974) annealed previously deformed phlogopite crystals at 10 kB. No recrystallization was observed below 1050 °C after 4 hours. The first sign of recovery was the migration along the (001) planes of the KBBs which assumed a serrated shape. The migration sometimes led to the disappearance of some kink bands; nucleation of new grains followed above 1050 °C. Similarly prepared samples of biotite and muscovite did not recrystallize after several hours at temperatures just below the breakdown point.

Etheridge and Hobbs reach the following conclusions:

Although nucleation is restricted to zones of high strain (especially in biotite crystals), it cannot be explained only by strain-induced processes such as the ones described in § 4.5.2; free energy differences resulting from small chemical fluctuations should also be taken into account.

There is some evidence that mobile serrated kink boundaries correspond to coincidence orientations (see § 4.5.3).

5.8. PLAGIOCLASES

5.8.1. Crystallography

The plagioclase feldspars ranging in composition from $Na[AlSi_3O_8]$ to $Ca[Al_2Si_2O_8]$ have a framework of linked (Si, Al)O_4 tetrahedra (Fig. 5.19).

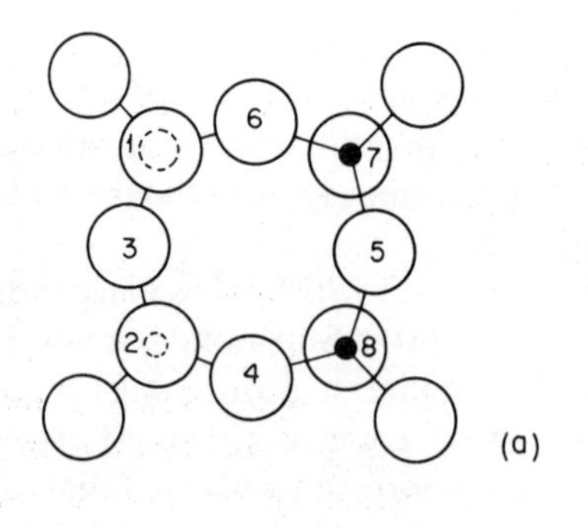

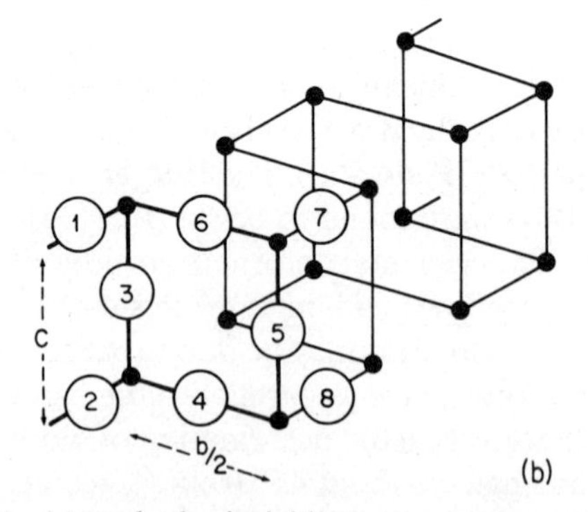

Fig. 5.19. Framework of linked tetrahedra in feldspars:
(a) Single link: eight oxygen atoms lie in the plane of the paper, two (Nos. 1 and 2) above and two (Nos. 7 and 8) below the plane
(b) Diagram of the chain, only the shared oxygens in the single link (a) are represented. The silicon and proxy-aluminium are shown in black
(after Hatch et al., 1972)

The lattice has a triclinic symmetry; this makes the crystallography quite complex, and we therefore refer the reader to recent books for details (Christie, 1962; Deer, Howie and Zussman, 1963; Marfunin, 1966).

The crystallographic habit and optical constants are summed up in Fig. 5.20.

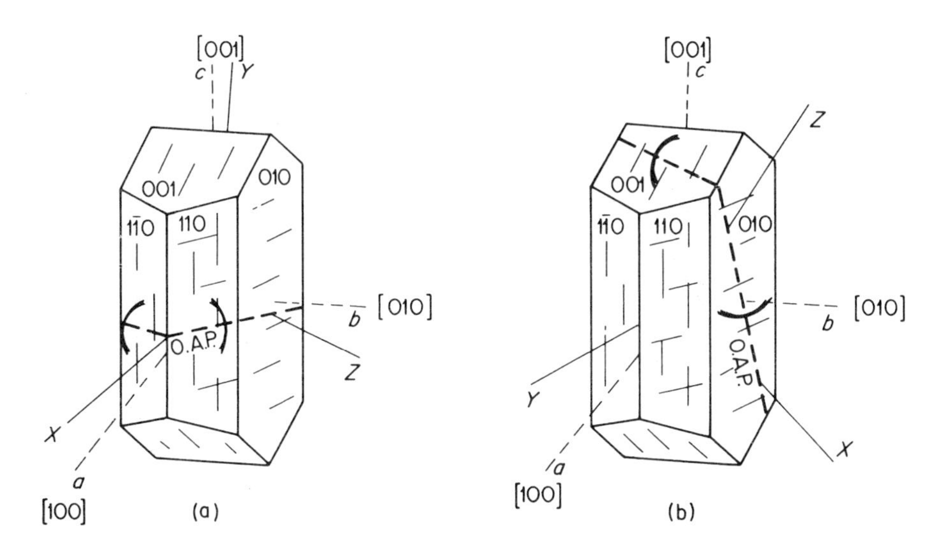

Fig. 5.20. Crystallographic habit and optical directions of:
(a) Low albite
(b) Anorthite
(after Tröger, reproduced by permission of E. Schweizerbart'sche, Stuttgart)

For cumulus plagioclases from stratiform complexes, Wager and Brown (1967) give the following $a:b:c$ aspect ratios:

An_{60}	1 : 0·2 : 1
An_{40}	1 : 0·3 : 0·5
Mafic plagioclases	1 : 0·3 : 1 to 1 : 0·08 : 1

5.8.2. Slip systems

Although feldspars are the most abundant minerals in crustal rocks, their plastic properties have been very little investigated; in particular the plasticity of alkali feldspars is completely unknown. This is why we will consider only plagioclases, which have been mainly studied by Borg and Heard (1969, 1970). These authors found that, at a given temperature (in the range 25–800 °C) and confining pressure (in the range of 5–10 kB), all members of the plagioclase series have similar plastic properties (Fig. 5.21). Slip systems have generally been inferred from the study of kink bands.

Seifert (1965) observed (010) slip in naturally deformed albite. Borg and Heard (1969, 1970) reported that at 800 °C $\dot{\varepsilon} \simeq 2.10^{-5}\,s^{-1}$ and about 10 kB confining pressure, plagioclases of practically all compositions deform by slip

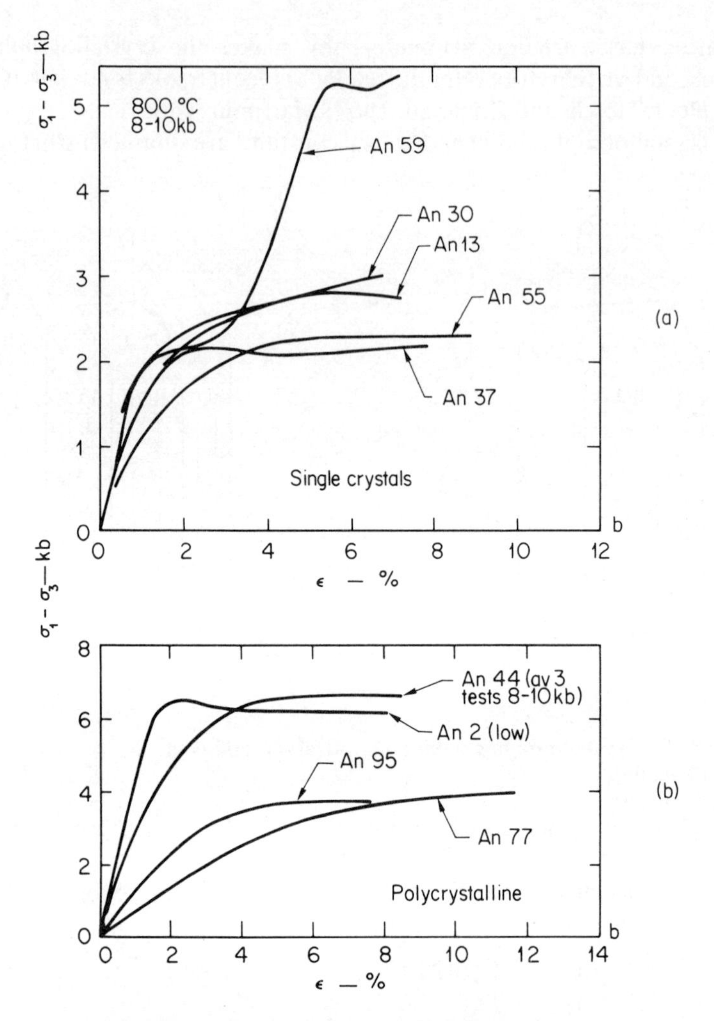

Fig. 5.21. Stress–strain curves for plagioclases:
(a) $T = 800\,°C$, confining pressure 8–10 kB, single crystals of oligoclase (An_{13}), andesines (An_{30} and An_{37}) and labradorites (An_{55} and An_{59}). The crystal An_{59} was disordered, all others were ordered
(b) $T = 800\,°C$, confining pressure 8–10 kB, Polycrystals of low albite (An_2), andesine (An_{44}), bytownite (An_{77}) and anorthite (An_{95})
(after Borg and Heard, 1970, reproduced by permission of Lawrence Livermore Laboratory, California)

and twinning. Under these conditions, (010) slip was specifically observed in single crystals of low albite and peristerite (An_{13}) unfavourably oriented for twinning; the slip direction in (010) was found to lie parallel to the irrational twinning direction associated with mechanical albite twinning. (001) slip was never observed although there was a high resolved shear stress on (001) in the direction [010].

5.8.3. Mechanical twinning

Reversible elastic twinning is well known in feldspar but the first evidence of permanent twinning was given by Mügge and Heide (1931) in anorthite. Borg and Handin (1966) and Borg and Heard (1969, 1970) showed that experimental deformation of plagioclases took place by permanent twinning along with slip.

The only known (but not necessarily the only existing) mechanical twinning systems operating during experimental deformation correspond to the albite and pericline law* (see Starkey, 1964) (Fig. 5.22) (Table 5.4).

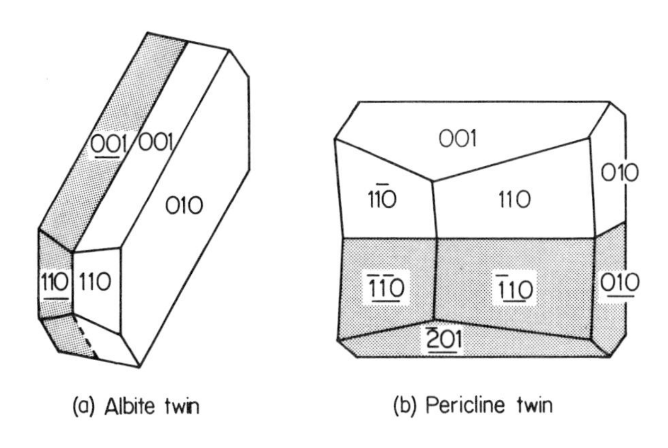

Fig. 5.22. Albite (a) and pericline (b) twins

Table 5.4

Twin law	Twin plane	Twinning direction	Shear (An_{55})
Albite (normal law)	$K_1 = (010)$	η_1 = irrational (see Fig. 5.23)	$S = 7° 50'$
Pericline (parallel law)	K_1 = irrational, near (001)	$\eta_1 = [010]$	$S = 7° 50'$

Borg and Heard observed both albite and pericline twinning for all plagioclases in the range An_{30}–An_{95} but not for An_{44}.

* Mechanical twins have the same geometry as growth twins. They can be distinguished by using the criteria given in the case of diopside (§ 5.5).

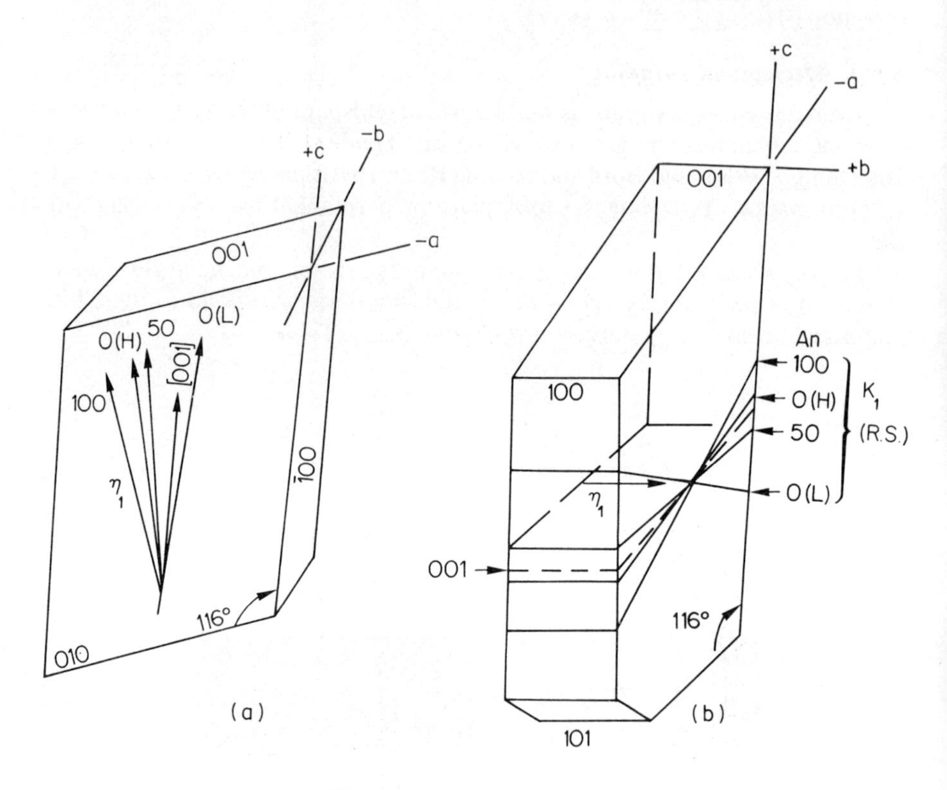

Fig. 5.23. (a) Orientation of η_1 and (+) positive sense of twinning in albite twinning. Arrows = η_1 of plagioclases of different An content. (010) = K_1 twinning plane

(b) Orientation of $\eta_1 = [010]$ and (+) sense of twinning in pericline twinning. Position of twinning plane K_1 indicated for plagioclases of different An content. K_1 = rhombic section (RS)

5.9. QUARTZ

5.9.1. Crystallography

Quartz, SiO_2, has a framework structure of SiO_4 tetrahedra. It can exist in several allotropic forms according to the (P, T) domain considered (Fig. 5.24).

(i) α quartz (low quartz) crystallizes in the trigonal system (Fig. 5.25) with an enantiomorphous symmetry (right- or left-handedness) corresponding to the handedness of helical chains of SiO_4 tetrahedra along the 3 fold axis. It is the quartz variety commonly found in the crust. The typical crystal shape is represented on Fig. 5.26.

(ii) β quartz (high quartz) is the form stable at high temperature (above 573 °C at atmospheric pressure). It is only found in geological domains where the geotherm was steep enough for the temperature to be higher than 650 °C at a depth of 10 km and higher than 800 °C at 30 km (Fig. 5.24). β quartz has a hexagonal symmetry (Fig. 5.27); the inversion from α to β on heating is a displacive transformation corresponding to minor atomic movements without any rupture of bonds. By slight rotations of the tetrahedra, the a axes of α quartz lose their polarity and the 3-fold axis becomes a 6-fold axis. The handedness of α quartz is preserved in β quartz.

(iii) Coesite is the high-pressure form of SiO_2. It has a monoclinic symmetry. The inversion from α quartz to coesite is not displacive and involves atomic diffusion.

5.9.2. Slip systems

(a) Experimental evidence

Early studies on experimental deformation of quartz and quartz-bearing rocks failed to produce definite plastic flow features and resulted in recrystallization (Griggs, 1940; Griggs et al., 1960). The first unequivocal evidence of extensive plastic deformation of quartz was obtained by Carter et al. in 1961. Slip systems were soon identified (Carter et al., 1964; Christie and Green, 1964; Griggs and Blacic, 1964) and detailed investigations followed (Baëta and Ashbee, 1967, 1969, 1970; Heard and Carter, 1968; Tullis, 1970; Blacic, 1971; Tullis and Tullis, 1972; Avé Lallemant and Carter, 1971; Hobbs et al., 1972).

Slip in quartz requires breaking the strong covalent Si–O bonds and should be very difficult. Indeed, dry quartz deformed at 500 °C exhibits a strength approaching the theoretical one (Griggs, 1967) and deforms only by twinning; no dislocations are observed in TEM (McLaren et al., 1967). However, slip becomes possible when traces of water are present in the crystal. This also causes 'hydrolytic weakening' at high temperatures (see § 5.9.4).

Plastic flow by slip is initiated, under atmospheric pressure, at temperatures above 600 °C ($\dot{\varepsilon} = 10^{-4}\ s^{-1}$). The easiest slip system is $(0001)\langle 11\bar{2}0\rangle$, (*Basal slip* in the **a** directions). At temperatures of about 700 °C *Prismatic slip* becomes equally easy:

$\{10\bar{1}0\}[0001]$ 1st order prism in the c direction
$\{10\bar{1}0\}\langle 1\bar{2}10\rangle$ 1st order prism plane in the a direction
$\{10\bar{1}0\}\langle 1\bar{2}13\rangle$ 1st order prism plane in the $(a+c)$ directions

Blacic (1971) reports that in dry quartz, $(0001)\langle 11\bar{2}0\rangle$ slip is the most active below 750 °C ($\dot{\varepsilon} = 10^{-5}\ s^{-1}$) and $\{10\bar{1}0\}[0001]$ the most active above 750 °C; in wet quartz above the weakening temperature, $\{10\bar{1}0\}[0001]$ tends to be supplanted by $\{11\bar{2}0\}[0001]$ and basal slip becomes important again.

Prismatic slip can be accompanied by slip on low-index pyramidal planes. The same slip systems have been identified in β quartz by Baëta and Ashbee (1969).

The principal slip systems are given in Table 5.5.

Table 5.5. Principal slip systems in α and β quartz (+: common, −: uncommon)

Basal	$(0001)\langle 11\bar{2}0\rangle^+$	Low T, high $\dot{\varepsilon}$
1st ord. prismatic	$\{10\bar{1}0\}[0001]^+$	High T, low $\dot{\varepsilon}$
	$\{10\bar{1}0\}\langle 1\bar{2}10\rangle^+$	High T, low $\dot{\varepsilon}$
	$\{10\bar{1}0\}\langle 1\bar{2}13\rangle^-$	High T, low $\dot{\varepsilon}$
2nd ord. prismatic	$\{11\bar{2}0\}[0001]^-$	High T, low $\dot{\varepsilon}$
2nd ord. pyramidal	$\{11\bar{2}2\}\langle 11\bar{2}3\rangle^-$	High T, low $\dot{\varepsilon}$

Slip systems have been mostly identified by the observation of slip lines on the surface of deformed samples, of KBB planes in thin sections and of deformation lamellae, remarkably abundant both in experimentally and naturally deformed specimens. (The use of deformation lamellae to determine slip systems raises a number of problems which will be briefly reviewed in the next paragraph.) Direct observation of dislocations by TEM (Plate 6(a)) has been very helpful in ascertaining the slip systems (McLaren and Phakey, 1965; McLaren et al., 1967, 1970; Baëta and Ashbee, 1968; McLaren and Retchford, 1969; McLaren and Hobbs, 1972); dislocations with *a* and *c* Burgers vectors have been positively identified (Ardell et al., 1974).

(*b*) *Deformation lamellae*

Deformation lamellae (defined in Chapter 9) were first investigated by Christie et al. (1964) and Carter et al. (1964). Avé Lallemant and Carter (1971) classified the lamellae in a (P, T) diagram according to their orientation with respect to basal and prism planes (Fig. 5.28): The plane of most natural lamellae makes an angle of 16°-30° with (0001) (sub-basal I type). The lamellae tend to be parallel to the basal plane (Plate 3(b)) at lower temperatures and faster strain rates and parallel to prism planes at higher temperatures.

The influence of strain on the lamellae orientations has also been investigated (Tullis, 1971).

The structure of deformation lamellae has been investigated by TEM (McLaren et al., 1967, 1970; McLaren and Hobbs, 1972; White, 1973b). It now seems clear that identical optical features characteristic of deformation lamellae may arise from several physical processes leading to widely differing dislocation structures. Some lamellae parallel to active slip planes probably correspond to dislocation arrays in the slip planes, while others have been interpreted as thin Brazil twins (McLaren et al., 1967); in many cases the lamellae can be associated with arrays of elongate subgrains and tangled sub-boundaries in whose formation recovery has clearly taken part (McLaren and Hobbs, 1972; White, 1973b).

Christie and Ardell (1974) found that many different submicroscopic structures may give rise to optical features classified as deformation lamellae. An important consequence of these observations is that optically visible lamellae

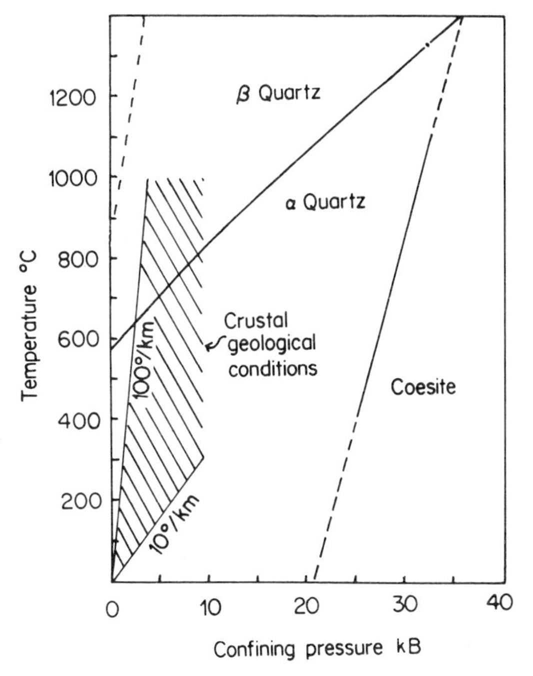

Fig. 5.24. (P, T) diagram of phase transformations of quartz

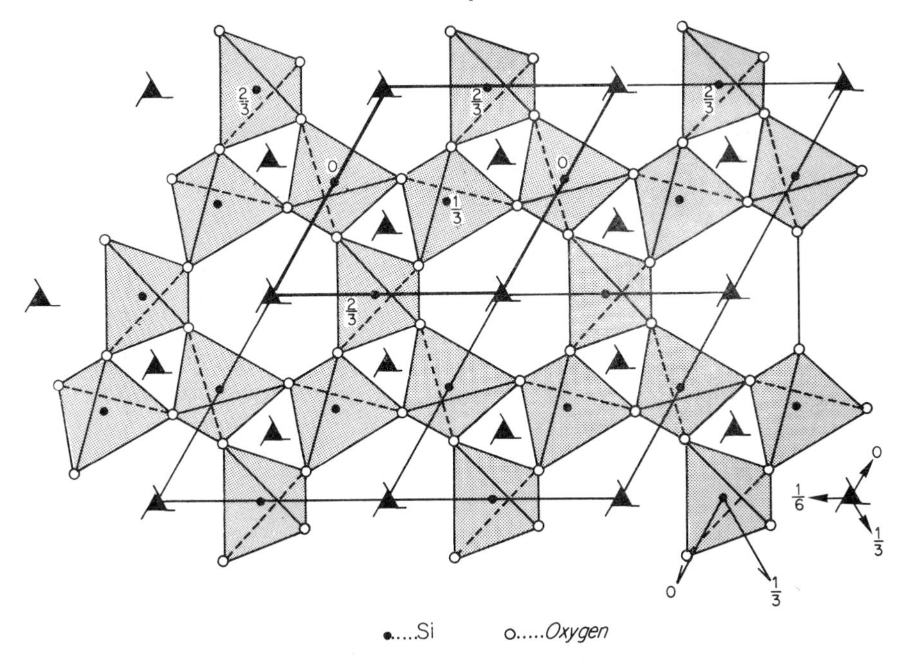

Fig. 5.25. Crystal structure of α-quartz projected on (0001) (after Wei, 1935, reproduced by permission of Akademische Verlagsgesellschaft, Frankfurt)

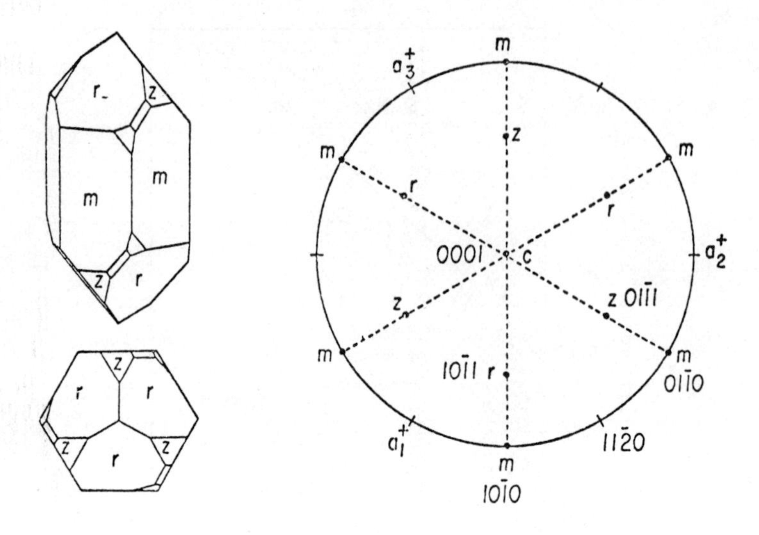

Fig. 5.26. Crystal faces of α-quartz. 0^+ orientation, often used in experiments, is such that the shear stress coefficient resolved parallel to a in the basal plane is equal to $0{\cdot}5$

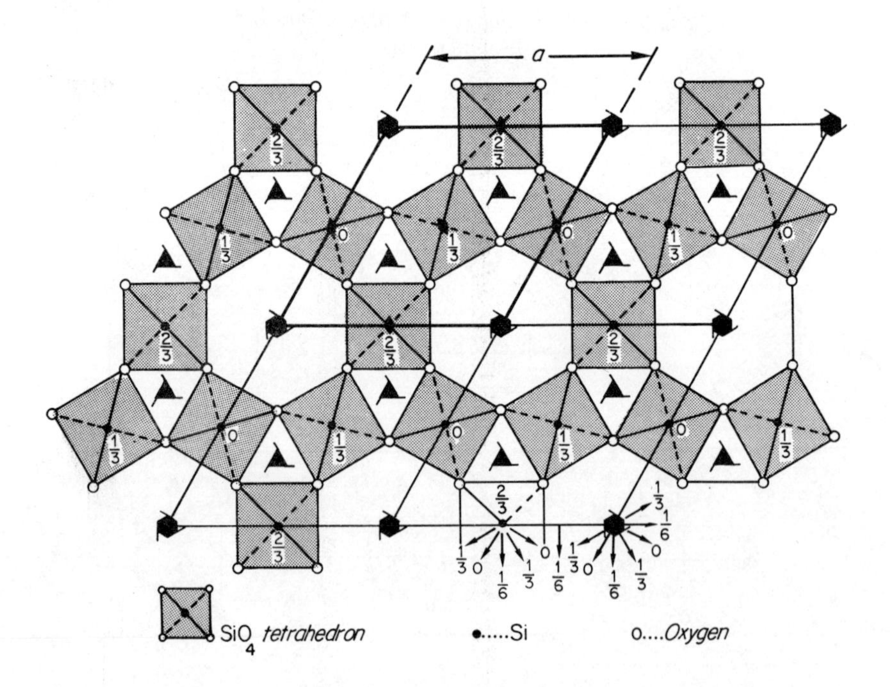

Fig. 5.27. Crystal structure of β-quartz projected on (0001) (after Deer, Howie and Zussman, 1963, reproduced by permission of the Longman Group Ltd.)

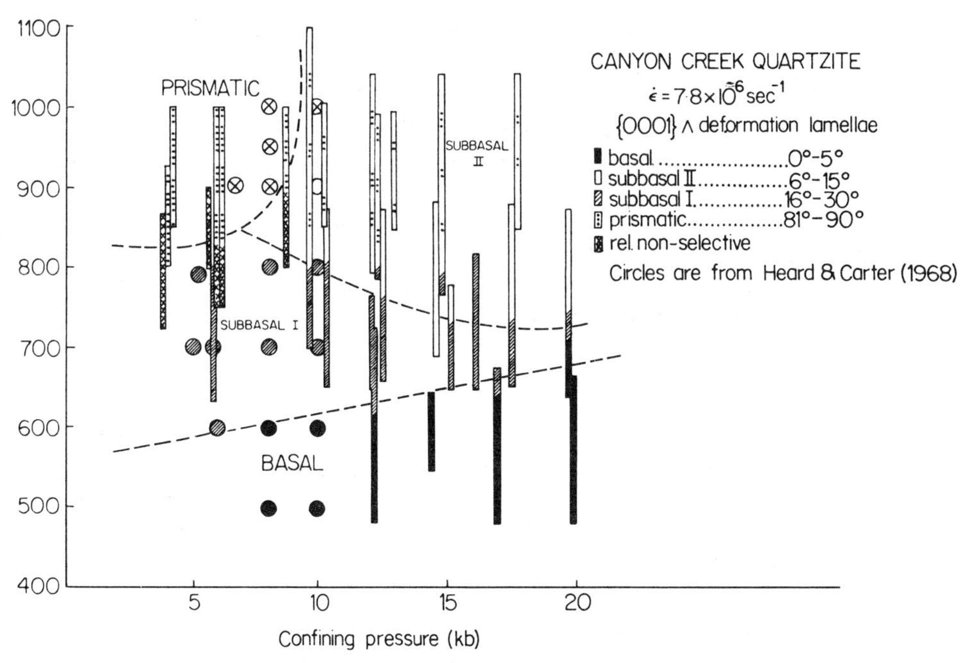

Fig. 5.28. Orientation of deformation lamellae (angle with (0001)) according to temperature and confining pressure for an experimentally deformed quartzite. (Reproduced by permission of The American Journal of Science, from Avé Lallemant and Carter, *Amer. J. Sci.*, **270**, 218, Fig. 7 (1971))

do not necessarily define active slip planes and must not be used indiscriminately as slip indicators.

5.9.3. Mechanical twinning

(a) Dauphiné twinning

Dauphiné twins are characterized by a 180° rotation about the c-axis (Fig. 5.29(a)); the crystal axes remain parallel but the polarity of the a-axis is

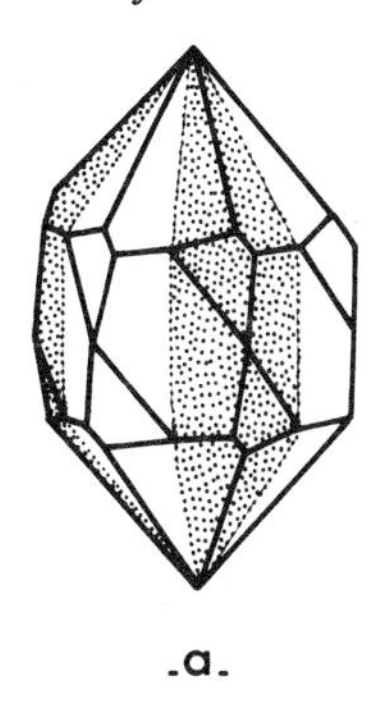

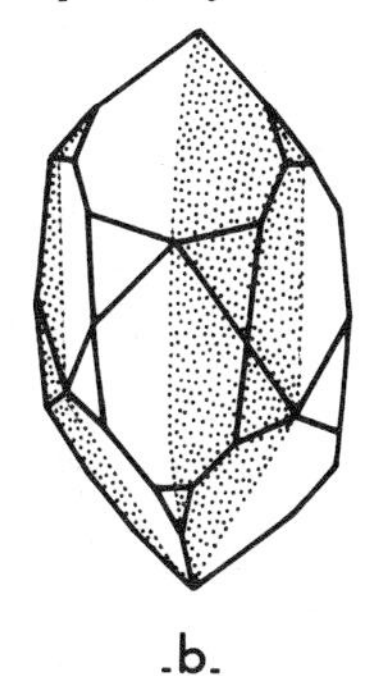

Fig. 5.29. (a) Dauphiné twin, reverses the polarity of the a-axes (b) Brazil twin, reverses the handedness of the crystal. Twinning planes are stippled

reversed; the composition planes are parallel to $(10\bar{1}1)$ and $(10\bar{1}0)$. Twinning of this type is common in natural quartz but cannot be detected by optical means. It disappears on inversion to β quartz, since the twin operation is then included in the point group symmetry of high quartz.

Mechanical Dauphiné twinning was investigated for its bearing on quartz-preferred orientation (see Chapter 6) by Tullis (1970, 1971) and Tullis and Tullis (1972). In axially stressed single crystals, twinning brings the direction of greater compliance (normal to r) parallel to the stress axis: a single crystal compressed normal to z completely changes its orientation, whereas it is unchanged if compressed normal to r (Tullis, 1971).

(b) Brazil twinning

Brazil twinning changes the handedness of the crystal by mirror reflection in a $\{11\bar{2}0\}$ prism plane; the c-axis orientation remains unchanged (Fig. 5.29(b)).

Brazil twins were observed by TEM in single crystals of natural quartz deformed at a relatively low temperature (500 °C) and fast strain rates, with a high resolved shear stress on (0001); they may look like lamellae in optical microscopy (McLaren et al., 1967).

5.9.4. Deformation mechanisms of wet quartz, hydrolytic weakening

In 1965 Griggs and Blacic discovered that a synthetic quartz crystal containing water molecules ($8{\cdot}8\times10^{-3}$ H/Si) could be deformed above a critical temperature of 380 °C, under stresses 10 to 20 times lower than a natural 'dry' crystal (Fig. 5.30). This anomalous weakness (Griggs and Blacic, 1964), was soon termed *hydrolytic weakening* (Griggs, 1967) and attributed to the enhanced mobility of dislocations by hydrolysis of the Si–O bonds. It then became evident that no experimental study of the deformation mechanisms of quartz could be made without some knowledge of the specimen's water content (determined by infrared absorption and expressed as the ratio of the number of hydrogen atoms to the number of silicon atoms). Indeed, the mechanism of hydrolytic weakening probably conditions all glide deformation mechanisms at high and low temperatures as long as the crystals contain some water. Griggs (1967) considers a crystal containing 0·0015 wt % H_2O (10^{-4} H/Si) as 'dry'.

It was predicted that hydrolytic weakening could be a controlling mechanism in the deformation of all silicates, and this was verified by Blacic (1971) in the case of olivine. In view of the importance of the problem of the plastic deformation of silicates in the deep crust and upper mantle, it does not seem out of place to review at some length the evolution of the ideas on the mechanisms of deformation of 'wet' quartz in light of recent well-documented compression experiments as well as TEM observations.

The first experiments on water weakening are summed up by Griggs (1967). They show that water weakening is reversible; a crystal first heated at 600° and then cooled, deforms only for a high normal stress; this proves conclusively that the weakening is not due to H_2O molecules migrating to some structural sites where they would cause intrinsic weakness; but rather that the deformation in

PLATE 1

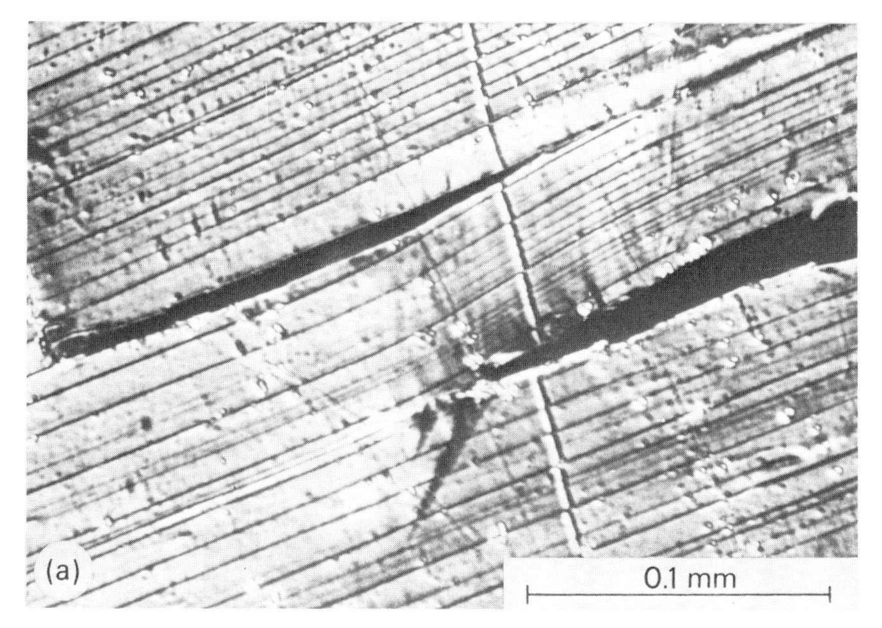

(a) Basal cleavage cracks in a beryllium single crystal. Note the curvature of basal planes between the cracks. (Photograph by J. M. Dupouy)

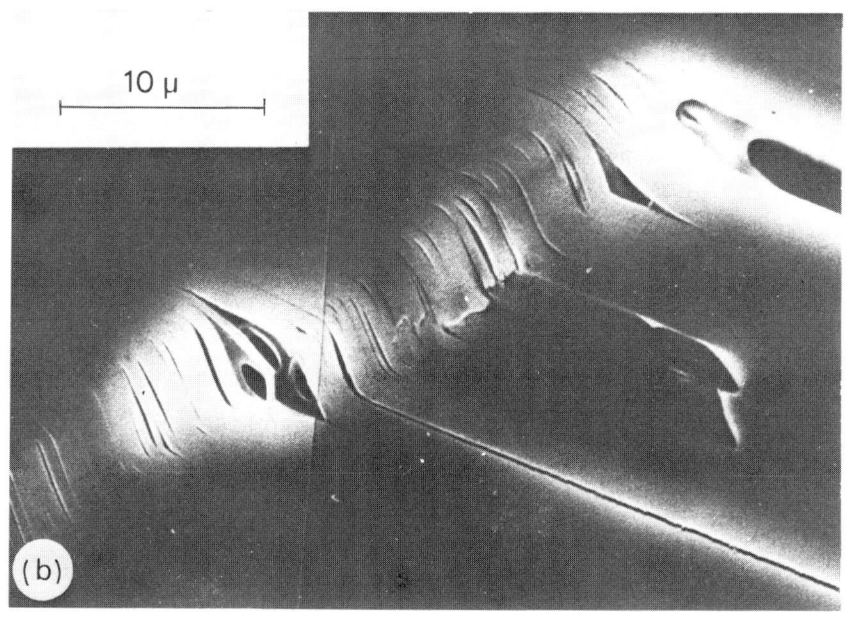

(b) Basal cleavage cracks and kink bands in biotite. Westerly granite deformed at $T = 25\,°C$, $P = 35$ bar, $\sigma = 3{\cdot}5$ kbar. (Photograph by P. Taponnier and W. F. Brace)

[*facing page 206*

PLATE 2

(a) Deformed diopside crystal in a peridotite mylonite (Lanzo massif). Exsolution lamellae underline the curved planes. Note the concentration of curvature into kink bands.

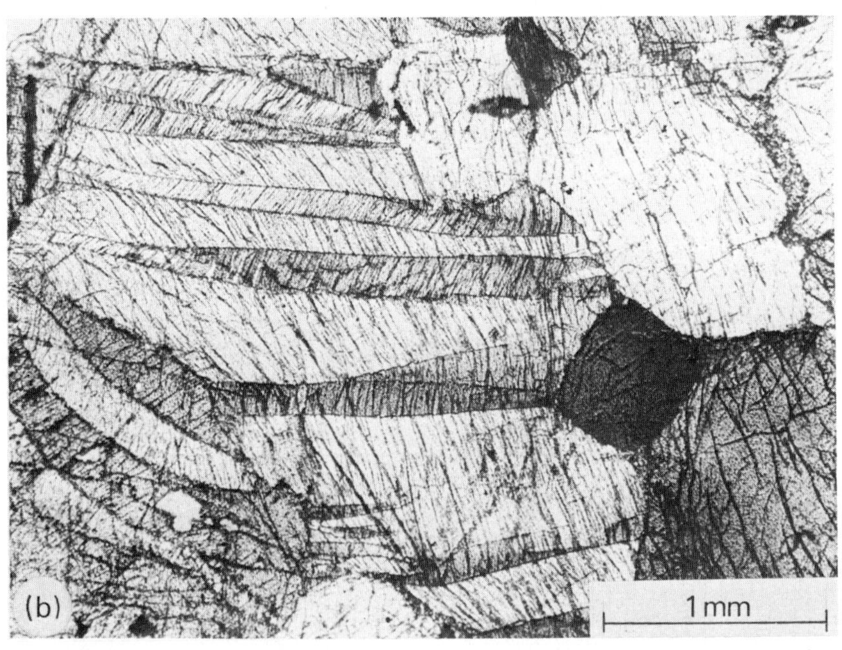

(b) Kinked enstatite crystal in a naturally deformed pyroxenite (Balmuccia massif). The kink bands are horizontal. Exsolution lamellae change orientations at kink-band boundaries.

PLATE 3

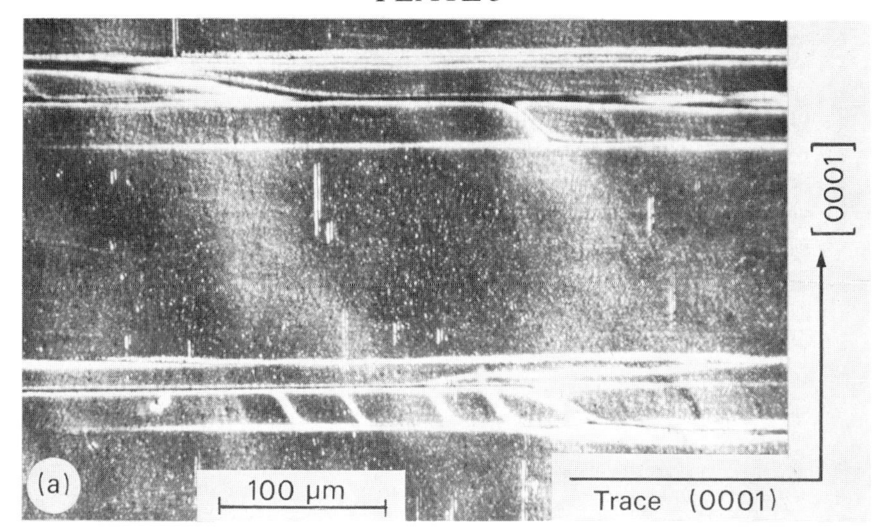

(a) Cross-slip in a beryllium single crystal. Horizontal lines are basal slip lines on the surface of the sample. They are connected by slip lines, traces of cross-slip planes on the surface. (Photograph by R. Le Hazif)

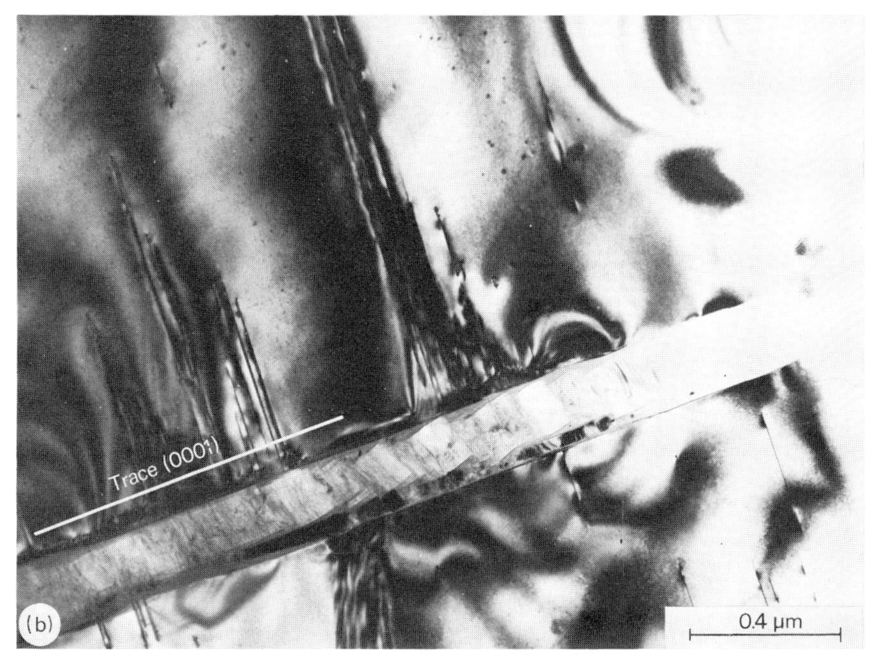

(b) Basal deformation lamella in a single crystal of Brazilian quartz. Compressed $\perp$ $(10\bar{1}1)$ at $T = 750\,°C$, $P = 15$ kbar, $\dot{\varepsilon} = 10^{-6}\,s^{-1}$, $\varepsilon = 3\,\%$. (TEM)
Note the rotated fragments of quartz in glassy phase. (Photograph by A. J. Ardell)

PLATE 4

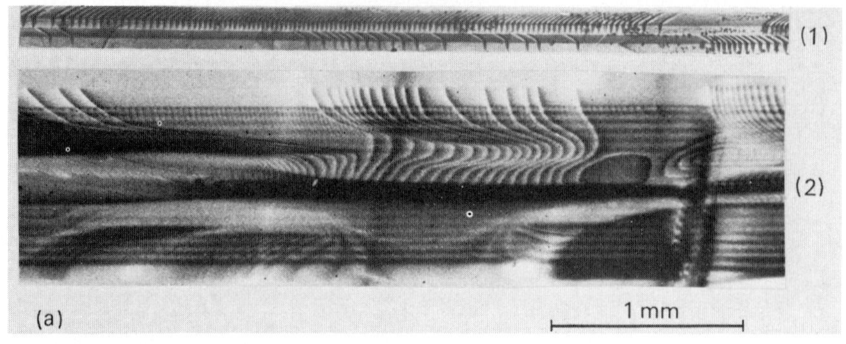

(a) Twinning dislocations in calcite
(1) Emergence points of twinning dislocations shown by etch-pits along the twin boundary (horizontal line viewed from above the sample).
(2) Lang projection topogram showing twinning dislocations in the twin boundary (plane of the figure). The dislocations are bowed out in the sense of propagation of the twin. (Courtesy M. Sauvage and A. Authier and *Bulletin Soc. Min. Fr.*)

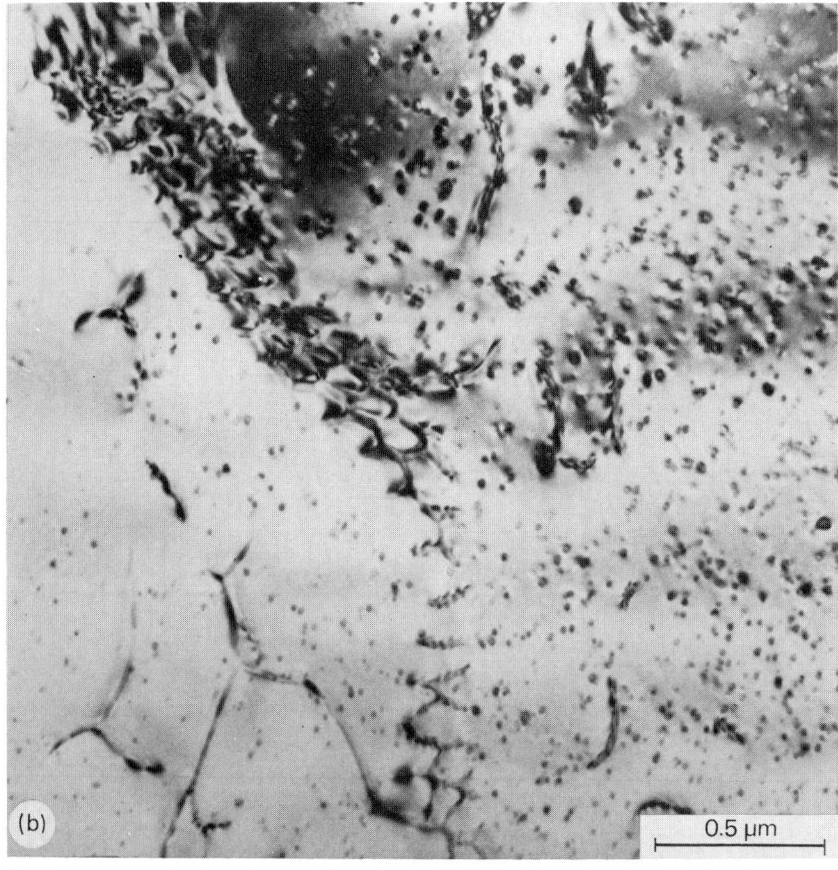

(b) Electron-irradiation damage in quartz (TEM). After 2-h irradiation by 100 kV electrons, nuclei of glassy phase (black dots) are created. Total vitrification may eventually occur. (Photograph by A. J. Ardell)

PLATE 5

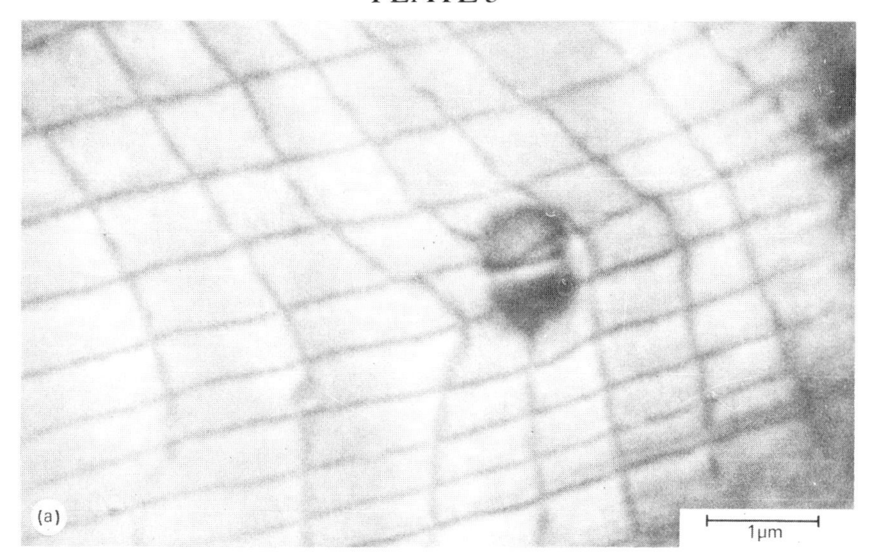

(a) Network of dislocations in the basal plane of muscovite (TEM). The dislocations are not visibly dissociated. They interact elastically with a bubble (round black feature). (Photograph by P. Delavignette)

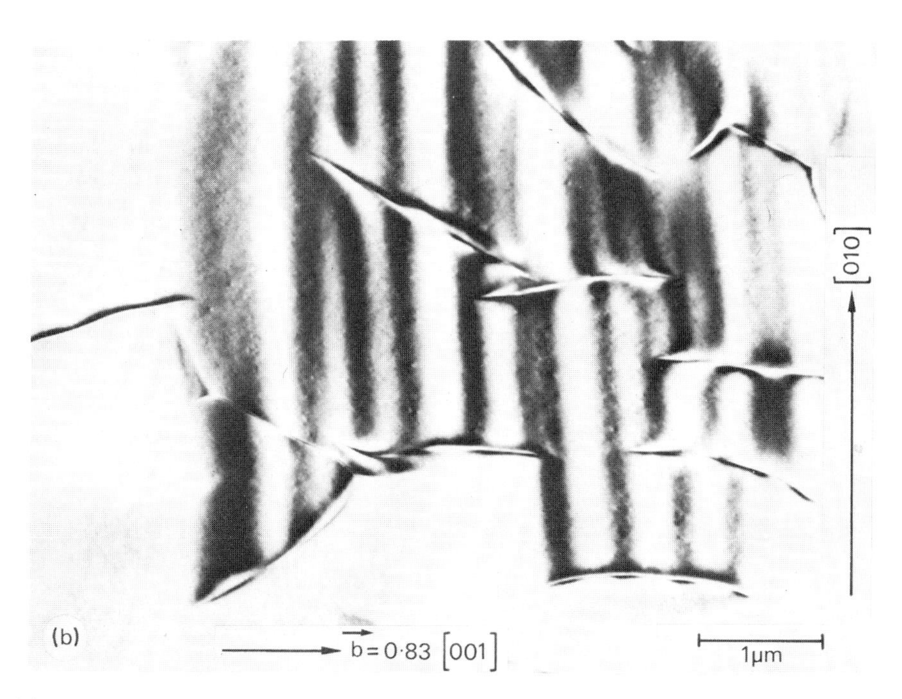

(b) Ortho-clino-enstatite inversion. Orthobronzite deformed at $T = 800\,°C$, $\dot{\varepsilon} = 10^{-4}\,s^{-1}$, $P = 15$ kbar. TEM shows stacking faults parallel to (100) (fringed contrast). Unit dislocations dissociate following:

$[001] \rightarrow 0{\cdot}83[001] + 0{\cdot}17[001]$

$\mathbf{b} = 0{\cdot}83[001]$ transforms ortho to clino-enstatite. (Photograph by S. H. Kirby)

PLATE 6

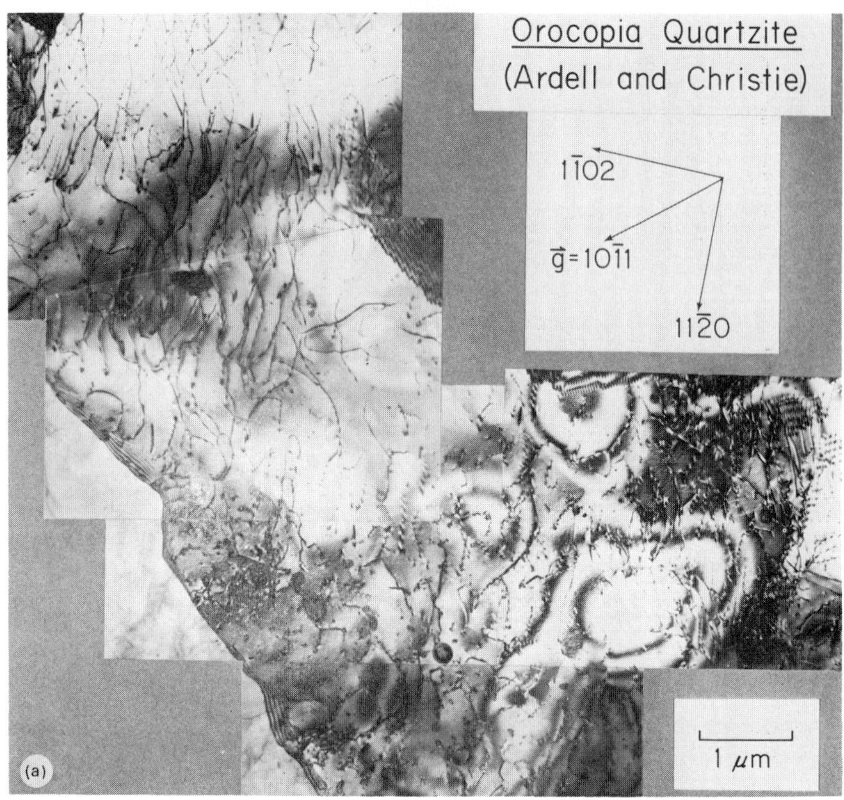

(a) Naturally deformed Orocopia quartzite (TEM). (Photograph by A. J. Ardell and J. M. Christie)

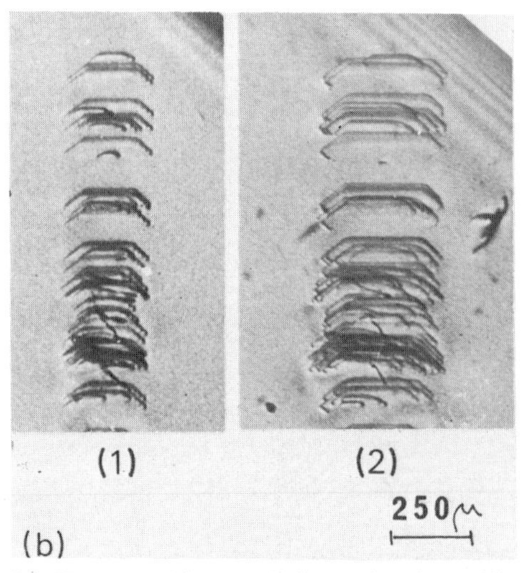

(1) The loops are nucleated at a scratch on the surface and correspond to (111) [1$\bar{1}$0] slip system. The open extremities of the half hexagons are pinned at the surface. The segment parallel to the surface is purely screw. Note the straight dislocations segments due to a strong Peierls force.

(2) After a short stress pulse, the dislocation loops have expanded.

(b) Hexagonal loops of dislocation in a silicon single crystal (Lang projection topogram). (Photograph by A. George)

PLATE 7

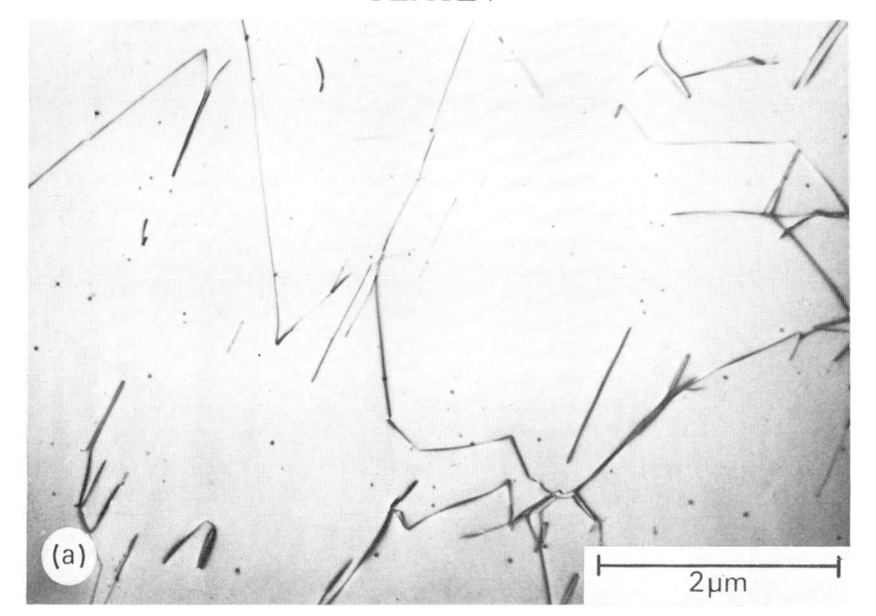

(a) Easy basal glide in a beryllium single crystal (TEM). Formation of debris and edge dipoles. The plane of the foil is (0001).

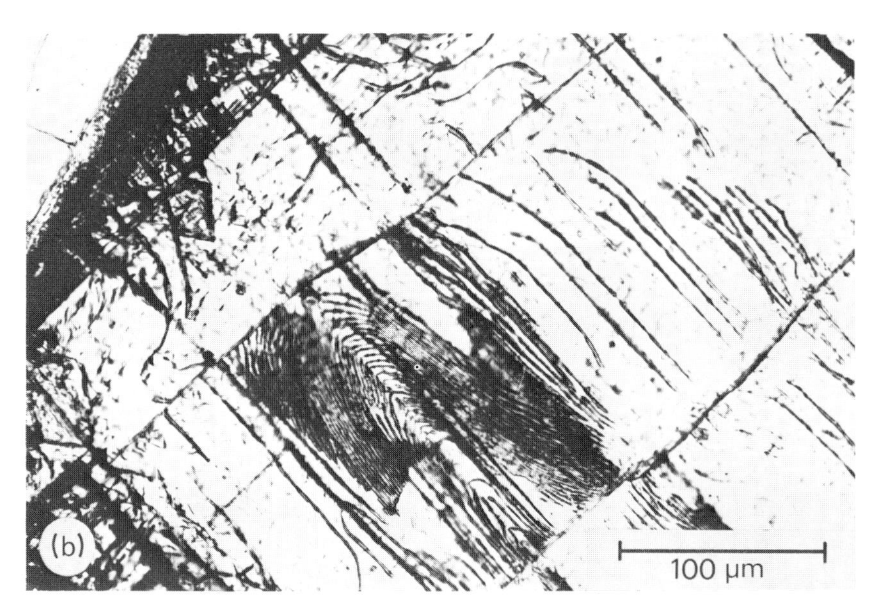

(b) Naturally decorated dislocations in olivine grains of a troctolite xenolith from Hawaii. (Optical micrograph.) The dislocations are decorated by iron oxides. Note the straight screw dislocations between walls and a procession of bowed out dislocations in a glide plane.

PLATE 8

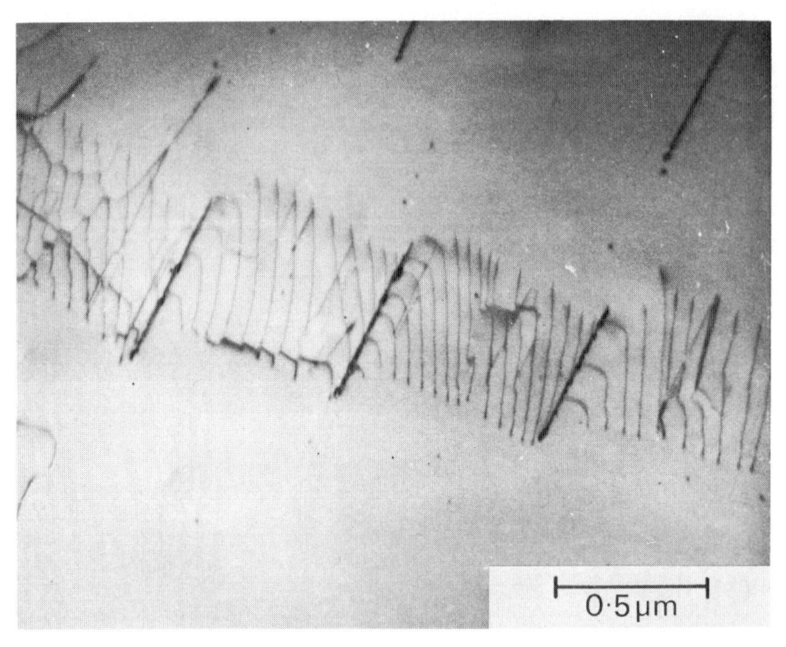

Tilt dislocation wall in beryllium formed by a edge dislocations, **c** dislocations in basal planes are visible as bold straight lines (TEM). (Photograph by R. Le Hazif)

PLATE 9

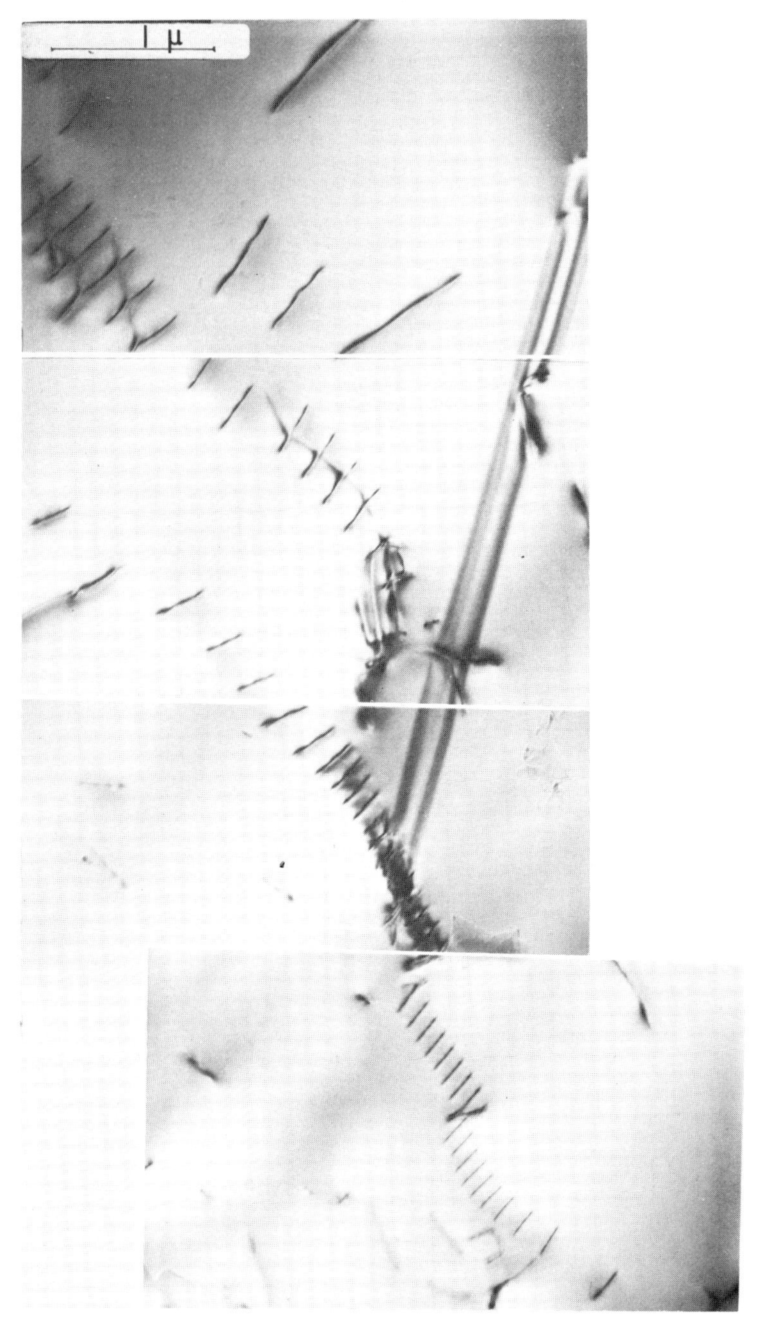

Tilt dislocation wall in an enstatite grain of a peridotite nodule from kimberlites, equilibrated between 1200 °C and 1400 °C (TEM). Note stacking fault ribbons with fringed contrast. (Photograph by Y. Gueguen)

PLATE 10

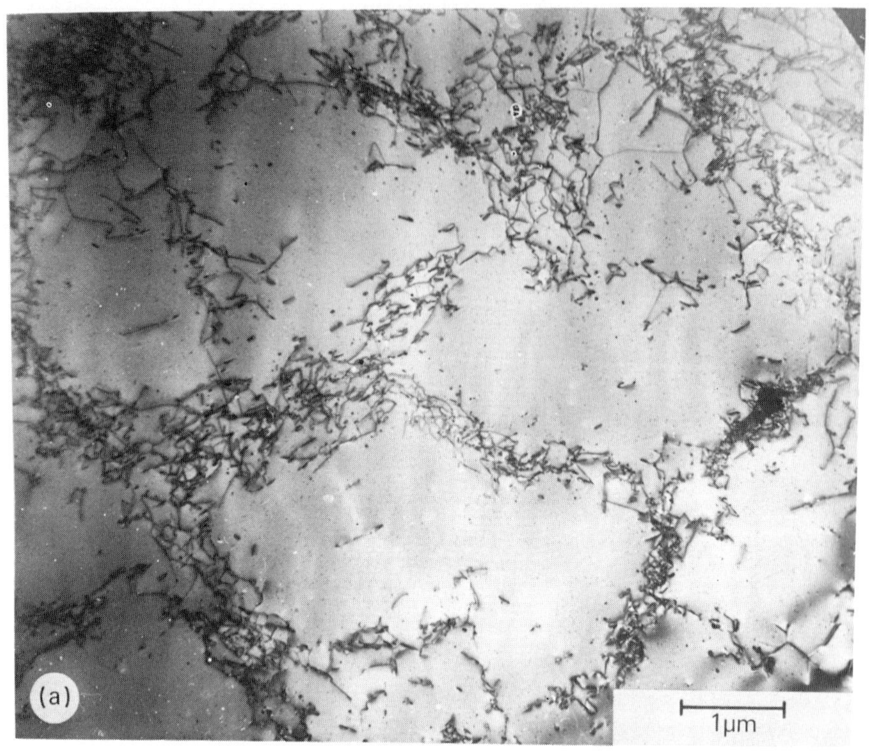

(a) Cellular dislocation structure in beryllium deformed at 175 °C (cold work) (TEM). The plane of the foil is (0001). (Photograph by J. Antolin)

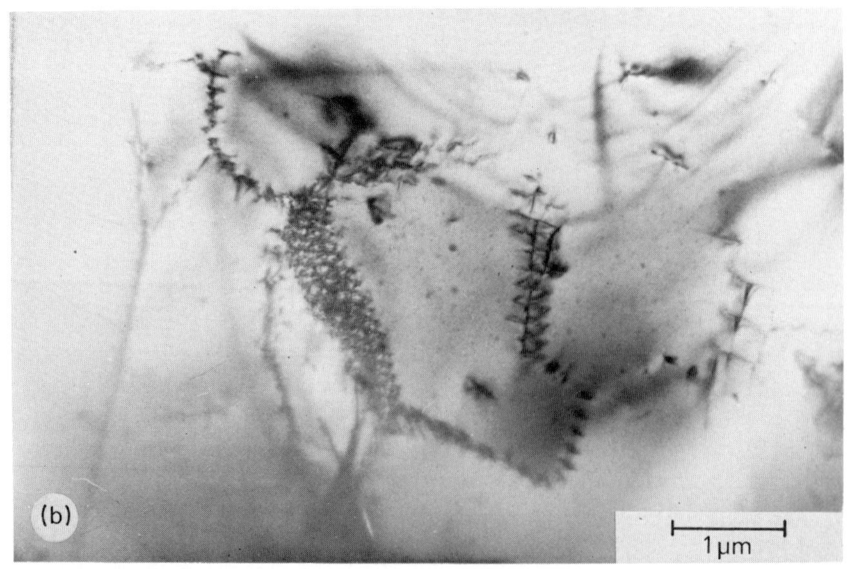

(b) Dislocation wall enclosing a subgrain in an orthoclase crystal. Granite naturally deformed ($T \simeq 500$ °C, $P \simeq 2$ kbar) (TEM). (Photograph by M. Gandais, C. Guillemin, C. Willaime)

PLATE 11

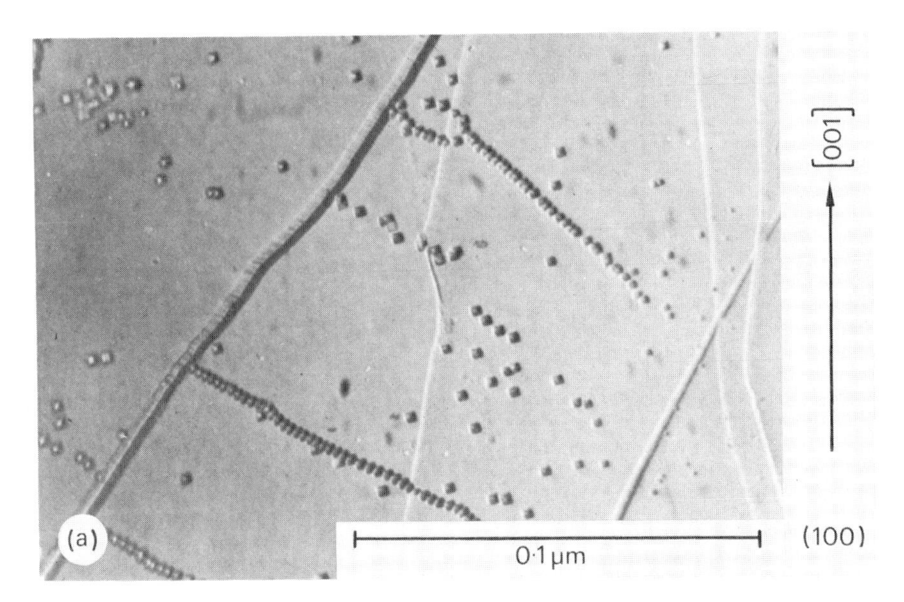

(a) Tilt subboundaries revealed by etch-pits on the (100) cleaved surface of halite single crystal. Creep deformation at $T = 780\,°C$, $\sigma = 35\ g/mm^2$, $\varepsilon = 24\ \%$.

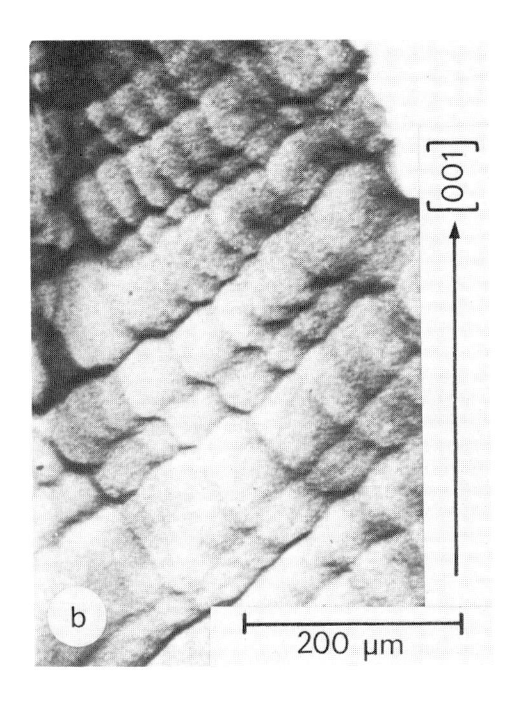

(b) Subgrains revealed by Berg-Barrett topogram in a spinel single crystal deformed by creep at $T = 1450\,°C$, $\sigma = 5\ kg/mm^2$, $\varepsilon = 7\ \%$ Surface (100). (Photograph by R. Duclos)

PLATE 12

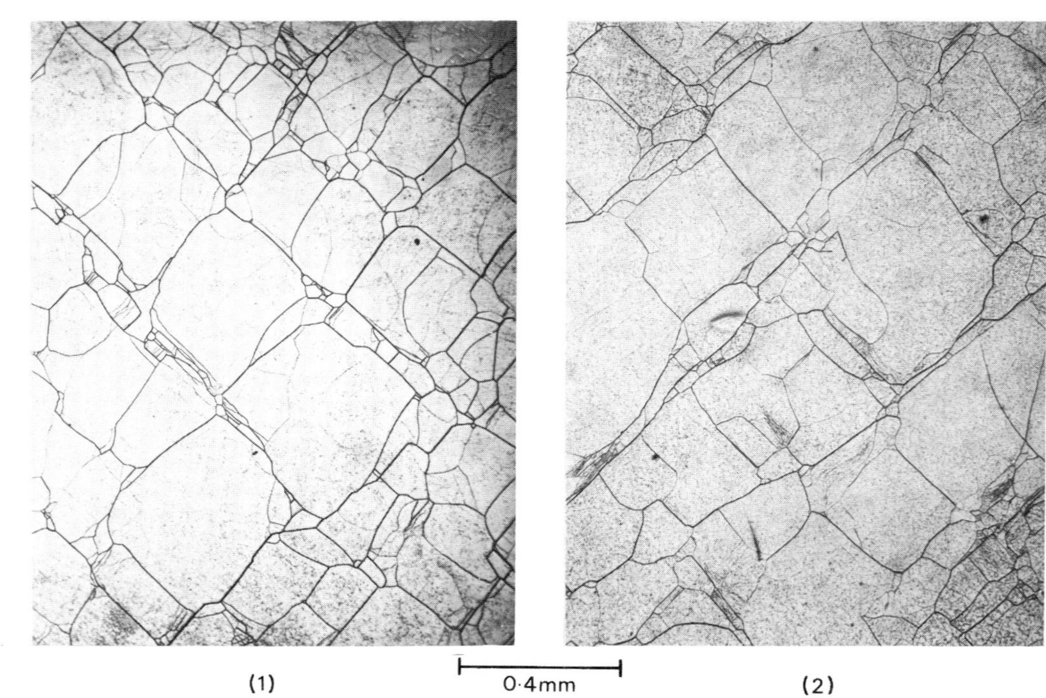

Subgrains revealed by photolytic silver decoration in a silver chloride single crystal deformed by creep at $T = 308\ °C$:

(1) Sample crept successively at $\sigma = 40\ g/mm^2$, $\sigma = 60\ g/mm^2$, $\sigma = 40\ g/mm^2$. Mean subgrain size $\bar{d} = 154\ \mu m$.

(2) Sample crept only at $\sigma = 40\ g/mm^2$. Mean subgrain size $\bar{d} = 186\ \mu m$. Note the dependence of $\bar{d}$ on the highest applied stress.

(Courtesy V. Pontikis)

PLATE 13

Subgrains revealed by electron diffraction orientation contrast in a gold single crystal deformed by creep at $T = 755\,°C$, $\sigma = 250\ g/mm^2$, $\varepsilon = 15\ \%$. (Photograph by P. Dorizzi)

PLATE 14

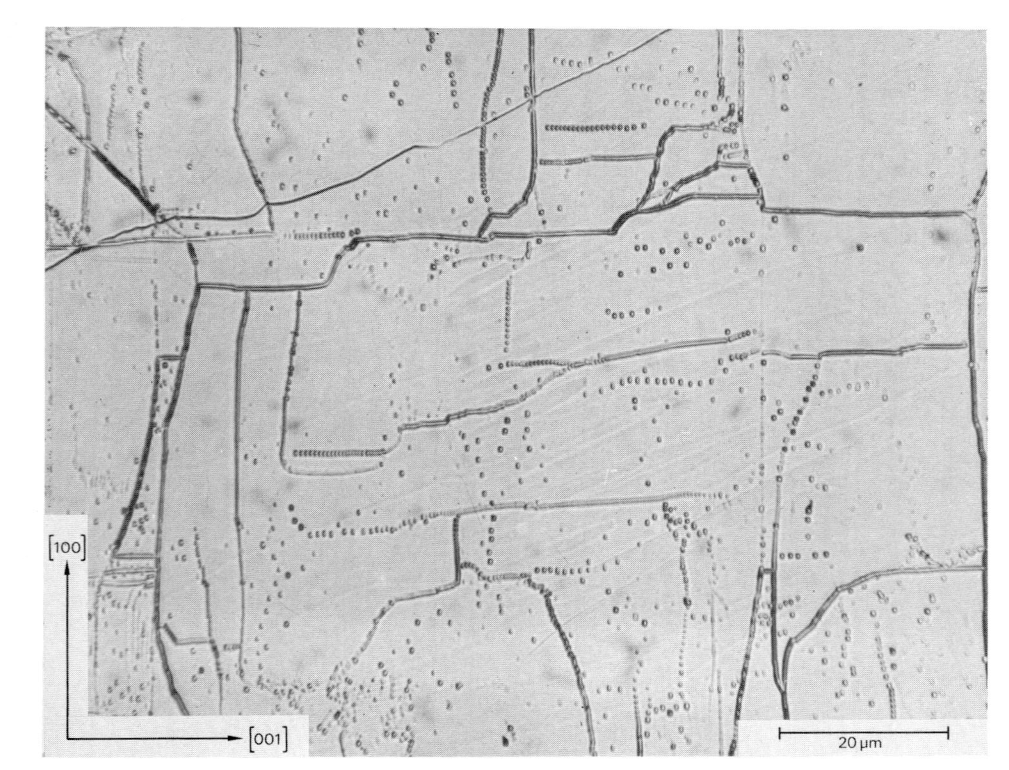

Subgrains revealed by etch pits in olivine from San Carlos, Ariz. lherzolite. Cleaved (010) surface. (Photograph by M. Wegner and S. H. Kirby)

PLATE 15

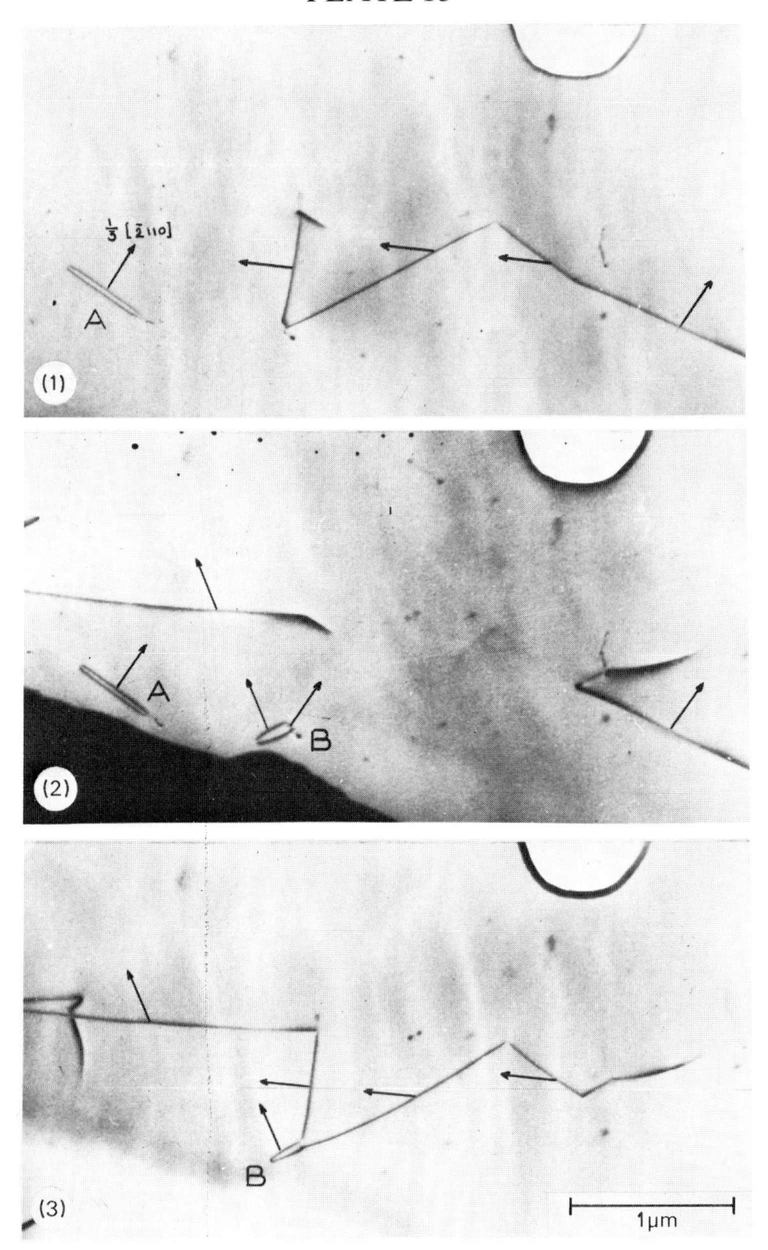

Determination of the direction of the Burgers vector of dislocations by TEM in beryllium. Same area with 3 different reflecting planes
(1) $10\bar{1}0$, (2) $\bar{1}100$, (3) $0\bar{1}10$

Note that dislocations are seen only on two out of the three micrographs. Dipole A seen on (1) and (2) and invisible on (3) has its Burgers vector **b** in a plane perpendicular to the reflecting plane $\mathbf{g} = [0\bar{1}10]$ and in the basal plane hence $\mathbf{b} = \frac{1}{3}[\bar{2}110]$ ($\mathbf{g} \cdot \mathbf{b} = 0$). The same reasoning applies to dipole B and other dislocation lines. The Burgers vectors are indicated by arrows

PLATE 16

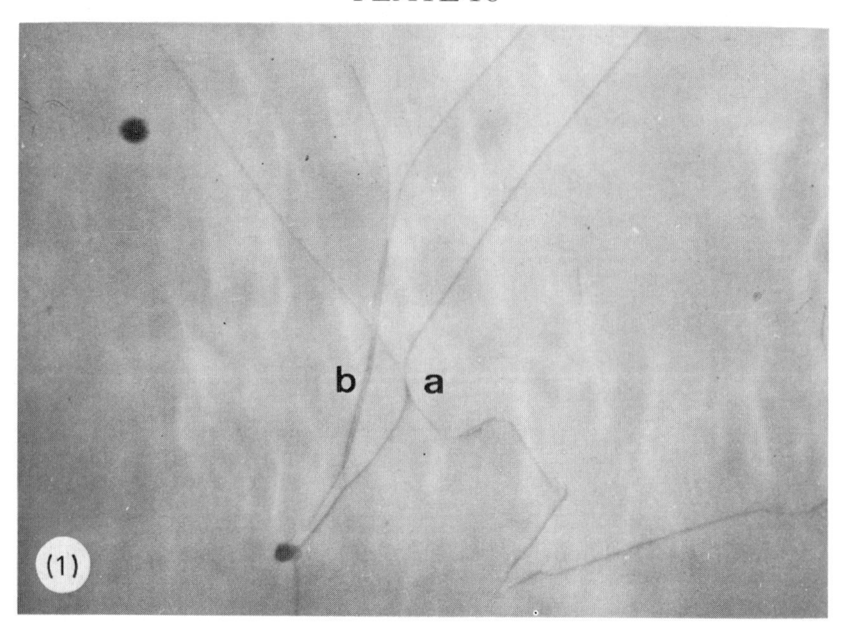

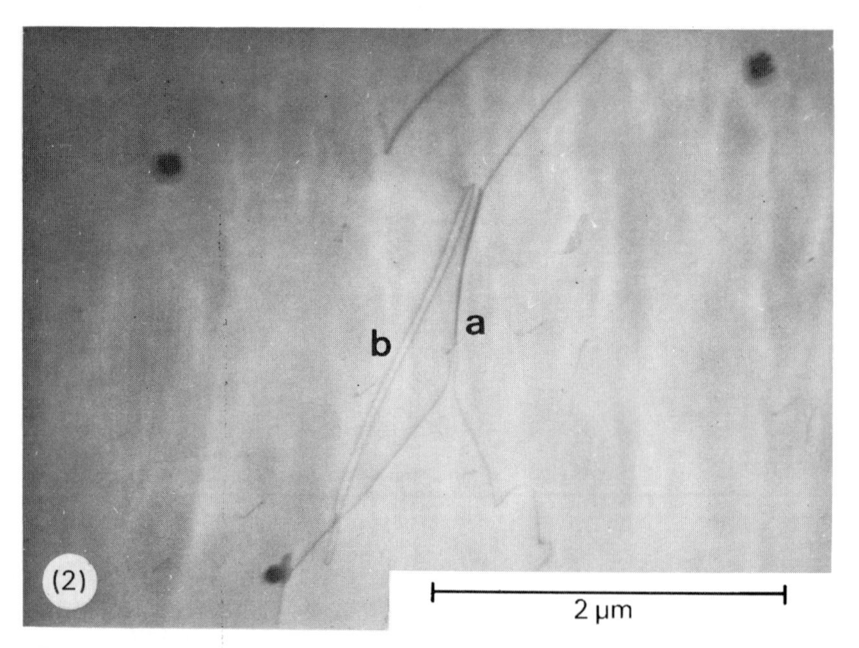

Formation of elongated dipoles in basal glide of beryllium (TEM). The plane of the foil is (0001).

(1) Two dislocations with opposite sign Burgers vectors gliding in parallel planes attract and tend to align in parallel positions (a). Two others are already parallel (b).

(2) A few seconds later, dislocations (a) are now aligned, dislocations (b) have formed a dipole by cross-slipping at each end.

PLATE 17

On Plates 17–19 the foliation and banding planes are normal to the section and the elongation direction, within the section, is parallel to their trace.

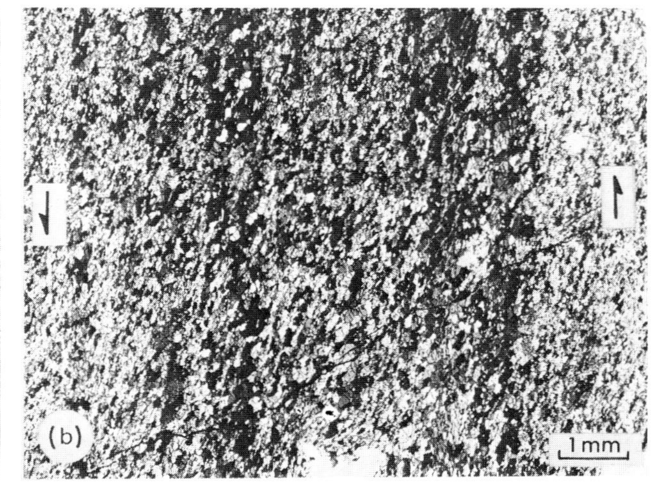

(a) Deformation gradient in an orthogneiss. Starting from a granite, the deformation grades into the banded mylonite seen on the photograph, within only one meter. The augen are formed of alkali feldspar and the matrix, of quartz, plagioclase, biotite and accessory minerals. (Nantes vicinity, France, Lasnier et al., 1973; photograph by Marchand and Lasnier)

(b) Micrograph of a finely banded peridotite from Ronda peridotite massif. The banding is due to alternate layers with finer and larger grain size. In both types of bands, olivine and enstatite have strong shape and lattice preferred orientations. As visible on the picture, the trace of shape preferred orientation (foliation) is oblique to the trace of the bands; it is more oblique in the coarse grained bands and less in the finer grained ones. The preferred orientations of [100] olivine and [001] enstatite (slip directions) follow a similar pattern with a smaller angle to the trace of the bands. In fact, on this micrograph taken between crossed polarizers, the fine grained bands appear, thanks to their extinction, for an orientation nearly parallel to their trace. *These relations strongly suggest a non-homogeneous sinistral shear in the direction of the banding*; the finer bands correspond to a larger strain with, as a result, an increased rotation of the shape and lattice preferred orientations toward the shear direction. (Photograph and interpretation by M. Darot)

PLATE 18

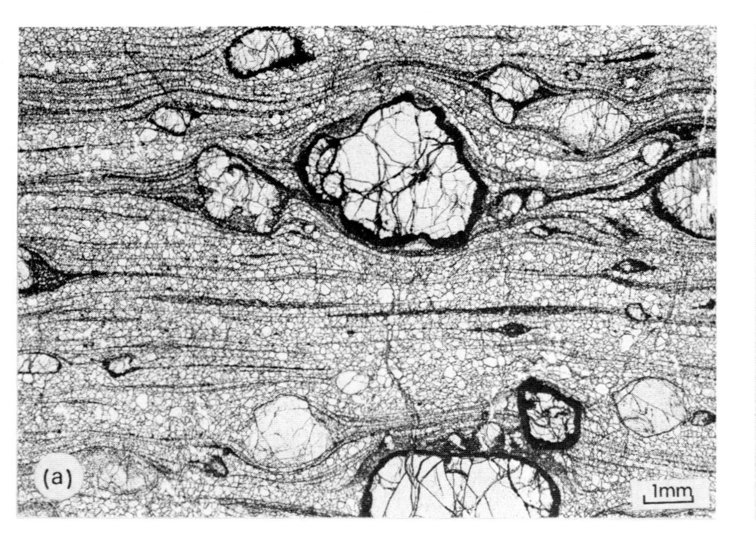

(a) Micrograph of an enstatite banding in a garnet lherzolite xenolith from kimberlite (South Africa). This banding prolonged by fractures in the olivine matrix has been attributed to a superplastic flow of 10 μm enstatite grains generated by syntectonic recrystallization in enstatite porphyroclasts (see Plate 22(a)). (Photograph by A. M. Boullier)

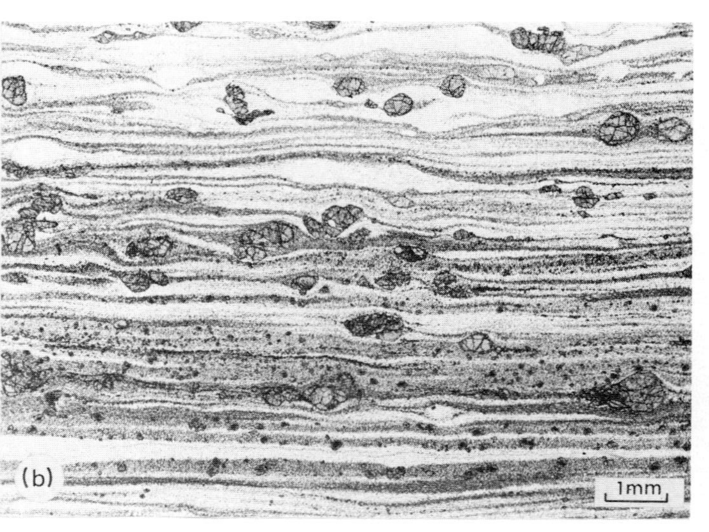

(b) Micrograph of a banded mylonite in an amphibole-bearing anorthosite from Sognjotun Nappe, Norway. The dark banding is formed by a mixture of plagioclase and amphibole; it is attributed to superplastic flow. (Photograph by A. M. Boullier and Y. Gueguen, 1975)

PLATE 19

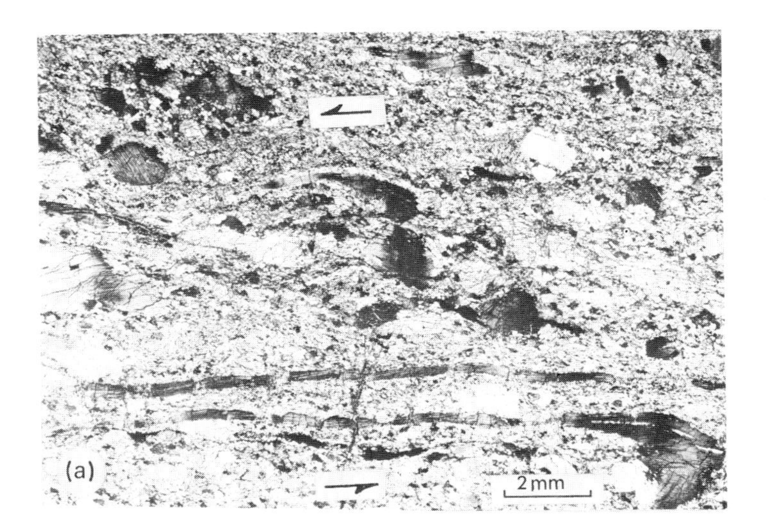

(a) Micrograph showing retort-shaped enstatite grains in a peridotite mylonite from Lanzo. The elongated enstatite blades forming the neck of the retorts determine on the specimen a lineation like that of Plate 27(a). Slip in the retort necks, once the (100) slip plane (shown by exsolution) has been reoriented, indicates a sinistral sense; some boudinage is also observed. The olivine matrix is nearly completely recrystallized. (Photograph by A. M. Boullier)

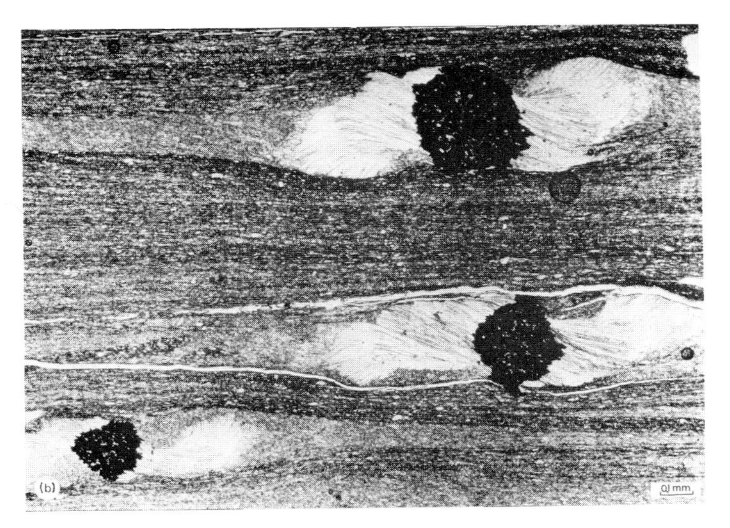

(b) Micrograph showing a sigmoid crystallization in the pressure shadow around pre-tectonic crystals of pyrite in a phyllite from Lourdes (Pyrénées). Rotational increments of strain are deduced from such observations. (Photograph by Choukroune, 1971)

PLATE 20

Plates 20–22. Micrographs in *XZ* sections

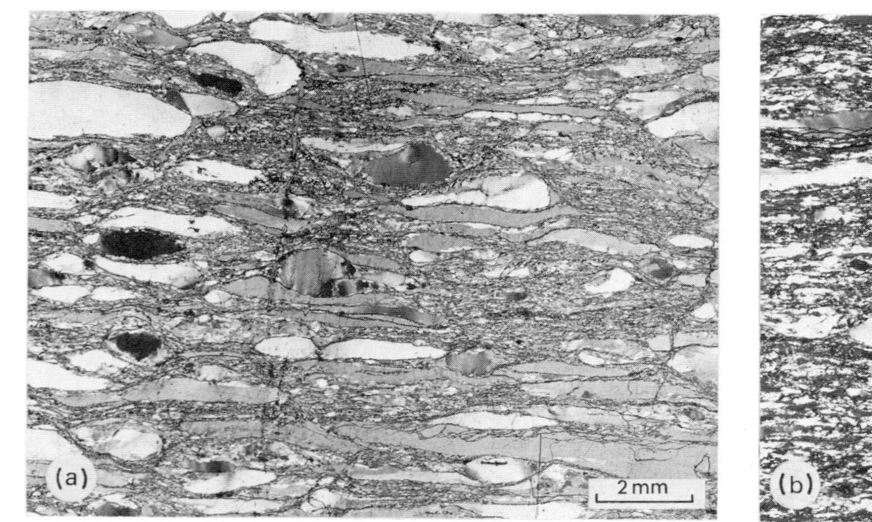

(a) Quartzite within low-grade metamorphic slates, Angers, France. The quartz porphyroclasts derive by plastic flow from equant grains. Some aspects of their substructure are shown in (b) and Plate 21(a). In this highly deformed facies, a strong preferred orientation is developed; basal slip is considered as dominant. (Photograph by J. L. Bouchez, 1975)

(b) Weakly misoriented subgrains with prismatic walls in a quartz porphyroclast oriented with the basal plane E.W. and normal to the section (undulose extinction). (Photograph by J. L. Bouchez)

PLATE 21

(a) Kink bands with prismatic boundaries in a quartz porphyroclast oriented with the basal plane N.S. and at a high angle to the section. In contrast with the substructure shown in (b), the rotation from one band to another is over 20°, inducing recrystallization in ribbons. The origin of these two distinct types of substructure in relation with the flowage is discussed in the case of olivine in § 6.2.1 and 10.3.2. (Photograph by J. L. Bouchez)

(b) Enstatite displaying either slip polygonization or kink banding in adjacent grains depending on the orientation of the (100) slip plane with respect to the E.W. trending foliation. The trace of (100) is underlined by diopside exsolutions which are thicker in the vicinity of KBBs. Enstatite-bearing pyroxenite, Balmuccia, Italy. (Photograph by F. Boudier)

PLATE 22

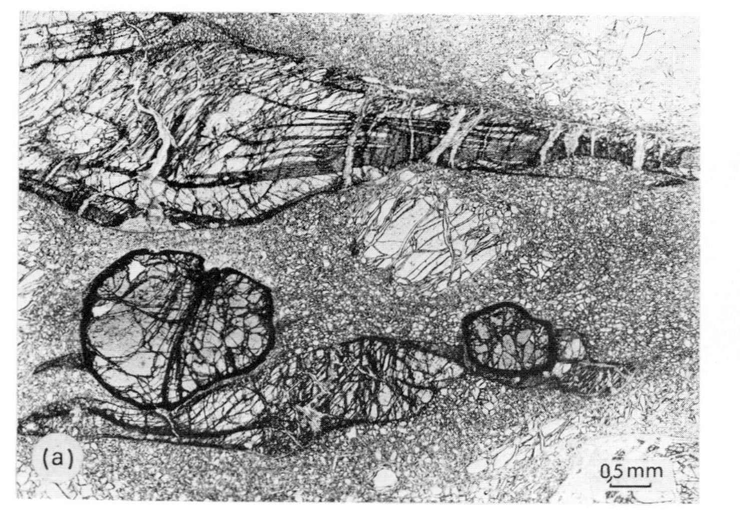

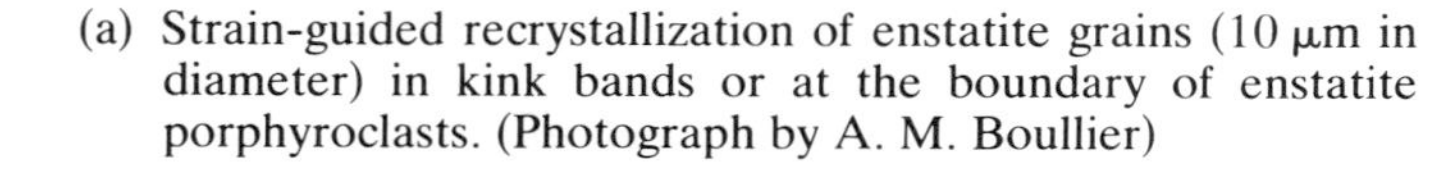

(a) Strain-guided recrystallization of enstatite grains (10 μm in diameter) in kink bands or at the boundary of enstatite porphyroclasts. (Photograph by A. M. Boullier)

(b) Cold working texture in the Baldissero peridotite: the wavy extinctions in the olivine porphyroclasts and the absence of recrystallization (except tiny grains in some boundaries) are typical of this regime. The (100) subboundaries stand statistically at a high angle to the E.W. trending foliation, thus indicating a strong olivine preferred orientation. (Photograph by F. Boudier)

PLATE 23

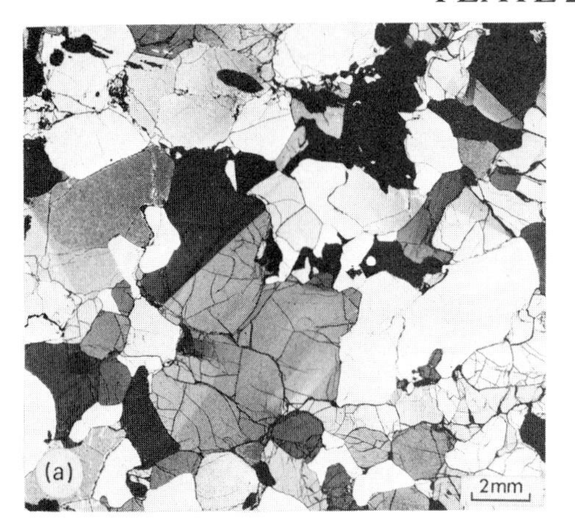

Plates 23 and 24. Micrographs from basalt xenoliths illustrating the successive stages of increasing deformation and recrystallization. The sections are normal to the foliation plane with the spinel lineation parallel to the long side, except on Plate 23(a) where no structure could be detected on the sample (photographs by J. C. Mercier and A. Nicolas, 1975)

(a) Protogranular texture. Vermicular grains of black spinel associated with pyroxenes can be seen in the centre and upper part of the micrograph. Salt Lake, Hawaii

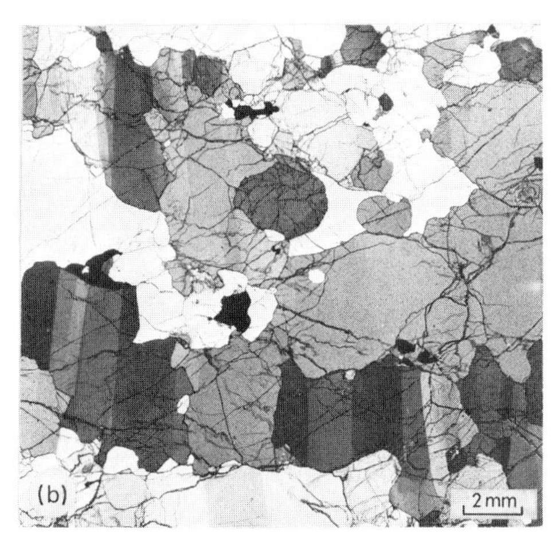

(b) and (c) Porphyroclastic textures. Salt Lake, Hawaii

(b) Moderate deformation. The olivine and enstatite grains are slightly elongated, thus defining a weak foliation on the specimen. The grain boundaries are more ragged than in (a) and the substructure more developed; no recrystallization in neoblasts

(c) Strong deformation. The olivine constitutes porphyroclasts and neoblasts. The porphyroclasts display a well developed substructure and a large aspect ratio. Note the obliquity of the subboundaries in a single direction. The neoblasts are equant or tabular; they are optically strain free

PLATE 24

(a) Equigranular tabular texture. Note the spinel inclusions in olivine tablets. Dreiser Weiher, Germany

(b) Equigranular equant texture. Spinel inclusions in olivine grains as in (a). Puy Beaunit, France

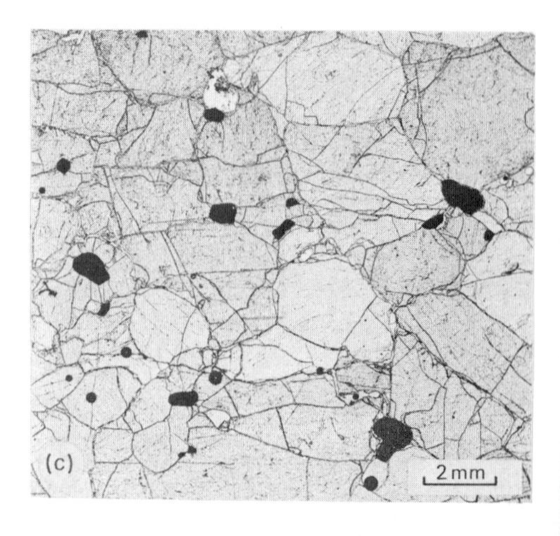

(c) Secondary protogranular texture (plain light). The characteristic difference with Plate 23(a) is that here the spinel forms blebs in the olivine grains or at triple junctions. Puy Beaunit, France

PLATE 25

Plates 25–27. Field structures in peridotite massifs. (a) and (b) Foliation marked by feldspar lenses cutting obliquely a former pyroxenitic layering. In (a) a strain-slip type cleavage superimposed upon the foliation is responsible for the saw teeth intersections with the layering, suggesting a sinistral sense of shear.

In (b), probably due to large deformation, the layering tends to vanish. (a): photograph by A. Nicolas et al., 1972, Lanzo lherzolite massif; (b): Liguria.

PLATE 26

(a) and (b) Small folds from the hinge area of a large scale fold in Lanzo massif. In (b), the foliation is at some angle to the layering; both the saw-teeth pattern and the rotation suggested by the folds indicate a sinistral sense of shear (photographs by A. Nicolas and F. Boudier, 1975)

PLATE 27

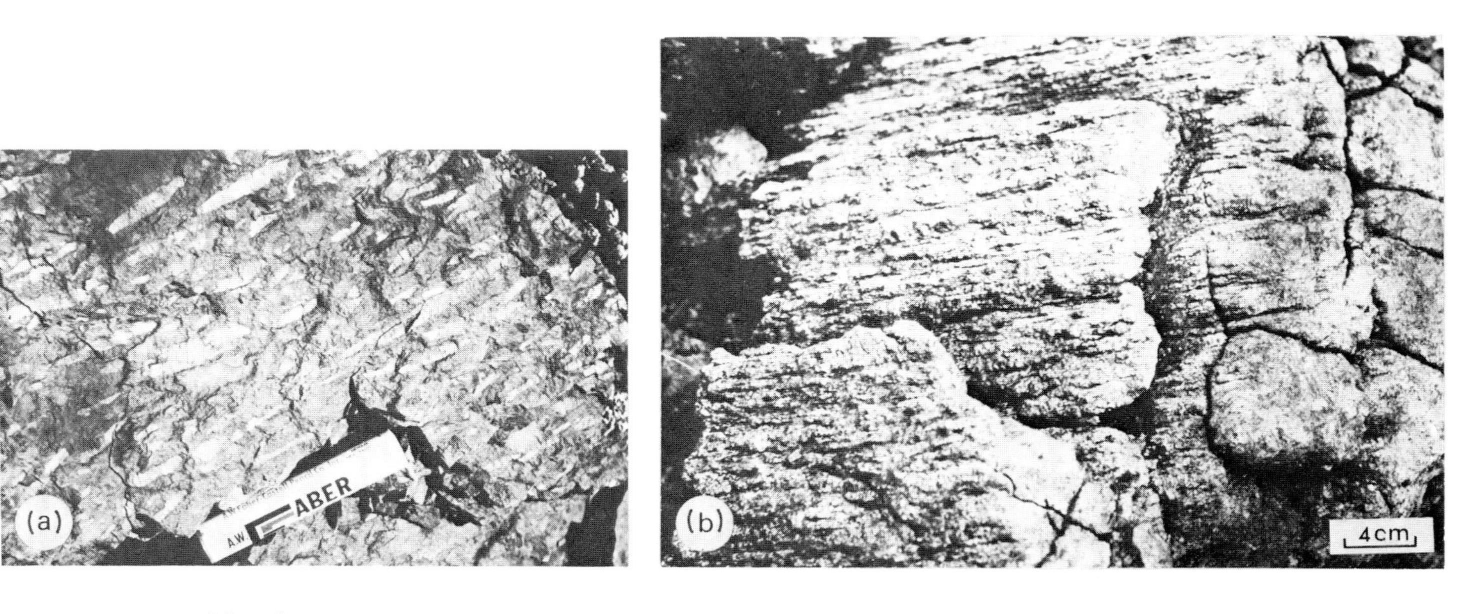

Lineations:

(a) Enstatite lineation due to extensive slip; Plate 19(a) shows such crystals under the microscope in a XZ section (W contact in Ronda massif, Southern Spain, photograph by M. Darot).

(b) Feldspar and spinel aggregate lineation (Lanzo massif)

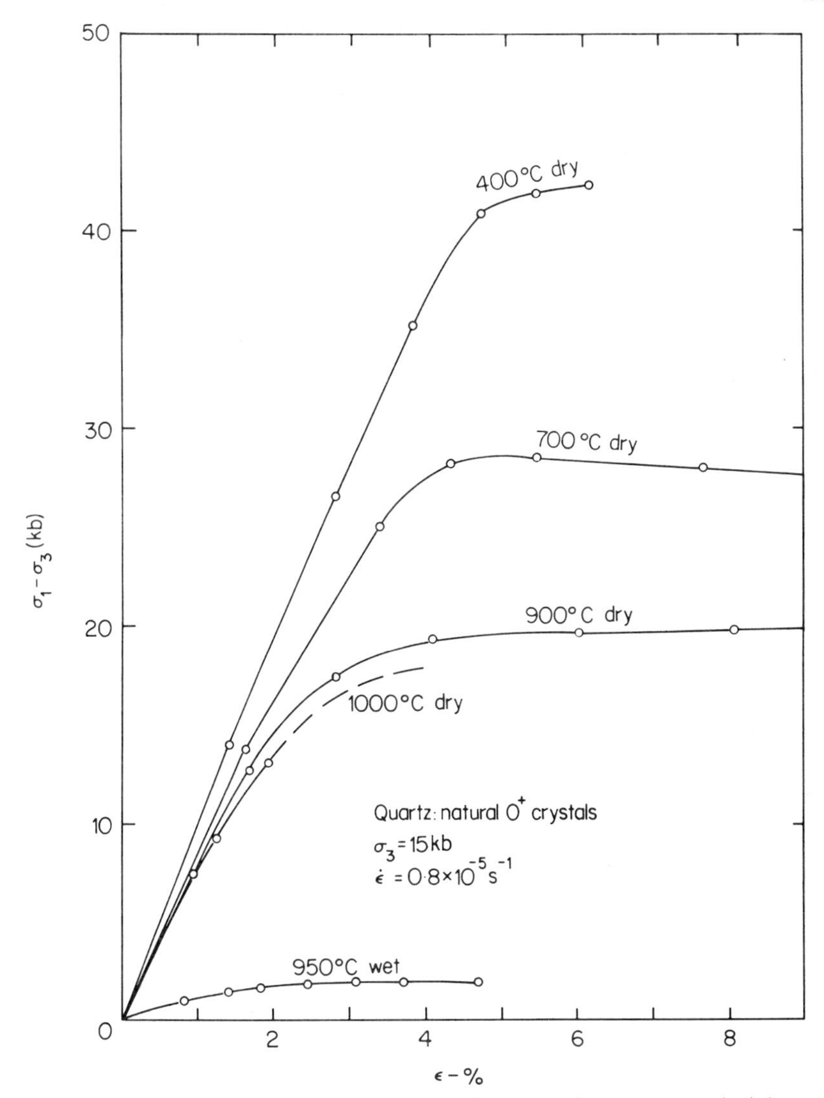

Fig. 5.30. Stress–strain curves of wet and dry natural quartz crystals (after Griggs, 1967, reproduced by permission of Blackwell Scientific Publications Ltd.)

the weakened condition is *thermally activated.* The thermal activation was investigated by a stress relaxation technique for crystals with different water contents: At a given moment in a constant $\dot{\varepsilon}$ compression test, the piston is stopped and the stress relaxing at constant total strain is recorded; during the relaxation the temperature is increased by 50 °C every 20 minutes. The fractional stress drop $\Delta\sigma/\sigma$ during each interval is then plotted against T. Weakening is apparent by a change of slope at a critical temperature T_c which depends on the water content.

If $\bar{C}=H/Si$ is the total average water content, an Arrhenius plot ln $\bar{C}$ vs $1/T_c$ shows that the critical temperature for weakening and the water content are related by

$$\bar{C}=7{\cdot}4\times10^{-8}\exp\left(\frac{E}{RT_c}\right)$$

the activation energy E revised value (Griggs, 1974) is:

$$E=15{\cdot}7\ \text{kCal/mole}$$

It was also noted that T_c increased as the strain rate was increased. TEM observations (McLaren et al., 1967) showed without a doubt that the deformation was due to dislocation glide. The Frank–Griggs hypothesis (Griggs, 1967) explained the weakening in the following way: For dislocations to move in a dry quartz crystal Si–O covalent bonds have to be broken; this requires a high energy and the Peierls stress is accordingly high. However, if H_2O can diffuse to dislocations, hydrolysed Si–OH HO–Si bridges form. These bridges may very easily be cut at their weakest link: the hydrogen bond (Fig. 5.31). The Peierls

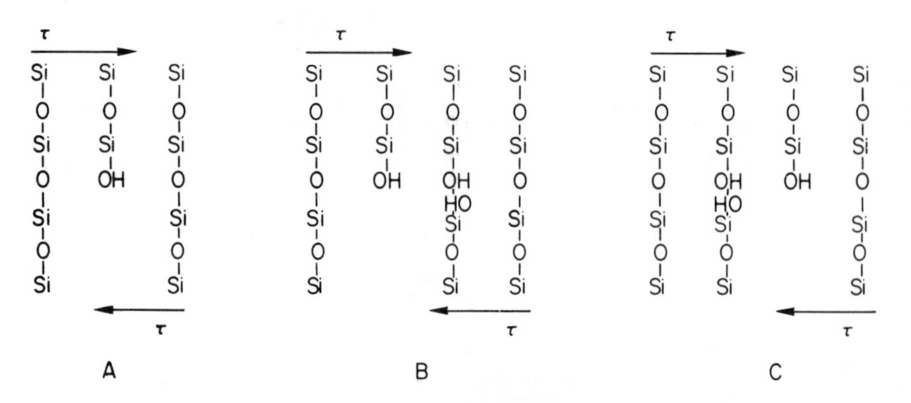

Fig. 5.31. Schematic model of the Frank–Griggs mechanism for hydrolytic weakening:
(A) Sessile hydrolyzed edge dislocation with anhydrous neighbours
(B) The right-hand side neighbouring Si–O bond has been hydrolyzed by arrival of a H_2O molecule and a hydrogen bond is formed
(C) The hydrogen bond can easily be broken under shear stress and reforms with the OH at the dislocation edge. As a result the dislocation moves to the right
(after Griggs, 1967, reproduced by permission of Blackwell Scientific Publications Ltd.)

stress is greatly lowered and dislocations can move under lower applied stresses. The diffusion of OH groups would therefore control the propagation of dislocations which in turn would control deformation.

Using TEM, McLaren and Retchford (1969) investigated samples of 'wet' quartz deformed at $\dot{\varepsilon}=$const. between 300 °C and 900 °C. In samples deformed at $T=300$ °C, below the critical temperature for weakening, they found a very high dislocation density ($\rho \simeq 10^{11}$–10^{12} cm^{-2}) for low strains. This was strong evidence that there was an important strain hardening in the first stages of deformation of the crystals below T_c, and that multiple slip systems were already operating in what had been previously thought to be the elastic domain.

On the other hand, samples deformed at higher temperatures, above T_c, showed a dislocation structure typical of recovery by climb. Accordingly, McLaren and Retchford proposed that hydrolytic weakening occurred when the recovery rate exceeded the work hardening rate; the recovery rate, in turn, would depend on the influence of the water concentration on the dislocation climb rate and could be written:

$$r = \bar{C}A \exp - \left(\frac{E}{kT}\right)$$

where A is a constant and E is the sum of the formation and migration energies of the H_2O molecules and of the activation energy for dislocation climb.

Thus, hydrolytic weakening would take place for

$$r = h\dot{\varepsilon}$$

(using the notations in § 4.1) and at $T = T_c$

$$\bar{C} = \frac{h\dot{\varepsilon}}{A} \exp\left(\frac{E}{kT_c}\right)$$

in agreement with the relation proposed by Griggs.

For McLaren and Retchford, therefore, water must play a role not only in promoting slip but also in promoting recovery by climb.

Baëta and Ashbee (1970) compressed variously oriented quartz single crystals at atmospheric pressure. At $T > 550$ °C and $\dot{\varepsilon} < 10^{-4}$ s^{-1} they could deform 'wet' quartz containing $\simeq 163$ p.p.m. OH groups at low stresses whereas 'dry' quartz (< 47 p.p.m. OH) did not yield. These authors discovered that for low strain rates ($\dot{\varepsilon} = 10^{-5}$ s^{-1}) the stress–strain curve exhibited a *yield point* (the applied stress reaches a value called upper yield stress, while little or no deformation is taking place, and falls suddenly as the sample yields easily in a heterogeneous manner).

The value of the yield stress depended on the orientation of the crystal and the magnitude of the stress drop increased as temperature was increased. TEM observations showed a high dislocation density ($\rho \simeq 10^{10}$ cm^{-2} at 800 °C). The yield drop was attributed to a high multiplication rate of dislocations. (It can be seen, by Orowan's formula $\dot{\varepsilon} = \rho_m bv$, that if ρ_m increases suddenly, the dislocations need not be driven so fast to maintain constant $\dot{\varepsilon}$, and the stress drops.) The nucleation of fresh disclocations would be made easier by the presence of hydrolyzed Si–O bonds. As Baëta and Ashbee performed their experiments with no confining pressure ($P = 1$ atm), their low temperature experiments were carried out on α quartz whereas at high temperature, α quartz changed to

β quartz. The authors reported different deformation behaviour in different temperature domains in relation with the phase change.

Hobbs, McLaren and Paterson (1972) discovered independently from Baëta and Ashbee that the presence of a yield point is a characteristic feature of 'wet' quartz stress-strain curves and they investigated it very thoroughly. They compressed at constant strain rate ($\dot{\varepsilon} = 10^{-5}\ \mathrm{s}^{-1}$) single crystals of quartz containing 700 to 6000 p.p.m. H/Si, with various orientations, under a 3 kB confining pressure. The temperature ranged from 350 to 900 °C.

A typical σ–ε curve is given in Fig. 5.32, it can be divided into 3 stages: the yield region, a region of low work hardening (Stage I) and a region of high work hardening (Stage II).

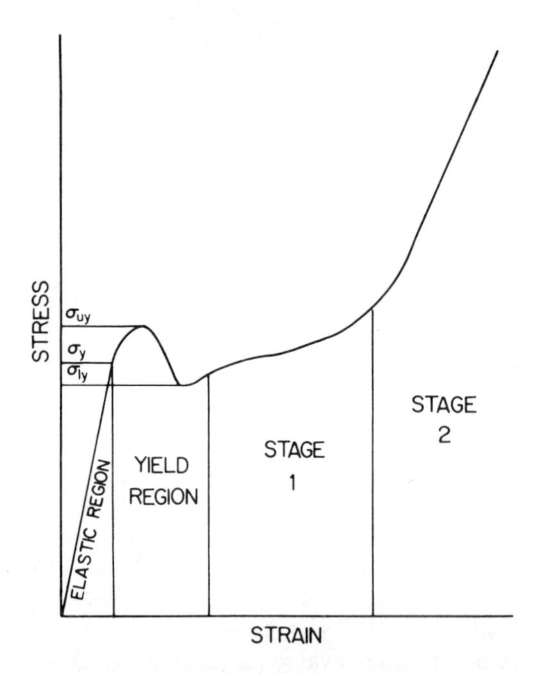

Fig. 5.32. Typical stress–strain curve for wet quartz showing the yield point. σ_y = yield stress at which departure from elastic behaviour is first observed; σ_{uy} = upper yield stress; σ_{ly} = lower yield stress (after Hobbs et al., 1972)

The orientation of the crystals and their water content change the shape of the curve. The initial yield stress and the strain hardening characteristics are very sensitive to the strain rate. In contradiction with Baëta and Ashbee's results, Hobbs et al. did not find significant changes in the σ–ε curve, when going through the α–β transition. TEM and optical microscopy observations were made in the different regions of σ–ε curve.

In the yield region, deformation lamellae develop throughout the specimen. $m=\{10\bar{1}0\}$ is the predominant slip plane in samples compressed $\perp r$.

In Stage I, lamellae are abundant in broad deformation bands. The dislocation density inside the lamellae is high ($\rho > 10^{10}$ cm^{-2}). Dislocation tangles are evidence of multiple slip.

In Stage II, the dislocation density is very high ($\rho > 10^{11}$ cm^{-2}). There must be a dislocation generating mechanism capable of rapidly producing such a high density.

The yield stress is thermally activated and varies with T according to the law

$$\sigma_y = \sigma_0 \exp \frac{Q}{kT}$$

σ_o and Q depend on the orientation and water content.

Hobbs, McLaren and Paterson noticed the striking similarity between the σ–ε curves of wet quartz and the σ–ε curves of the diamond cubic crystals (silicon, germanium) or ionic crystals such as LiF. In these crystals the initial dislocation density is very low and the yield point is due to a rapid increase in mobile dislocation density for a stress equal to the upper yield stress. Similarly, the yield point in quartz is attributed to the multiplication of dislocations from a low initial density (10^3–10^4 cm^{-2}) to a density capable of producing the required strain rate under stresses lower than the yield stress. In consequence, the microdynamical theory of yielding proposed by Alexander and Haasen (1968) for the diamond cubic structure, was applied to quartz. This theory proposes that the occurrence of the yield point is dependent on the relative strengths of the tendencies for the stress to fall, as smaller velocities are required of the great number of moving dislocations and to rise as the increasing dislocation density hardens the crystal.

Balderman (1974) systematically investigated the effect of temperature and strain rate on the yield point of single crystal quartz containing 8515 p.p.m. H/Si. The experiments were performed at 500 °C at strain rates ranging from 7×10^{-3} to 5×10^{-8} s^{-1} and at $\dot{\varepsilon}=6\times10^{-6}$ s^{-1} for temperatures between 250 and 500 °C, under a 5 kB confining pressure (Fig. 5.33). Balderman found the following empirical relations:

$$\dot{\varepsilon} = 5{\cdot}8\times10^{-7}\,\sigma_y^n$$

$$\sigma_y = \left[A \exp - \left(\frac{31\,600}{RT}\right)\right]^{1/n}$$

with

$$n = 3{\cdot}64$$

These relations were, without much justification, compared with Weertman's law for high temperature steady-state creep.

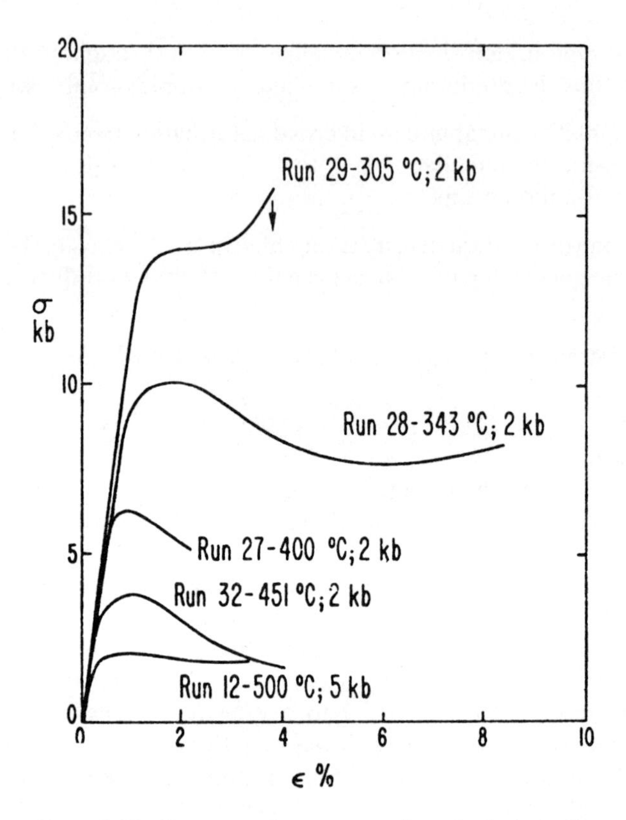

Fig. 5.33. Stress–strain curves for hydrolytically weakened synthetic quartz. $\dot{\varepsilon} = 6 \times 10^{-6} s^{-1}$. Temperatures and confining pressures are noted (after Balderman, 1974)

In a recent paper (Griggs, 1974), Griggs made a synthesis between the new experimental data and his model (Griggs, 1967) for hydrolytic weakening.

In his recent model, Griggs modified Alexander and Haasen's theory to take recovery into account. Multiplication of dislocations is thought to be the key process which can account for the nature of the stress–strain behaviour; here, multiplication is synonymous with 'dislocation growth', or increase in the length of dislocation line by expansion of loops. In Grigg's own terms: 'a dislocation can only grow in size as enough HOH diffuses to the growing segment to saturate the newly created core and to develop a cloud of HOH interstitials around the new dislocation core segment in the potential field created by its elastic stress'. (As in the former model, the role of HOH is to hydrolyse the Si–O bonds, enabling the dislocation to move freely.)

Dislocation growth is therefore the rate-limiting process and is allowed by the diffusion of HOH to the dislocation, in bulk (radial diffusion) or along the line (core diffusion).

The model rests on two main equations:

$$\dot{\varepsilon} = \tfrac{1}{2}\rho_m b(v_1 + v_2) \tag{5.1}$$

$$\frac{d\rho_m}{dt} = \dot{\rho}_m = \rho_m[\delta(v_1 + v_2) - \beta(v_3 + v_4)] \tag{5.2}$$

Equation (5.1) is Orowan's equation (where the factor $\frac{1}{2}$ allows the expression of the plastic shear strain in terms of the compressive strain). v_1 and v_2 are, respectively, the dislocation velocities corresponding to radial and core diffusion of HOH

$$v_1 = C_0\left(\frac{\sigma_{\text{eff}}}{\mu}\right)^n \frac{B_1(T)}{T} \tag{5.3}$$

$$v_2 = \bar{C}\left(\frac{\sigma_{\text{eff}}}{\mu}\right)^m \frac{B_2(T)}{T} \tag{5.4}$$

these are equivalent to Haasen's original equation for the velocity of a dislocation as a function of stress, modified to take into account the effect of diffusing HOH.

Haasen's equation has the form:

$$v = \left(\frac{\sigma_{\text{eff}}}{\mu}\right)^m B(T) \tag{5.5}$$

with

$$B(T) = B_0 \exp -\left(\frac{Q}{RT}\right)$$

$$\sigma_{\text{eff}} = \sigma - \sigma_i = \sigma - A\sqrt{\rho} \tag{5.6}$$

$$\left(A = \frac{\mu b}{2\pi(1-\nu)}\right)$$

σ_{eff} is the effective stress on the dislocation, the difference between the applied stress and the internal stress due to the other dislocations (see § 4.1).

$\bar{C}$ is the average concentration of HOH and

C_0 the concentration of HOH far from the dislocation.

C is approximated by

$$C_0 = \left[1 - \tanh\frac{\alpha\rho}{\bar{C}}\right]\bar{C} \tag{5.7}$$

(5.2) derives from Haasen's equation

$$\frac{d\rho_m}{dt} = \dot{\rho}_m = \rho_M v\delta \tag{5.8}$$

where the multiplication rate of dislocations is assumed to be proportional to the dislocation density, the dislocation velocity and a multiplication factor δ proportional to the effective stress.

$$\delta = K\sigma_{\text{eff}} \tag{5.9}$$

The important modification consists in the introduction of a term of dislocation annihilation by recovery

$$-\beta\rho_m(v_3 + v_4) \qquad \text{(see § 3.4.9)}$$

v_3 and v_4 are climb velocities of dislocations, v_3 has the applied stress and v_4 the internal stress of the other dislocations for driving force. Both are assumed to be proportional to the average concentration of HOH, $\bar{C}$.

$$v_3 = \left(\frac{\sigma}{\mu}\right) \bar{C} \frac{B_3(T)}{T} \tag{5.10}$$

$$v_4 = \left(\frac{A\sqrt{\rho}}{\mu}\right) \bar{C} \frac{B_4(T)}{T}$$

Equations (5.1) and (5.2) plus two other equations describing the behaviour of the system sample apparatus during the stress relaxation following the yield point, are solved by computer to give calculated σ–ε curves. There are twelve adjustable parameters fitted on Balderman's results.

There is a good agreement between the calculated curves and Balderman's experimental curves, especially if one considers the wide variety of possible shapes.

It can be noticed in equation (5.2) that $\dot{\rho}_m = 0$ if $\delta(v_1 + v_2) = \beta(v_3 + v_4)$; the dislocation density and the plastic strain rate at constant stress remain constant: this is a case of steady-state creep which can be attained if the temperature is high enough. It is possible only because a recovery term has been introduced.

5.9.5. Recovery and recrystallization

(a) Diffusion in quartz

(i) The diffusion coefficient for the oxygen ion has been measured in 'dry' and 'wet' conditions, by ^{18}O isotopic exchange.

Haul and Dümbgen (1962) performed diffusion anneals in oxygen and found the following expression for the self-diffusion coefficient:

$$D_{Sd} = 3{\cdot}7 \times 10^{-9} \exp\left(-\frac{55\ 000}{RT}\right)$$

Choudhury et al. (1965) measured the diffusion coefficient by means of anneals in ^{18}O enriched steam. They found a diffusion anisotropy (as could be expected) and a much higher value for the diffusion coefficient than in dry oxygen. The results are summarized in Table 5.6.

Table 5.6

T	P	D cm^2 s^{-1}	Ref.
667 °C	820 bars H_2O	$8{\cdot}1 \times 10^{-14}$ ($\perp c$) $4{\cdot}1 \times 10^{-12}$ ($\parallel c$)	Choudhury et al. (1965)
1000 °C 1200 °C	0·12 bar O_2	$1{\cdot}3 \times 10^{-18}$ $2{\cdot}5 \times 10^{-17}$	Haul and Dümbgen (1962)
600 °C	1000 bars H_2O	$1{\cdot}2 \times 10^{-17}$	Gilleti et al. (1975)

(ii) Kats, Haven and Stevels (1962) measured the diffusion coefficient for hydrogen deuterium exchange by heating crystals containing OH in D_2O vapour at various temperatures. The time taken to replace all H^+ by D^+ was measured. (The replacement of H^+ by D^+ was checked by infrared absorption.)

The diffusion coefficients have different activation energies in two temperature domains:

$$D = 5{\cdot}0 \exp -\left(\frac{42\,000}{RT}\right) \qquad \text{for } T > 700\,°\text{C}$$

$$D = 5 \times 10^{-5} \exp -\left(\frac{19\,000}{RT}\right) \qquad \text{for } T < 620\,°\text{C}$$

(*b*) *Experimental evidence of recovery and recrystallization*

Typical recovery substructures (subgrains, dislocation walls) (§ 4.2.2) have been frequently observed in natural quartzes by TEM (Hobbs, 1968; McLaren and Hobbs, 1972; White, 1973) and could be correlated with optical features; in particular, undulatory extinction can often be associated with the existence of long narrow prismatic subgrains misoriented by less than 1°; these subgrains are enclosed in larger subgrains with greater misorientation, which are responsible for the optically visible banded substructure (White, 1973).

Dynamic or static recrystallization has been investigated in quartzites and flints. In quartzites, recrystallization occurs above 850 °C ($\dot{\varepsilon} = 10^{-5}\ \text{s}^{-1}$); for $\dot{\varepsilon} = 10^{-7}\ \text{s}^{-1}$, it begins at 650 °C at grain boundaries and is total above 800 °C (Tullis, 1971). During dynamic recrystallization, the dislocation density falls from about 10^9 disl. cm^{-2} to 10^6 disl. cm^{-2} (McLaren and Retchford, 1969). Recrystallization in flints occurs at 500 °C ($\dot{\varepsilon} = 0{\cdot}8 \times 10^{-7}\ \text{s}^{-1}$) (Green et al., 1970).

In single crystals (Hobbs, 1968), recrystallization does not occur unless trace amounts of OH^- are present; the initial dislocation density, higher than 10^{10} disl. cm^{-2} is reduced to 10^8 disl. cm^{-2} by stress annealing for 1 hour at 700 °C, with local rearrangement into sub-boundaries. The growth orientations and mineral preferred orientations are discussed in § 6.4.3.

5.9.6. Inversion to coesite during deformation experiments

Inversion of both α and β quartz to coesite has been repeatedly observed in the course of experimental deformation (Hobbs, 1968; Green et al., 1970; Baëta and Ashbee, 1970; Tullis, 1971; Tullis et al., 1973). It is remarkable that coesite which is the high-pressure polymorph of quartz (Fig. 5.24) should nucleate and grow during experiments performed at low confining pressures or even at atmospheric pressure. Inversion was always found to occur in low T, high $\dot{\varepsilon}$ experiments for large strains, when recrystallization did not take place. Baëta and Ashbee (1970) noting that the transformation had occurred at deformation nodes tentatively attributed it to the existence of stress concentrations able to raise the hydrostatic pressure locally to about 7 kB. Green (1972), following Hobbs (1968), refutes this interpretation and considers that coesite is not in equilibrium and has grown metastably since the free energy of unstrained coesite is lower than that of quartz in high deformed regions with a high dislocation density (up to 10^{12} disl. cm^{-2}). Green calls attention to possible departures from phase equilibria in highly strained rocks and concludes that care must be taken in inferring natural or experimental confining pressures from polymorphic transitions.

5.10. CARBONATES

The two principal rock-forming carbonates are *calcite* $CaCO_3$ and *dolomite* $CaMg(CO_3)_2$. These will be studied simultaneously and no other carbonate will be considered here. No data have been found concerning aragonite plastic properties, although this calcite polymorph can be present in very high pressure metamorphism (above 8–10 kB at 400 °C).

5.10.1. Crystallography

Calcite and dolomite both have a trigonal symmetry. The structure is commonly described by a 4-molecule rhombohedral cell whose axes [100], [010], [001] coincide with the edges of the perfect cleavage rhomb. The 2-molecule unit cell is also rhombohedral and twice as long as the cleavage cell in their common [111] direction (see Deer et al., 1963 and Fig. 5.34). The

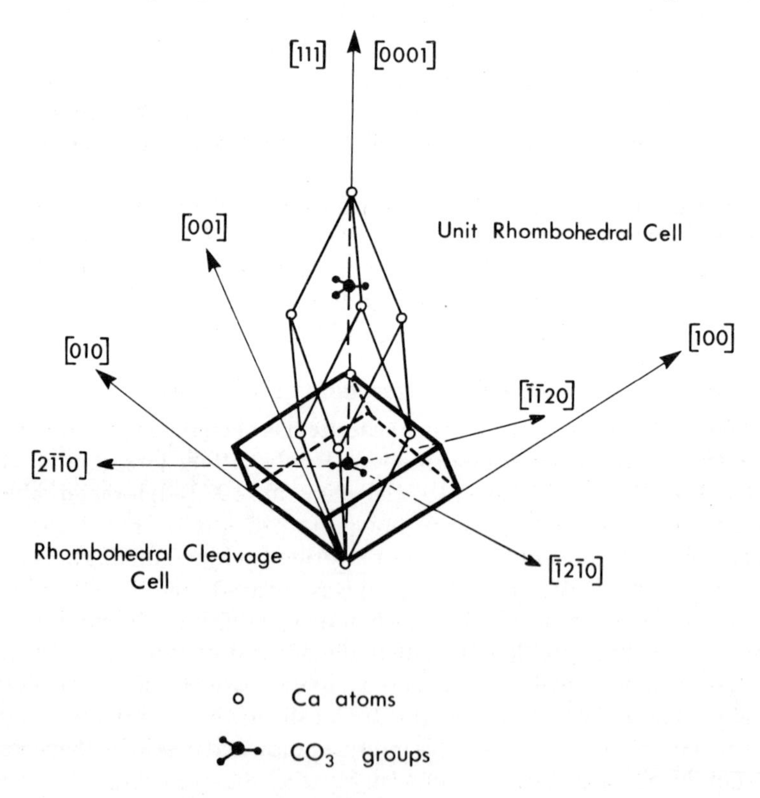

Fig. 5.34. Crystal structure of calcite. Relation between rhombohedral cleavage cell and unit cell. (Reproduced by permission of Société francaise de Minéralogie et de Cristallographie)

structure can also be described in hexagonal axes corresponding to the cleavage cell, with the following conventions:

$c\,[0001]$ corresponds to $[111]$	(trigonal symmetry axis)
$(10\bar{1}1)$ corresponds to (100)	(cleavage plane)

Each plane can therefore be labelled with four Bravais–Miller indices $(hkil)$ or three Miller indices (HKL)* (Fig. 5.35).

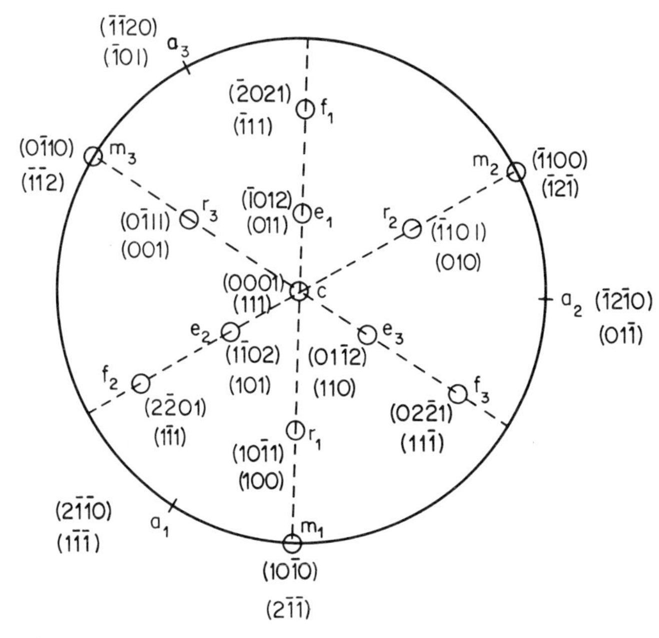

Fig. 5.35. Crystallographic planes in calcite. For each plane Miller indices (HKL) and Bravais–Miller indices (hkil) are given relative to the cleavage cell

Dolomite is structurally similar to calcite except that Ca and Mg ions alternate along the trigonal symmetry axis [0001].

Optically, calcite and dolomite are uniaxial of negative sign. As for quartz, the only information obtained by optical means on the indices ellipsoid is the C-axis orientation, but here, additional elements needed to define uniquely the

* Miller indices (HKL) referring to the *cleavage cell*, and Bravais–Miller indices $(hkil)$ are related in the following way:

$$\begin{cases} H = 2h + k + l \\ K = k - h + l \\ L = -2k - h + l \end{cases} \qquad \begin{cases} h = \frac{1}{3}[H - K] \\ k = \frac{1}{3}[K - L] \\ i = \frac{1}{3}[L - H] \\ l = \frac{1}{3}[H + K + L] \end{cases}$$

lattice orientation are provided by the presence of twin lamellae ($\{01\bar{1}2\}$ for calcite, $\{02\bar{2}1\}$ for dolomite) and $\{10\bar{1}1\}$ cleavage faces. Turner and Weiss (1963, pp. 237–239 and 244–245) give a complete account of the procedure used to determine the orientation optically.

5.10.2. Mechanical twinning

(a) Calcite

Twinning in calcite has been extensively studied (Turner and Chi'h, 1951; Griggs et al., 1951, 1953, 1960; Turner et al., 1954a, 1956; Paterson and Turner, 1970; see Klassen–Neklyudova, 1964 for Sovietic work references).

The principal twinning systems are listed in Table 5.7.

Table 5.7

Twinning plane*	Twinning direction	Sense	Number of equivalent systems	Comments
e_1 $(\bar{1}012)$ (011)	$[e_1 : a_2]$ $[(\bar{1}012) : (\bar{1}2\bar{1}0)]$ $[100]$	+	3	Very common at all temperatures. Very large strain. Complete twinning possible
r_1 $(10\bar{1}1)$ (100)	$[r_1 : f_2]$ $[(10\bar{1}1) : (2\bar{2}01)]$ $[011]$	+	3	Not uncommon for $T \leqslant 300\,°C$. Strain insignificant
f_1 $(\bar{2}021)$ $(\bar{1}11)$	$[f_1 : a_2]$ $[(\bar{2}021) : (\bar{1}2\bar{1}0)]$ $[211]$	−	3	Rather rare. Strain insignificant

* The twinning planes labelled by letters (see Fig. 5.35) are also indicated by their four Bravais–Miller indices and their Miller indices relative to the cleavage cell.

The twinning directions in the hexagonal cell are given as the intersection of planes. The positive sense corresponds to a displacement upward of the upper part of the crystal with respect to the lower part relative to the twinning plane, if the trigonal axis [0001] is oriented upward.

Twinning on $e = \{\bar{1}012\}$ planes is the most common mode in calcite; it is readily induced at room temperature by a compression normal to [0001]. An e_1-twin thus produced can be deduced from the matrix by a shear on the (011) plane parallel to the [100] direction, in the positive sense as defined by Turner et al. (1954) (Fig. 5.36) (see Table 5.7); the cleavage planes (010) and (001) remain parallel to themselves during twinning.

The shear has a high value:

$$S = 2 \tan 19°\, 9' = 0{\cdot}694$$

it corresponds to a maximum possible strain of 29% in experiments where twinning is achieved by compression normal to the C-axis and 41 % in

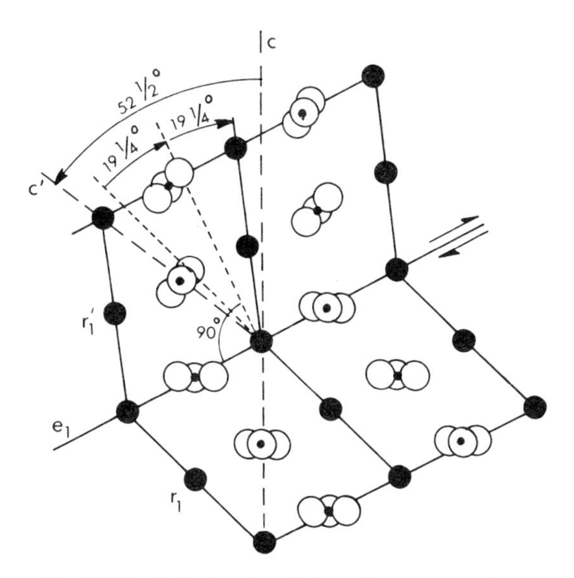

Fig. 5.36. Twinning in calcite. The sense of shear is indicated by arrows. Closed circles = calcium ions. (Reproduced by permission of Gordon and Breach Science Publishers Ltd.)

experiments where it is achieved by traction parallel to the C-axis. The C-axis rotates through 52·5° (Handin and Griggs, 1951).

Below about 500 °C, twin lamellae tend to be thin and sharply bounded by planar surfaces extending across the full width of the deformed grain. As strain increases, the number of lamellae increases until they finally coalesce and the crystal becomes completely twinned. At higher temperatures, especially at 700 °C to 800 °C, lamellae are few, thick, with irregular boundaries, and sometimes lensoid in outline. At 800 °C complete twinning is never attained even in the most favourably oriented crystals; according to Turner (1963) this is due to a notable lowering of the critical resolved shear stress for slip, which enables the crystal to deform simultaneously by slip and twinning.

The twinning dislocations responsible for e-twinning were observed by means of etch-pit techniques by Bengus et al. (1961) and Bengus (1963). The correspondence between etch-pits and dislocations was verified by Sauvage and Authier (1965) who obtained images of twinning dislocations by X-ray topography (Plate 4(a)).

(*b*) *Dolomite*

Mechanical twinning was first identified in naturally deformed dolomite crystals by Fairbairn and Hawks (1941) who showed that the prevailing system is f-twinning with a negative sense of shear (Fig. 5.37) (Table 5.7). The shear is:

$$S = 2 \tan 16^\circ\, 22' = 0{\cdot}587$$

The f-twin lamellae in naturally deformed dolomite are neither so profusely nor so widely developed as e-twin lamellae in naturally deformed calcite.

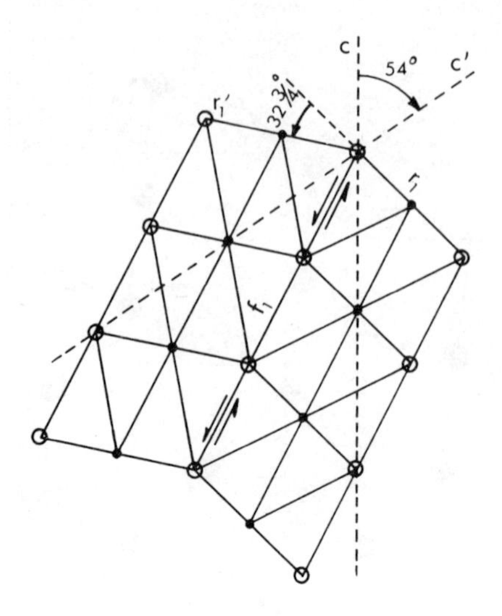

Fig. 5.37. Twinning in dolomite. The sense of shear is indicated by arrows

Handin and Fairbairn (1955) experimentally deformed a dolomite rock by uniaxial traction or compression, parallel and perpendicular to the natural foliation; they found f-twinning at 300 °C but not at room temperature, the flow stress for a few percent deformation did not depend on the orientation.

Higgs and Handin (1959) experimentally deformed single crystals of dolomite in uniaxial compression and tension between 25 and 500 °C under 5 kB confining pressure ($\dot{\varepsilon} = 1{\cdot}6 \times 10^{-4}\,\mathrm{s}^{-1}$). For all orientations they found f-twinning in a negative sense at temperatures higher than 400 °C but never at $T < 400$ °C. Samples unfavourably oriented for slip failed in a brittle fashion for $T < 400$ °C.

5.10.3. Slip systems

(a) Calcite

The slip systems in calcite single crystals and in marble have been studied in great detail, principally by Turner, Griggs and their co-workers (Turner and Chi'h, 1951; Turner et al., 1954, 1956; Paterson and Turner, 1970; Weiss and Turner, 1972; Griggs et al., 1951a,b, 1953, 1960; Keith and Gilman, 1960; Thomas and Renshaw, 1967; Borg and Handin, 1967; Braillon et al., 1972; Mugnier, 1973; Braillon et al., 1974). Table 5.8 lists the known slip systems.

The two main slip systems in calcite are

$$r = (100)\,[011]$$

and

$$f = (\bar{1}11)\,[101]$$

they were established by Turner et al. (1954) on single crystals experimentally deformed under confining hydrostatic pressure at various temperatures, by considering the internal rotation of passive markers like early formed *e*-twin lamellae. *r*-slip is the most common in the temperature range 20–300 °C. At higher temperatures (500–800 °C), Griggs et al. (1960) report that *f*-slip tends to be dominant; at lower temperatures *f*-slip has a higher critical resolved shear stress than *r*-slip and is mainly observed in experiments where the sample is constrained and where high shear stresses may be locally created.

Keith and Gilman (1960) investigated the slip systems of calcite by the etch-pit technique, and their results generally agree with those of Turner et al. (1954). However, their experiments on single crystals in 3-point bending showed *f*-slip to be much more extensive than *r*-slip over the temperature range from room temperature to 600 °C.

Braillon et al. (1972) investigated the deformation on single crystals of calcite compressed along the *C*-axis (anti-twinning orientation) with no confining pressure; for temperatures lower than 100 °C the crystals are brittle and fail by fracture, for higher temperatures up to 600 °C the samples deform by *r*-slip only. For temperatures higher than 200 °C the stress–strain curves exhibit three stages (Fig. 5.38). The critical resolved shear stress decreases as *T*

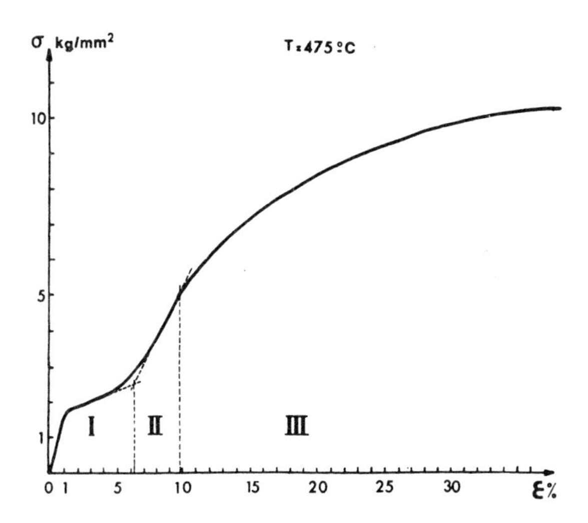

Fig. 5.38. Typical stress–strain curve for single crystalline calcite, compressed along *C*-axis in the anti-twinning orientation at 475 °C. Note the three stages: I, II, III (after Braillon et al., 1972)

increases, from 22 kg/mm^2 at 25 °C to 0·75 kg/mm^2 at 600 °C. In the range from 100 to 375 °C it follows a law of the form:

$$\sigma^{\mathrm{kg/mm^2}} = \sigma_o - 2{\cdot}8\sqrt{T}$$

These authors observed no *f*-slip in the whole temperature range (25–600 °C).

Table 5.8.

Slip plane*	Slip direction	Sense of observed slip	Number of equivalent systems	Comments
r_1 $(10\bar{1}1)$ (100)	$[r_1 : f_2]$ $[(10\bar{1}1) : (2\bar{2}01)]$ $[011]$	–	3	Common. Strain may be extensive
f_1 $(\bar{2}021)$ $(\bar{1}11)$	$[r_2 : f_3], [r_3 : f_1]$ $[(\bar{1}101) : (02\bar{2}1)], [(0\bar{1}11) : (\bar{2}021)]$ $[101], [01\bar{1}]$	–	6	
a_2 $(\bar{1}2\bar{1}0)$ $(01\bar{1})$	$[r_1 : f_2]$ $[(10\bar{1}1) : (2\bar{2}01)]$ $[011]$	Neutral	3	Very rare

* The slip planes labelled by letters (see Fig. 5.35) are also indicated by their four Bravais–Miller indices and their Miller indices relative to the cleavage cell.

The slip directions in the hexagonal cell are given as the intersection of planes. The sense of slip is defined using the same conventions as for twinning (see Table 5.7).

r-slip versus mechanical twinning. It can be seen on Table 5.8 that r-slip has been observed experimentally mostly in the anti-twinning (negative) sense. This can be attributed to the fact that, when twinning is possible, it is activated under a shear stress much lower than the critical resolved shear stress for slip. Samples solicited in the twinning sense (positive) therefore deform by twinning rather than by slip and slip can usually be observed only if twinning is impossible (negative sense); Fig. 5.39 shows stress–strain curves for samples deformed by slip in the anti-twinning sense and by twinning. It can be seen that the stress for slip is much higher than the stress for twinning. (However, this comparison gives no information about the relative magnitude of the stresses for twinning and slip in the twinning sense (positive)). Weiss and Turner (1972) succeeded in producing r-slip in calcite, along with e-twinning, and they suggested that although r-slip is more difficult than twinning, it is easier in the twinning sense than in the anti-twinning sense. Thus, r-slip would be asymmetrical, with different critical resolved shear stresses in the positive and negative senses. Although this seems inconsistent with the views on slip exposed in Chapter 2, the asymmetry of slip is, however, a well-documented fact in body-centred-cubic metals: for instance, in pure iron single crystals, the CRSS is about 14 % lower for slip on a {112} plane (twinning plane) in the twinning sense than for slip in the anti-twinning sense (Spitzig and Keh, 1970). This was explained in terms of dislocation dissociation.

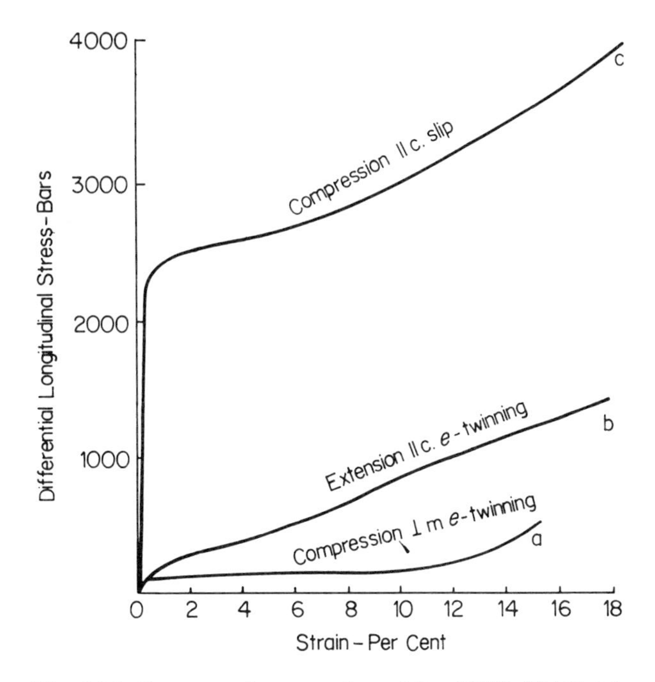

Fig. 5.39. Stress–strain curves for calcite, 20 °C, 10 kB. (a) (b) = *e*-twinning possible, (c) = *e*-twinning impossible (after Turner et al., 1954, reproduced by permission of The Geological Society of America)

(b) Dolomite

The only known slip system in dolomite is basal slip (0001) ⟨11$\bar{2}$0⟩. First reported by Johnsen (1902), it was more completely investigated in single crystals by Higgs and Handin (1959) at temperatures between 25 and 500 °C and 5 kB confining pressure ($\dot{\varepsilon} = 1{\cdot}6 \times 10^{-4}\ s^{-1}$). In samples favourably oriented for slip (stress axis parallel to $[f_1 : f_3]$ or perpendicular to r), only (0001) ⟨11$\bar{2}$0⟩ slip occurs at temperatures lower than 400 °C, f-twinning appears at about 400 °C and entirely supersedes basal slip above 500 °C, when the stress for twinning becomes lower than the stress for slip (Fig. 5.40).

Weiss and Turner (1972), from studies of kink band formation by basal slip, conclude that the yield stresses are probably equal for slip in twinning and anti-twinning senses. These authors also report slip on (0001) along ⟨10$\bar{1}$0⟩ directions.

5.10.4. High temperature creep

Calcite and marble have probably been the most intensely studied geological materials as far as flow properties are concerned, starting with Griggs' (1936, 1938) experiments.

Heard (1963) and Heard and Raleigh (1972) more specifically considered steady-state flow in Yule marble. These authors obtained steady-state creep in

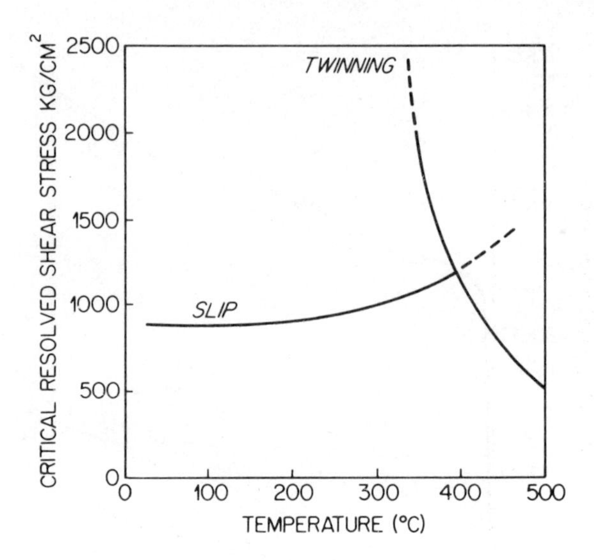

Fig. 5.40. Shearing stresses for f-twinning and basal slip in dolomite as functions of temperature (after Higgs and Handin, 1959, reproduced by permission of The Geological Society of America)

the 500–800 °C temperature range at 5 kB pressure, for low enough strain rates ($\dot{\varepsilon} = 10^{-7}\,s^{-1}$ at 400 °C, $\dot{\varepsilon} = 10^{-4}\,s^{-1}$ at 500 °C) (Fig. 5.41). The creep curves can be fitted by a Weertman equation

$$\dot{\varepsilon} = \dot{\varepsilon}_0 \sigma^n \exp -\left(\frac{\Delta H}{kT}\right)$$

the activation enthalpy and stress exponent are nearly identical when the tension axis is parallel and perpendicular to the marble foliation:

$$\dot{\varepsilon}_{0\parallel} \simeq 1{\cdot}2 \times 10^{-5}\,s^{-1} \qquad \dot{\varepsilon}_{0\perp} \simeq 6{\cdot}3 \times 10^{-2}\,s^{-1}$$

$$\Delta H_{\parallel} = 60{\cdot}9 \pm 2{\cdot}9\ \text{kCal/mole} \qquad \Delta H_{\perp} = 62{\cdot}5 \pm 3\ \text{kCal/mole}$$

$$n_{\parallel} = 7{\cdot}7 \pm 0{\cdot}5\ (\sigma \text{ in bars}) \qquad n_{\perp} = 8{\cdot}2 \pm 0{\cdot}4\ (\sigma \text{ in bars})$$

Studies of the polygonization substructure by optical microscopy and etch-pits show that it is well developed for temperatures higher than 500 °C. The coincidence between the onset of the formation of subgrains by dislocation climb and the onset of steady-state flow supports the view that dislocation climb is the controlling mechanism for high temperature creep. Recrystallization is commonly observed at 600 to 800 °C ($10^{-7} < \dot{\varepsilon} < 10^{-3}\,s^{-1}$) especially at grain boundaries.

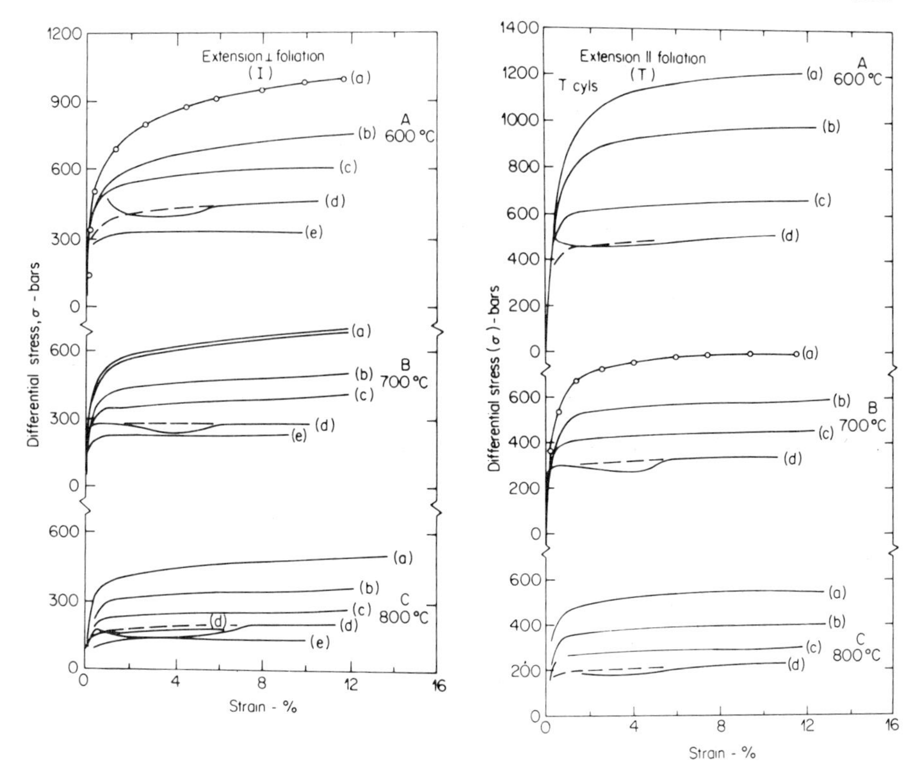

Fig. 5.41. High temperature flow in Yule marble (after Heard and Raleigh, 1971):
(*Left*) Differential stress–strain curves for samples extended normal to the foliation at different temperatures and strain rates
(*Right*) Differential stress–strain curves for samples extended parallel to the foliation at different temperatures and strain rates
Strain rates (a) $2\times10^{-3}\,s^{-1}$, (b) $2\times10^{-4}\,s^{-1}$, (c) $2\times10^{-5}\,s^{-1}$, (d) $2\times10^{-6}\,s^{-1}$, (e) $2\times10^{-7}\,s^{-1}$
(Reproduced by permission of The Geological Society of America)

5.11. HALITE

5.11.1. Crystallography

Halite, or sodium chloride: NaCl, represents the simplest class of ionic crystals. It crystallizes in the cubic system, each ion Na^+ or Cl^- is surrounded by six nearest neighbour ions with the opposite charge. This structure is easily described by a face-centred-cubic cell with a Na^+Cl^- motif (Fig. 5.42). The edge of the FCC cell is taken as the lattice parameter $a = 5{\cdot}6402$ Å.

The planes {100} are perfect cleavage planes. Halite is optically isotropic but becomes birefringent under non-hydrostatic stresses; glide bands are therefore visible in polarized light through their locked-in stresses.

5.11.2. Slip systems

(*a*) *Generalities*

Halite is one of the minerals in which slip is easiest, but it was not recognized

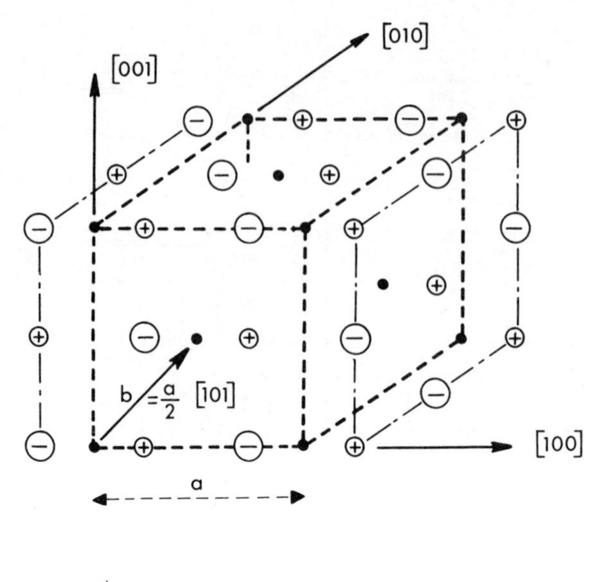

Fig. 5.42. Crystal structure of NaCl

as such until relatively recently. Although several slip systems had been recognized at an early date, many authors attributed plastic deformation mainly to twinning or kinking. Pratt (1953) proposed on the contrary that slip was the dominant deformation mode of halite as it is for metals. Since then, extensive studies of deformation by slip have been performed on halite and the closely related alkali halides (LiF, KCl, NaBr, KBr, etc.). No reliable evidence of twinning has been reported.

At low temperatures, NaCl slips on the $\{110\}$ planes in the $\langle 1\bar{1}0 \rangle$ direction. The Burgers vector of the active dislocations is $\mathbf{b} = a/2\langle 1\bar{1}0 \rangle$ and is equal to the vector linking two nearest neighbour identical ions. There are six $\{110\}$ planes, each of them containing only one slip direction; it can be shown (Groves and Kelly, 1963) that there are only two independent slip systems. The Von Misés criterion for ductility is therefore not met and polycrystalline halite is brittle. Slip on the more densely packed $\{001\}$ and $\{111\}$ planes, in the $\langle 1\bar{1}0 \rangle$ direction, is more difficult than on $\{110\}$ planes for electrostatic reasons. However, there is evidence of slip on these systems. Recently, Gutmanas and Nadgornyi (1970), using etch-pit techniques, found clear evidence for $\{001\}\ \langle 1\bar{1}0 \rangle$ and $\{111\}\langle 1\bar{1}0 \rangle$ slip at room temperature; they reported that the critical resolved shear stress for $\{001\}$ slip is ten times as high as the CRSS for $\{110\}$ slip. Carter and Heard (1970) deformed single crystals of halite, in tension under a confining pressure of 2 kB, at temperatures below 500 °C and strain rates in the range 10^{-8}–10^{-1} s^{-1}. They found that, according to their orientation, the single crystals deformed mostly by $\{110\}\langle 1\bar{1}0\rangle$, $\{001\}\langle 1\bar{1}0\rangle$ or $\{111\}\langle 1\bar{1}0\rangle$ slip. They found that the yield stresses for crystals deforming entirely by {100} slip or mainly by {111} slip were much higher than for crystals deforming by {110} slip (Fig. 5.43).

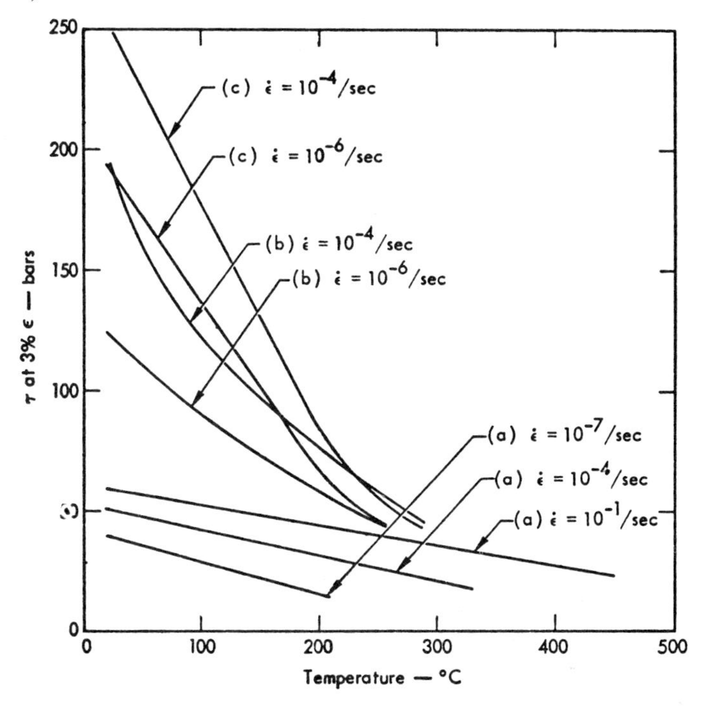

Fig. 5.43. Slip systems of halite single crystals. Resolved shear stress (3 % strain) for observed slip systems versus temperature: (a) $\{110\}\langle 1\bar{1}0\rangle$, (b) combined $\{111\}\langle 1\bar{1}0\rangle$ and $\{110\}\langle 1\bar{1}0\rangle$, (c) $\{100\}\langle 110\rangle$. (Reproduced by permission of The American Journal of Science, from Carter and Heard, *Amer. J. Sci.*, **269**, 193, Fig. 7 (1970))

In any case, for high enough temperatures, cross-slip of screw $a/2\langle 1\bar{1}0\rangle$ dislocations becomes possible on {001} and {110} plane and brings these secondary slip systems into operation. This is evidenced by the presence of wavy slip lines and the apparition of aggregate ductility. For polycrystalline halite tested in tension at constant strain rate ($\dot{\varepsilon} = 5\times 10^{-5}$ s^{-1}), the brittle ductile transition occurs at temperatures between 150 and 200 °C at atmospheric pressure; polycrystals exhibit an increasing amount of plastic deformation as temperature and grain size increase (Stokes, 1966). In fact, high temperature plasticity due to the occurrence of cross-slip and dislocation climb is so important as to warrant a comparison with the plasticity of FCC metals. We will now briefly summarize the extensive corpus of studies on deformation by slip on $\{110\}\langle 1\bar{1}0\rangle$ systems.

(*b*) *Slip on* $\{110\}\langle 1\bar{1}0\rangle$

(i) *Stress strain curves.* Most recent experiments were carried out in compression at constant strain rate on single crystals, under atmospheric pressure and at temperatures between room temperatures and 100 °C.

The stress–strain curve exhibits three stages and closely resembles the stress–strain curve of single crystalline FCC metals deforming by slip on $\{111\}$ planes (Davidge and Pratt, 1964; Hesse, 1965; Frank, 1970) (Fig. 5.44). Carter

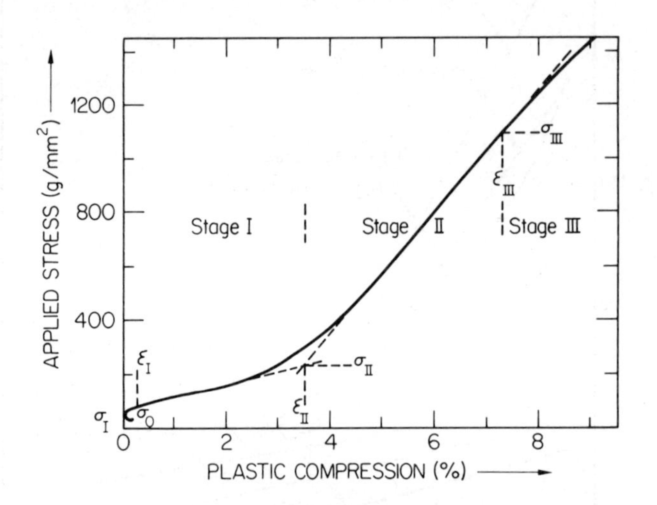

Fig. 5.44. Three-stage stress–strain curve for halite single crystals (after Davidge and Pratt, 1964, reproduced by permission of Physica Status Solidi, Berlin)

and Heard (1970) also found a three-stage curve for single crystals deformed on $\{110\}\langle 1\bar{1}0\rangle$ systems under 2 kB pressure at temperatures between 25 and 500 °C.

Stage I: For stresses greater than the CRSS on the $\{110\}$ slip plane with the higher Schmid factor, glide occurs on this slip plane alone. This stage of the σ–ε curve has a low work-hardening coefficient ($d\sigma/d\varepsilon \simeq 10^{-3}\ \mu$). During this easy glide stage, plastic strain is concentrated in glide bands which nucleate and widen until they fill the crystal.

Stage II: After a plastic strain of about 5 to 10 % a linear hardening stage appears with a higher work-hardening coefficient. This stage is characterized by the coming into operation of a second $\{110\}$ slip plane and an intense *glide polygonization* by which edge portions of dislocations held up at secondary slip bands gather into deformation bands perpendicular to the primary slip plane. These *deformation bands* are analogous to narrow kink bands (Fig. 5.45). They have been thoroughly investigated using etch pits and stress

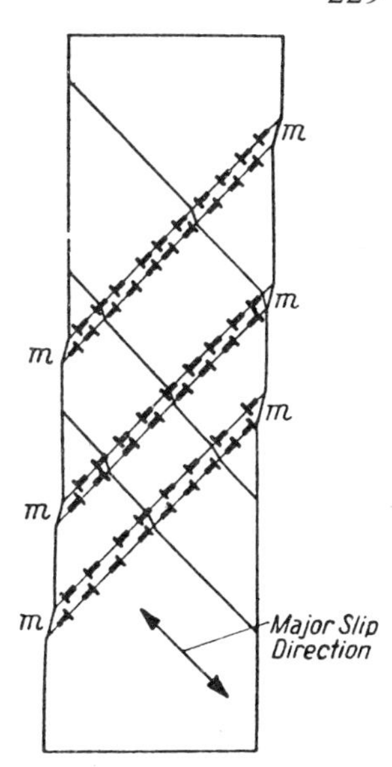

Fig. 5.45. Idealized model for the formation of dislocation walls by glide polygonization (after Davidge and Pratt, 1964, reproduced by permission of Physica Status Solidi, Berlin)

birefringence (Mendelson, 1962; Davidge and Pratt, 1964) or X-ray topography (Strunk, 1972).

Work hardening during Stage II can be attributed to the interaction between primary and secondary slip systems, evidence for which can be found in the presence of the deformation bands and a characteristic dislocation structure (dislocation dipoles and loops) found by TEM (Hobbs, Goringe, 1970).

Stage III: The linear Stage II may be followed by a parabolic stage with decreasing work hardening (Stage III). This stage usually occurs for total strains of about 20 %; the flow stress at the beginning of Stage III, σ_{III} (or τ_{III}), decreases as temperatures increases at constant strain rate and as strain rate decreases at constant temperature (Fig. 5.46). This behaviour is characteristic of a thermally activated phenomenon; in this case, cross-slip of screw dislocations (see § 4.1.2). In the case of NaCl, screw $a/2\langle 110\rangle$ dislocations lying in $\{110\}$ planes cross-slip in $\{100\}$ or $\{111\}$ planes containing their Burgers vector, and slip lines become wavy. As will be seen in the next paragraph, σ_{III} also decreases as hydrostatic pressure increases.

(ii) *Critical resolved shear stress.* The analogy with FCC metals concerning the geometry of glide and the various stages of the σ–ε curve ends when we consider the mechanisms which control the value of the CRSS and its variation with temperature, strain rate and especially impurity content. Indeed, due to the fact that dislocations can be electrically charged in ionic crystals and can therefore interact with various electrically charged point defects, the problem of understanding the causes of the resistance to motion of dislocations is a

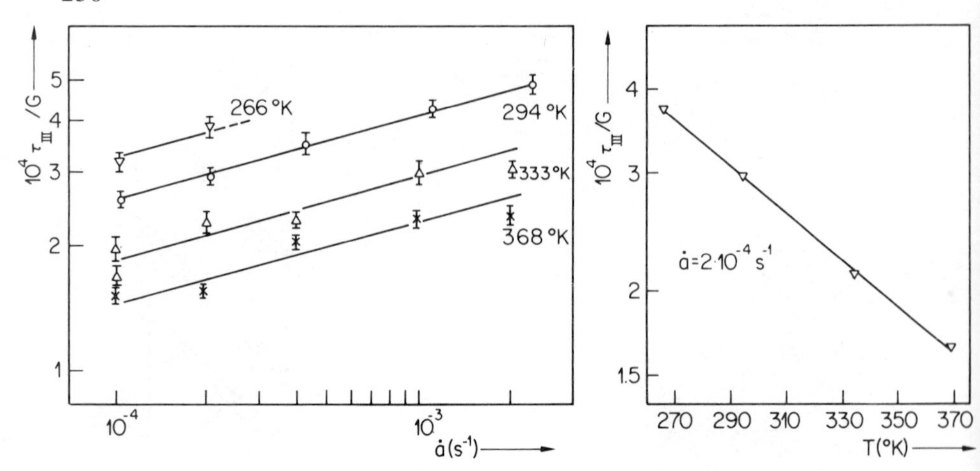

Fig. 5.46. (a) Variation of the flow stress of single crystalline halite at the beginning of Stage III as a function of strain rate for different temperatures
(b) Dependence in temperature of τ_{III}/G for constant strain-rate
(after Hesse, 1965, reproduced by permission of Physica Status Solidi, Berlin)

formidable one; even in the simplest case of NaCl, exposing the theoretical models purporting to explain a small number of reliable experiments clearly falls beyond the scope of this book; we will therefore only outline briefly the principal physical mechanisms that may intervene (see, for instance, Smoluchowski, 1966).

Two types of point defects can be considered in NaCl: negatively charged cationic vacancies and positively charged anionic vacancies (see § 3.3). Vacancies are formed at jogs of edge dislocations. Now $a/2\langle 110\rangle$ edge dislocations in NaCl have two extra half-planes of ions (Fig. 5.47) and jogs may be electrically neutral, positive or negative. (The charge of a jog may change by the absorption or emission of a vacancy.) The number of positive jogs is not usually equal to the number of negative ones, so that there is a net electrical charge on the dislocation (Eshelby et al., 1958); the charge depends on the temperature and on the concentration and charge of the impurity ions present (Ca^{++}, for instance). In order to screen the electric field of the dislocation, a cloud of oppositely charged vacancies surrounds the dislocation and exerts a force opposing its motion on it. The situation, in practice, is even more complicated, since there are usually several impurities present and since, according to the temperature range, they can form various complexes with the intrinsic cationic and anionic vacancies.

This situation leads to the following consequences:

The value of the CRSS depends in a very sensitive fashion on the concentration and nature of the impurities present and on whether they are isolated, associated with vacancies or aggregated.

The temperature dependence of the CRSS is usually quite complex.

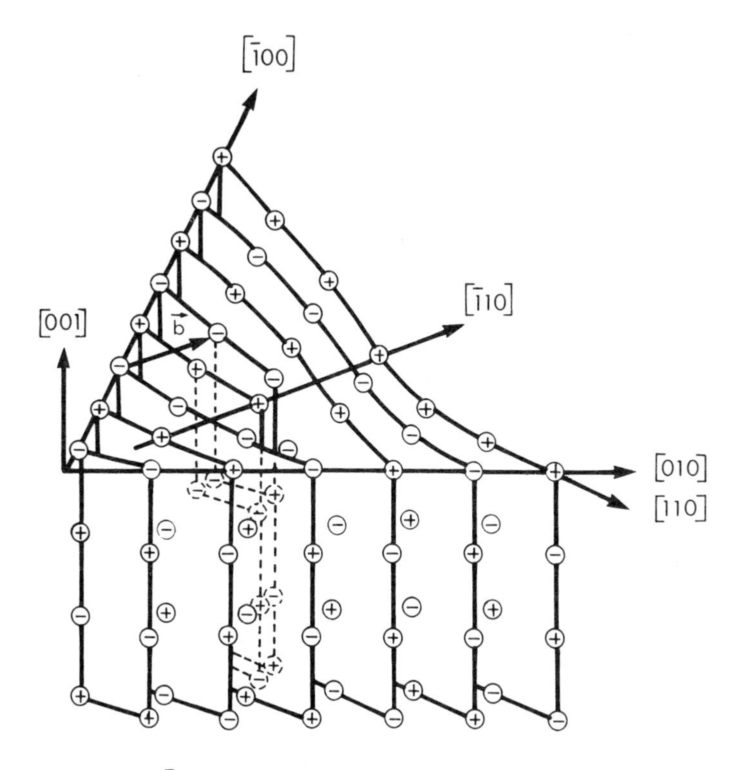

Fig. 5.47. [$\bar{1}$10] Edge dislocation in halite. The dislocation line is vertical and comports a jog. Note the double extra half-plane. (Reproduced by permission of Taylor & Francis, Ltd., from Poirier, *Phil. Mag.*, **26**, 701, Fig. 7 (1972))

5.11.3. Influence of pressure on cross-slip

It has seemed worth developing in some detail the mechanisms of the softening effect of pressure on halite, since this effect may not be restricted to halite only.

Hesse (1965) found that σ_{III}, the flow stress for which Stage III begins, is thermally activated and varies as:

$$\log \frac{\sigma_{\mathrm{III}}}{\sigma_0} = \frac{kT}{A} \log \frac{\dot{\varepsilon}}{\dot{\varepsilon}_0}$$

where σ_0, $\dot{\varepsilon}_0$ and A are constants (see Fig. 5.46).

By comparison with the models for thermally activated cross-slip in metals, this suggests that the screw dislocations are dissociated in $\{110\}$ planes and that their deviation into $\{100\}$ planes is controlled by their stress-assisted, thermally activated recombination. Fontaine (1968) showed theoretically that a stacking fault can exist in the $\{110\}$ planes and that the screw dislocations $a/2\langle 110\rangle$ can dissociate into two partial dislocations separated by a stacking fault ribbon:

$$\frac{a}{2}[110] \rightarrow \frac{a}{4}[110] + \frac{a}{4}[110]$$

Fontaine found that the stacking fault should be accompanied by an important local dilation of the lattice:

$$\varepsilon_0 = \frac{\delta d_{110}}{d_{110}} \simeq 0{\cdot}3$$

(where d_{110} is the distance between $\{110\}$ planes).

The dissociation of a screw dislocation therefore works against an applied hydrostatic pressure.

Fontaine and Haasen (1969) calculated the width of a split dislocation and found that the effective stacking fault energy increases and thus that the width of a split dislocation decreases as hydrostatic pressure increases. They accordingly predicted that cross-slip should become easier under hydrostatic pressure. This was experimentally verified on single crystalline NaCl by Davis and Gordon (1969) who found that at room temperature and under a pressure of 4·3 kB, σ_{III} was lowered by:

$$\frac{\Delta\sigma_{\text{III}}}{\sigma_{\text{III}}} = -0{\cdot}5 \pm 0{\cdot}1$$

Further experiments were carried out by Aladag et al. (1970). These authors deformed single crystals and polycrystals of NaCl at a constant strain rate $\dot{\varepsilon} \simeq 10^{-4}\ \text{s}^{-1}$ and at room temperature, under hydrostatic pressures from 1 atm to 10 kB. They found that the flow stress at the beginning of Stage II was not affected by pressure, but that σ_{III} decreased linearly as pressure increased with the following rate:

$$\frac{\text{d}\sigma_{\text{III}}}{\text{d}p} \simeq 7 \text{ to } 10\ \% \text{ per kbar}$$

(Fig. 5.48). The value of $\text{d}\sigma_{\text{III}}/\text{d}p$ was calculated by Haasen et al. (1970) and agrees reasonably well with the experimental results.

At the same time, the extent of Stage II decreases as pressures increases and extrapolation of the results shows that Stage II should vanish at about 10 to 13 kB.

The effect of pressure on the deformation of polycrystals at room temperature is also quite important: the flow stress is considerably lowered and the strain to failure is increased. (Under 10 kB the flow stress at $\varepsilon = 5\ \%$ is 20 % lower than under 1 atm and the samples fail at strains of over 30 % instead of 10 to 17%.)

It is therefore clear that *by favouring cross-slip, hydrostatic pressure increases the plasticity of polycrystalline halite.* The softening effect of pressure is rather

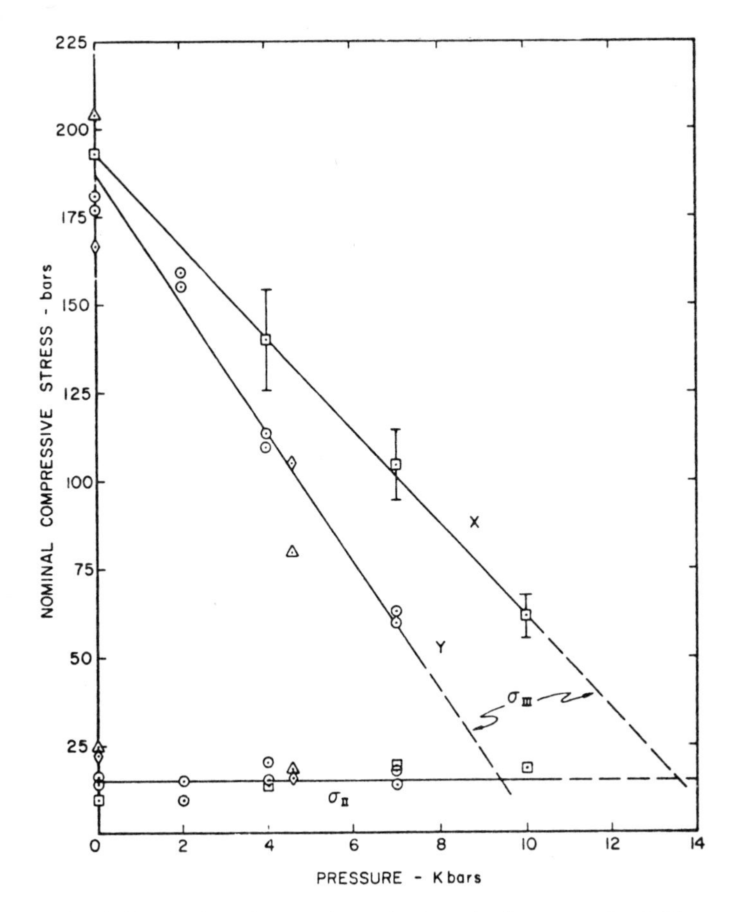

Fig. 5.48. A plot of σ_{II} and σ_{III} the (nominal) stresses for initiation of Stages II and III, respectively, versus pressure for single crystal NaCl. Lines X and Y correspond to data from samples cleaved from two different parent crystals. σ_{II} is approximately pressure-independent for all samples. If the linear extrapolation of σ_{II} and σ_{III} to higher P is valid, Stage II should be eliminated in the vicinity of 10 to 13 kB. The error bars indicate an uncertainty of ± 10 % in σ_{III}. (Reproduced by permission of Taylor & Francis, Ltd., from Aladag et al., *Phil. Mag.*, **21**, 469, Fig. 2 (1970))

remarkable since the usual effect of pressure is to harden solids by increasing the elastic constants or decreasing the diffusion coefficients in the case of diffusion-controlled deformation.

5.11.4. High temperature creep

In recent years many authors have investigated steady-state flow of polycrystalline or single crystalline halite, either by constant-stress creep experiments at atmospheric pressure or by constant strain-rate experiments under a confining pressure at high temperature or low strain rates, when work-hardening is balanced by recovery. Polycrystalline halite with a grain size smaller than 140 μm was found to deform by Nabarro–Herring creep at 740 °C (Kingery and Montrone, 1965). Most of the time, however, the creep law can be written in the form:

$$\dot{\varepsilon} = \dot{\varepsilon}_0 \sigma^n \exp -\left(\frac{\Delta H}{RT}\right)$$

and the samples were found to deform by glide on {110}, controlled by dislocation climb as proposed by Weertman (see Chapter 4). Some recent results are listed in Table 5.9.

Table 5.9.

S = Single crystal P = Poly crystal	Confining pressure bars	T °C	n	ΔH kCal/mole	Reference
S	1	550–800	4	57·5	Blum and Ilschner, 1967
P	1	365–530	5	38	Burke, 1968
		530–740	5	48	Burke, 1968
S	2000	300–400	7	33	Carter and Heard, 1970
S	1	750–780	4	57·5	Poirier, 1972
P	2000	100–400	5·5	23·5	Heard, 1972

We can see that the observed activation energies of the creep rate fall into two domains: For $T < 550$ °C the values are much lower than for $T > 550$ °C.

The high temperature values of the activation energies of creep rate are comparable with the values of the activation energies of self-diffusion of the slower anion Cl^- by Cl^- vacancies ($\Delta H_{Cl^-} \simeq 48$ kCal/mole) or of self-diffusion of chlorine ions by neutral vacancy pairs ($\Delta H \simeq 57$ kCal/mole) (Poirier, 1972).

Considering that the highest values of the activation energies of the creep rate were found in experiments on very pure crystals, well in the intrinsic regime (see § 3.3) (Blum and Ilschner, 1967; Poirier, 1972) it is reasonable to think that creep in this regime is controlled by dislocation climb, in turn controlled by the diffusion of the chlorine ion.

The problem is much more complicated at lower temperatures for less pure crystals which are certainly in the extrinsic regime, if not in the regime where impurities are associated with vacancies. Under these conditions it is practically impossible to determine a controlling diffusion mechanism merely from knowing the activation energy for creep. Schuh, Blum and Ilschner (1970) found that in the range 260–760 °C, crystals doped with 0 to 1300 p.p.m. of Ca^{++} ions exhibited a pronounced and complex temperature dependence of the activation energy for creep which could not be interpreted in terms of known diffusion activation energies.

It is safe to say that steady-state creep is diffusion-controlled but the detailed mechanism depends, in quite a sensitive way, on the nature and concentration of impurities and the temperature range. For this reason it is impossible to assess the effect of pressure on the creep rate by comparing experiments carried out on different crystals. Le Comte (1965) alone performed creep experiments under various confining pressures from 1 bar to 1 kB; he found that at 104 °C and with an axial stress of 69 bars the creep rate was 4 times lower at 1 kB than at atmospheric pressure. Such an effect is qualitatively consistent with the fact that pressure should decrease the diffusion coefficient through the activation volume (see Chapter 3).

5.12. ANHYDRITE

5.12.1 Crystallography

Anhydrite $CaSO_4$ crystallizes in the orthorhombic system. The crystallographic habit is shown in Fig. (5.49).

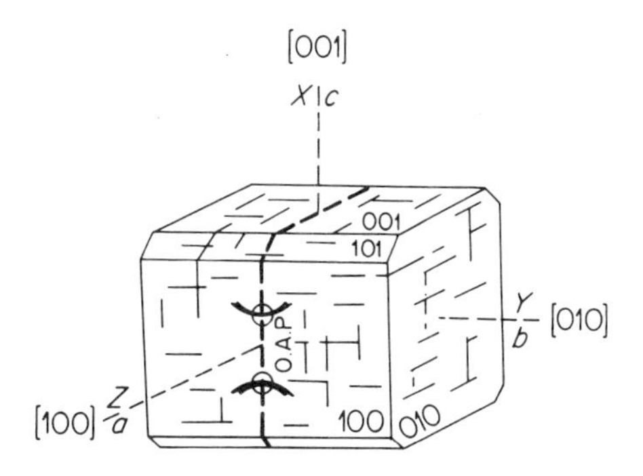

Fig. 5.49. Crystallographic habit and optical directions of anhydrite (after Tröger, reproduced by permission of E. Schweizerbart'sche, Stuttgart)

Anhydrite is cleaved easily along the (010) and (001) planes and with more difficulty along (100).

5.12.2. Mechanical twinning and slip systems

Slip and twinning in anhydrite were studied at room temperature and confining pressures up to 7 kB by Mügge (1898), Johnsen (1918) and Veit (1922). The slip systems reported were (001) [010] and {012} [100] or {012} $\langle 0\bar{2}1 \rangle$. Twinning was reported on the {101} $\langle \bar{1}01 \rangle$ system. More recently, the slip and twinning systems were again investigated by Ramez (quoted by

Müller and Siemes, 1974). The only documented slip systems (partly invalidating Veit's results) are:

$$(001)\,[010]$$
$$(012)\,[\bar{1}\bar{2}1] \quad \text{and} \quad (012)\,[1\bar{2}1]$$

The last two systems on (012) are active in the positive sense of shear (upper part of the crystal displaced in the direction of positive [001]). The twinning system (101) $[\bar{1}01]$ operates only in the negative sense.

5.12.3. Stress–strain curves

Following Handin and Hager (1957, 1958), the plastic behaviour of polycrystalline anhydrite was investigated by Müller and Siemes (1974) between 25 and 300 °C at a constant strain rate $\dot{\varepsilon} \simeq 0{\cdot}5 \times 10^{-3}\ \mathrm{s}^{-1}$ and under confining pressures ranging from 500 bars to 5 kB.

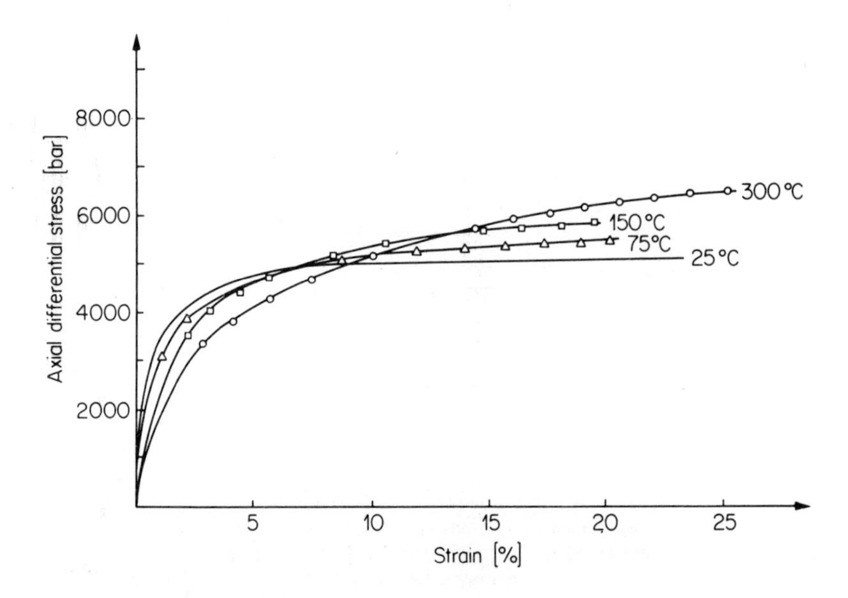

Fig. 5.50. Stress–strain curves of anhydrite at 2 kB confining pressure and different temperatures (after Müller and Siemes, 1974)

At room temperature anhydrite is brittle and deforms by cataclastic flow for confining pressures lower than 1 kB. For pressures higher than 1 kB important deformations may be attained. The flow stress increases with increasing pressure. For temperatures between 25 and 300 °C (Fig. 5.50) the flow stress decreases with increasing temperature for a strain $\varepsilon < 5$ % and increases with increasing temperature for $\varepsilon > 15$ %.

CHAPTER 6

Development of Textures and Preferred Orientations by Plastic Flow and Recrystallization: Results of Experiments and of Computer Simulation

6.1. INTRODUCTION

The relations between flow, strain and the development of preferred orientations and special textures in metamorphic rocks are complex and still being debated. They will be investigated in depth only in the most favourable case of peridotites (see Chapter 10). For the other rocks, some light will be cast on the problem of these relations by the presentation of the experimental contributions in the simpler case of monomineralic rocks deformed by axial compression. The experiments have dealt with rocks formed by olivine, quartz, micas, calcite and dolomite and, to a lesser extent, by halite and anhydrite.

The hot creep apparatuses, either with solid confining medium (Griggs, 1967) or gas confining medium (Paterson, 1970), have been designed for axial compression or extension tests except Kern and Karl's (1968) apparatus which realizes conditions of real triaxial stresses. In these apparatuses, it is only incidentally that shear regimes are generated. No significant results have been obtained in the few experiments purposely designed to produce shear conditions (Griggs et al., 1960). The lack of information on the development of textures and preferred orientations in shear flow is regrettable, considering the importance of shear conditions in natural deformation. Elements of answer will be found in computer simulation, although the crudeness of this simulation limits its applicability to real materials.

6.2. EXPERIMENTAL DEFORMATION OF OLIVINE AGGREGATES

The development of textures and preferred orientations of olivine in olivine aggregates by plastic flow and recrystallization has been experimentally investigated by Raleigh (1963), Avé Lallemant and Carter (1970), Carter and Avé

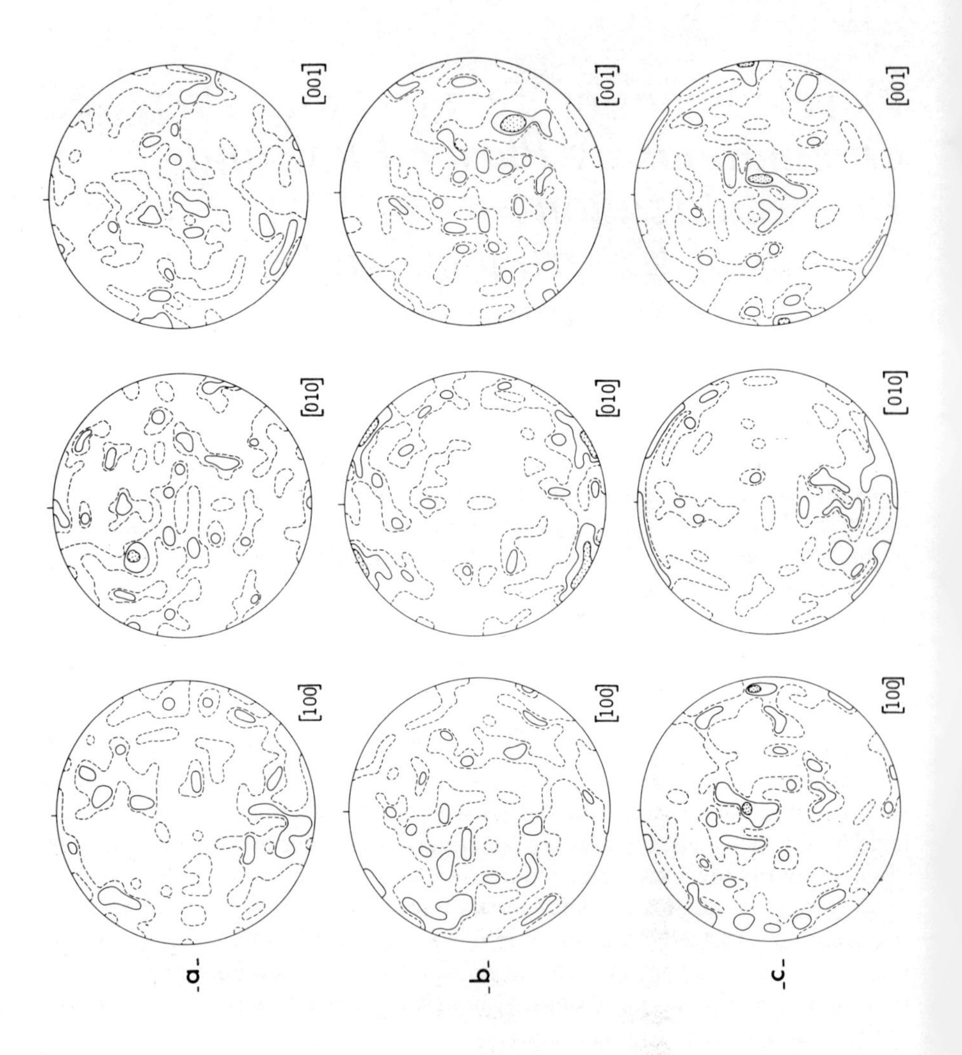
[001]
[010]
[100]
a
b
c

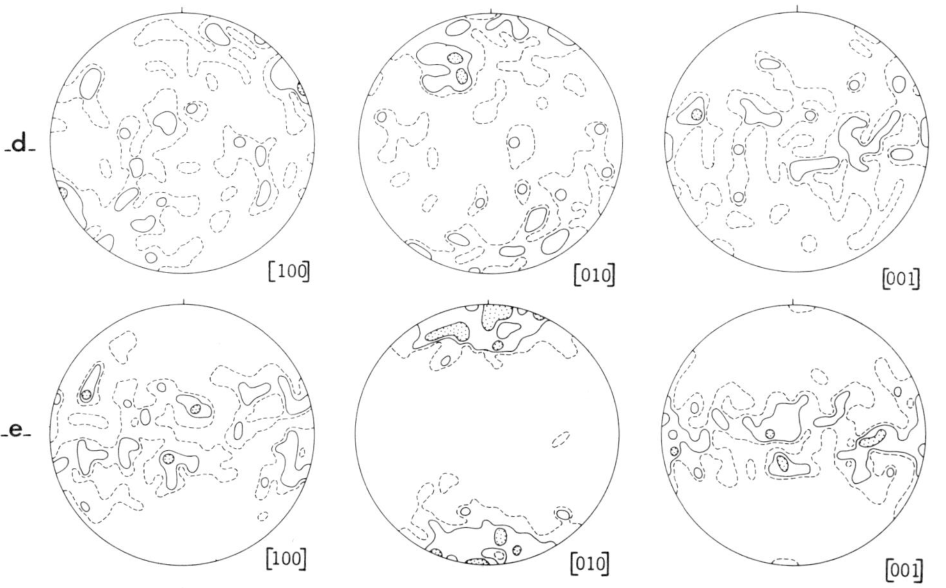

Fig. 6.1. Preferred orientations of olivine porphyroclasts in experimentally deformed synthetic dunite, σ_1 is oriented N.S. Equal area projections on the lower hemisphere. Measurements on 100 grains. Contours are at 1 % (dashed line); 2 % (continuous line); 4 and 8 % (decorated areas), per 0·45 % area:
(a) Starting material, cored parallel to the deformed samples
(b) 20 % strain, 1250 °C, 13 kB, 10^{-5} s^{-1}
(c) 33 % strain, 1250 °C, 13 kB, 10^{-5} s^{-1}
(d) 44 % strain, 1300 °C, 14 kB, 10^{-4} and 10^{-5} s^{-1}
(e) 58 % strain, 1300 °C, 14 kB, 10^{-4} and 10^{-5} s^{-1}
(Reproduced by permission of The American Journal of Science, from Nicolas et al., *Amer. J. Sci.*, **273**, 853, Fig. 1 (1973))

Lallemant (1970) and Nicolas et al. (1973). The experiments performed with Griggs' solid pressure apparatus were oriented in different ways: Avé Lallemant and Carter's, to obtain syntectonic recrystallization; Nicolas and his coworkers, to obtain plastic flow by slip. The results exposed below have been applied to the understanding of the origin of preferred orientations and modes of flow in natural peridotites (see Chapter 10).

6.2.1. Plastic flow

Nicolas et al.'s experiments (1973) on a dry synthetic dunite composed of pure forsterite were performed in the following conditions: $T = 1250$–$1300\,°C$,* $\dot{\varepsilon} = 10^{-4}\,s^{-1}$, P confinement = 13–14 kB. One of the principal requirements to develop successfully textures and preferred orientations by intracrystalline slip is to impose large strains; in these experiments strain currently exceeded 50 %. The sequence of changes in preferred orientations of large plastically deformed crystals (porphyroclasts) with increasing strain is illustrated in Fig. 6.1. Starting with a random orientation (Fig. 6.1(a)), migration of [010] towards the direction of compressive stress σ_1 can be observed, and conversely a migration of [001] and [100] towards directions normal to σ_1 in the foliation plane of the deformed specimen, that is, in the plane of mineral elongation (Fig. 6.1(e)). Due to the axial symmetry ($\sigma_2 = \sigma_3$), there is no maximum of [100] or [001].

The role of intracrystalline slip in the development of these textures and preferred orientations is demonstrated by utilizing as strain markers spherical inclusions of periclase (10 μm) enclosed in the olivine crystals. Assuming that the inclusion passively records the strain in the enclosing olivine, for a given shear angle there is a theoretical relation connecting the elongation ratio of the inclusion to the angle between the elongation direction and the shear plane (see § 2.2). By measuring these quantities with the universal stage in the (001) plane, it was possible to verify this relation with a good approximation and therefore to check the slip mechanism and incidentally to confirm (010) [100] as the dominant slip system in these experimental conditions (Fig. 6.2). It was also observed that the strain in single crystals is not homogeneous. The highest value of strain recorded in porphyroclasts by the inclusions was 400 %, corresponding to a shear angle around 75 %.

Fig. 6.2. Relationship between the elongation ratio r of the deformed inclusions in the (001) plane of olivine and the angle α between the elongation direction and the [100] direction. The line represents the theoretical relationship for (010) [100] slip. Dots and crosses: Inclusions in specimens with, respectively, over 33 % and under 33 % strain. Open circles: Inclusions in crystals with [100] close to σ_1. (Reproduced by permission of The American Journal of Science, from Nicolas et al., *Amer. J. Sci.*, **273**, 853, Fig. 6 (1973))

* For similar results in natural dunite, the necessary temperatures should be lowered by 50–100°.

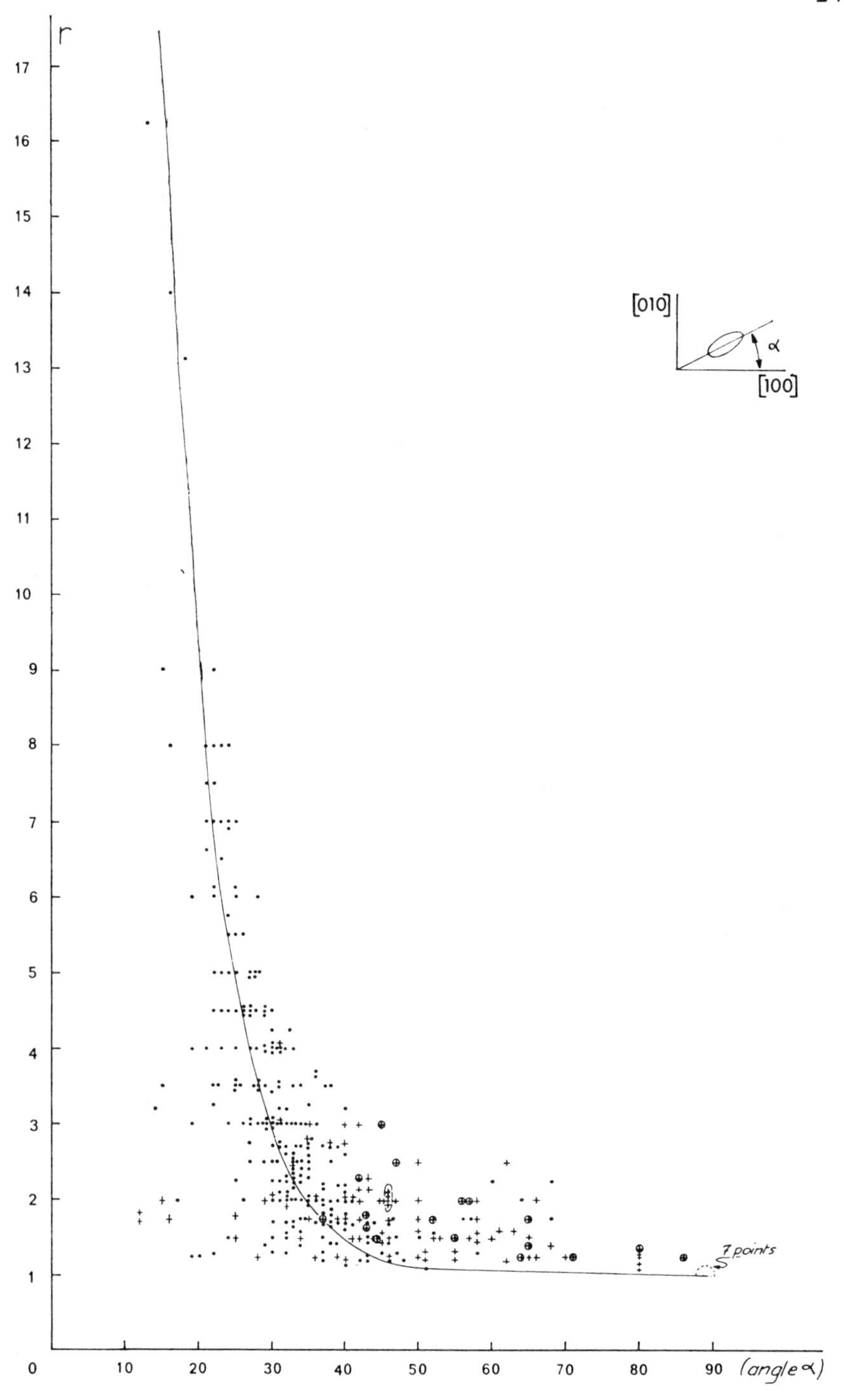
r
17
16
15
14
13
12
11
10
9
8
7
6
5
4
3
2
1
0
10
20
30
40
50
60
70
80
90
(angle α)
[010]
[100]
α
7 points

Another result obtained using the spherical inclusion as strain markers is the clear distinction between kink banding, produced by compression in olivine grains oriented with [100] close to σ_1 and slip polygonization, produced by lattice bending and twisting during slip, in olivine grains with [100] at a high angle to σ_1. As seen above, in the latter case, the strain in the [100] direction is elongation and can be very large; in the kink banded grains, the strain is shortening along [100] with measured values of 20 %. This shortening is a mean effect of slip in opposite directions from one band to the other as verified by observation of individual inclusions elongated in opposite direction from one band to the other.

The development of textures and preferred orientations in porphyroclasts with increasing strain was interpreted as a two-stage process:

(1) From 0 % to around 30 % strain the mechanism seems to be mainly external rotation (as defined in § 6.9.4) of the initially anisometric grains accompanied by strong kinking and noticeable shortening in grains oriented with their slip direction close to σ_1. This mechanism implies a good deal of grain boundary sliding which was not directly evidenced.

(2) With over 30 % strain intracrystalline slip becomes dominant as shown by the strain markers. This can be explained by considering that at the initiation of this stage a great number of grains are in a favourable orientation to slip (high resolved shear stress in the [100] direction (Fig. 6.1(b) and (c)). Internal rotations in crystals due to slip cause elongation in directions normal to σ_1 and induce external rotations and slip in surrounding crystals so that their glide system progressively rotates away from σ_1. This is necessary to minimize grain boundary sliding (for a more complete mechanical treatment see § 6.9). Undulose extinctions in elongated crystals (slip polygonization, see § 4.2.1) record the bending of lattice during slip.

The resulting fabric should be a [010] small cone around σ_1, or alternatively a [100] open cone almost normal to σ_1. The [010] migration toward σ_1 with increasing strain is limited by the fact that the resolved shear stress on the slip plane, assuming here it is (010), simultaneously tends toward zero. The existence of porphyroclasts with [010] parallel to σ_1 (Fig. 6.1(c)) can be explained by Tullis et al.'s (1973) interpretation of 'augen'. The 'augen', called here porphyroclasts, have a slip plane lying in the foliation plane of plastically deformed quartzites and have remained relatively undeformed. Tullis and her co-workers consider that these crystals were initially oriented in a way inappropriate for slip and that they would not recrystallize because they store less deformation than others (see below and § 4.5). This interpretation has been partially confirmed in the case of the synthetic dunite. Indeed, the absence of significant deformation recorded by the strain markers has been checked in 20 % of the porphyroclasts represented in Fig. 6.1(e), indicating that they had not slipped or if so, very little (Nicolas et al., 1973). A new observation reinforces Tullis and co-workers interpretation in the case of the undeformed augen: comparing the preferred orientations of the porphyroclasts in relation with their strain, Nicolas et al. have checked that the undeformed ones fall in the 'forbidden domain' for slip, that is in close coincidence with σ_1 where the resolved shear stress is nearly zero. The strained porphyroclasts are more oblique, with orientations which do not preclude slip.

(3) With over 40 % strain, the syntectonic recrystallization becomes marked. It progresses mainly at the grain-boundaries or in highly strained domains of porphyroclasts. This is why the elongation ratios of porphyroclasts and enclosed strain markers differ noticeably for large strains. This also suggested to the authors that the recrystallization, which depends on the amount of strain at a given T and $\dot{\varepsilon}$, is driven by the strain energy.

6.2.2. Syntectonic recrystallization

Avé Lallemant and Carter performed about 300 experiments on natural dunites and lherzolites and on a powdered dunite compressed into compact cylinders; the results are reported in their 1970 publications. They operated in the following range of experimental conditions: $T = 300\text{–}1400\,°\mathrm{C}$; $\dot{\varepsilon} = 10^{-3}\,\mathrm{s}^{-1} - 10^{-8}\,\mathrm{s}^{-1}$; $P = 5\text{–}30$ kB. The T, $\dot{\varepsilon}$ conditions in these experiments and in Nicolas and co-workers' cannot be directly compared because of differing starting materials; they cover, however, the same range. Therefore the marked difference between their respective results (important to complete recrystallization versus limited to absent recrystallization) is to be sought, in Avé Lallemant and Carter's experiments, in the presence of externally released water and in the lack of constraint on the sample in all the experiments that they performed with talc sleeves. However, they also performed some tests with the strong AlSiMag sleeves later employed by Nicolas et al.; in these tests the only water present was that released by dehydration of the small quantities of serpentine initially in the specimen. Finally, Avé Lallemant and Carter applied limited strains on their samples, usually below 20 %.

The experimental conditions necessary for the onset of recovery and recrystallization in olivine have been described in § 5.2.3. Avé Lallemant and Carter have studied the preferred orientations of recrystallized olivine grains found in the three following environments:

(i) New grains totally enclosed by host grains. The preferred orientations of these recrystallized grains are represented in Fig. 6.3; in Fig. 6.3(a) they are related to the host crystal orientation and in Fig. 6.3(b) to the stress. Avé Lallemant and Carter conclude that the orientations of the new grains do not seem related in a very simple way to the orientation of the stress but appear to be most closely related to the orientations of the host crystals. However, no simple crystallographic relation between host and newly recrystallized grains was found.

(ii) New grains along host grain-boundaries. Fig. 6.3(c) shows the

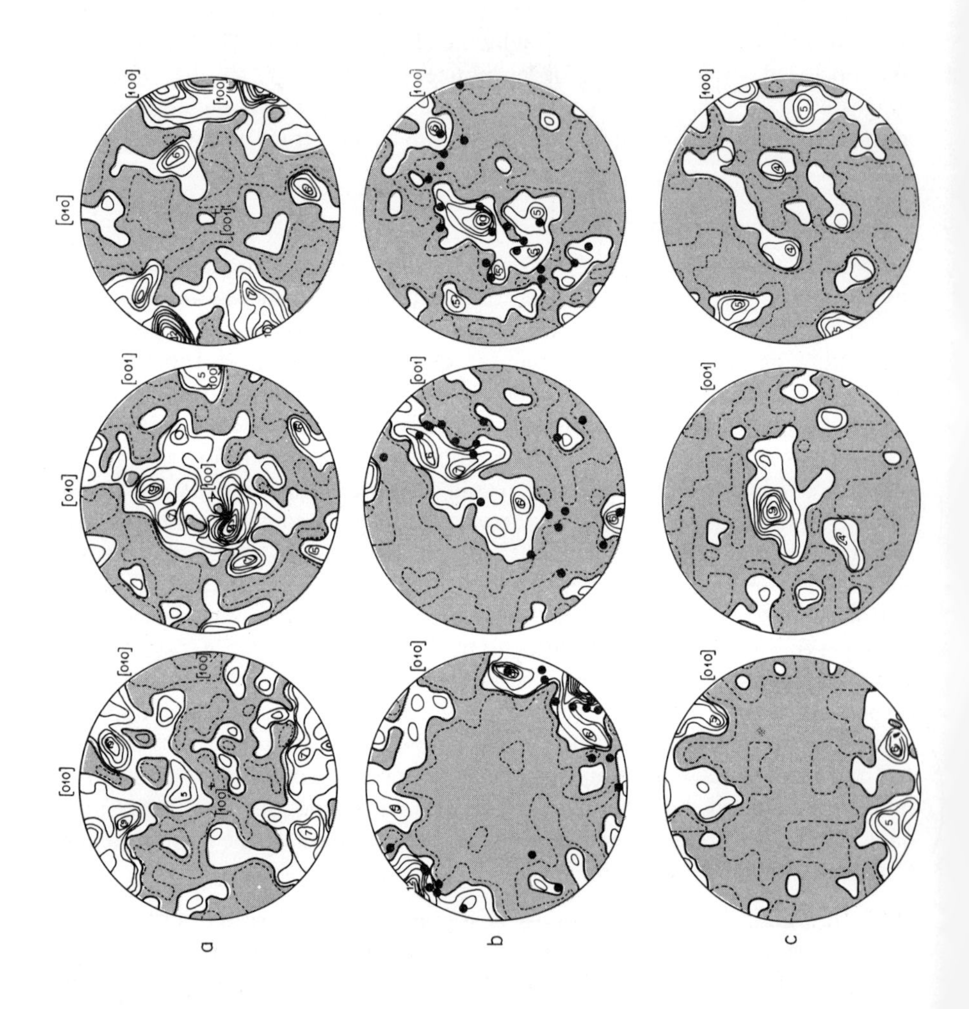
[010]
[100]
[001]
a
b
c

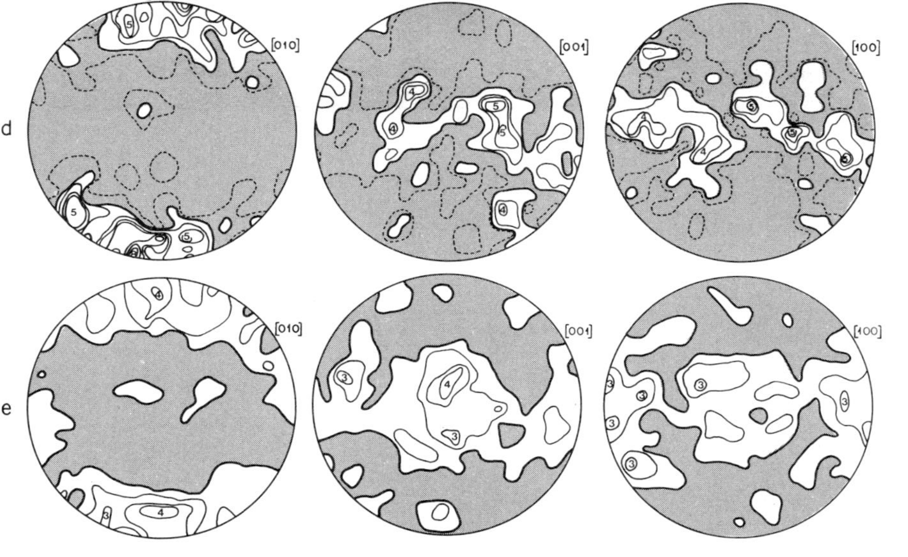

Fig. 6.3. Preferred orientations of experimentally recrystallized olivine from natural dunites. Except in (a) σ_1 is oriented N.S. Equal area projections on the lower hemisphere. Contours in the diagrams (a), (b), (c), (d) and (e) are, respectively, at 0·67 %, 0.8 %, 1 % and 1 % intervals per 1 % area; shaded areas contain less than 1·33 %, 1.6 %, 2 %, 2 %, and 2 % per 1 % area:

(a) 150 neoblasts enclosed in 30 porphyroclasts rotated in such a way that the axes of the porphyroclasts coincide. 1225 °C, 15 kB, 10^{-5} s^{-1}, wet conditions
(b) 80 neoblasts enclosed in 20 porphyroclasts (dots), not rotated
(c) 100 neoblasts at host grain boundaries
(d) 100 neoblasts in the totally recrystallized central (hottest) part of a specimen. 1100 °C, 15 kB, 10^{-6} s^{-1}
(e) 350 neoblasts, 1100 °C, 13 kB, $7{\cdot}8 \times 10^{-7}$ s^{-1}

(after Avé Lallemant and Carter, 1970, reproduced by permission of The Geological Society of America)

orientations of these grains with respect to the stress, which is N.S. Their lattice orientations are symmetrically related to the axial or nearly axial stress.

(iii) The preferred orientations of new grains in completely recrystallized areas are shown in Fig. 6.3(d). They exhibit the same relation to the stress as in the previous case. In order to eliminate the orientation effects of the large host grains, Avé Lallemant also syntectonically recrystallized a powdered dunite. The results shown in Fig. 6.3(e) compare closely with those of intergranular and total recrystallization (Fig. 6.3(d)) and prove that such a fabric can be achieved irrespective of any control by host grains.

Avé Lallemant and Carter envisage the following process leading to complete recrystallization with increasing temperature and decreasing strain rate:

(1) Nucleation of new grains at host boundary.

(2) Growth of these nuclei and nucleation of new grains within strained host crystals. The orientations of new grains within the host are controlled predominantly by the host orientation, and those at grain boundaries by the stress.

(3) Advancement of grain-boundary recrystallization until the host and new grains oriented unfavourably with respect to the stress are consumed.

[010] olivine is oriented in this process parallel to σ_1 and [100] olivine would be parallel to σ_3. This orientation in syntectonic recrystallization conforms to Kamb's (1959) thermodynamic theory for solids in the presence of an intercrystalline fluid as applied to olivine by Hartman and Den Tex (1964). Kamb's theory has been examined in § 4.4.2 and it was concluded that it does not provide a satisfactory way of explaining preferred orientations in flowing solids. A more reasonable explanation is the mechanism of strain-driven recrystallization invoked by Nicolas et al. (1973) or the mechanism by subgrain rotation proposed by Poirier and Nicolas (1975) (see § 4.5.4). The elementary mechanism of flow is glide and climb of dislocation and not stress-controlled diffusion. The grains become oriented in the stress field by the slip and rotation processes examined in the previous section but, beyond a certain amount of strain, they recrystallize to new strain free grains. Their preferred orientation is guided by the host orientation as shown by Avé Lallemant and Carter in the case of new grains enclosed in a host. In their turn these new grains plastically deform, opening the way to a new generation of recrystallization.

The strong [010] maximum in fabrics in the σ_1 direction is comparable with that found by Nicolas and his co-workers (Fig. 6.1). It would ultimately derive from the same origin, that is slip on the dominant [100] (010) system which, as it happens, is the system dominant at the high temperatures and slow strain rates favourable to recovery and recrystallization (Carter and Avé Lallemant, 1970).

6.3. EXPERIMENTAL DEFORMATION OF MICA AGGREGATES

6.3.1. Situation of the problem

The question of the development, during deformation, of preferred orientations of the basal plane in sheet silicates is extremely important in investigating the origin of slaty cleavage, foliation or schistosity, which are the most striking features of metamorphic rocks. Efforts to understand the origin of these planar structures include early experiments on clay with platy rigid inclusions of various natures (Sorby, 1853; Daubrée, 1879), along with theoretical models predicting the behaviour during homogeneous strain of rigid and deformable particles imbedded in a viscous medium (March,* 1932; Jeffery, 1922) (see also § 7.2).

We will restrict ourselves to giving an account of the recent experimental studies on the development of preferred orientations in mica aggregates (Means and Rogers, 1964; Means and Paterson, 1966; T. Tullis, 1971; Etheridge and Hobbs, 1974; Etheridge et al., 1973). This matter is approached differently in mica aggregates and in the case of the other polycrystals examined in this chapter. In high temperature runs, mica usually is not the unique phase present in the aggregate, and fluids comprise at least 20 % of the total volume (Means and Paterson, 1966; Etheridge et al., 1973). This derives from the technique consisting in making the mica *crystallize*, from a mixture of oxides and water during or before the deformation process. Mineral crystallization during the process is indeed a major difference with the other cases studied in this chapter. The nature of some orienting mechanisms may also be distinct in mica aggregates as a result of the usual strong aspect ratios of the mica crystals possibly combined with the effect of a matrix of contrasting ductility.

T. Tullis (1971), in his review of the possible mechanisms responsible for preferred orientations, opposes 'mechanical' orienting process to crystallization or recrystallization processes. This age-old argument has not been completely settled by the experimental studies: Means and Paterson (1966) and T. Tullis (1971) favoured the mechanical process of rotation which is now partly questioned by Etheridge et al. (1973). The contribution of the mechanical processes can be studied independently at low temperatures since in these conditions crystallization or recrystallization does not operate in micas. We will therefore deal with the case of low temperatures before considering the more controversial case of high temperature deformations.

6.3.2. Low temperature deformation

Means and Paterson (1966) compressed samples mainly composed of phlogopite or talc at room temperature, 3 kB confining pressure and 10^{-3} s^{-1} strain rate. The pholopite and the talc had been crystallized in a previous stage at 500–600 °C from appropriate mixtures of oxides and water. T. Tullis (1971)

* A summary of March's analysis is given by Means and Paterson (1966).

operated on pellets of fluorophlogopite and muscovite. The tests on fluorophlogopite were conducted at room temperature and pressure and were designed to investigate the development of preferred orientations during *compaction*; those on muscovite, at 200 °C and 15 kB confining pressure, to investigate the development of preferred orientations due to extensive compressive strain on already compacted samples.

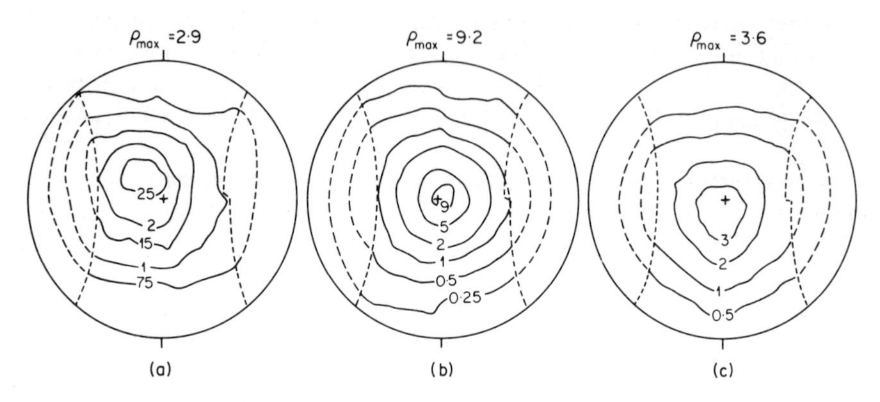

Fig. 6.4. Preferred orientations of [001] in micas at low temperature. Uniaxial compression with σ_1 in the centre of the diagrams. Equal area projection, contours in multiples of a uniform distribution:
(a) and (b) Compaction of a fluorophlogopite powder at room temperature and pressure. Though already compacted (a) is taken as a reference; in (b) $\varepsilon = 0{\cdot}80$, as defined in § 2.9
(c) Compression of a muscovite aggregate at 200 °C and 15 kB, $\varepsilon = -68$ %
(after T. Tullis, 1971)

These experiments succeeded in producing preferred orientations of the (001) plane. The (001) mean orientation is always normal to the direction of greatest shortening (which does not necessarily coincide in each region of the sample with that of the maximum compressive stress). This result which also holds true for the high temperature experiments in the recrystallization domain was demonstrated by T. Tullis using built-in strain markers. Preferred orientations for T. Tullis compaction and compression experiments at constant or nearly constant volume are reported respectively in Fig. 6.4(b) and (c). The strength of the preferred orientation, estimated by the maximum concentration ρ_{max} of poles of (001), is plotted against the strain in Fig. 6.5(a) for the compaction experiments. This proves that, at least in this particular case, the strength of the preferred orientation increases with strain. The straight line in the figure represents March's predictions (1932) of preferred orientations of planar markers passively deformed in a perfectly homogeneous deformation. These predictions or Jeffery's ones for rigid particles (1922) can be equally accepted here, because they do not differ significantly for particles with aspect ratios between four and ten, as in the case of Tullis' micas. The good

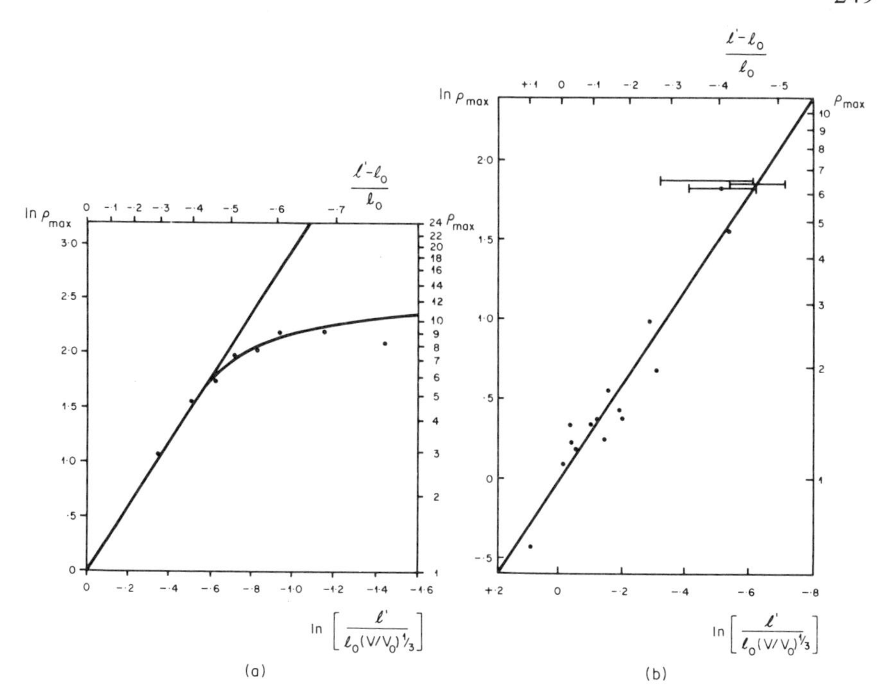

Fig. 6.5. Intensity of preferred orientation (ln $\rho_{max.}$) of [001] in micas versus magnitude of deviatoric strain, account being taken of volume changes (see, Tullis, 1971). The departure point of the curve from the straight line is not fundamentally important; it depends upon many parameters. Straight line: March theoretical model:
(a) Compacted fluorophlogopite powder
(b) Crystallized and recrystallized specimens
(after T. Tullis, 1971)

coincidence between predictions and experiments for moderate deformations led Tullis to the conclusion that rotation of the mica flakes is the orienting mechanism. For greater deformations in compaction the discrepancy is attributed to increasing interactions between grain-boundaries which reduce the effectiveness of the shape-induced rotation. Means and Paterson's results (1966) which are now questioned by Etheridge et al. (1973) (see below) agree well with Tullis and lead these authors to the same conclusion. More recently Oertel (1974) brings convincing indirect evidence, from naturally deformed slates, in favour of this interpretation by mechanical reorientation at low temperature. This author finds an excellent agreement between the finite strain directly measured by strain gauges and that computed from the strength of the preferred orientation, assuming mechanical reorientation is the operating process. Tullis and Wood (1975) found a good correlation between finite strain estimated from deformed reduction bodies in natural slates and preferred orientation of mica; their observations are once more in good agreement with the predictions of the March model.

T. Tullis discards slip on (001) as being the orienting mechanism because the patterns of preferred orientations show a point maximum usually parallel to the compression direction as predicted by March, and not small girdles centred on the compression direction as would be expected if slip were active. However, in § 6.2.1 and 6.4.1 it has been shown that in experimental deformation, this theoretical prediction concerning slip fabric is not entirely fulfilled. More convincing are all the descriptions of the coarsest textures which provide no evidence that extensive slip occurred. The crystals are strongly bent and consequently exhibit marked undulatory extinction. However, kinking is not common. In Means and Paterson's and in Tullis' experiments at constant volume, numerous shear zones crossing the sample in orientations of high shear stress are observed. The mica flakes are reoriented in these zones, but the (001) planes are never found to lie parallel to the shear surface, as Etheridge et al. (1973) also reported. This is further evidence against slip as the orienting mechanism. Tullis concludes that these shear zones appeared when increasing strain made it too difficult for the aggregate to deform homogeneously by grain rotation because of the strong preferred orientations of these grains. The analogy between these shear zones and strain slip cleavage is noted by the authors.

6.3.3. High temperature deformation

High temperature syntectonic crystallization and recrystallization in micas have been investigated by Means and Paterson (1966), T. Tullis (1971), Etheridge et al. (1973) and Etheridge and Hobbs (1974). The latter authors have not been interested in the development of preferred orientations as much as in the mechanisms controlling syntectonic recrystallization. With chemical data supporting their interpretation, they propose that nucleation and growth in deformed mica of recrystallized grains is promoted by free energy differences resulting from differences in both chemical composition and stored strain energy. They observe, however, that new grains are often oriented with their (001) plane parallel to kink band boundaries. Recalling that kink band boundaries form at high angles to the principal finite shortening direction, they conclude that this is a potential way to orient the (001) plane of new grains in the flattening plane of metamorphic rocks.

Means and Paterson (1966) conducted experiments involving crystallization of phlogopite from stoichiometric mixtures of oxides at 500–600 °C, 3 kB confining pressure, $10^{-3}\ s^{-1}$ and $10^{-4}\ s^{-1}$ strain rates and 20 % strain. T. Tullis (1971) conducted his crystallization experiments on phlogopite at 775 °C and 15 kB along with recrystallization experiments on pellets of finely ground micas: muscovite recrystallized at 700 °C, 15 kB; fluorophlogopite coarsely recrystallized at 1250–1300 °C, 13 kB; biotite coarsely recrystallized at 850 °C, 14 kB and was deformed at that temperature, $10^{-6}\ s^{-1}$ strain rate. The crystallization experiments were either static and followed by deformation, or syntectonic with or without subsequent annealing at the same temperature.

Well-oriented textures are reported in all these experiments with no mention of any significant discrepancy which could be ascribed to the crystallization-deformation sequences. The (001) plane of micas tends to be normal to the shortening direction (Fig. 6.6). The grains have mean aspect ratios of 4. Their

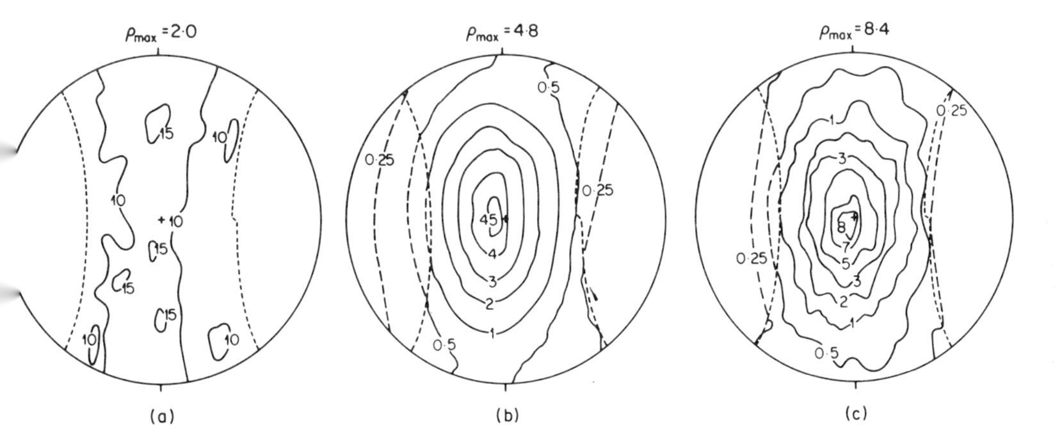

Fig. 6.6. Preferred orientations of [001] in micas at high temperature. Graphical conventions as in Fig. 6.4:
(a) Hydrostatic recrystallization at 1300 °C in fluorophlogopite. This sample does not display any significant change relative to the starting pellet and is the reference state for (b) and (c)
(b) Deformation without recrystallization at 800 °C, 10^{-4} s^{-1}, $\varepsilon = 43$ %
(c) Syntectonic recrystallization at 1300 °C, 10^{-5} s^{-1}, $\varepsilon = 45$ %
(after T. Tullis, 1971).

internal strain features depend on the temperature: for instance, the fluorophlogopite grains deformed at 1300 °C, 10^{-5} s^{-1}, are almost strain free, contrasting with their strong undulose extinctions and kink bands in the tests conducted at 800 °C, 10^{-4} s^{-1}. Two striking features also reported by Etheridge et al. (1973) are that, in the same experimental conditions, mica grows coarser in tests where strain is imposed than in hydrostatic tests and that the larger aspect ratios in mica are associated with the larger strains.

T. Tullis notes that at all strains the strength of the preferred orientation observed in his high temperature runs matches the strength of the preferred orientation predicted by the March model (Fig. 6.5(b)). He concludes that rotation of tabular crystals during deformation (on which the March model is based) is the dominant orienting mechanism, as it is at low temperature. The contrast with the low temperature tests where a divergence was observed for large strain between the experimental results and the prediction (Fig. 6.5(a)) suggests to him that recrystallization evidenced at high temperature plays an active role in reducing the interferences between grains responsible for the divergence at low temperature.

Etheridge et al. (1973) with materials similar to those of Means and Paterson (1966) question the relevance of some results at low temperature in which the observed strong preferred orientations had led to their accepting the interpretation of rotation as the dominant mechanism. They reconsider this interpretation at high temperature when crystallization and recrystallization are active for various reasons in particular because of the increase in size, aspect ratio and preferred orientation normal to σ_1, in the phlogopite crystals with increasing strain. However, a model presented by Oertel (1970) (Fig. 6.7) explains how

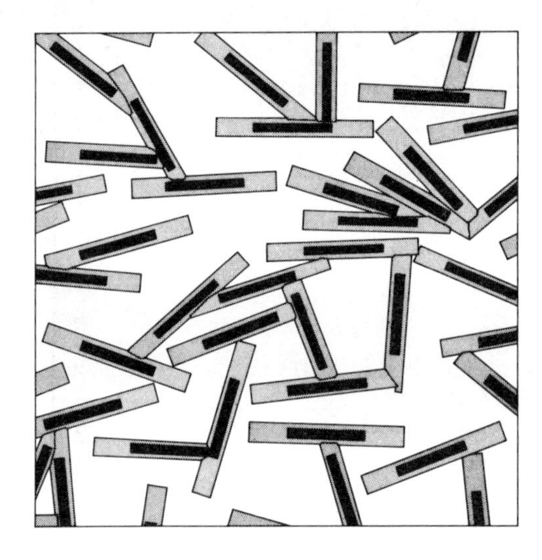

Fig. 6.7. Enhancement of preferred orientation by post-tectonic crystal growth. Black, crystals with a mechanically developed preferred orientation; Gray, the same after growth to twice the original size. Impingement stops more crystals that deviate strongly from the plane of preferred orientation than crystals that are nearly parallel to it (after Oertel, 1970, reproduced by permission of The Geological Society of America)

mica's aspect ratios and preferred orientation can be enhanced by post-tectonic crystal growth: crystals that deviate strongly from the plane of preferred orientation impinge more frequently during growth than crystals which are nearly parallel to it.

Recalling that 20–25 % of the total volume is composed of fluids, Etheridge and his co-workers conclude that the preferred orientation of phlogopite may be controlled by the interaction of several factors. One general factor which may operate in combination with any other one is that the growth rate of mica crystals in the presence of a fluid is the greatest in directions parallel to the (001) plane. Another factor may be an anisotropic pore structure created in the specimen by the strain which would also result in an anisotropic permeability. Provided that the strain has developed some preferred orientation of (001) planes parallel to the flattening orientation of the pores, preferential growth in the pores, possibly aided by the anisotropic permeability, will enhance the existing fabric. Another cause may be an orientation-dependent 'pressure solution' (see § 4.4). The compressive stress acting on a crystal face normal to it will increase its solubility compared with the solubility of a face subject only to the pressure in the fluid. A difference in solubility also arises from the elastic anisotropy in the crystal; in the case of phlogopite the elastic anisotropy effect will contribute to favouring the fastest growing (001) faces which happen to be more stable than others in stressed specimen, whereas the stress-dependent solubility effect will contribute to generating flat shapes normal to the compressive stress. In these conditions, crystals already oriented with (001) normal to the compressive stress will have their flattening markedly increased; crystals inclined to the compressive stress will suffer from competition between the rapid growth parallel to (001) and the pressure solution effect which will tend to inhibit it.

Etheridge et al. (1973) investigations complement those of T. Tullis (1971). At low temperatures where recrystallization does not operate, the dominant orienting mechanism is by rotation of platy sheet silicate. When syntectonic crystallization or recrystallization can occur, this mechanism is complemented or superseded by growth of micas controlled by the strain features in the matrix (flattened pores) or by the local stress field (pressure solution).

6.4. EXPERIMENTAL DEFORMATION OF QUARTZ AGGREGATES

The observation of recrystallized textures in many quartz tectonites and the fact that the first successful creep experiments in quartz developed either recrystallized textures at high temperature (Griggs, 1940, 1941; Griggs et al., 1960) or cataclastic flow at low temperature (Borg et al., 1960) contributed to accredit syntectonic recrystallization as a mode of flow, dominant in the case of quartz rocks. It also explains why in recent years, many important studies, which will be examined in the second part of this section, have been devoted to recrystallization during and after flow (Carter et al., 1964; Raleigh, 1965a; Green, 1966, 1967; Hobbs, 1968; Green et al., 1970). However, at the same time the plastic flow mechanisms in quartz were analysed (see § 5.9) and their contribution to flow in quartz aggregates, clearly shown (Carter et al., 1961; Christie et al., 1964; Heard and Carter, 1968). Blacic and Griggs (1965) were the first to produce preferred orientations in plastically deformed quartzites which they attributed to grain-boundary sliding. We will present here the contributions of J. Tullis (1971) and Tullis et al. (1973) which demonstrated the importance of slip in the development of strong preferred orientations in flowing quartz aggregates.

A detailed account of the different pattern of preferred orientations of quartz in naturally deformed quartz-bearing rocks is given by Turner and Weiss (1963, pp. 425–430). The interpretations proposed at that time are often to be reconsidered in the light of the recent experimental contribution discussed here (see the conclusions in the papers of Tullis et al., 1973, and Green et al., 1970), and of the large number of more recent studies carried out on naturally deformed rocks.

6.4.1. *Plastic flow*

In J. Tullis' experiments, the starting material was intentionally chosen among quartzites with a grain size large enough (around 0·1 mm) in order to evaluate with the optical microscope the response of individual grains to the imposed deformation. The *c*-axis-preferred orientations were determined optically and by an X-ray technique giving the complete orientation of the lattice. More than 200 axial compression experiments in the solid-confining medium apparatus designed by Griggs were performed, mainly in the α-quartz field at temperatures between 400 and 1110 °C, strain rates of $10^{-5}\,s^{-1}$ to $10^{-7}\,s^{-1}$, confining pressure of 15 kB. Some samples were shortened by as much as 75 %. Water was present in the system above 820 °C due to the outer talc sleeve dehydration; this occurred above 750 °C in long-term experiments. Coesite nucleates and grows metastably in most of the samples deformed at lower temperatures and higher strain rates (see § 5.9.5.). In these experimental conditions J. Tullis prospected the complete (T, $\dot{\varepsilon}$) range of conditions in quartzites flow, from cold-working with limited plastic flow before the onset of fracturation, to hot-working with continuous recrystallization.

From Tullis (1971) and Tullis et al. (1973) descriptions, two stages can be envisaged depending on the types of microstructures present in the quartzites.

(*i*) *Plastic flow without recrystallization*

Below 800 °C at $10^{-5}\,s^{-1}$ (or other equivalent (T, $\dot{\varepsilon}$) values) (see Fig. 6.8), the samples work-harden continuously and very high differential stresses are required to achieve even moderate strains (roughly 20 kB at 10 % strain on the samples described below). Deformation takes place mainly by intracrystalline slip as shown by the presence of deformation lamellae and bent planes. Split cylinder experiments according to Raleigh's technique (1968) prove that grain-boundary sliding is a minor occurrence. For moderate strains the sub-basal orientations of lamellae and the geometry of deformation bands suggest that basal slip is dominant; rotations in the bands bring the *c*-axis closer to σ_1. For higher strains, grain flattening perpendicular to σ_1 is more and more conspicuous. Except for 'augen' generally oriented with their *c*-axis parallel or perpendicular to σ_1, the original grains are no longer recognized as such. This is the result of inhomogeneous distribution of slip within the grains, the strain

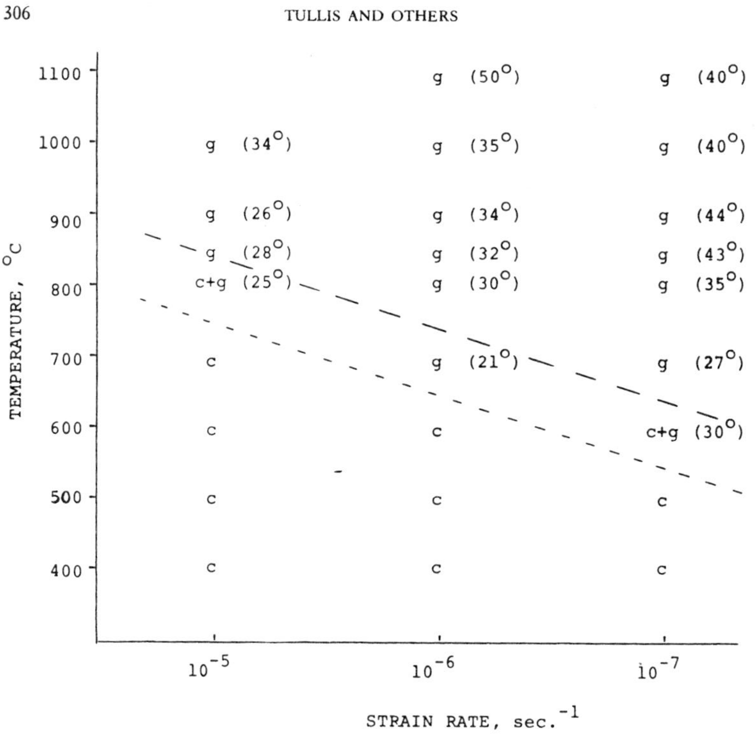

Fig. 6.8. Variation of the preferred orientation of quartz c-axes with temperature and strain rate in quartzites; c indicates a maximum of c-axes parallel to σ_1, g indicates a small girdle of c-axes about σ_1. The number in parentheses indicates the average opening half-angle of the girdle. The transition between the two patterns occurs in the same interval in which a marked decrease in the flow stress (hydrolytic weakening) occurs (short-dashed line). Recrystallization first appears (long-dashed line) at slightly higher temperature (after J. Tullis et al., 1973, reproduced by permission of The Geological Society of America)

concentrating in the regions of the boundaries. Ribbon-like grains, fragmented from original grains, enhance the foliation, they exhibit sharp and clear boundaries, believed to arise when extreme local deformation promotes recovery. The deformation lamellae are no longer basal or subbasal. This is attributed to the activation of non-basal slip systems, but in the light of the most recent TEM observations on lamellae, this interpretation may need a reexamination.

The preferred orientations for these lower temperatures or higher strain rates are shown in Fig. 6.9. The U-stage determination of c-axes shows a maximum parallel to σ_1, stronger as the strain increases. This is explained by basal slip with the accompanying external rotations of c toward σ_1, as in the

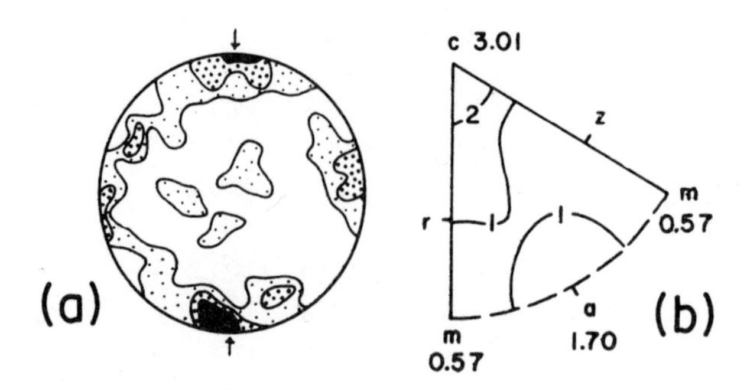

Fig. 6.9. Pole figure (a) and inverse pole figure (b), illustrating the quartz c-maximum type of preferred orientation:
(a) 200 grains; contours 0·5, 1, 2 and 3 % per 1 % area. 500 °C, 15 kB, 10^{-6} s^{-1}, 30 % strain
(b) 600 °C, 15 kB, 10^{-6} s^{-1}, 40 % strain
(after J. Tullis et al., 1973, reproduced by permission of The Geological Society of America)

case of the [010] index in the plastic flow of olivine studied in a preceding section.

The X-ray goniometer allows to determine statically the orientations of the other crystallographic elements (Fig. 6.9(b)), like those of the first- and second-order prism planes and those characterizing positive and negative forms. The poles of the second-order prism planes $\{11\bar{2}0\} = a$ show a significantly higher concentration parallel to σ_1 than do the poles of the first-order prism planes $\{10\bar{1}0\} = m$. This tendency is independent of the type of c-axis preferred orientation and of the presence or absence of recrystallization, thus indicating that these various weak preferred orientations are due to mechanical reorientation. Prismatic slip on $\{11\bar{2}0\}$ in the various directions defined in § 5.9.2 is most probably responsible for them. The relative contributions of m and a slip systems cannot be separated.

The strong asymmetry of the inverse pole figures (Fig. 6.9(b)) relative to the ac line indicates that the poles of $r = (10\bar{1}1)$ positive form have a much stronger tendency to lie parallel to σ_1 than the poles of the corresponding $z = (01\bar{1}1)$ negative forms. This component of the fabric has been ascribed to mechanical Dauphiné twinning (Tullis, 1970; Tullis and Tullis, 1972).

(*ii*) *Plastic flow with recrystallization*

At temperatures of the order of 700 °C and strain rates of 10^{-7} s^{-1}, recovery and recrystallization become important and tend to balance work-hardening: high strains can be achieved with low to moderate differential stresses (roughly 6 kB at 10 % strain). Observation of the textures indicates that plastic flow is

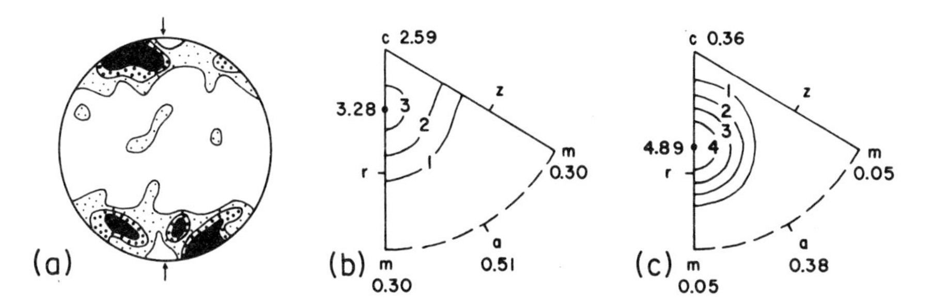

Fig. 6.10. Pole figure (a) and inverse pole figures (b) and (c), illustrating the quartz small-circle girdle type of preferred orientation:
(a) 200 grains; contours 0·5, 1, 2 and 3 % per 1 % area. 1000 °C, 15 kB, $10^{-5}\ s^{-1}$, 60 % strain, 10 % recrystallization
(b) 700 °C, 15 kB, $10^{-6}\ s^{-1}$, 75 % strain, no recrystallization
(c) 900 °C, 15 kB, $10^{-7}\ s^{-1}$, 70 % strain, 80 % recrystallization
(after J. Tullis et al., 1973, reproduced by permission of The Geological Society of America)

still important. With increasing strain, the original grains are more and more flattened in a foliation plane perpendicular to σ_1. They exhibit smooth and fairly continuous undulatory extinction and their deformation lamellae are mostly near-basal. The amount of recrystallization also increases with strain. Small and inequant recrystallized grains are first observed at the grain-boundaries, then, progressively, recrystallization invades the strained domains in the original grains.

At even higher temperatures or lower strain rates (800 °C at $10^{-7}\ s^{-1}$), very low stresses (~1 kB) are required to deform the quartzite. At these temperatures water is released in the samples and weakens them. Recrystallization is extensive or complete and internal deformation features such as lamellae, uncommon. The lamellae are mainly near-basal and to a lesser extent near-prismatic. The recrystallized grains and polygonal subgrains in preserved original grains are of the same general size (roughly 30 μm).

The undeformed augen with their c-axes parallel or perpendicular to σ_1 are still present, even in the highly strained samples. With increasing strain the size of recrystallized grains decreases and they become inequant, (10×20 μm) creating a definite foliation perpendicular to σ_1.

The main change in preferred orientations of the original grains with the onset of recrystallization is that the c-axis maximum parallel to σ_1 characterizing the lower T, higher $\dot{\varepsilon}$ domain, now spreads out, first in a narrow girdle about σ_1 which opens to a maximum angle of 45 ° with increasing temperatures or decreasing strain rates and consequently increasing amounts of recrystallization (Figs. 6.8 and 6.10). The opening angle does not change with increasing strain at a particular temperature and strain rate.

The girdle patterns of preferred orientation of the original grains are considered as stable when flow is due to a combination of basal and prismatic slip. A computed model shows that the ratio of the strain due to basal slip to that due to prismatic slip, determines the steady-state angle by which the c-axes of grains of all initial orientations will rotate for an infinite strain. In the model the steady-state angle increases with an increase in the proportion of prismatic slip: for example, the angle will be 22 ° assuming the ratio of shear on the basal plane to shear on the prism planes is 6/1, and 35 ° with a ratio of 2/1. This correlates with the larger girdle opening angles which are observed in samples deformed at higher temperatures when prismatic slip predominates over basal slip (§ 5.9.2).

As already stated, the $(T, \dot{\varepsilon})$ conditions do not modify the preferred orientations of first- and second-order prisms planes. The positive and negative forms are unchanged except in the case of some samples deforming with recrystallization at high temperature or low strain rates.

6.4.2. Syntectonic recrystallization

The most extensive studies on the textures and preferred orientations developed by syntectonic recrystallization have been carried out by Green (Green, 1968; Green et al., 1970), they come after Carter et al. (1964) and Raleigh's (1965) investigations on the textures created in quartz aggregates by syntectonic recrystallization. They are well complemented by Hobbs' (1968) experiments on recrystallization of single crystals which open the way to understanding the recrystallization mechanisms and their bearing on the development of preferred orientations.

Green's experiments were carried out on flint and novaculite samples with a grain size of 1 to 5 μm and a water content of 1·2 to 0·1 %; most were conducted in axial compression and a few in axial extension. The range of conditions were: temperatures from 300 to 1300 °C, that is in the β-quartz field above 950 °; strain rates from $10^{-4}\ s^{-1}$ to $10^{-7}\ s^{-1}$; confining pressures (solid medium) from 5 to 25 kB, strains up to 80 %. The data on preferred orientations were obtained mainly with the X-ray goniometer technique devised by Baker et al. (1969) and, when possible, the c-axis orientations were determined optically.

(a) *Compression tests in the α-quartz field*

Fig. 6.11 shows the influence of temperature and strain rates on the onset of recrystallization in Green's fine-grained aggregates. For comparison sake, the corresponding conditions in Tullis's quartzites have also been indicated (Tullis, 1971).

The water present in the starting material or released during the experiments has a determining influence on the temperature at which recrystallization is initiated and on the way in which it evolves with increasing temperature or decreasing strain rates (Hobbs, 1968; McLaren and Retchford, 1969; Green et al., 1970; Tullis, 1970). This fact, along with the influence of the grain size of the starting material on recrystallization (Green et al., 1970; Tullis, 1971), can explain the discrepancy observed on the diagram of Fig. 6.11 between Green's

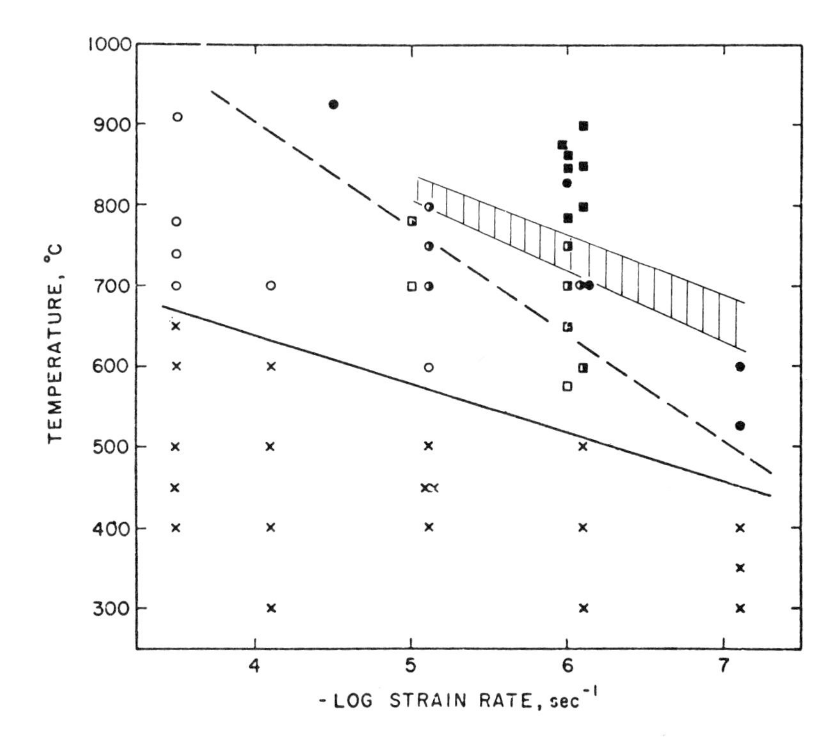

Fig. 6.11. Variation of the preferred orientation of quartz c-axes with temperature and strain rates in flints. Open symbols: c-maximum fabric; closed symbols: r-maximum fabric; half-closed figures: composite fabrics; Xs: specimens that did not recrystallize and did not develop measurable preferred orientation. (Squares: X-ray and universal stage determinations; circles: flat-stage estimations.) The hatched area represents the incipient recrystallization in quartzites (see Fig. 6.8.) (after Green et al., 1970 (modified), reproduced by permission of Springer-Verlag, Heidelberg)

and Tullis's results. In this brief review of the physical factors initiating and activating recrystallization, the effect of strain (Hobbs, 1968; Tullis, 1971), also evidenced in peridotites (Nicolas et al., 1973), should not be omitted.

Above the threshold for recrystallization, in compression tests, at lower temperatures and higher strain rates the grains are flattened normal to the compressive stress σ_1. The shape anisotropy generally increases with increasing strain and can reach ratios as high as 8 to 1. The best evidence that recrystallization is effective is that the grain size is increased to dimensions of the order of 10 μm. Intragranular strain features are abundant and the grain boundaries are diffuse and irregular. This textural domain is also characterized by a *c-maximum fabric* parallel to σ_1 (Fig. 6.11). On the inverse pole figure corresponding to this type of preferred orientations, the concentrations of σ_1 are somewhat greater when parallel to the poles of the positive trigonal forms than when parallel to the poles of their negative counterparts (Fig. 6.17(a)).

With increasing temperature and decreasing strain rates, the grains become progressively larger (up to 100 μm) and, at the highest temperatures in slower experiments, they are more equant. Intragranular strain features become less prominent and grain boundaries more sharply defined and polygonal but without any tendency to develop common morphological faces. This high temperature, low strain rate, domain is also that of *r-maximum fabric* parallel to σ_1 (Figs. 6.11 and 6.12). The inverse pole figures (Fig. 6.17(c)) clearly show

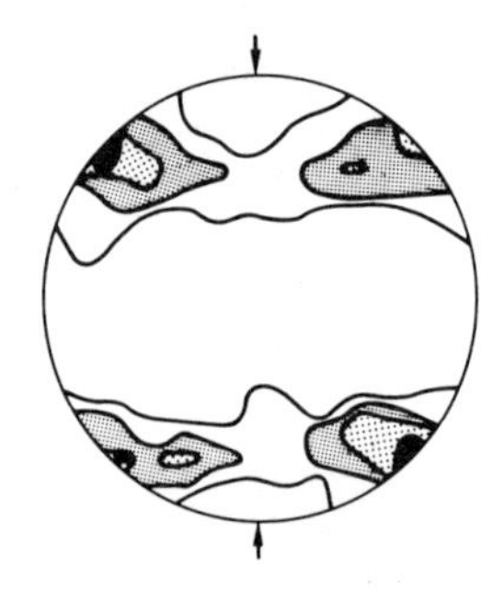

Fig. 6.12. c-axes pole figure from a recrystallized flint specimen exhibiting the r-maximum preferred orientation. 516 grains. Contours: 1, 2, 3, 4 % per 1 % area. 800 °C, 15 kB, $10^{-6}\,s^{-1}$, 29 % strain (after Green et al., 1970, reproduced by permission of Springer-Verlag, Heidelberg)

the $r_1 = (10\bar{1}1)$ maximum parallel to σ_1, contrasting with a $z = (01\bar{1}1)$ minimum.

Composite preferred orientations are met in the transition zone from c-maximum to r-maximum fabrics in Fig. 6.11. On inverse pole figures they display the typical aspects of both (Fig. 6.17(b)), that is two maxima, one at c and one near the pole of r. With increasing temperature the sequence is first the broadening of the c maximum, next the development of a minimum ~20° to σ_1 and finally the disappearance of the c-maximum.

Crossed-girdle preferred orientations were obtained in the r-maximum field in two samples exhibiting a pronounced lack of axial symmetry. The crossed girdles of c-axes are presented in the diagram of Fig. 6.13 with the inferred orientations of σ_2, σ_3 and ε_2, ε_3.

(b) *Compression tests in the β-quartz field*

In the β-quartz field, the grain-flattening perpendicular to σ_1 is weak or non-existent. Porphyroblasts tend to grow in the matrix; they are bounded by

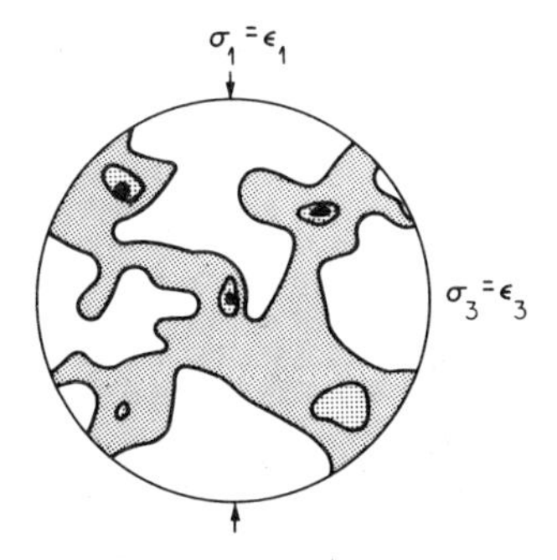

Fig. 6.13. c-axes pole figure of a specimen exhibiting a crossed-girdle pattern. 597 grains. Contours: 1, 2, 2·5 % per 1 % area. 850 °C, 16 kB, 10^{-6} s^{-1}, 28 % strain (after Green et al., 1970, reproduced by permission of Springer-Verlag, Heidelberg)

$\{10\bar{1}1\}$ planes. The coarser grained β-quartz specimens at higher temperatures have planar or near-planar boundaries, symmetrically inclined to the c-axis, probably approximating $\{10\bar{1}1\}$ planes. The preferred orientations combine a c-axis maximum parallel to σ_1 with a concentration normal to σ_1 (Fig. 6.14).

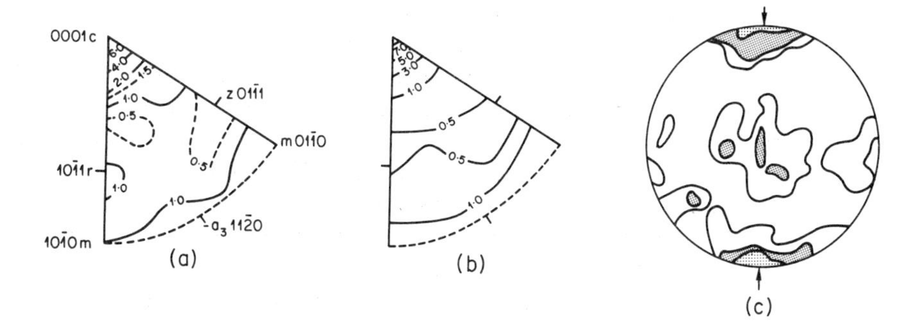

Fig. 6.14. Pole figure (c) and inverse pole figures (a) and (b) from specimens compressed in the β-quartz field:
(a) 750 °C, 6 kB, 10^{-5} s^{-1}, 34 % strain
(b) 800 °C, 4 kB, 10^{-4} s^{-1}, 37 % strain
(c) 254 grains. Contours: 1, 2, 4 % per 1 % area. 1100 °C, 15 kB, 10^{-6} s^{-1}, 16 % strain
(after Green et al., 1970, reproduced by permission of Springer-Verlag, Heidelberg)

(c) *Extension tests*

There is a tendency for grains to be elongated in the axial extension direction (σ_3). Qualitative estimations in the α-quartz field indicate that c-axes are concentrated at high angles to the extension direction. Fig. 6.15 shows the results of c-preferred orientation measurements in the β-quartz field: a weak girdle is present, normal to the extension direction.

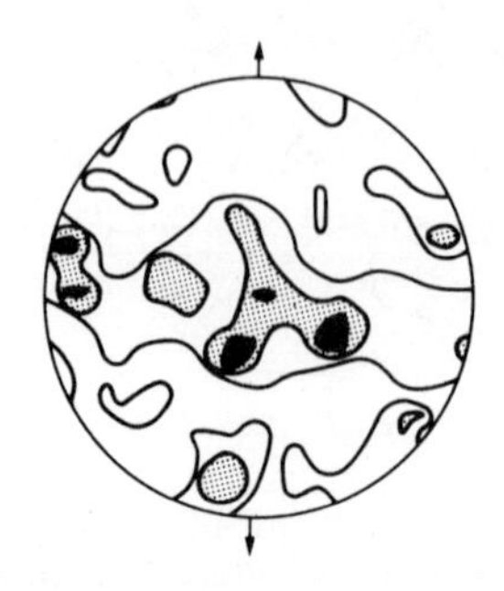

Fig. 6.15. c-axes pole figure from a specimen deformed, in extension, in the β-quartz stability field. 150 grains. Contours: 1, 2, 3 % per 1 % area. 950 °C, 15 kB, $10^{-6}\,s^{-1}$, 10 % strain (after Green et al., 1970, reproduced by permission of Springer-Verlag, Heidelberg)

(d) *Single crystals syntectonic recrystallization*

Hobbs (1968) shortened single crystals of both natural and synthetic quartz up to 50 % at 10–15 kB in the range of 400–950 °C and with strain rates of $10^{-5}\,s^{-1}$ to $10^{-6}\,s^{-1}$. The recrystallization did not take place before 800 °C so that the process occurred in the β-quartz field.

In these conditions, for small strains, a dislocation substructure is first formed within wide deformation bands. The subgrains are delimited by {0001} and {11$\bar{2}$0} walls and in some cases by other planes at high angles to {0001}. Misorientations across walls are of the order of 5°. As the strain increases the misorientation between subgrains in the deformation band also increases to 20–30°. At high strains new grains are observed in the deformation bands; they are strain-free with sharp boundaries but their shape and size, identical to that of subgrains in which they grade progressively, show that the new grains derive directly from the subgrains. Consideration of the diagrams of Fig. 6.16 leads to the following conclusions:

(i) Large rotations of the c-axis in the host crystal are achieved by cumulative subgrain rotations (Fig. 6.16(a), (d), (g), (j)) with a progressive migration toward σ_1, except in Fig. 6.16(g), in which the starting c orientation is normal to σ_1.

(ii) The recrystallized grains orientations are scattered within 30–50° of those of the subgrains from which they derive.

This recrystallization process by progressive misorientation of subgrains, which has since been more explicitly studied (Poirier, 1974; Poirier and Nicolas, 1975, see §4.5.4), is clearly demonstrated here in the case of experimentally deformed quartz. It has also been verified in the case of naturally

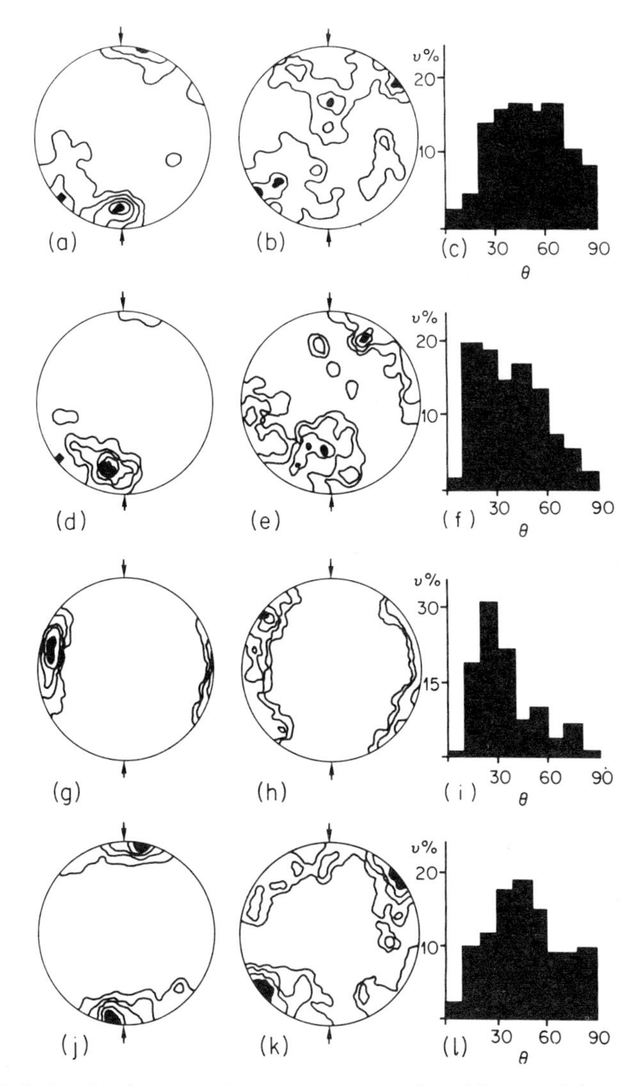

Fig. 6.16. Relationship between the quartz *c*-axes of neoblasts and the *c*-axes of their host crystal in syntectonic recrystallization experiments. 900 °C, 10 kB. The full square represents the initial host orientation of *c*. (a), (d), (g), (j) *c*-axis orientations in the deformed host crystal loaded in various orientations; (b), (e), (h), (k), *c*-axis orientations in neoblasts adjacent to the host; (c), (f), (i), (l), histograms showing frequency, ν, of angles θ, between *c*-axes of neoblasts and adjacent host crystal: (a), (b), (c) 200 *c*-axes for the host loaded in the 0^+ orientation (a): Contours: 1, 5, 10, 15, 20 % per 1 % area; (b): Contours: 1, 3, 5 %. (d), (e), (f) 96 *c*-axes for the host loaded in the *r* orientation. (d): Contours: 1, 5, 10, 15, 20 % per 1 % area; (e) Contours: 1, 2, 4, 6 %. (g), (h), (i) 112 *c*-axes for the host loaded in the $\perp$ *m* orientation. (g) Contours: 1, 5, 10, 16, 20 % per 1 % area; (h) Contours: 1, 3, 6, 9, 12 %. (j), (k), (l) 155 *c*-axes for the host loaded in the $\|$ *c* orientation. (j) Contours: 1, 5, 10, 15, 20 % per 1 % area; (k) Contours: 1, 2, 4, 6 % (after Hobbs, 1968, reproduced from *Tectonophysics* by permission of Elsevier Scientific Publishing Company, Amsterdam)

deformed quartz by Ransom (1971); although Hara and Paulitsch (1971) do not directly refer to this mechanism they bring strong evidence for it. As shown in Fig. 6.16. the *orientation of the new grains is controlled by that of the host crystal.* A puzzling orientation for the new grains is, however, present in Fig. 6.16(k), where they tend to be at 45° to the host and its subgrains.

Locally, the existence of larger new grains with serrated boundaries suggests *that strain-induced boundary migration* also takes place.

6.4.3. Annealing recrystallization

The two principal studies on annealing in quartz are those of Green et al. (1970) and of Hobbs (1968), the former on fine-grained aggregates, the latter on single crystals. Green has only considered the case of hydrostatic annealing, unlike Hobbs who has also studied stress-relaxation (or stress annealing).

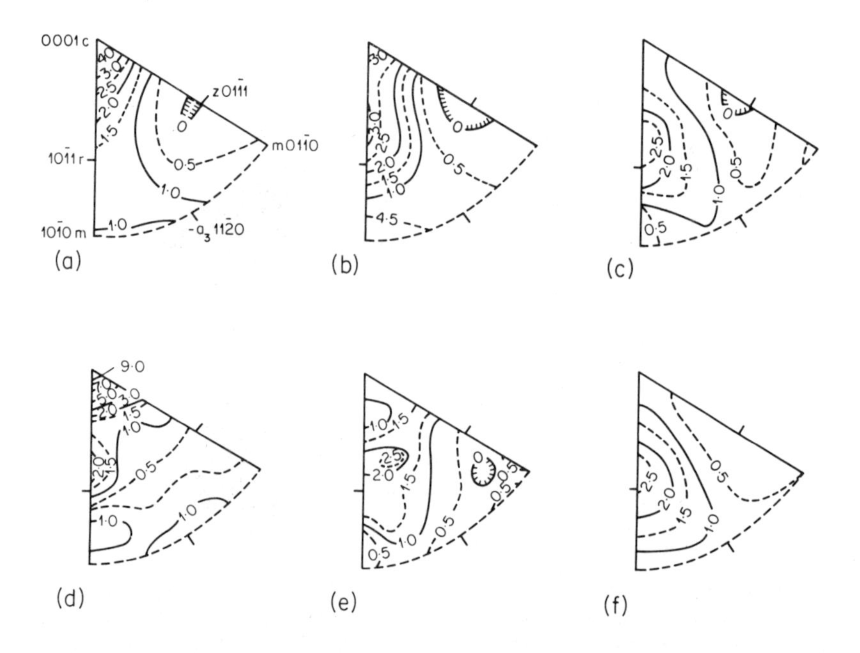

Fig. 6.17. Inverse pole figures of three annealed specimens (d–f) of flints and their non-annealed counterparts (a–c):

(a) *c*-maximum preferred orientation (700 °C, 16 kB, 10^{-5} s^{-1}, 34 % strain) and (d) annealed version (700 °C, 70 hours)

(b) Composite preferred orientation (780 °C, 16 kB, 10^{-5} s^{-1}, 41 % strain) and (e) annealed version (900 °C, 48 hours)

(c) *r*-maximum preferred orientation (780 °C, 15 kB, 10^{-6} s^{-1}, 34 % strain) and (f) annealed version (900 °C, 48 hours)

(after Green et al., 1970, reproduced by permission of Springer-Verlag, Heidelberg)

Annealing of fine-grained aggregates

In Green's experiments the temperature and duration of the anneals vary greatly (*T*: 450–1110 °C; *t*: 20 to 74 hours). Anneals in both the α- and β-quartz fields result in a large coarsening of the texture, up to 8–10 times; the flattening of the grains and the intragranular strain features characteristic of the syntectonic recrystallization have disappeared. In the β-field, the square grains described in the syntectonic recrystallization textures are well represented and are often bounded by $\{10\bar{1}1\}$ planes.

The changes caused by annealing in the preferred orientations developed during syntectonic recrystallization in the α-quartz field are recorded in Fig. 6.17. The c-maximum fabric is strengthened; the composite fabric tends to lose its c-maximum parallel to the compression axis and to reinforce its 30–40 ° girdle component; the r-maximum fabric is not significantly changed. It seems that samples which had not previously recrystallized and had developed a weak c-axis maximum fabric parallel to the cylinder axis acquire by annealing a distinct mean orientation with c-axes at 70–80 ° to the cylinder axis.

In the β-quartz field, annealing produces a great strengthening of the $c \| \sigma_1$ fabric component. When the grain size before anneal is very fine, the strenthening is extreme and the $c \perp \sigma_1$ component is completely obliterated (Green, 1967; Fig. 6.18). This strengthening is strongest in regions where the grain

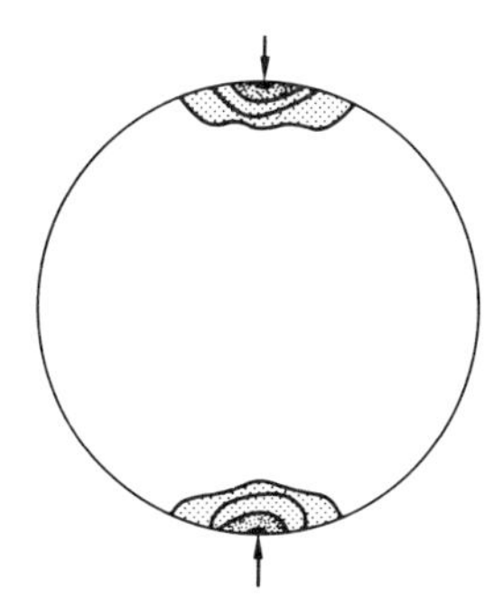

Fig. 6.18. c-axes pole figure from a specimen deformed and annealed in the β-quartz field at 900 °C, for 48 hours, after compression at 750 °C, 6 kB, $10^{-5}\,s^{-1}$, 34 % strain. 200 grains. Contours: 1, 10, 30, 50 % per 1 % area. Maximum concentration is 60 % (after Green et al., 1970, reproduced by permission of Springer-Verlag, Heidelberg)

growth was greatest during annealing. The regions are not necessarily those where the grain size is the coarsest after annealing, but those where there has been the greatest reduction in the number of grains during the process (reduction in the ratio of 500/1). Green (1967) attributes this special effect to the fact that growth of the grains with c parallel to the cylinder axis is much faster than that of the grains in other directions.

Annealing of single crystals

Hobbs' (1968) experiments on annealing and stress annealing (mainly in the α-quartz but in some cases also in the β-quartz field) lead to very comparable results as far as textures and preferred orientations are concerned. A significant amount of recrystallization appears mainly in 'wet' experiments.

New quartz grains nucleate along the approximately basal lamellae, or at the boundaries of deformation bands, and grow faster parallel to the boundary than normal to it, so that elongated grain-shapes tend to form. When the grains are large, euhedral prisms tend to be best developed but after the grain size has

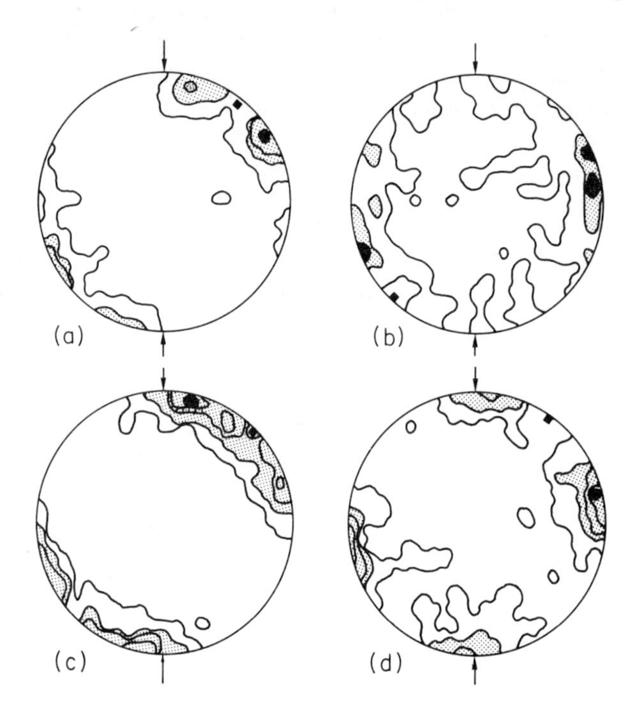

Fig. 6.19. c-axes pole figure of quartz neoblasts formed by annealing of single crystals compressed in the 0^+ and $\perp r$ orientations (full square: c-axis of the host crystal):
(a) $\perp r$ orientation. 210 grains. Contours: 1, 5, 9, 13 % per 1 % area. Annealing at 1280 °C for 61 min, wet
(b) $\perp r$ orientation. 300 grains. Contours: 1, 3, 4 % per 1 % area. Annealing at 1550 °C for 62 min, wet
(c) 0^+ orientation. 250 grains. Contours: 1, 3, 5, 7, 9 % per 1 % area. Annealing at 900 °C for 2 h, dry
(d) 0^+ orientation. 228 grains. Contours: 1, 3, 5, 7, 9 % per 1 % area. Annealing at 900 °C for 2 h, wet. Nucleation occurs along kinks developed during compression approximately parallel to $\{11\bar{2}0\}$
(after Hobbs, 1968, reproduced from *Tectonophysics* by permission of Elsevier Scientific Publishing Company, Amsterdam)

increased to the extent where all grains are touching, the grain shape becomes polygonal. Hobbs has also investigated the kinetic aspects of nucleation and grain growth. At temperatures above 650 °C, the nucleation of all new grains is completed within 30 seconds; it takes days at 300 °. Grain growth leads to complete recrystallization in 90 seconds at 900 °C. At 600 °C the volume fraction recrystallized never increases above 5 % even after heating for $12\frac{1}{2}$ hours. At 300 °C no more than 1 % is recrystallized after 7 days.

Fig. 6.19 illustrates the preferred orientations of c-axes in annealing experiments. The host control over the orientation of the new grains appears immediately but as in the case of syntectonic recrystallization, the new grains are oriented at 20–40 ° to the host. This result is independent of the host orientation with respect to σ_1. The grains in the host orientation remain small and are progressively consumed by grains in other orientations. This conflicts with some recrystallization experiments conducted by Green et al. (1970) on bicrystals with different orientations in the β-quartz field. They concluded that the new grains nucleated and were oriented with their c-axes parallel to that of the host by growing at the bicrystal boundary. Some ambiguity may perhaps arise when it comes to deciding whether a given grain belongs to one particular host or to the other since they are at 45 ° to each other, it is thus possible that Hobbs' conclusion still holds here.

Another result of Hobbs' annealing experiments is that the presence or absence of stress has no appreciable influence on the pattern of preferred orientations developed during annealing. This conforms with the general conclusions of Chapter 4.

The similarity of the annealing results with those of syntectonic recrystallization also contributes to ruling out stress as a factor of development of preferred orientations in the latter case. This similarity should also call for identical recrystallization mechanisms. Now, for annealing, Hobbs, according to the evidence presented above, suggests a process controlled by nucleation of grains with preferred orientations in the kink bands followed by a preferred growth of grains in certain orientations. In syntectonic recrystallization, along with nucleation and grain-boundary migration, he mentioned progressive misorientation of subgrains as a possible mechanism.

6.5 EXPERIMENTAL DEFORMATION OF CALCITE AGGREGATES

Plastic properties in calcite aggregates, essentially Yule marble, were studied very early because large strain can be imposed at the low temperatures and pressures which were accessible to the first experimentalists. Starting with Griggs (1936), a vast programme of experimental investigation on the deformation of Yule marble in the 1950s and early 1960s led to an excellent grasp of that particular question and of the plastic flow mechanisms in rocks, also inviting comparison with metals (Griggs and Miller, 1951; Handin and Griggs, 1951; Turner and Chi'h, 1951; Griggs et al., 1951, 1953; Borg and Turner,

1953; Turner et al., 1956; Griggs et al., 1960a and b). The findings of this period of investigation are summarized by Turner and Weiss (1963, pp. 339–355). From our point of view the relevant studies carried out since then are those of Karl and Kern (1968) who achieved triaxial strain experiments and those of Wenk et al. (1973) who presented an extensive study of uniaxial deformation in fine-grained limestones. The latter authors also attempted to correlate their experimental results with observations made on naturally deformed limestones.

6.5.1. Textures and preferred orientations developed in uniaxial experiments

Griggs, Turner and their coworkers deformed Yule marble either by compression or by extension. These conditions were obtained in specimens highly constrained by a liquid or a solid medium in a piston cylinder apparatus. The textures and preferred orientations with respect to the deviatoric stress orientation compare well in both cases. The range of experimental conditions was 20 to 150 °C at 10 kB and 150 to 800 °C at 5 kB. Wenk et al. (1973) operated only by compression from 25 to 1000 °C, with strain rates from 10^{-4} s^{-1} to 10^{-7} s^{-1} and confining pressures from 5 kB to 22 kB, usually around 10 kB. They synthetized aragonite in its stability field, above 500 °C. The inverse relation between temperature and strain rates has been more thoroughly investigated in the case of marble by Ferreira and Turner (1964) who showed that lowering the strain rate from 10^{-4} s^{-1} to 10^{-7} s^{-1} was equivalent to increasing the temperature by 100 °C.

Textures

With increasing strain, the calcite grains are more and more flattened or elongated depending on the compressive or extensive nature of the experiment. As already mentioned in the study of calcite plasticity (§ 5.10.2), at low temperatures swarms of thin *e* lamellae appear in the grains on those *e* planes for which the CRSS is large and the sense of gliding positive. Slip on *r*, in the grains unfavourably oriented for twinning, is deduced from irrational rotation of lamellae inherited from earlier stages. Undulatory extinctions are also reported with a 20 % strain, the *e* lamellae become very numerous, coalesce and give almost completely twinned grains with a few clouded relict grains. With increasing temperature the grains are less clouded; the lamellae grow thicker and lensoid in aspect and at 700 to 800 °C, in spite of high strains, they seldom occupy more than half of the grains surface. In all experiments except those at such high temperatures, there is a profuse development of extremely thin lamellae superimposed on each other and oriented at a high angle to the compressive stress, that is in the direction of low CRSS. They are thought to have developed during the experiment relaxation. For this reason, they will not be discussed further.

Remarkably enough, in hot working conditions, recrystallization is very weak in Yule marble. Griggs et al. (1960b) report the occurrence of local clusters of new grains generated at 600–800 °C along the boundaries of highly strained primary grains.* This process is increasingly effective between 400 and 600 °C and decreases above this to become insignificant at 800 °C. In limestones, grain growth is obvious above 700 °C (Wenk et al., 1973) and at 900–1000 °C the texture is entirely recrystallized to a mosaic-shaped assemblage with a marble like texture. The first evidence of recrystallization in limestones is observed at 350 °C.

In syntectonic recrystallization of marble (Turner et al., 1956) incipient recrystallization seems to appear at 400 °C, but in annealing (Griggs et al., 1960a) it occurs only at 500 °C.

Preferred orientations

At low temperatures, even for moderate strains, a strong preferred orientations of calcite c-axes is generated. Beyond 20 % strain, the preferred orientation does not increase in a comparable manner (compare Fig. 6.20(b) and (c). As seen on Fig. 6.20(d), the final fabric reflects the influence of the initial one in its symmetry.

At higher temperatures the strength of the preferred orientations decreases (Fig. 6.21). For large strains (50 % shortening, 100 % elongation), the c-axes tend to be inclined at 20 to 30 ° to σ_1 in compression and at 60 to 70 ° to σ_1 in extension. Turner et al. (1956) regard this as a stable orientation for calcite grains deforming by e twinning and r slip. This is illustrated by the texture and preferred orientation in the neck of a sample after an extension experiment conducted at 500 °C, 5 kB, the strain having reached 118 % locally (Fig. 8.10). The a-axes in this experiment, as in others where large strains have been attained, present preferred orientations which correlate with those of the corresponding c-axes. They are characterized by three point maxima (X, Y, Z) 60 ° apart which are also considered as defining a stable orientation of calcite (Fig. 6.22). Inverse pole figures obtained by the X-ray goniometer technique (Wenk et al., 1973) complement these data by showing the preferred orientations of other crystallographic directions (Fig. 6.23).

At low temperatures a shoulder of σ_1 orientation extends from c toward the low angle (0kil) rhombohedra (negative forms) (Fig. 6.23(b) and (c)) and with very large strains a second maximum, comparable with that in c may develop in the vicinity of e (Fig. 6.23(d)).

At higher temperatures (500 to 600 °C) the inverse pole figures are almost axially symmetrical (Fig. 6.23(f) and (g)) although, as in other cases, there is an absolute minimum close to f form. At still higher temperatures a shoulder developing towards the positive forms (Fig. 6.23(k)) ultimately centres on r at 1000 °C in completely recrystallized textures (Fig. 6.23 (l) and (m)). Low

* In this paper, they also report the existence of intragranular recrystallization forming elongated lensoid domains close to $\{02\bar{2}1\} = f$ of the host. The c-axis in these domains tended to be aligned with σ_1, this interpretation is now dismissed in favour of kinked structures with external rotation of c-axis toward σ_1 (Turner, in Wenk et al., 1973). Comparable structures with c-axis in the vicinity of σ_1 are described by Kern and Karl (1968), although in this case they are $\{01\bar{1}2\}$ lamellae. It is suggested that attributing them to recrystallization may need the same revision.

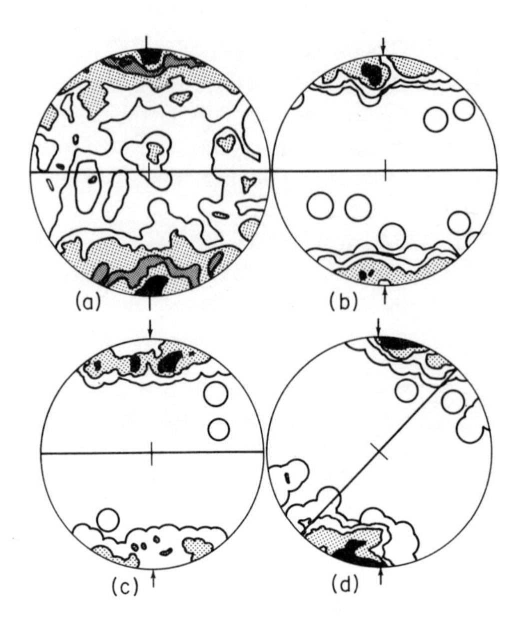

Fig. 6.20. c-axes pole figures of calcite in Yule marble. E.W. straight line: foliation trace in the starting material. Compression axis N.S.:
(a) 172 grains in the undeformed specimen. Contours 0·3, 1, 3, 5, 7 % per 1 % area
(b) and (d) 100 grains in specimens shortened 19 % at 300 °C, 5 kB. Contours: 1, 3, 5, 10 % per 1 % area
(c) 100 grains in a specimen shortened 39·9 % at 400 °C, 3 kB. Contours: 1, 5, 10 % per 1 % area
(after Turner and Weiss, 1963, reproduced by permission of McGraw-Hill Book Company, New York)

pressures seem to produce a subsidiary maximum of the high angle $(h0\bar{h}1)$ rhombohedra (positive forms) (compare Fig. 6.23(h) and (m) with (g)).

The technique used here makes it impossible to distinguish the contribution to the total fabric of the plastically deformed and recrystallized grains when they are simultaneously present. Measurements with the U-stage avoid this problem. Griggs et al. (1960b) show that the c-axis preferred orientation of the new grains generated at the boundary of the highly strained grains lie at 80–90° to the extension direction and that individually they diverge from the c-axis of their host by 20–30° (Fig. 6.24). However, those new grains which obviously grew during a late stage of the deformation process may even postdate it (Turner, in Wenk et al., 1973).

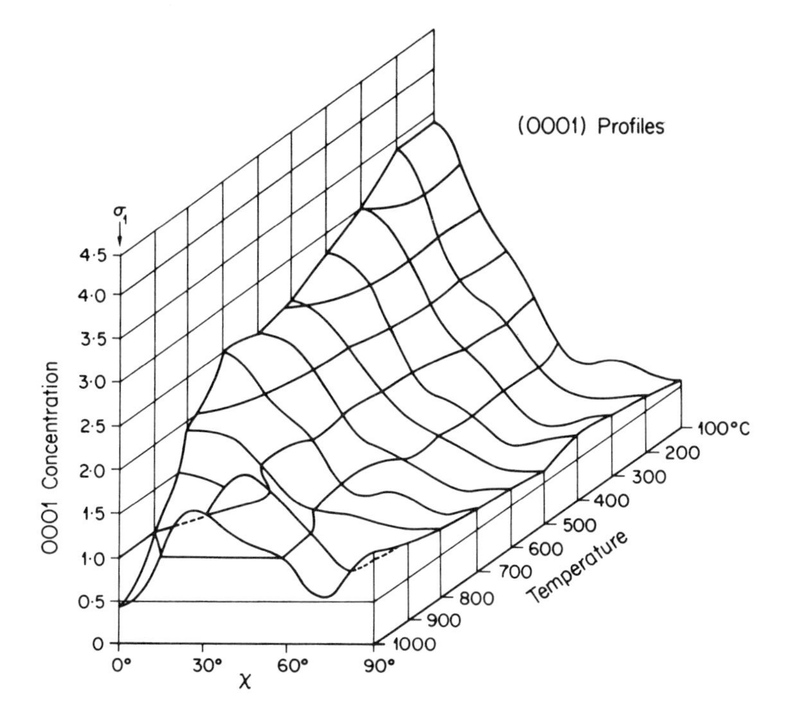

Fig. 6.21. c-axes profiles of calcite in fine grained limestones, from parallel to normal to the compression axis as a function of temperature (confining pressure 9 to 11 kB) (after Wenk et al., 1973, reproduced by permission of Springer-Verlag, Heidelberg)

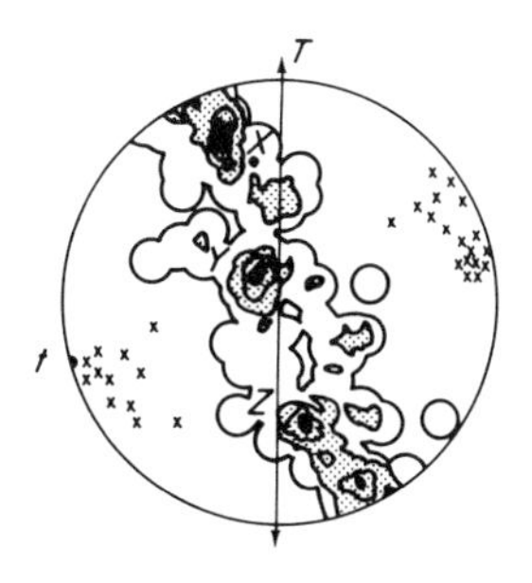

Fig. 6.22. a-axes pole figure of calcite in Yule marble. N.S. line is the trace of the foliation in the starting material (Fig. 6.20a). It is parallel to the extension axis. 90a-axes in 30 grains whose c-axes are shown by crosses. Contours: 1, 3, 5·5, 9 % per 1 % area (after Turner et al., 1956, reproduced by permission of The Geological Society of America)

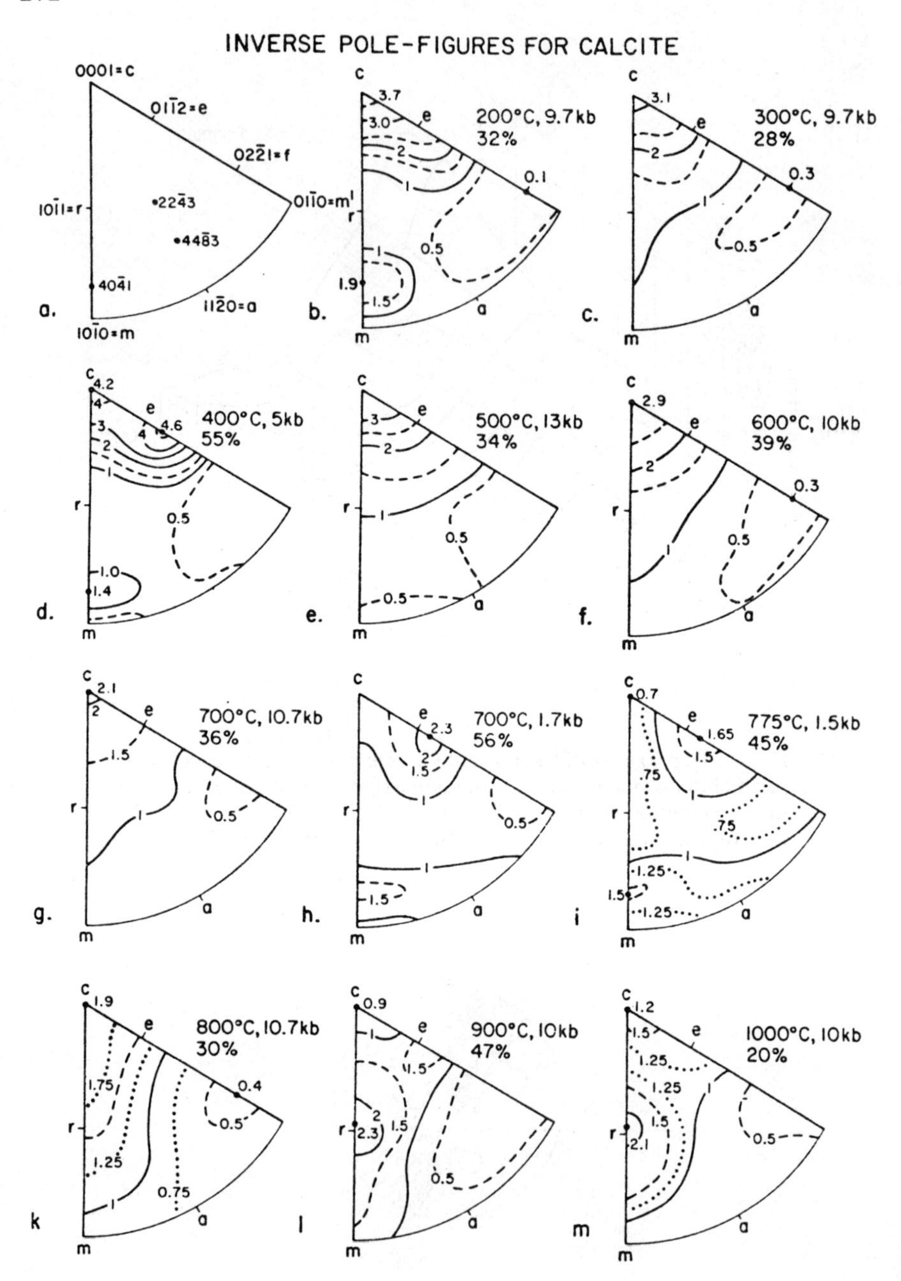

Fig. 6.23. Inverse pole figure of calcite in fine-grained limestones. (a) Crystallographic orientation of calcite; (b–m) Specimens deformed in conditions specified on the figure (after Wenk et al., 1973, reproduced by permission of Springer-Verlag, Heidelberg)

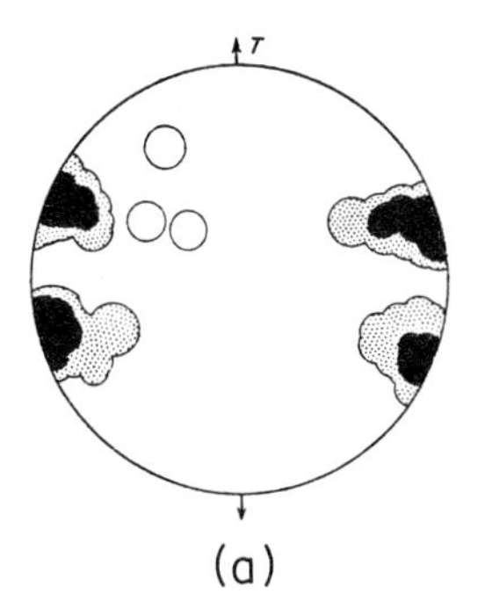

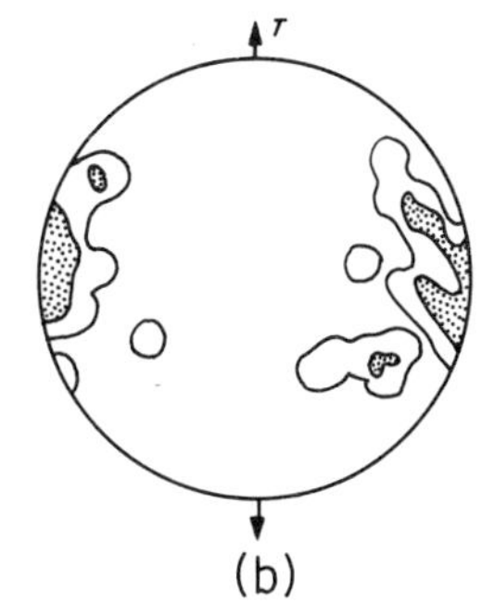

Fig. 6.24. c-axes pole figure of calcite in Yule marble deformed by N.S. extension at 600 °C, 3 kB, 590 % strain:
(a) 38 grains across neck. Contours: 3, 8 % per 1 % area
(b) 46 larger recrystallized grains. Contours: 2, 6 % per 1 % area

(after Griggs et al., 1960)

6.5.2. Interpretation

Plastic flow

The way in which strong preferred orientations are produced in calcite aggregates by plastic flow is well known largely because, in contrast with quartz, calcite presents internal tracers of strain, i.e. twin lamellae and cleavages. Turner and his co-workers (1954) took advantage of this in their geometrical analyses. Calcite can deform, mainly by e twinning and r slip, on several independent systems, and meets fairly well the Von Misés criterion. Thus, it is possible to apply Taylor's analysis (1938) for predicting preferred orientations with increasing strain, although it is strictly valid only for a homogeneous deformation (i.e. where each grain deforms in the same way as the aggregate). This was achieved by Handin and Griggs in 1951 and yielded predictions which strikingly fitted the experimental results. Admitting that the strain is approximately homogeneous in any grain means moreover that the stable orientations of calcite in various homogeneous flow regimes can be calculated. Calnan and Clew's method (1950, 1951a and b) was successfully applied by Wenk et al. (1973) in a similar attempt to explain low temperature preferred orientations.

Reorientation during strain depends on e twinning, which is a powerful mechanism since the c-axis is rotated through 52° toward σ_1, and on r slip which is more sluggish and proceeds by external rotation as in the cases already studied (§ 6.2.1, 6.4.1, 6.9). Both mechanisms act simultaneously but in different proportions according to the strain and the temperature.

At low temperatures, all the crystals oriented favourably for e twinning (possible only in the positive sense, see § 5.10.3) do twin. Consequently, a strong preferred orientation of c-axes close to σ_1 rapidly develops below 20 % total strain. The maximum strain possible in a crystal by complete twinning is 29 % by compression (Handin and Griggs, 1951), thus suggesting that at 20 %

total strain all the crystals that could twin have completely twinned. For larger strains the grains will continue to deform by *r* slip. Henceforth, they behave like other grains whose initial orientation prevented *e* twinning at the onset. In this new stage, the preferred orientation changes comparatively little per unit strain. For 50 % shortening, experiments and calculations indicate that the stable orientation consists of an *e* plane and an *a*-axis approximately normal to σ_1, with the *c*-axis of the principal lattice at 26° to σ_1. In the elongation experiments, the stable orientation is the same, as shown on Fig. 6.22 (T position).

Wenk's calculations also predict that *e* twinning operating alone would produce a broad *c*-axis concentration parallel to σ_1. When combined with *r* slip, it favours an *e* concentration close to σ_1; if *r* slip operates alone, *e* and σ_1 coincide. These predictions are in agreement with the conclusions just presented and with the experimental findings: the *c*-axis maximum observed for moderate strain (Fig. 6.23(c)) would derive from the predominance of *e* twinning and the *e* pole maximum observed for large strain (Fig. 6.23(d)), from a combination of *e* twinning and *r* slip.

In plastic flow at high temperature, *r* slip competes more efficiently with *e* twinning, as shown by incomplete twinning in most grains. This may explain why the preferred orientations are weaker; there is also a possibility that other slip systems may be activated. The apparition of a new fabric induced by recrystallization as observed by Wenk and his co-workers may contribute to the activation of new systems.

Recrystallization

Here the results and interpretation of Griggs et al. (1960b) conflict with those of Wenk et al. (1973). In Griggs' experiments on Yule marble, limited recrystallization is observed with new orientations close to those of the host grains. The strain-induced mechanism (§ 4.5), with an orientation controlled by the host, is invoked. In this framework, the observation of the progressively decreasing amount of recrystallization above 600 °C is explained by the decrease in the recrystallization driving strain energy, which in turn is itself a consequence of the decrease in the marble strength.

Recrystallization in Wenk and co-workers fine-grained limestone is more fully documented. Due to the large angular difference the *r*-shoulder and *r*-maximum at 1000 °C (Fig. 6.23(k), (l) and (m)) can hardly be attributed to a host-controlled recrystallization. This is probably why, following Kern's reasoning (1971), Wenk et al. explain it by MacDonald's thermodynamic theory (1960) for a stable orientation in a non-hydrostatic stress field. This theory based on the assumption that the stable orientation is that which maximizes the strain energy has been shown by Kamb (1961) to be invalid. As a possible orienting mechanism, it also suffers from the drawbacks inherent to Kamb's theory itself (§ 4.4). For these reasons, although at present no other explanation can be proposed for the preferred orientation of recrystallized calcite grains, this interpretation does not seem to be satisfactory.

6.5.3. Preferred orientations developed in triaxial experiments

Triaxial deformation was performed by Karl and Kern. Their first experiments dealt with marbles deformed at room temperature (Karl and Kern, 1968) and at 650° (Kern and Karl, 1968). The lamellae and c-axis orientations were subsequently studied by optical means. Kern's more recent experiments

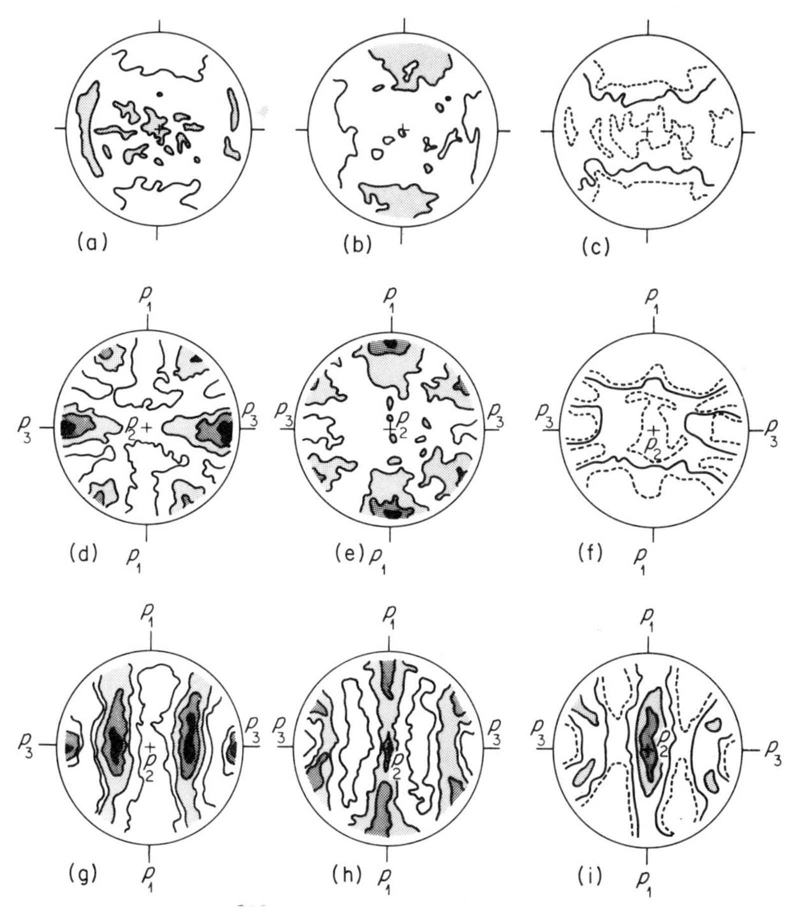

Fig. 6.25. Preferred orientation of calcite in triaxial deformation a, d, g: r-rhombohedron orientations; b, e, h: a-prism orientations; c, f, i: e-rhombohedron orientations:

(a), (b), (c) Initial preferred orientation. X-ray relative intensity: $>2-1(-0{\cdot}5)-0$

(d), (e), (f) Plane strain at 650 °C; $p_1 = 8{\cdot}24$ kB, $p_2 = 4{\cdot}12$ kB, $p_3 = 2{\cdot}74$ kB, $\varepsilon_1 = 16{\cdot}6$ %, $\varepsilon_2 = 0{\cdot}65$ %, $\varepsilon_3 = -10{\cdot}7$ %. X-ray relative intensity: $>4-3-2-1(-0{\cdot}5)-0$

(g), (h), (i) Triaxial strain at 600 °C; $p_1 = 6{\cdot}87$ kB, $p_2 = 4{\cdot}58$ kB, $p_3 = 2{\cdot}29$ kB; $\varepsilon_1 = 23{\cdot}7$ %, $\varepsilon_2 = 9{\cdot}2$ %, $\varepsilon_3 = -16{\cdot}1$ %. X-ray relative intensity: $>4-3-2-1(-0{\cdot}5)-0$

(after Kern, 1971, reproduced by permission of Springer-Verlag, Heidelberg)

(1971) applied to fine-grained Solnhofen limestone deformed between 20 and 650 °C and studied for preferred orientations by X-ray texture analysis techniques. In all the room temperature tests the total strain is of the order of a few percent in the three main directions, which is insufficient to cause any significant c-axis concentration, even considering that the predominant orientating mechanism is e twinning. Appreciable strain, that is in excess of 10 %, in the shortening direction, is only attained at high temperatures, 600 °C and 650 °C. Point maxima are found in experiments which approximate plane strain (Fig. 6.25(d), (e) and (f)) and girdle patterns around σ_3 and ε_2 are found in other triaxial experiments (Fig. 6.25(g), (h) and (i)).

Fig. 6.26(b) represents the results of a uniaxial test at 650 °C with 19·3 % shortening ($\sigma_1 = 6{\cdot}87$ kB). The c-axis preferred orientation, in particular the girdle at 90° to σ_1, does not conform to the results presented above for

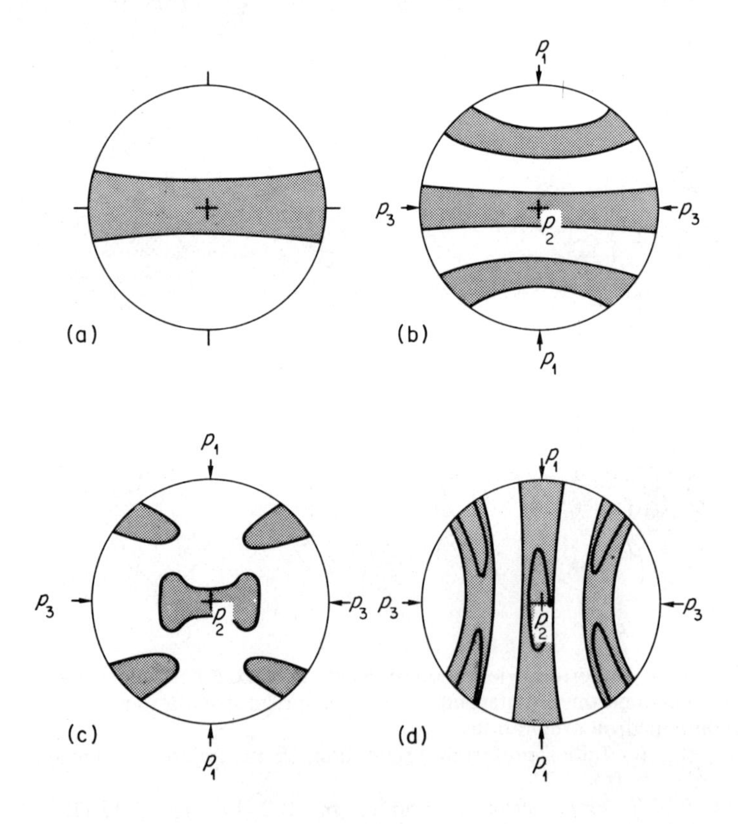

Fig. 6.26. c-axes pole figures of calcite in uniaxial and triaxial deformation. Experimental conditions as in Fig. 6.25:
(a) Initial preferred orientation, (b) uniaxial strain, (c) plane strain, (d) other triaxial strain
(after Kern, 1971, reproduced by permission of Springer-Verlag, Heidelberg)

limestones (Fig. 6.21). Comparing it with the fabric of the starting material (Fig. 6.26(a)), we suggest that the strain has been insufficient to obliterate it. The initial fabric may be reflected too in the triaxial experiments for which the strain is comparable (Fig. 6.26(c) and (d)). We are inclined to resort to the same explanation in the case of similar experiments in a moderately anisotropic marble from Auerbach. It was deformed at 650 °C with stress ratios 9 : 6 : 1 and 9 : 3 : 1 and shortening between 12 and 15 % and elongations between 9·5 and 11 %. The resulting preferred orientations reflect only a very weak migration of c-axes toward σ_1. This hypothesis is reinforced by the fact that in a uniaxial experiment with 30·5 % shortening a stronger c-axes concentration is observed.

6.5.4. Annealing

In Yule marble deformed at low temperature and subsequently annealed, incipient recrystallization occurs at 500 °C and is complete at 800 °C provided the strain was sufficient during cold working (Griggs et al., 1960a). For strain of the order of a few percent both the incipient and complete recrystallization temperatures are higher. This shows the influence of strain on annealing. Probably the same explanation has to be invoked in the annealing of single crystals after equal shortenings achieved in one case by r slip and in the other by e twinning at room temperature. In the former case annealing at 800 °C produces a complete recrystallization to an aggregate, whereas in the latter it results only in incipient recrystallization. This could be due to the fact that r slip induces more dislocations in the lattice than e twinning.

Solnhofen limestone recrystallizes at 800 °C without prior strain, illustrating the mechanism of recrystallization induced by surface energy (§ 4.5.3). Granoblastic textures, occasionally containing porphyroblasts, are produced by annealing. The preferred orientations are much weaker than in the preceding cold-worked aggregate. Observation of deformed crystals suggests that they recrystallize to a cluster of 5 to 10 new grains with their c-axes inclined at angles of less than 40° to c of the host grain.

6.6. EXPERIMENTAL DEFORMATION OF DOLOMITE AGGREGATES

6.6.1. Plastic flow

Plastic flow was experimentally investigated in coarse-grained dolomite aggregates by Turner et al. (1954b), Handin and Fairbairn (1955) and Griggs et al. (1960b) and in fine-grained aggregates (5–20 μ) by Wenk and Shore (1974, in press). Unfortunately, in the two former studies, strains larger than 10 % could not be achieved before failure; in Griggs et al.'s experiment, although it was conducted at a higher temperature (500 °C), only 13 % strain was obtained without any clearly developed plastic features. Strains up to 30 % were obtained by Wenk and Shore in the plastic flow domain.

In the section on plasticity in dolomite crystals (§ 5.10) the dominant mode of flow below 400 °C was said to be basal slip in the *a* directions; above 400 °C *f* twinning away from the *c*-axis predominates due to a smaller CRSS (Fig. 5.41). From this it can be predicted that at lower temperatures, when *c*-slip predominates, large strains would be required to produce strong preferred orientations whereas at higher temperatures, *f* twinning through a 54° rotation of the *c*-axis would readily produce a strong preferred orientation even for a moderate strain, as in the case of *e* twinning in calcite, at low temperature (Handin and Fairbairn, 1955). In this respect, calcite and dolomite aggregates have contrasting plastic behaviour.

Actually, the experiments conducted at room temperature with less than 10 % strain did not modify the preferred orientation in the starting dolomite specimen; nor is there an appreciable development of *f* lamellae (Handin and Fairbairn, 1955). However, at 300 °C the lamellae develop profusely in the grains favourably oriented for twinning, i.e. those having their *c*-axes inclined at 25° or less to the compression axis. Again, in Turner et al.'s experiments with 9·4 % strain, at 380 °C no definite change is registered in the initial preferred orientation. In Handin and Fairbairn's experiments with comparable strain at 300°C in compression, the *c*-axes possibly migrate slightly away from σ_1 as predicted.

In recent experiments in compression, Wenk and Shore (1975) report that below 700 °C, where no recrystallization seems to occur, the preferred orientations of σ_1 on the inverse pole figure are weak and complex (Figs. 6.27 and 6.28): the principal maximum is near *e*, a secondary maximum near a high-angle positive rhomb and the principal minima are at *c* and *f*. When analysed with Calnan and Clew's (1950, 1951) method, this inverse pole figure is shown to be consistent with *f* twinning and *r* slip, which was not previously described. These two mechanisms counteract each other, explaining why under these conditions preferred orientation in dolomite is much weaker than in calcite aggregates. Below 100 °C the minimum at *c* is less pronounced suggesting that *c*-slip may be active, but no firm conclusion can be reached because only small strains can be induced in these very strong specimens.

6.6.2. *Syntectonic and annealing recrystallization*

Neumann (1969) optically investigated the preferred orientations developed in dolomite rocks during syntectonic recrystallization and annealing. The experimental conditions for syntectonic recrystallization were: $T = 1000$ °C, $P = 15$ kB, $\dot{\varepsilon} = 8 \times 10^{-7}$ s^{-1}. In the coarser grained specimens, the old grains are replaced by several new grains, one-tenth their diameter, with a new granoblastic texture. A tendency for *c*-axes to be at a low angle to σ_1 is observed. A study of the orientation of new grains relative to their hosts suggests that the *c*-axes in the new grains are inclined at angles of 30 and 60° to the *c*-axis in the host. Annealing experiments, conducted at 1000 °C, 15 kB for 13 and 24 hours after 25 and 16 % strain at 200 °C result in the same preferred orientation of *c*-axes close to σ_1.

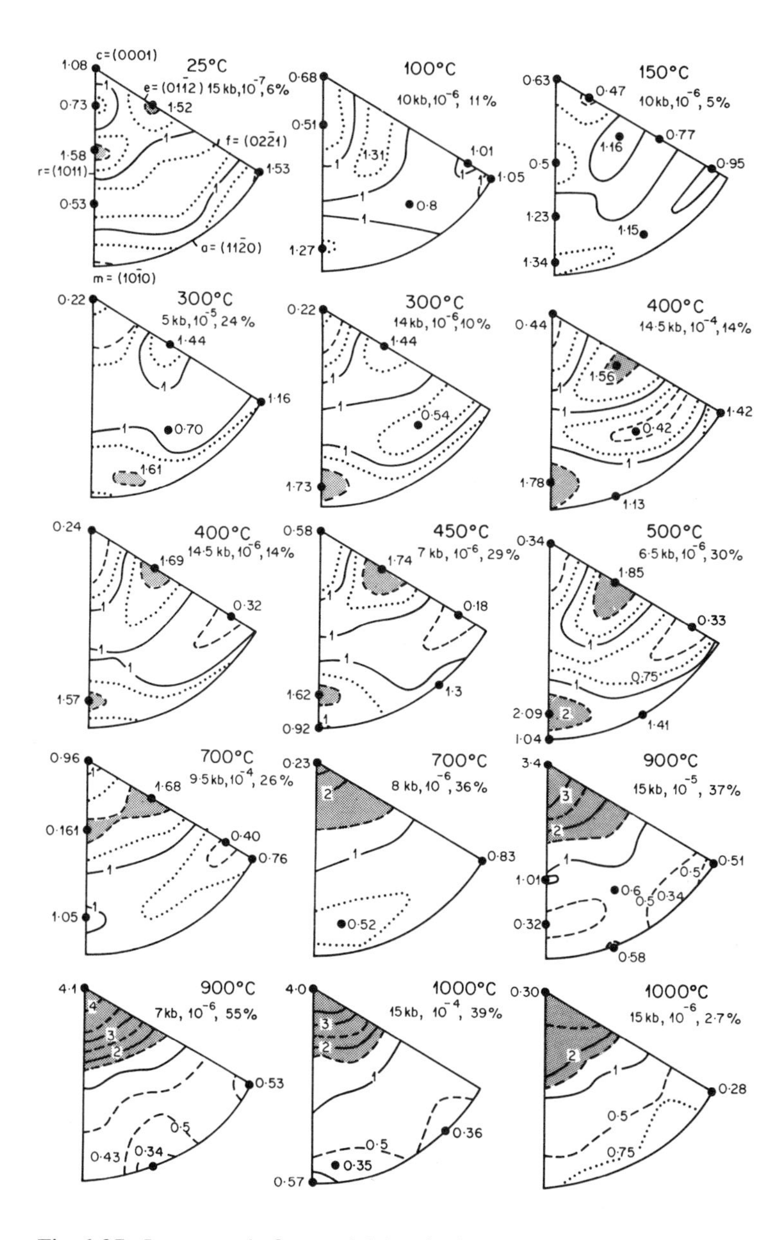

Fig. 6.27. Inverse pole figure of dolomite in uniaxial compression tests, experimental conditions are indicated. Ordered with increasing temperature. Contours in multiples of a uniform distribution: 0·5, 0·75, 1, 1·5, 1·75, 2, 2·5, 3, 3·5, 4 (after Wenk and Shore, 1975, reproduced by permission of Springer-Verlag, Heidelberg)

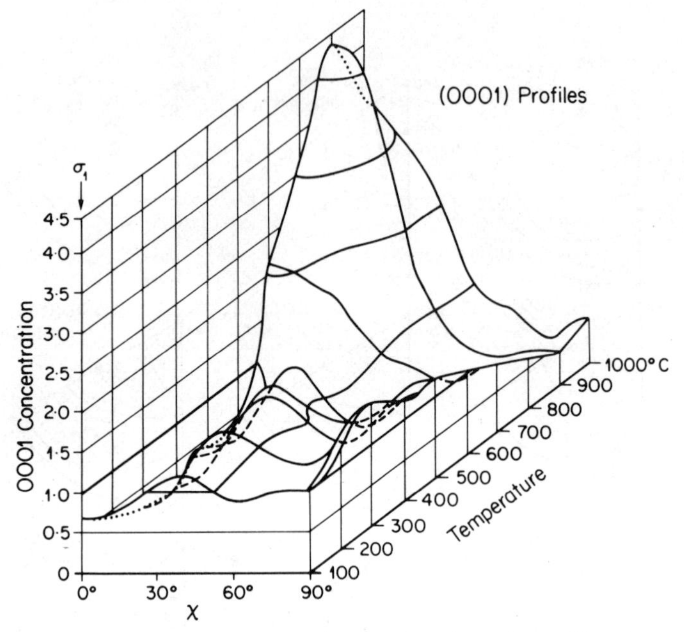

Fig. 6.28. Block diagram with normalized c-axis profiles from parallel to normal to the compression axis as a function of temperature. All experiments performed at $10^{-6}\ s^{-1}$, 10 kB, shortening from 15 to 40 % (after Wenk and Shore, 1975, reproduced by permission of Springer-Verlag, Heidelberg)

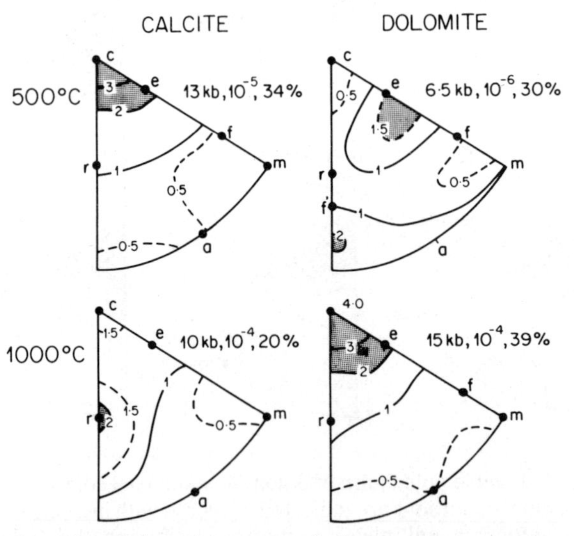

Fig. 6.29. Selected inverse pole figures representative for low and high temperature preferred orientations in calcite and dolomite. Shaded areas indicate maxima (after Wenk and Shore, 1975, reproduced by permission of Springer–Verlag, Heidelberg)

Wenk and Shore observed no increase in the grain size of their dolomite rocks, even at 1000 °C at low strain rates. From the colour change to pure white above 700 °C, they conclude that grain-boundary diffusion becomes important above that temperature. This 700 °C limit is strongly reflected in the preferred orientation patterns: above, a good *c* axis maximum parallel to σ_1 contrasting with the weak e/f maxima found below. The *c* maximum can be explained by basal slip being the dominant slip system at high temperature. Unfortunately in these fine-grained rocks the necessary textural evidence is lacking.

Fig. 6.29 summarizes the large differences between calcite and dolomite orientations in both cold and hot working conditions. As pointed out by Neumann, it is therefore puzzling that in naturally deformed dolomites and limestones or marbles, similar *c*-axis preferred orientations are often observed normal to the foliation or forming a girdle normal to the lineation in tectonites dominated by this structural feature. This important topic requires further studies, a conclusion reached at by Wenk and Shore who also report the case of a natural deformation in which dolomite and calcite display distinct fabrics.

6.7. EXPERIMENTAL DEFORMATION OF HALITE AGGREGATES

The experimental deformation of halite aggregates was investigated by Kern and Braun (1973) on artificial fine-grained specimens with a random fabric. They used an apparatus producing triaxial stresses normal to the faces of cubic specimens, as in the case of calcite aggregates (§ 6.5.3). The temperature range was 20–200 °C, the maximum compressive stress σ_1 below 2 kB, and σ_3 maintained at 0·25 kB. The preferred orientations of $\langle 110 \rangle$, $\langle 111 \rangle$ and $\langle 100 \rangle$ were determined with an X-ray texture goniometer. The specimens were studied in uniaxial compression (Fig. 6.30(a), (b) and (c)), triaxial compression, plane strain (Fig. 6.30(d), (e) and (f)), and uniaxial extension (Fig. 6.30(g), (h) and (i)). The uniaxial tests resulted in axially symmetrical pole figures and the triaxial ones in pole figures with an orthorhombic symmetry, which also combined features of both kinds of axial tests. These experimental results are compatible with Schwerdtner's predictions (1968)* of fabrics using Calnan and Clew's (1950, 1951a and b) and Dillamore and Roberts (1964) theoretical methods, which support $\{110\}$ slip as the dominant plastic mechanism responsible for the flow and the development of fabrics.

6.8. EXPERIMENTAL DEFORMATION OF ANHYDRITE AGGREGATES

This question was investigated by Muller and Siemes (1974) in axial compression experiments up to 5 kB and 300 °C, at a constant strain rate of

* A comparison with naturally deformed rocks from salt domes is also incorporated in this paper and in a more recent one (Schwerdtner and Morrison, 1973).

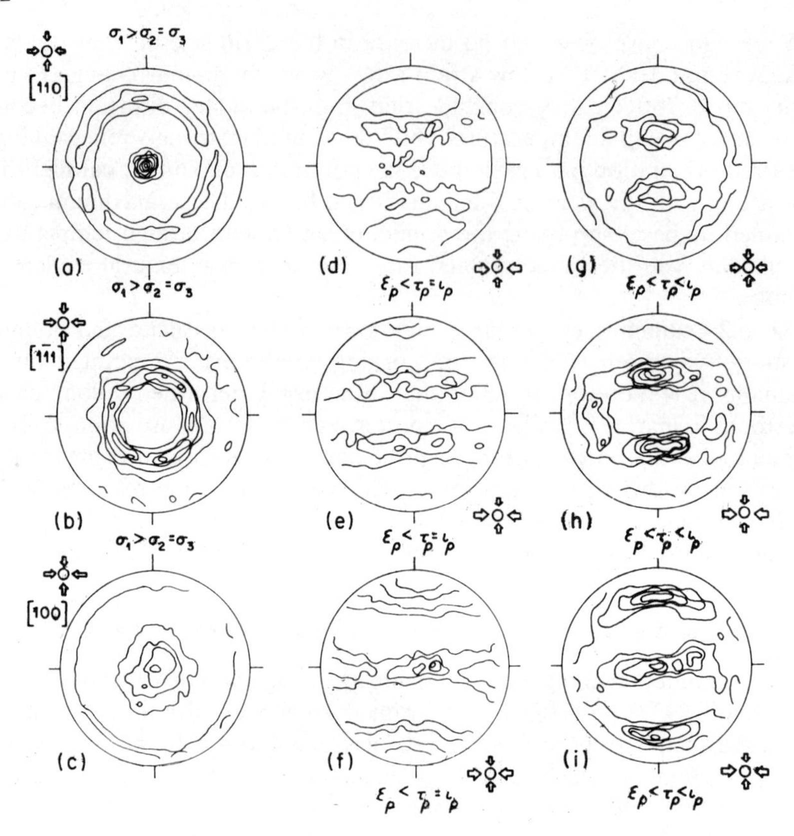

Fig. 6.30. Preferred orientation of halite in uniaxial and triaxial deformation at 100 °C. The maximum compressive stress σ_1 is in the centre of the diagrams. (a), (d), (g): [110] orientations; (b), (e), (h): [111] orientations; (c), (f), (i): [100] orientations. Contours: 1, 1·2, 1·3, etc., %:
(a), (b), (c) Uniaxial compression. $\sigma_1 = 0{\cdot}95$ kB, $\sigma_2 = \sigma_3 = 0{\cdot}25$ kB; $\varepsilon_1 = 21{\cdot}4$ %, $\varepsilon_2 = \varepsilon_3 = -9{\cdot}7$ %
(d), (e), (f) Plane strain. $\sigma_1 = 1{\cdot}57$ kB, $\sigma_3 = 0{\cdot}25$ kB; $\varepsilon_1 = 18{\cdot}34$ %, $\varepsilon_2 = -1{\cdot}03$ %, $\varepsilon_3 = -15{\cdot}78$ %
(g), (h), (i) Uniaxial extension. $\sigma_1 = \sigma_2 = 1{\cdot}69$ kB, $\sigma_3 = 0{\cdot}25$ kB; $\varepsilon_1 = \varepsilon_2 = 12{\cdot}5$ %, $\varepsilon_3 = -22{\cdot}45$ %
(after Kern and Braun, 1973, reproduced by permission of Springer-Verlag, Heidelberg)

$0{\cdot}5\ 10^{-3}\ s^{-1}$. Strong preferred orientations were developed and analysed with an X-ray texture goniometer. For technical reasons the pole figures correspond to the (002/020)† and (210) reflections which are in no way related to the slip

† These two reflections cannot be separated on account of the fact that the interreticular distances are too close.

systems (see § 5.12). Since the tests are axially symmetrical, the preferred orientations of these lattice elements can be analysed on the profiles of Fig. 6.31.

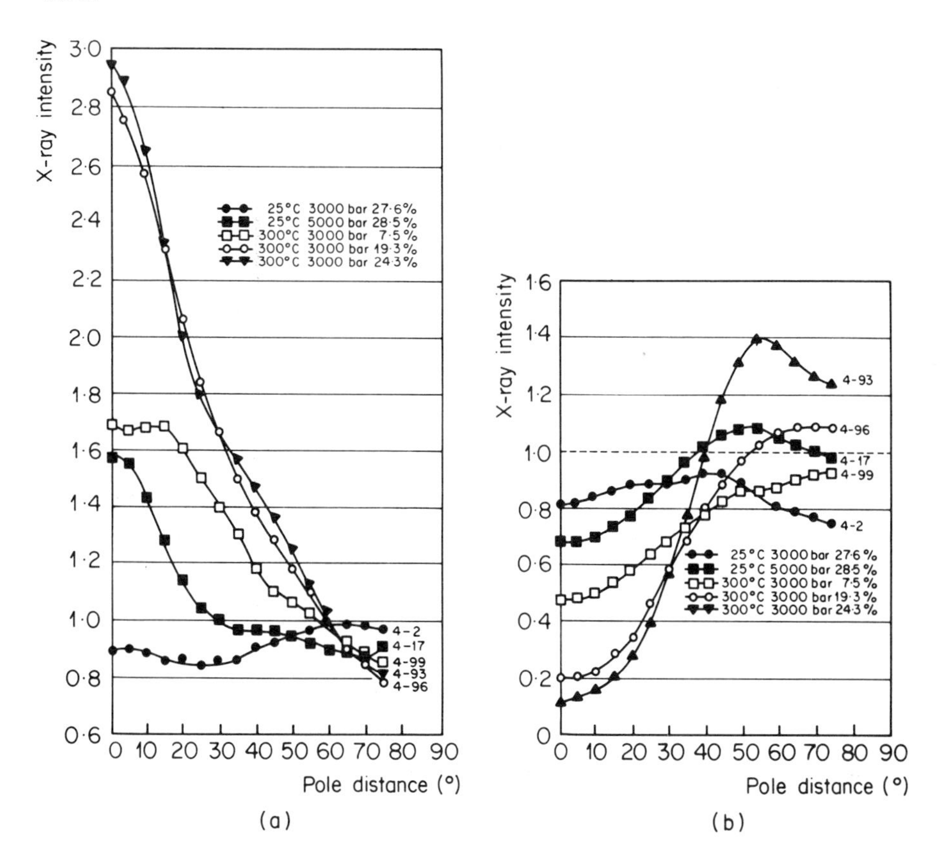

Fig. 6.31. Mean intensity profiles in anhydrite from parallel to normal to the compression axis in uniaxial tests:
(a) Profiles of the pole figures of the (210) reflection
(b) Profiles of the pole figure of the (002/020) reflection
(after Muller and Siemes, 1974)

At room temperature below 3 kB no preferred orientation appears even for fairly large strains. Above 3 kB, moderate preferred orientations are developed (solid square lines on Fig. 6.31) which are assigned to intragranular slip.

At 300 °C, preferred orientations are progressively stronger with increasing strain, but above about 20 % strain they do not change significantly whilst no more strain hardening is recorded by the stress strain curves.

Schwerdtner (1970) reports that, in naturally deformed anhydrite rocks, [010] anhydrite axes are parallel to the lineation and [001] often normal to the foliation or form girdles normal to the lineation. He finds this orientation incompatible with intragranular slip as the orienting mechanism and favours Kamb's theory of syntectonic recrystallization. We disagree with his analysis which relies on unwarranted assumptions about the stress deviator orientation relative to the fabric elements and on their direct correlation. The reported natural preferred orientations are indeed compatible with (001) [010] as the dominant flow mechanism (compare with olivine in peridotites, Chapter 10). However, it must be noted that the experimental preferred orientations just described, which were also attributed to intracrystalline slip, cannot have been produced by the action of (001) [010] alone.

6.9 A COMPUTER SIMULATION OF PLASTIC FLOW

6.9.1. Object

Mineral preferred orientations developed by slip in compressive or extensive strain, rolling or other deformation regime, can be predicted using Boas and Schmid's (1931) and Taylor's (1938) methods as applied in the case of calcite and dolomite aggregates, respectively by Handin and Griggs (1951) and by Handin and Fairbairn (1955) (§ 6.5.2 and 6.6.1). Calnan and Clew's simpler method (1950, 1951a and b) was also applied in the case of quartz aggregates by Bhattacharya and Pasayat (1968), in that of limestone by Wenk et al. (1973) (§ 6.5.2) and in that of halite by Schwerdtner (1968) who also introduces the more recent method of Dillamore and Roberts (1964) (§ 6.7) and gives a brief account of the principles underlying these modelling methods.

In these predictions on preferred orientation, nothing is known about the development of the oriented texture with increasing strain.

It is, however, a very important matter in our general object of kinematic analysis of textures and preferred orientations in naturally deformed rocks (§ 7.2). It would complement the information obtained on experimentally deformed mineral aggregates because, on the necessarily simple model, the orienting mechanisms and their bearing on textures and preferred orientations are more easily elucidated.

For technical reasons, experimental deformations have not been carried out in shear conditions with significant results. The absence of information on the development of textures and preferred orientations in these conditions is critical since we consider that shear (rotational shear flow) is largely responsible for tectonite fabrics in many deep-seated rocks (see Chapters 10 and 11).

These objects are partially fulfilled by Etchecopar's (1974) model, presented below, which simulates the development of oriented textures and preferred orientations under various conditions.

6.9.2. Method

By a method of finite increments, a computer simulation of large deformations by pure shear, simple shear and a combination of both has been carried out. The prime interest of the pure shear simulation was to check the validity of the method by comparing its predictions with what has been learnt from experimental deformation. This comparison encourages acceptance of the predictions made concerning simple shear.

The simulation models are two-dimensional and suppose that the slip directions of individual cells are in the plane of deformation and remain there. The cells have a hexagonal contour (this does not seem critical), and a single slip direction. Other restrictive conditions are that in a cell the slip must be homogeneous and no bending allowed. Finally, because dynamical or rheological considerations, like critical shear stress for slip, viscosity, etc., have not been retained, the cell is perfectly fluid in the slip direction and undeformable in other directions.

For each finite strain increment, in a first stage, the contours of each hexagonal cell are modified to fit perfectly the new shape imposed on the whole assemblage (no voids, no overlaps). In a second stage, returning to the original cells, each one is required to fit its new perfect shape by combining slip, rotations and translations: voids and overlaps appear. Corners common to three adjacent cells in the perfect assemblage have now split in three distinct points since in the second stage, each cell has deformed independently of its neighbours. The aim of the third stage is to minimize voids and overlaps by interactions between neighbouring cells. For this the corners split from an initial triple point are required to tend toward a common point. An iterative method is used. At the end of this operation there are still voids and overlaps. If they exceed a certain percentage, in surface, a fracturation programme can be introduced through which the cells with the largest misfits fracture in two or several new cells which subsequently behave like independent individuals.

6.9.3. Statistical results on the assemblage of cells

Some typical textures of the cell assemblage are presented on Figs. 6.32 and 6.33. Fig. 6.32 corresponds to increasing strain in pure shear conditions and Fig. 6.33, to simple shear conditions. The initial texture is also represented on Fig. 6.33. The consideration of those textures and of preferred alignments of slip lines for both pure and simple shears (Fig. 6.34) suggests two general conclusions:

(i) In both regimes there are two peaks of preferred orientations symmetrically inclined to the general elongation L at least for weak to moderate strains. With increasing strain they converge toward L. For large strains in simple shear, the peak closer to the shear direction widens at the expense of the other peak. This tendency is increased by the fracturation of the cells. Contrary to expectation, *the large peak is not centred on the shear line* but its maximum is closer to this direction than to the elongation direction;

(ii) With increasing strain, a continuous increase of voids and overlaps occurs in pure shear, whereas in simple shear there is no further increase beyond a threshold (Fig. 6.35).

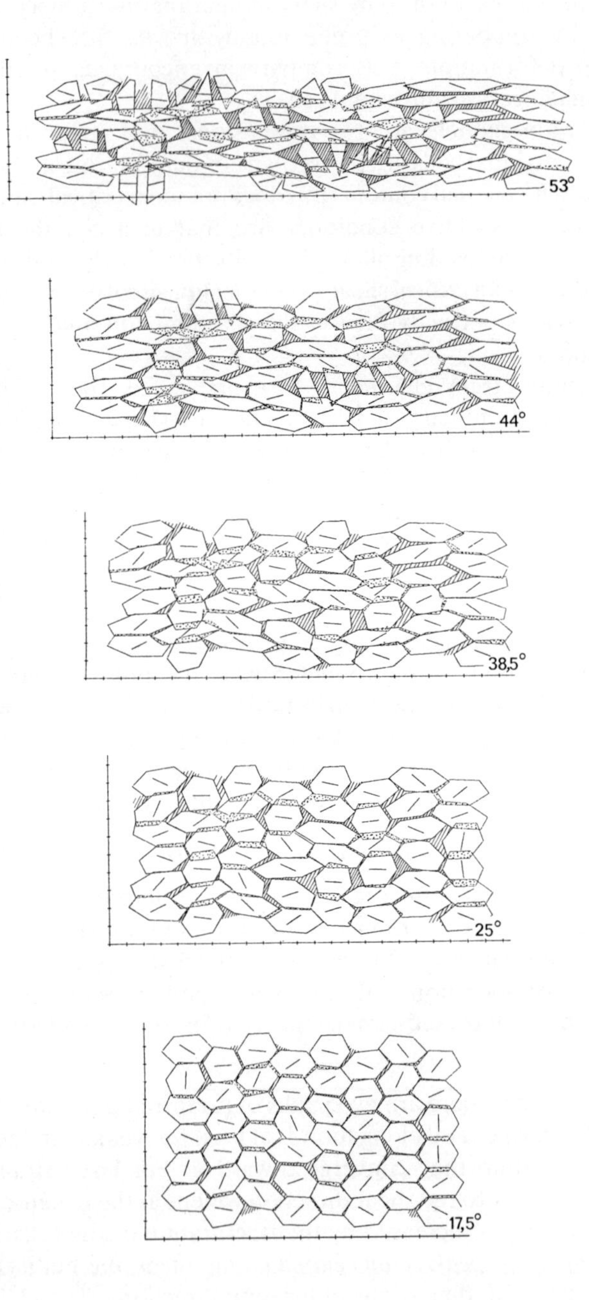

Fig. 6.32. Textural changes with increasing pure shear of a planar network of hexagonal cells (see Fig. 6.33). Voids and overlaps are decorated. The line inside the cells is the slip line.

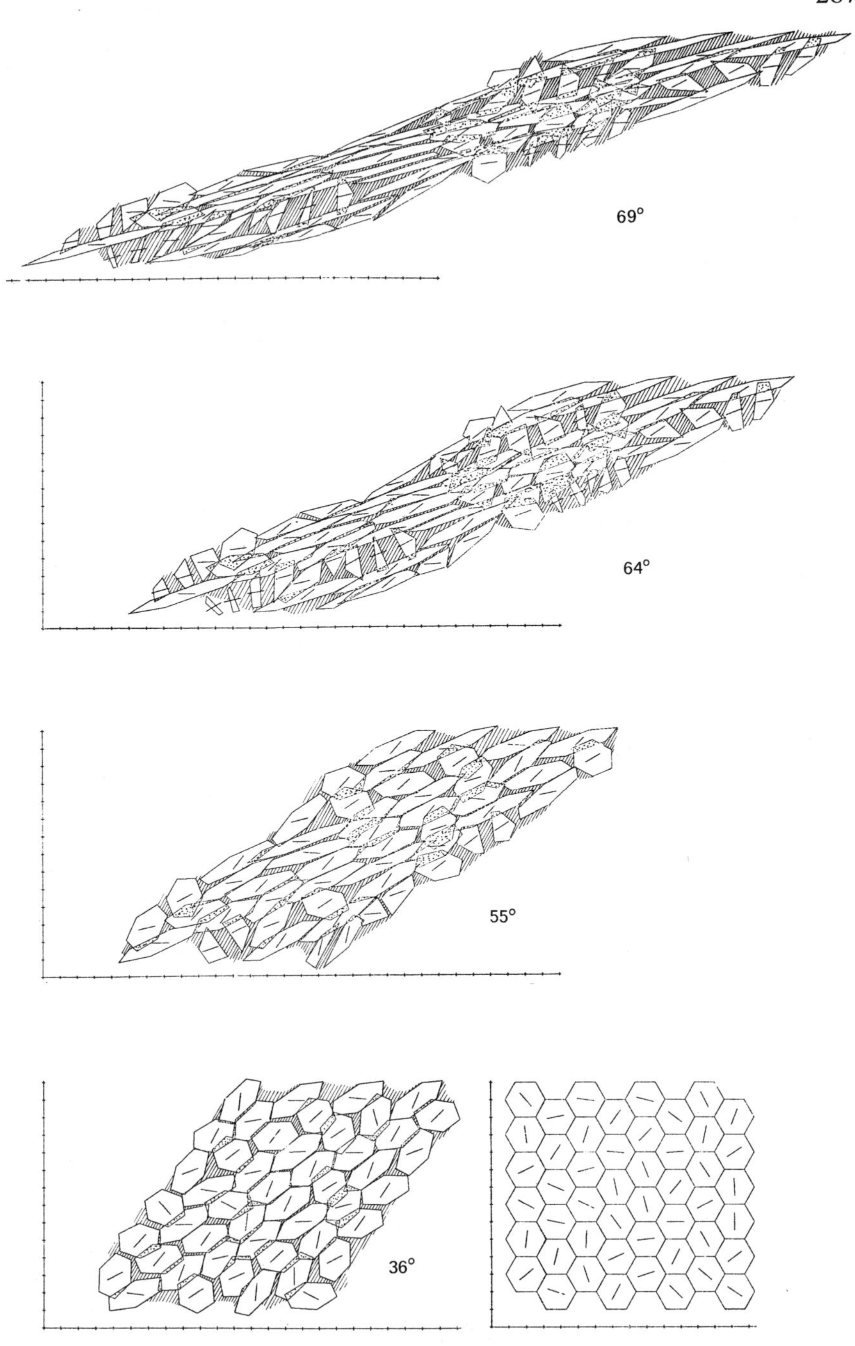

Fig. 6.33. Textural changes with increasing simple shear of a planar network of hexagonal cells. Graphical conventions as in Fig. 6.32

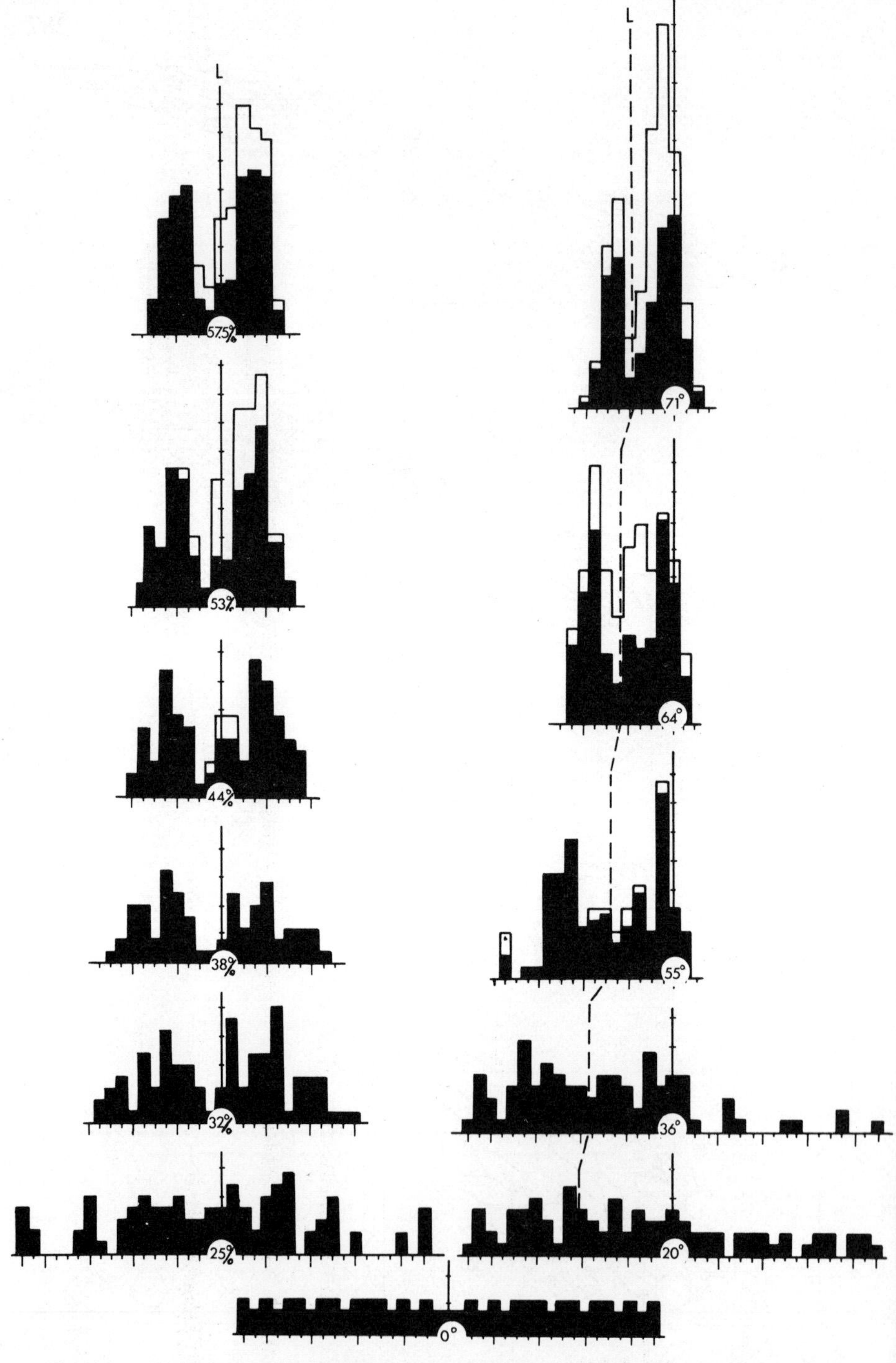

Fig. 6.34. Slip lines preferred alignment with increasing pure shear (left column) and simple shear (right column), starting from a random orientation (bottom). Shaded: unfractured cells; unshaded: fractured cells. The graduated zero line at left is normal to the shortening direction; at right it is the shear direction. The dashed line L at right is the mean elongation direction in the deformed aggregate

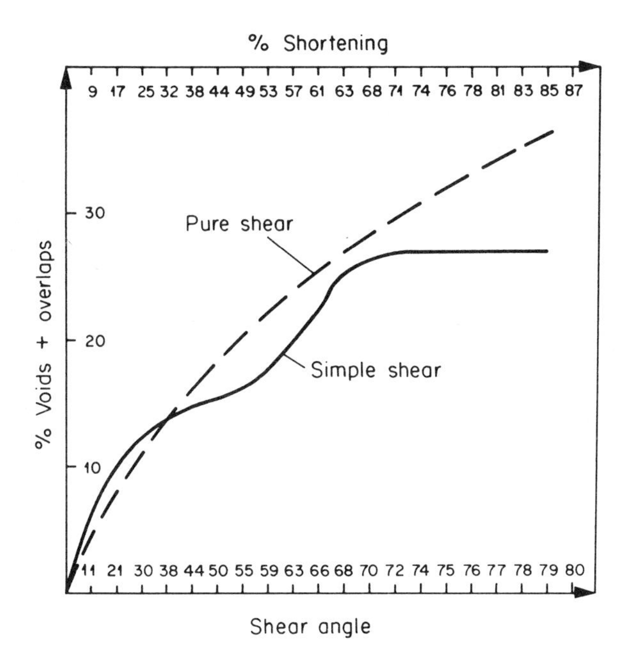

Fig. 6.35. Shear angles and percent of shortening in equivalent units (see Etchecopar, 1974) versus cumulated % of voids and overlaps. Full line: simple shear, dashed line: pure shear

6.9.4. Observations on isolated cells

Before proceeding it is useful to recall some concepts about rotations and slip initially developed by Sander (1948). *External rotation* is related to axes external to the body; here it is simply referred to as rotation and is measured by the rotation of the slip direction in a given cell during strain increments. *Internal rotation* is related to axes internal to the body and is induced by slip; for instance here, internal rotations are those of the contours of a given cell relative to the slip line when slip is taking place. Turner et al. (1954) clearly expose these concepts and apply them to the deformation of single crystals of calcite which are constrained to remain more or less parallel and coaxial to the axis of the applied stress. They show that the external and internal rotations are in opposite senses: if a crystal is slipping in a sinistral sense, its internal rotation is counterclockwise and its external rotation, by lattice curvature, is clockwise (Fig. 6.36). In a crystal, the internal rotation is responsible for the change in shape and the external rotation for the change in orientations. In aggregates the latter accounts for the development of preferred orientation by slip, both by lattice curvature (Turner and Weiss, 1963, p. 348) and by bodily rotation as exposed here (see also § 6.2.1).

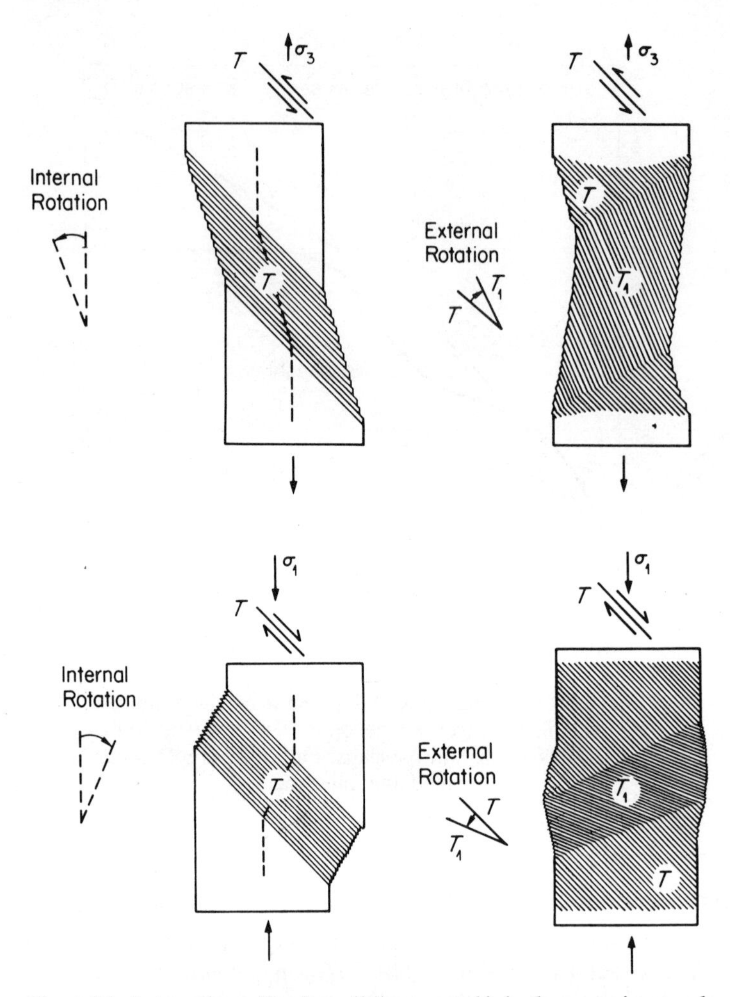

Fig. 6.36. Internal rotation in a gliding crystal is in the sense imposed by gliding and external rotation (lattice curvature) in the opposite sense (after Turner and Weiss, 1963 (modified), reproduced by permission of McGraw-Hill Book Company, New York)

Pure shear

The cells are classified in three groups, following the initial orientations of their slip direction relative to the extension line (Figs. 6.37 and 6.38).

40°, 80° group: the cells undergo slip and rotations larger than the average strain until they have rotated beyond the 40° limit of the group. For a given cell the senses of internal and external rotations are opposed.

80°, 90° group: no slip or rotation takes place until the total strain reaches 25 %, beyond this, the corresponding cells enter the 40–80° group and rapidly slip and rotate.

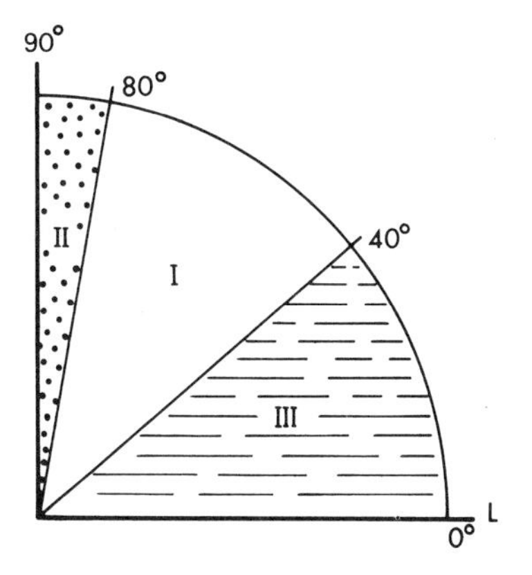

Fig. 6.37. Classification of the cells as a function of their initial orientation in pure shear (0° = L = extension direction)

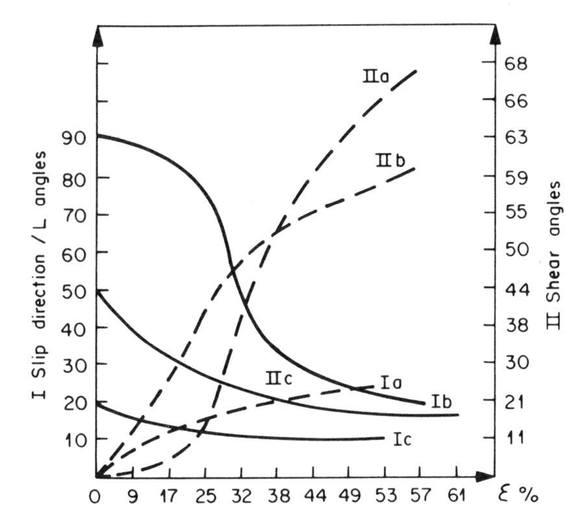

Fig. 6.38. Rotations (curves I) and shear (curves II) of individual cells of differing initial orientations with increasing strain during pure shear. Curves I: angle between the slip line in the cell and the extension direction L versus total strain ε %. Curves II: shear angle of the cell versus total strain ε %. Cells initial orientations: a, 90°/L; b, 50°/L; c, 20°/L

0°, 40° group: slip and rotations are more limited than the average strain; the smaller the angle of the slip direction in a cell with the L elongation direction, the smaller slip and rotation in that cell.

Simple shear

When a shear stress is applied to a body, a torque appears which tends to rotate it. If the shearing is dextral the resulting torque is clockwise. Accordingly, in the kinematic analysis presented here, the clockwise rotations generated in cells by the dextral applied shear will be counted as positive and the counterclockwise ones as negative. The cells are classified in three groups following their initial orientation relative to the shear line (Fig. 6.39).

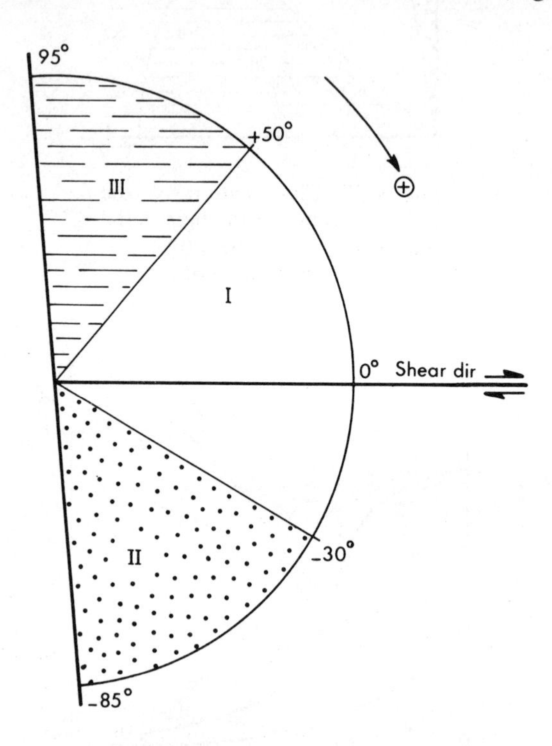

Fig. 6.39. Classification of the cells as a function of their initial orientation in simple shear. Dextral sense, along the 0° line.

+50°, −30° group: the cells close to 0° are affected by intense dextral slip accompanied by weak positive and negative rotations towards 0°. The cells closer to the group-limiting angles present less slip and more rotations toward 0°. This results in orienting them in the low angles range appropriate to intense slip. The rotations of the slip direction (external rotation) have the same sense as the clockwise internal rotations (Fig. 6.40(a)).

−30°, −85° group: the cells are submitted to a shear in the sinistral sense and rotation in the positive sense. The initially strong shear and rotation progressively decrease with increasing total strain until a locked orientation is reached.

In this orientation the cells are elongated parallel to the L elongation of the assemblage and their slip directions are at some 10° above L, that is in the vicinity of +50°. With increasing applied shear, they rotate and slip in proportion to the stretching and rotation of L, unless they are unlocked (see below). If fracturing is allowed, these cells are readily fractured and the fragments rotate and join the +50°, −30° −30 ° group (Fig. 6.40(b)).

+50°, +95° group: the cells are first submitted to a shear in the sinistral sense and a rotation in the positive sense. The larger the initial angle in the cells, the greater is the amount of slip and rotation. By the rotation, the cells progressively reach the +50° orientation in which they are locked. The cells initially in that orientation have remained practically locked in.

For applied shear angles in excess of 30 to 50°, many cells which were in the lockage orientation are now unlocked. Their sense of slip reverses and becomes dextral, the rotation remains positive. At a given total strain the initial shape is reconstructed and the rotation has been such that the cell joins the +50°, −30° group and behaves consequently (Fig. 6.40(c)). The only cells which remain locked are those oriented parallel to the elongation L: they are fractured.

Rotation and slip in pure and simple shears.

From the preceding observation it follows that:

(i) In pure shear, slip and rotation are always associated; in simple shear, not necessarily;

(ii) In pure shear, slip-induced internal rotation is in the opposite sense to the external rotation. In other words, external rotation is in the opposite sense to the torque accompanying slip. In the case of an aggregate, this is an extension of Turner et al.'s (1954) concept on opposed senses of internal and external rotations, initially evolved for single crystal deformations. This relation does not always stand for simple shear;

(iii) In pure shear, rotation and slip keep the same sense in a given cell throughout the whole deformation sequence; in simple shear, rotation keeps the same sense, imposed by the applied stress (clockwise for a dextral shear) but slip may change its sense in some cells during the deformational sequence.

6.9.5. Comparison with naturally and experimentally deformed polycrystals

Of course, comparisons ought to be sought only with polycrystals deforming with single slip system. High temperature flow in olivine aggregates and low temperature flow in quartzites are acceptable in this respect.

Pure shear

As already mentioned a comparison with experimentally deformed aggregates is mainly intended to justify the simulation models. The results of models

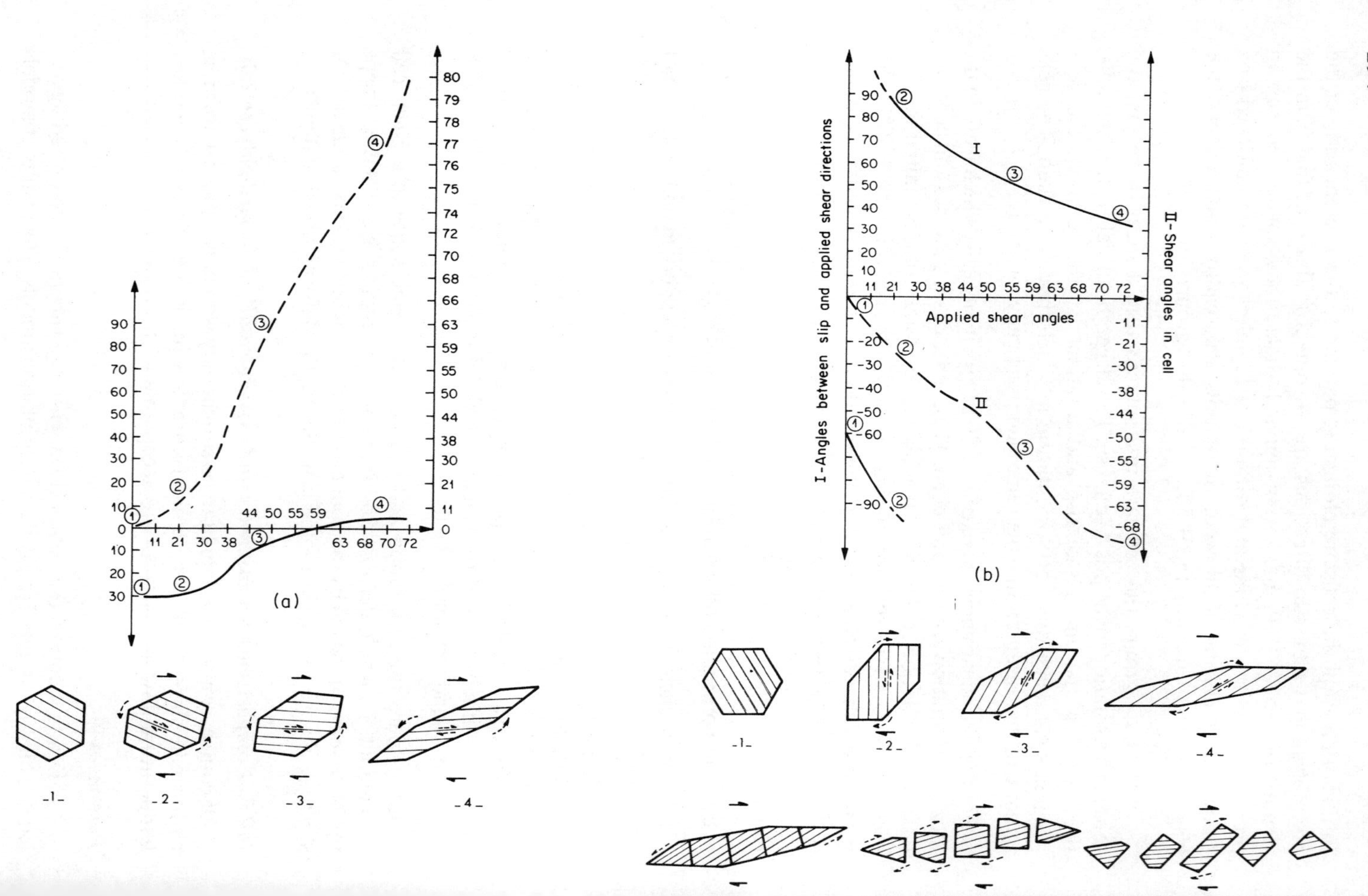

(a)
-1-
-2-
-3-
-4-
I-Angles between slip and applied shear directions
Applied shear angles
II-Shear angles in cell
I
II
(b)
-1-
-2-
-3-
-4-

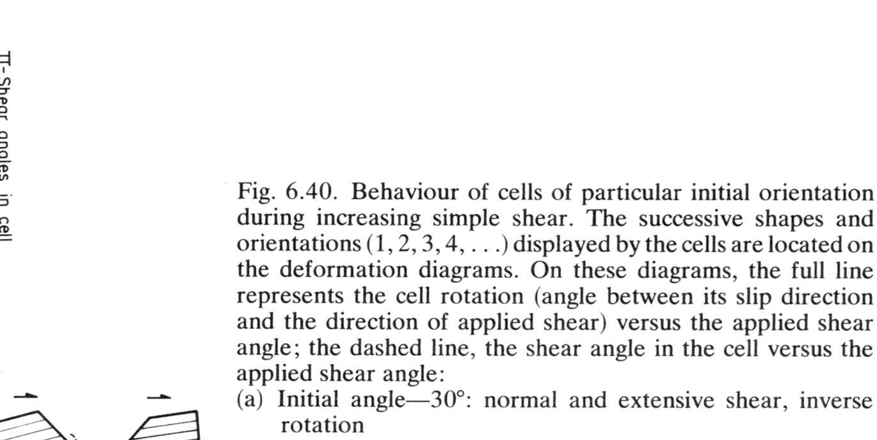

Fig. 6.40. Behaviour of cells of particular initial orientation during increasing simple shear. The successive shapes and orientations (1, 2, 3, 4, . . .) displayed by the cells are located on the deformation diagrams. On these diagrams, the full line represents the cell rotation (angle between its slip direction and the direction of applied shear) versus the applied shear angle; the dashed line, the shear angle in the cell versus the applied shear angle:

(a) Initial angle—30°: normal and extensive shear, inverse rotation

(b) Initial angle—60°: inverse shear, lockage, fracturation and normal shear in fragments, normal rotation

(c) Initial angle 90°: first inverse shear and rotation (Stage 2), next normal shear and rotation (following stages) are obviously in agreement with the general textures and preferred orientations described in the first part of this chapter, once the different conditions (plane versus axial symmetry, etc.) are taken into account. More interesting is the consistency of the models from the dynamic point of view:

Maximum slip in the 40–80° group which corresponds to orientations of large resolved shear stress;

Absence of slip in cells in the 90° and 0° orientations for which there is no resolved shear stress;

Increasing statistical importance of voids and overlaps for large strain when the slip direction in most cells has rotated to small angles with the elongation direction. This expresses the decreasing importance of slip for angles with decreasing resolved shear stresses.

Simple shear

In this case the simulation models are essentially predictive.

The dominant peak in the preferred orientation of slip directions for shear angles over 60° which is oblique to the elongation direction is characteristic of shear regime and indicates the sense of shearing. Its maximum does not coincide with the shear direction and therefore cannot be used for inferring the shear angle.

From the observations and the discussion in the preceding section it follows that the behaviour of some cells characterizes a shear regime and indicates the sense of the applied shear. The best criterion is the inversion in some cells in the sense of shear which operates first in the opposite sense to the applied shear and later in the same sense. This inversion is observed in natural occurrences and leads to a retort shape (Plate 19). Extreme slip in one sense would characterize the sense of the applied shear but it must be borne in mind that moderate shear is simultaneously present in the opposite sense. Rotations are also to be considered because it has been demonstrated that there is statistically a single sense of rotation in simple shear, compared with the two senses in pure shear. This results from the torque applied by shearing. The rotation sense is clockwise in dextral shear. In natural crystals somewhat elongated by slip the sense of rotation is indicated by bending at their ends.

Finally the fracturation of cells oriented on the extension direction in both pure and simple shear conditions and yielding thin polygons elongated perpendicular to this direction, is an observation which contributes to answering the intriguing question raised by the enstatite pull-apart lineation commonly observed in alpine type peridotites (see § 7.3.2).

In conclusion, attention should be drawn to the limits of applicability of this simulation principally due to the simplified conditions imposed for its achievement: two-dimensional model, only one slip direction, homogeneous slip and no bending within the cells. Another restriction to Etchecopar's simulation is that it is essentially geometrical and ignores the dynamic and mechanical aspects of flow; its predictions are, however, in general agreement with what is known about them in flowing solid.

CHAPTER 7

Kinematic Method and Strain Analysis of Field Structures

7.1. INTRODUCTION

In this chapter we first define the kinematic method, that is the sequence of operations which must be performed on the structures indicative of flowage at the successive stages of the investigation with a kinematic interpretation in view. Next, the field structures are described. We wish to recall that the structures under consideration are, in metamorphic terrains, those evocative of a large flowage and not contorted by several episodes of folding. (In the latter situation, a structural study would probably have distinct objects (Chapter 1).) The field structures are principally examined for the information they contain on the finite strain due to flowage. This is a prerequisite of further laboratory investigations (Chapters 8 and 9). A more complete discussion of the kinematic interpretations of the field structures is presented in Chapter 12, after the methods introduced here have been illustrated in the case of peridotite flowage (Chapters 10 and 11).

7.2. METHOD OF KINEMATIC ANALYSIS

The kinematic interpretation of the structure of a deformed rock specimen consists in deducing from the available structural evidence the principal orientations of the flow and its main geometrical properties. These, operating in the geological body from which the specimen originates, were responsible for the investigated structure. The kinematic analysis requires combining various pieces of information which can be almost entirely provided by means of the field study and measurements under the optical microscope, in the favourable cases. The main items of information are:

(i) *The orientation of the axes of the strain ellipsoid and, when possible, the estimation of their length.* The strain ellipsoid should not change in shape or orientation within the sample as we are dealing with fairly homogeneous structures. This piece of information is commonly provided by the study of the penetrative planar and linear structures (foliation, lineations). Measurements in the field yield their orientation, less commonly their relative magnitude.

Observations under the microscope are necessary to ascertain the nature of the field structures, making it possible to rule out penetrative structures related to other causes (e.g. annealing). They also afford new means for estimating the strain magnitude or the orientation of the strain axes, when the field studies have failed to produce the required information. The microscopic investigations are then conducted on the aspect ratios of anisometric minerals, their preferred orientations and their elongation in the corresponding directions. Such considerations of the dimensional or shape preferred orientations are contrasted with those on lattice preferred orientation, now examined.

(ii) *The lattice preferred orientations of the principal rock-forming minerals.* They can be estimated on properly orientated thin sections, but are best measured with the universal stage or with the X ray-goniometer.

(iii) *The nature of the mechanisms responsible for the lattice and shape-preferred orientations.* The different mechanisms responsible for the development of preferred orientations of minerals in stressed rocks, whether flowage has occurred or not, have been discussed in Chapter 4. As a result of metallurgical knowledge and recent experimental studies on geological material, it is concluded that plastic flow (rotation and slip in crystals due to generation and motion of dislocations) controls most of the preferred orientations in tectonites. This statement is mainly true in cold-working conditions provided that the flowage in the investigated rocks has been extensive, in order to delete preferred orientations inherited from a previous state, for instance during the course of a sedimentary or magmatic accumulation process. As a result of Turner et al.'s work on the Yule marble (1956), it seems, however, that the influence of an initially strong fabric is still recorded after a large strain.

We should like to emphasize also that in geological materials deforming or recrystallizing in the presence of fluids, like micas (§ 6.3.3), other orienting mechanisms are, or can be, operative. In hot-working conditions, the crystallographic controls of a plastically strained host grain on the orientation of the strain-free recrystallized grains, though existing, are not clearly understood and therefore the interpretation of preferred orientations, at least in totally recrystallized rocks, is still uncertain. Hot-working textures should retain attention only if the study can be carried out on the porphyroclasts which, in view of their large size, elongated habit and strained features, have been shaped by the plastic flow. It should also be recalled that some anneals may develop a new shape-preferred orientation, which clearly goes against the accepted criterion that the shape fabric provides the orientations of the strain ellipsoid axes. Indeed, this criterion is mainly valid for plastically deformed rocks. Incidentally, it also applies in the case of mica-bearing metamorphic rocks even though the preferred orientation of micas is not necessarily controlled by slip (T. Tullis, 1971; Etheridge et al., 1974).

Deciding whether the texture results from cold working or hot working is done, on correctly oriented thin sections observed under the optical microscope, by referring to the general criteria given in Chapter 4. One danger would be to assign to cold working a textural state developed during hot working and only slightly modified at a lower temperature or faster strain rate.

The provisional diagnosis regarding the orienting mechanism needs to be complemented by other pieces of evidence as shown in the concluding remarks below.

(iv) *The determination of the active slip systems in the crystals.* The most reliable techniques for determining the active slip systems in the minerals participating in the deformation, operate at the scale of individual dislocations; they are mainly transmission electron microscopy (TEM) and in certain cases decoration and etching (Chapter 9). The systems responsible for macroscopic slip in crystals can also be determined by optical methods using either internal rotations or pre-existing planar structures (Turner et al., 1954) or more commonly external rotation in adjacent kink bands (Christie et al., 1964). However, in the description of the last techniques (§9.1.2), emphasis is laid on the fact that it applies only to restricted cases and may be otherwise misleading.

Combining the preceding pieces of information (i), (ii) and (iv) may provide the means of verifying whether or not the preferred orientations were developed by slip (point (iii)). For instance, the [010] axis of enstatite preferentially oriented normal to the foliation in some pyroxenites is incompatible with an origin by slip because the [100] axis, pole of the unique slip plane, should have this preferred orientation. The [010] type of fabric possibly resulted from a mineral deposition in a magmatic cumulate (§ 5.4.1).

From this cross-checking it may be concluded that slip operating in the main rock-forming minerals on identified slip systems controlled the lattice and shape-preferred orientations; the control may be found to be indirect but in a way which can still be inferred. It is then possible in favourable cases to assess with a certain approximation how the main flow directions (flow line and flow plane) were oriented. A closer matching of (i) and (ii) yields new information about the geometry of the flow (irrotational versus rotational, sense of shear if rotational and, when the technique is calibrated, probably an estimate of the shear angle). If we now turn back to the penetrative structures (foliation and lineations) mapped in the field, the approximate relation connecting them with the flow elements opens the way to drawing a map of the flow structure (kinematic map) in the considered geological body. The non-penetrative structures mapped in the field may sharpen the kinematic picture. This was the case with folding in the peridotites from the Lanzo massif which was instrumental in showing that the flow was three-dimensional and not plane as first thought. It is discussed at length in Chapter 10.

7.3. FIELD STRUCTURES

In a metamorphic body with a strong and fairly homogeneous deformation, the penetrative planar and linear structures which are often ubiquitous are of primary importance. Evidently, no structural study can proceed unless these structural features have been carefully identified and measured. Field investigations will be complemented by considering the non-penetrative structures which are geometrically related to the penetrative ones.

The concept of penetrative and non-penetrative discontinuities in a geological body is linked with the notion of homogeneity which in turn depends on the scale of observation of the body (Turner and Weiss, 1963, p. 22; Dimitrijevic, 1967). A discontinuity is regarded as penetrative on a given scale of observation if it pervades the body uniformly and is present at any point. Such a structure is usually statistically homogeneous, that is, it becomes non-penetrative on a finer scale. The scale used here to decide whether a discontinuity is penetrative or not is that of the optical microscope, but very often field observations are discriminative enough. Furthermore, we are dealing primarily with strain-induced structures. In this respect, the penetrative structures are believed to result from homogeneous deformation on that scale applied on a homogeneous body and the non-penetrative from an inhomogeneous and discontinuous deformation applied on a homogeneous or inhomogeneous body.

Reference frames in structural analysis

The unique or dominant linear and planar penetrative structures define a natural reference system such that one axis is parallel to the lineation, another normal to it in the penetrative deformation plane and the last normal to this plane (Fig. 7.1). It seems natural to describe the structure and to pursue the

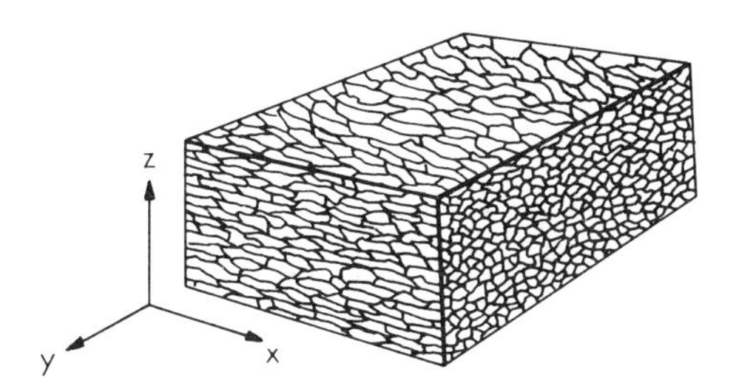

Fig. 7.1. Penetrative foliation and mineral stretching lineation and how they relate to the direction of the (XYZ) principal axes of the finite strain ellipsoid

analysis by referring to this framework, which in turn can be referred to the geographic coordinates. We will see in the next sections that the main axes of the penetrative deformation coincide with that of the finite strain, represented by an ellipsoid ($X \geqslant Y \geqslant Z$). This is true in most cases, possibly sometimes with a certain approximation. Serious confusions have arisen from equating this reference system with those describing the symmetry of an inhomogeneously deformed structure or the kinematics responsible for the deformed structure.

In this regard, b-axis in Sander's terminology could designate altogether the lineation trend, the axis of a fold and the direction normal to the flow in the flow plane. The danger of this practice has often been pointed out. Ramsay (1967, p. 333) distinguishes the three reference frames (*XYZ*) for the principal finite strain directions, (*ABC*) for the fabric symmetry axes and (*abc*) for the principal kinematic axes (Fig. 7.2). The two latter are accepted in a restricted

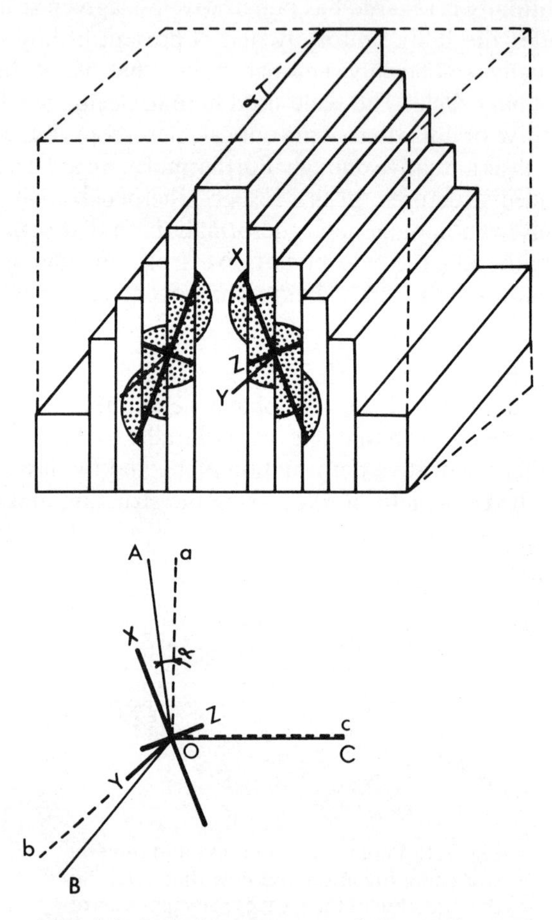

Fig. 7.2. Relations between finite strain axes (*XYZ*), geometrical or fabric symmetry axes (*ABC*) and kinematic axes (*abc*) illustrated in the case of a shear folding. The stippled areas correspond to initially circular domains which would be deformed into ellipses in the case of a penetrative shearing

sense (monoclinic fabrics and simple shear flow). Mattauer (1973, p. 104) extends the *ABC* system to other symmetries. Accordingly, the *abc* kinematic frame can be used in the most general case of flow, with a as the principal direction of flow, *b* as the direction normal to it in the flow plane and *c* as the direction normal to the flow plane. It seems that this usage is now widely accepted.

Finite strain estimations

In the kinematic investigation, estimations of finite strain carried out in the field and in the laboratory make a quantitative study on the flow possible. One theoretical method, which is in its first developments, depends on the relation between the finite strain and the degree of preferred orientation in minerals and also on observations carried at the level of individual deformation in minerals. It has been applied in the case of slates by Oertel (1974) (see § 6.3.2). The method should be used only if the texture and mineral-preferred orientation in the body before the onset of deformation and the operating flow mechanisms are sufficiently known. Methods akin to the continuum mechanics (see introduction) are in a more advanced stage of development. These use strain markers, like deformed ooids, nodules, pebbles, fossils, etc. Pressure shadows, fibre lineations and inclusion trails also yield information on the path of incremental strains (Choukroune, 1971; Elliott, 1972; Durney and Ramsay, 1973).

Estimation of finite strain using strain markers is a well-documented matter; therefore there is no need to develop here the appropriate techniques. They slightly differ in the case of fossils and in that of spherical or ellipsoidal objects. The strain markers can also have the same ductility as the matrix, thus recording faithfully the finite strain or have a ductility generally smaller than the matrix, thus recording only partially the finite strain. Cloos' (1947) study on strained oolitic limestones is a landmark in the progress of methods using ellipsoidal markers. Since then, Ramsay (1967) has presented an extended treatment. The theory and techniques have been further improved by Gay* (1968b), Dunnet (1969), Burns and Spry (1969), Elliott (1970), Dunnet and Siddans (1971) and Gosh and Sengupta (1973).

7.3.1. Planar penetrative structures

Slaty cleavage

In low-grade metamorphic rocks with a minute grain size such as slates, the penetrative plane is the slaty cleavage. It is a planar fabric produced by the shape preferred orientation of the minerals. This plane is remarkable in that it contains the XY section of the local finite strain ellipsoid (X, Y, Z respectively largest, mean and shortest axes) (Fig. 7.1) which is defined whenever strain gauges are available (Plate 20(a)). There has been a long-standing argument about the origin of cleavage, whether it is a shear plane or the XY plane of the strain ellipsoid. It seems now to be settled in favour of the second interpretation; overwhelming evidence has progressively been accumulated from strain estimations in objects of various sorts to confirm this (review in Siddans, 1972

* In a companion paper (1968a), Gay considers the theory of particle reorientation in a viscous fluid and sets up the relation with the mode of flow.

and Wood, 1974). However, the fact that slaty cleavage can originate in a macroscopic shear regime is in no way ruled out; in such a situation, the cleavage would simply not coincide with the shear plane (Fig. 7.3) (Ramsay and Graham, 1970; T. Tullis, 1971, § 12.2.3).

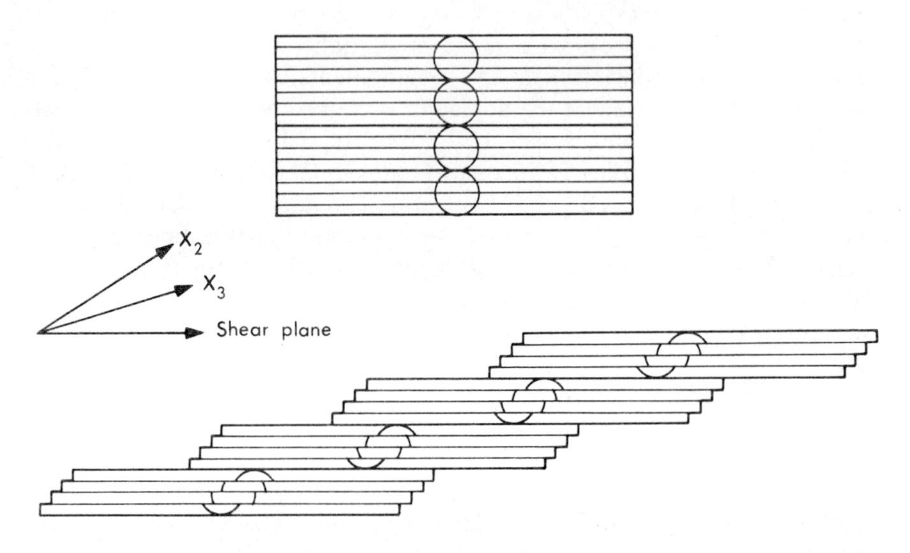

Fig. 7.3. Different states of finite strain on different scales under rigid conditions. Starting from the upper block, the shear at the smaller scale produces the X_2 direction of maximum finite extension. Discontinuous shear at the larger scale modifies the maximum finite extension into the X_3 direction (after Schwerdtner, 1973 (simplified), reproduced from *Tectonophysics* by permission of Elsevier Scientific Publishing Company, Amsterdam)

We have equated the slaty cleavage plane with the XY section of the strain ellipsoid but it is primarily the plane of preferred orientation of basal planes in layered silicates like chlorite and mica. Indeed the fissility is a consequence of this preferred orientation. In our kinematic interpretation, this does not necessarily imply that it is a shear plane since this lattice preferred orientation may have been produced by other mechanisms than plastic flow (Chapter 6). It is usually acknowledged that the XY section of the strain ellipsoid coincides with the preferred orientation of flaky minerals. Yet this has been argued and a slight obliquity between the two orientations has been postulated on the basis of observations defining the preferred orientations by means of the X-ray technique (Oertel, 1970, 1971) and theoretical reasoning (Ramsay, 1967, p. 93; Elliott, 1972; Bayly, 1974). In turn, the existence and the effective importance of this obliquity is questioned by Helm and Siddans (1971) and Borradaile (1974). We shall re-examine the problem in Chapter 12 and provisionally accept the conclusion that flaky minerals-preferred orientation and the XY section of the local strain ellipsoid are parallel. If a discrepancy of more than a few degrees were detected, we would then take into account the orientation of the XY section.

Accepting the slaty cleavage plane as the XY section of the strain ellipsoid is no longer true if there is a discontinuity plane (shear surface, strain slip cleavage) parallel to it. The deformation would be then partly discontinuous and, as concluded by Schwerdtner (1973), the slaty cleavage would no longer record the total finite strain but solely the fraction of it obtained in the stage of continuous strain (Fig. 7.3).

In coarse crystalline rocks, deformation in the low-grade metamorphism conditions tends to lead to non-penetrative shear structures (Plate 17(a)).

Schistosity, foliation

In the metamorphic conditions in which a conspicuous recrystallization has been induced, the grains usually display a preferred dimensional orientation in a plane called the *schistosity or foliation plane* (Plate 17). We purposely exclude from this definition any other property such as the coincidence with the lattice preferred orientation, the presence of fissility or of layering defined either by differences in composition or grain size. The term of schistosity has always been understood in this meaning, although it often refers to the planar disposition of micas in mica-bearing metamorphic rocks. The term of foliation originally indicated the occurrence of a layering (Harker, 1932, p. 203, after Darwin's definition). After a period when no precise meaning could be assigned to it (Schieferdecker, 1959; Turner and Weiss, 1963, p. 97; Whitten, 1966, p. 216; Spry, 1969, p. 207), it seems, in recent papers, that a tendency can be traced to use it in the restricted sense proposed above.

Having distinct terms to designate the plane of mineral flattening in low- and high-grade metamorphic rocks is fortunate for the flattening was probably produced by distinct processes. However, in both cases, with the aforementioned exception of particular anneals, we consider that the slaty cleavage and the schistosity planes correspond to the XY section of the specimen strain ellipsoid.

The plane of penetrative deformation is sometimes very indistinct in the field when the shape of individual minerals is not conspicuous and no jointing underlines it. Its critical importance justifies conceding a punctilious care to evidence it. In some peridotite bodies this is the case, when no other indicators than the spinel, forming a few percent of the rock, exist. It is then necessary to complement the field evidence by laboratory measurements on systematically collected oriented samples (Darot, 1973). The adequate techniques to orient a sample in the field are described by Turner and Weiss (1963, p. 85) (see also Fig. 9.1).

7.3.2. Linear penetrative structure—lineations

Once the plane of penetrative deformation has been identified, the next step consists in looking for the linear penetrative structures or lineations lying in it.

However, certain domains in the field are characterized by a pervasive lineation without any evidence of a flattening plane; the structure is then axially symmetrical, as in the case of rodding.

Among the lineation types defined by Cloos (1962), only two are of primary importance here; they are due to alignment either of minerals or of strain markers and strain shadows (Plate 27). As a matter of fact, some mineral lineations are strain markers as well. Strictly speaking, strain markers are seldom penetrative structures, at least on our scale of investigation. They are considered above under the more general heading of finite strain estimation. These lineations, when marked by the alignment of ellipsoids resulting from the stretching of initially more spherical particles, should not be confused with lineations due to the boudinage of thin layers in a flattening regime ($1 > K \geqslant 0$, Flinn, 1962). In that case, boudinage operates in two orthogonal directions, thus dividing the layer in aligned elongated objects. In contrast with the case of stretched particles, the main stretching direction is, in the foliation plane, normal to the long axis of the boudins.

The mineral lineations should be characterized in the field only by their mineralogical nature. Their structural meaning is ascertained by a careful investigation under the microscope using the distinctive criteria developed below. In this regard, the mineral lineations are due to the alignment of:

(i) *Tectonically inactive minerals.* They are passively aligned during flowage provided they had a large aspect ratio. Amphibole which is both brittle and needle shaped or prismatic can behave accordingly. March's (1932) mathematical analysis of the orientation of rigid rods on a flowing medium can be applied here; it predicts that the rods tend to be aligned with their long axes parallel to the longest axis X of the strain ellipsoid in the surrounding medium;

(ii) *Tectonically active minerals.* They are elongated and aligned parallel to the X-axis of the strain ellipsoid by slip (Fig. 7.4) and may simultaneously or

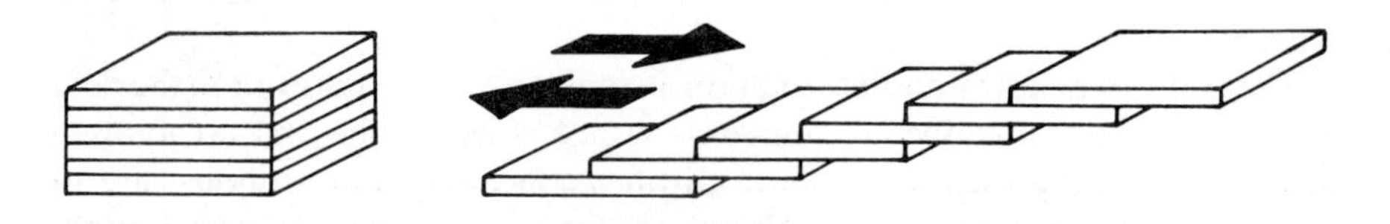

Fig. 7.4. Sketch illustrating the formation of a mineral lineation by large plastic flow (after Darot, 1973)

subsequently recrystallize in an aggregate lineation (Plate 27). They are as ductile as the flowing matrix unless they constitute the matrix thus offering, in favourable cases, the possibility of measuring the strain;

(iii) *Tectonically partially active minerals.* This category consists of minerals whose ductility is smaller than that of the matrix. It is also possible that in monomineralic aggregates some grains behave in a more brittle fashion throughout the deformational process, depending on their initial orientation (§ 6.9.4). Large equant crystals or clusters of crystals can be pulled apart in roughly disc-shaped grains, the axis of the disc being parallel to the X strain ellipsoid direction. This occurs in the moderate deformation of chromite clusters and enstatite crystals in peridotites (Fig. 7.5) (Plate 27(a)). Otherwise, minerals with a distinct initial aspect ratio and minerals weakly elongated during flowage are aligned parallel to the X strain ellipsoid direction with fracture or boudinage;

(iv) *Mimetic minerals.* After a mineral has been elongated and aligned in a direction during flow, it can recrystallize in an aggregate lineation or be replaced by a mineral belonging to a different species which inherits the preferred orientation, for instance static growth of amphiboles out of pyroxenes.

The last type of lineation is immediately recognized due to the absence of strain in minerals. The main distinction between (i) and (ii) is based on parallelism between elongation and crystallographic axes in (i) contrasting with obliquity* in (ii), and on the lack of relation between the mineral aspect ratio and the finite strain in (i) contrasting with some relation of this kind in (ii).

7.3.3. Non-penetrative structures

Non-penetrative cleavages

Unlike slaty cleavage and foliation, the structures known as *crenulation cleavage*, *strain slip cleavage*, *shear cleavage* or *fracture cleavage* are not penetrative on the microscopic scale. They form discontinuity surfaces, often very closely spaced. However, as pointed out by Williams (1972), the distinction between them and the penetrative slaty cleavage or foliation is not always that clear-cut (Fig. 7.6). Since we are dealing with otherwise homogeneous structures, attention will be focused on the cleavages, like *strain slip or shear cleavage*, which form straight parallel bands differing from the surrounding medium only by a finer grain size and possibly by a slight change in texture (Plate 25(a)); less attention will be paid here to *crenulation cleavage* which is related to minor folds. According to Ayrton and Ramsay (1974) it lies close to the XY plane of the finite strain ellipsoid.

The non-penetrative cleavages are often parallel or nearly so to slaty cleavage or foliation, suggesting that the non-penetrative deformation has been superimposed on the penetrative one during a single deformation process (Vialon, 1974). Whitten (1966, p. 260) recommends using fracture cleavage when the non-penetrative cleavage is not parallel to the penetrative one; it is then just a closely spaced jointing.

* In favourable cases it is possible to detect this obliquity even in the field on an outcrop surface normal to the foliation and parallel to the lineation. A statistical study of its sense yields direct information on the sense of shear (see § 6.9.3).

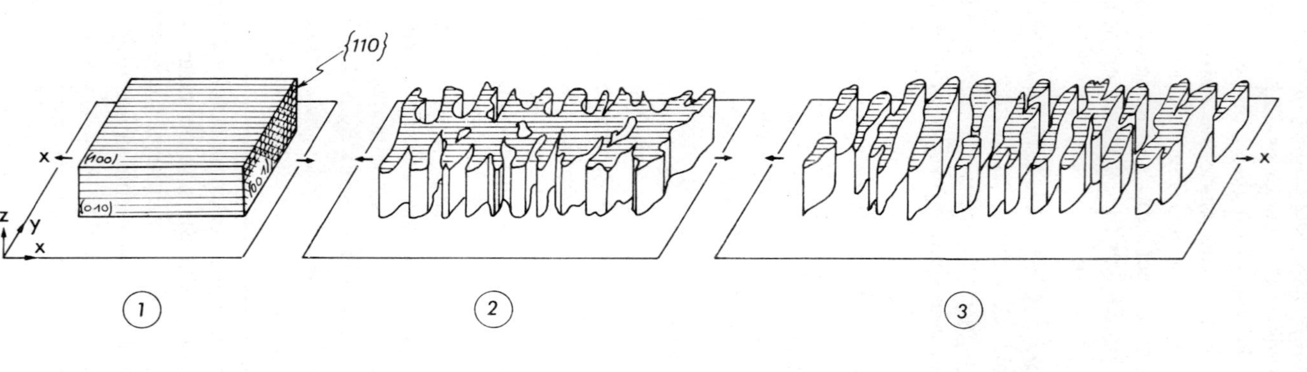

Fig. 7.5. Sketch illustrating the formation of a pull-apart mineral lineation during plastic flow (inspired from enstatite pull-apart lineation in peridotites; Nicolas et al., 1971)

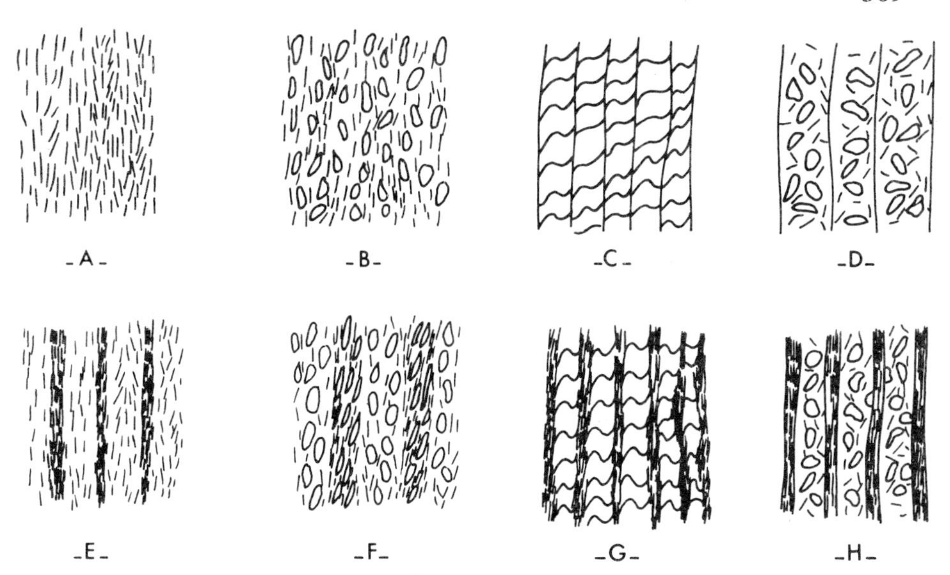

Fig. 7.6. Various types of penetrative and non-penetrative cleavages as they can appear together in a single area (inspired from Williams, 1972). (A) and (B): Slaty cleavages; (E), (F) and (H): Strain slip cleavages superimposed respectively on (A) and (B); (C) and (G): Crenulation cleavages, differentiated in the case of (G); (D): Fracture cleavage. (Reproduced by permission of The American Journal of Science, from *Amer. J. Sci.*, **272**, 1, Fig. 7 (1972))

Terms with a genetic connotation, like shear cleavage, have been introduced because, when markers are present simultaneously with these cleavages, close examination shows that they are planes of discontinuous displacement. When it has been established that the discontinuous flow was related to a continuous one, the discontinuous flow should be attentively considered on account of its contribution to the finite strain (Schwerdtner, 1973) and of its importance in materializing the flow plane. It has been seen (§ 6.9) that, in a shear flow regime, the orientation of the flow plane could be only approximately known; the information provided here is then particularly valuable. An example derived from the study of the peridotites in Ronda massif (Southern Spain) illustrates this point (Darot, 1973; Plate 17(b)).

Metamorphic banding

Metamorphic banding designates a planar mineral segregation which originates during a tectonic and/or metamorphic event and which is independent of any sedimentary or magmatic layering. It is usually a few minerals thick and constitutes a discrete structure like the cleavages just described above (Fig. 7.6). It is also called metamorphic layering; this term may be appropriate to designate fairly thick planar structures, presumably transposed from a former layering during metamorphism. Metamorphic banding accordingly may be reserved for planar structures only a few minerals thick, more compatible with an origin by a metamorphic or tectonic differentiation. The occurrence of a new banding independent of any pre-existing discontinuity has been explained by a variety of chemical and mechanical processes operating during metamorphism; they have been reviewed by Turner and Verhoogen (1960, p. 581). We are only interested here in banding formed in relation with mechanical processes.

As already noted by Talbot and Hobbs (1968), the banding is most commonly developed parallel to a crenulation cleavage or to the deformation plane in mylonites (Plate 18). When markers are present, it is seen to correspond to a discontinuous strain. Therefore it can be safely said that in many cases the banding underlines a plane of discontinuous motion. In this respect its kinematic meaning is the same as that of the crenulation cleavage into which it can progressively grade (Fig. 7.6). The association of banding and crenulation cleavage can be explained by the transposition of a former layering (Ayrton, 1969) or by a mineral segregation induced by different plastic behaviour during shear flow (Schmidt, 1925). The latter interpretation also put forward to explain the banding in mylonites, has been recently argued by Shelley (1974). In mylonites, another cause can be sought in the transposition of any kind of heterogeneity which is then flattened into a lens; in high temperature mylonites, Boullier and Guegen's interpretation (1974) of a differentiation induced by superplastic flow may be relevant.

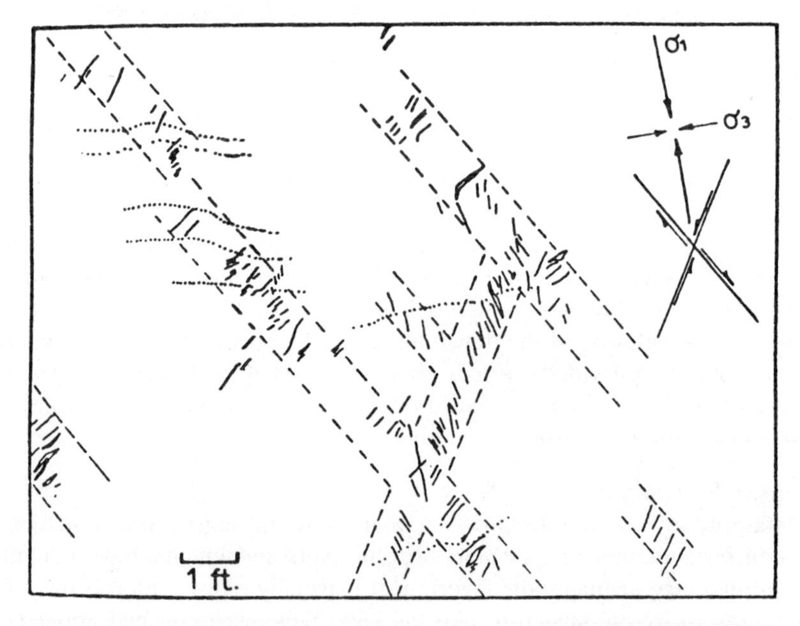

Fig. 7.7. En-echelon cracks in quartzite. Individual cracks parallel the complementary shear or σ_1 direction. Deviation from this is caused by rotation within shear zones. Dotted lines indicate bedding. (Drawn from a photograph) (after Roering, 1968, reproduced from *Tectonophysics* by permission of Elsevier Scientific Publishing Company, Amsterdam)

Tension gashes

Tension gashes usually form minor structures but they can range in size to that of dikes. Dikes are treated separately because they pre-date the flow unlike the gashes which display remarkable orientations relative to the flow, indicating that they opened during it. This criterion of contemporaneity is often complemented by petrogenetic considerations on the nature of the minerals filling the gash; they are either in equilibrium with the plastically flowing minerals or derived from them as in the case of pressure solution or partial melting. Experimentally it has been shown that tension gashes are planes containing the directions of maximum compressive and intermediate principal stresses. They also tend to be aligned in an en-echelon pattern, the alignment plane being a surface of high resolved shear (Fig. 7.7). The intersection of this plane with individual gashes coincides with the mean principal stress direction.

The feldspathic gashes in the peridotites from Lanzo massif were produced by the partial melting of the peridotites during their flowage (Boudier, 1972; Boudier and Nicolas, 1972) (Plate 25(a)). As shown on Fig. 7.8, they cut the

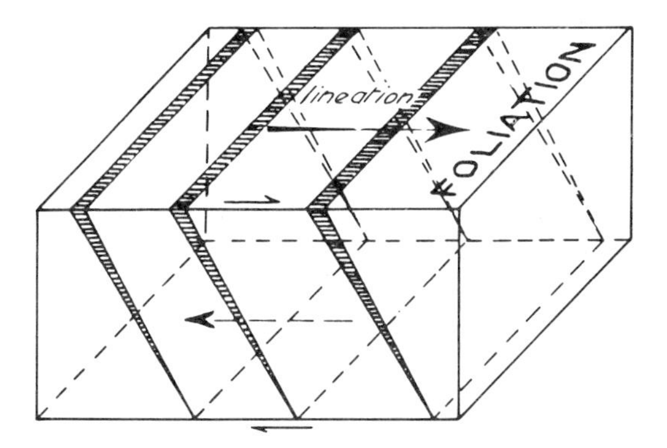

Fig. 7.8. Block diagram showing the preferred relationship between the en-echelon feldspathic gashes and the penetrative structure in the Lanzo peridotite massif. The kinematic pattern attached to these structures is discussed in Chapter 10. The feldspathic gashes result from partial melting during plastic flow (after Boudier and Nicolas, 1972, reproduced by permission of Verlag Leeman, Zürich)

foliation in a direction which is statistically normal to the lineation trend and the plane of the en-echelon pattern is the foliation or a close surface. The gashes are oblique to the foliation, thus indicating a maximum compressive stress inclined relative to the foliation and consequently a shear regime. In quartzites from Angers in France, strongly deformed under very weak metamorphism conditions, the gashes are either empty or quartz-filled fractures. The intersections with the foliation are the same as above but the gashes here are normal or nearly normal to the foliation, thus suggesting a dominant flattening regime. This is confirmed by the close symmetry of the petrofabric data relative to the foliation (Bouchez, oral com.).

Layering-dikes

Layers are characterized by stratigraphic changes in the proportion, grain size or composition of minerals. They are usually parallel. They pre-date the deformation otherwise they would fall into the category of metamorphic bandings which are often thinner too. Mapping the attitude of layers in a massif indicates whether or not the flow has been homogeneous on a large scale. An inhomogeneous flow would result in folding (see below). In rotational shear flow, the sense of shearing can sometimes be inferred in the field, considering the offsetting at the intersection between layers and non-penetrative cleavages present in the rock when the flow is somewhat discontinuous (Fig. 7.9). Dikes

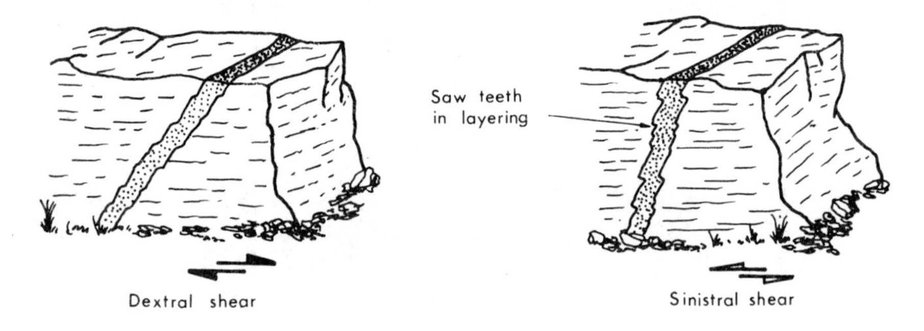

Fig. 7.9. Shear sense deduced from the observation of the offsetting of dikes in the case of a non-penetrative flow (after Darot, 1973)

emplaced before the flowage yield the same sort of information. They can also be used to estimate the finite strain by means of rotations from known initial orientations (Borradaile and Johnson, 1973; Borradaile, 1974) or measurements on their internal strain (extension characterized by boudinage; shortening, by folds) as a function of their orientation (Talbot, 1970).

Folds

We consider here the folds which have originated during the extensive flowage responsible for the penetrative structures under investigation; we exclude those formed during a subsequent tectonic event and those formed in a previous stage of the rock history. In fact, the latter would probably be erased by the flowage unless they had special orientations. In this case, they would be transposed and probably be indistinguishable from those formed during the flowage since the bases on which this contemporaneity is assessed are mainly geometrical: the fold axis is parallel or close to the direction of the penetrative lineation and the axial plane is parallel to the slaty cleavage or foliation. These criteria can be complemented by examining the nature and the orientation of the minerals in the folded layers. These observations would verify whether or not their nature and orientation are conformable to those determined by the flowage in the surrounding medium.

Careful mapping of the fold's geometrical elements may yield additional information on the flow regime. The study of folds in peridotites (§ 10.4.2) is instrumental in illustrating this point.

CHAPTER 8

Representation and Analysis of Structural Data

Written with the collaboration of Jean-Luc Bouchez and Michel Darot

8.1. INTRODUCTION

Our purpose in this chapter is to give a brief survey of the numerous techniques of representation and analysis of structural data. Many techniques are commonly used and are described in structural geology textbooks to which there will be considerable reference. Stress will be laid on general principles and only on those techniques for which extended descriptions have not been found.

Before embarking on any investigation, the homogeneity and representativeness of the numerous data collected inside the studied body must be examined. Once this has been accomplished, the analysis of the nature of the data leads one to consider them either as directional data (vectors) or as dimensional data (scalars). In structural geology we are mainly concerned with the former. Another consideration is whether or not the data need to be topographically located in a spatial coordinate system. The need arises when the structure is not statistically homogeneous. We will deal first with the case of non-located directional data, next that of located dimensional data and finally that of located directional data.

8.2. REPRESENTATIVENESS AND SIZE* OF THE DATA POPULATION†

At a time when diagrams of mineral preferred orientation bearing over 1000 measurements were fashionable, Turner (1942) stated: 'Where, as with

* Whitten (1966, p. 80) uses 'size' in relation to scale concepts (dimensions of objects) and does not conform with the meaning of it in statistics, where it is the number of attributes composing the population; the statistics definition is accepted here.

† In statistics, 'sample' means what we call here a population (Whitten, 1966, p. 71) for the obvious reason that sample already designates an object of the population in geological language.

oriented olivine fabrics, the degree of preferred orientation is high and the pattern is simple, it is more profitable to measure, in as many rocks as possible, the minimum number of grains* necessary to bring out the pattern of the fabric, than to investigate a few specimens in greater detail'. We will come back to this problem later in this section (size of the population). At present we think that it can be used to illustrate the common absence in geological investigations of a critical appraisal of the more general significance or representativeness in a larger body of local observations and measurements. 'Structural geology is essentially a statistical study, because a quantitative estimate . . . has to be made on the basis of a limited number of observations' writes Whitten (1966, p. 70) to call attention to the fundamental problem of a representative sampling.

It is indeed a fundamental problem and, in our opinion, if it had been correctly examined, in many instances this would have resulted in quashing the arguments between geologists or in raising them from the level of the raw data to that of the concepts. When it comes to selecting the material for study, often the geologist focuses his attention on a few objects, sometimes deliberately chosen as oddities, which will be studied in great detail. At other times he presents data on a large number of samples without any guarantee concerning their representativeness of the larger body to which the conclusions will be extended. It is mainly this second aspect which is discussed now.

Let us illustrate some of the difficulties involved in questions of representativeness by the following example. We want to investigate in a peridotite body the distribution of strain which is approximated by that of the grain size in the peridotites (see § 11.2.2). The grain size is an *attribute* attached to a given sample, an *object* in Whitten's terminology, which has many other attributes of directional and dimensional nature. Since the studies involve a consideration of different scale levels (in this case the scales of the thin section, the sample and the map), representativeness should be verified at each different level. Are the *measurements* on individual grains in thin section representative of the distribution in the thin section and in the sample? Is the hand specimen representative of the outcrop where it was collected? This may give rise to the question: Does the *nature* of the object or of the attribute in the object conform with the one designed? (For instance, a serpentinized peridotite sample does not, since the grain size in serpentine has little to do with that in the peridotite.) Finally, is the *population* of objects representative of the peridotite body? To answer the last question and because we already know some structural features of the body, we decide to trace a grid of an appropriate spacing and orientation and to collect the samples at the mesh points of the grid. Obviously, for practical reasons, the sampled population will differ from the one theoretically defined. These accumulated drawbacks limit the chance of the final result being as representative as is wished.

With classical statistical tests (variance analysis) it is, however, possible during the course of the study and *a posteriori* to know whether the attribute

* 50 grains in Turner's study.

population was representative or not, that is, whether the grid which was devised *a priori* was suitable or not. In the latter case, a new grid with a finer mesh and/or a new orientation may have to be used.

The empirical method suggested here can be improved in statistical mapping by a close treatment (Matheron, 1962, 1963, 1965). In our example, a variogram is built on the grid chosen *a priori*. This is a curve defining the variability of the studied property and is used to find its range (*portée*), that is the maximum distance within which the information carried by this property is valid. Then it should be possible to design the optimal grid.

This is just one of the aspects of the general problem of the size of the population which is clearly important. In studies on mineral preferred orientations, simple tests can be performed to ascertain how many individual measurements should be provided in order to have a fair representation of the fabric. Thus, measuring 1000 to 5000 grain orientations may be avoided when a tenth of these figures would be sufficient. This matter has been investigated by Stauffer (1966) in relation to considerations on types of preferred orientations and their densities. His estimations, calling for measurements of over a thousand grain orientations in unfavourable cases, could discourage the most dedicated or obstinate scholar! They seem to be excessive on account of the simple test devised by Hopwood (1968), in the somewhat different objective of comparing the degrees of preferred orientations. It consists in defining a coefficient R_N of preferred orientation for N individual measurements and in plotting this coefficient, versus N. The number of measurements is regarded as sufficient when R_N ceases to vary significantly (Fig. 8.1). Hopwood investigated the case of both strong and random preferred orientations; he concluded that 350 measurements are sufficient to characterize these fabrics. Stauffer sets the number of his measurements at 400 for a pattern of random preferred

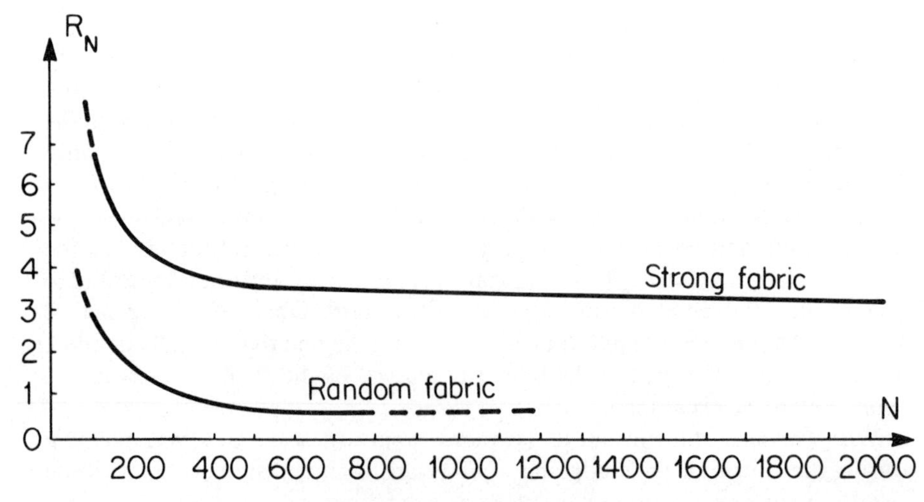

Fig. 8.1. Graph of the degree of preferred orientation coefficient R_N against the number of points N (after Hopwood, 1968 (simplified), reproduced by permission of The Geological Society of America)

orientation. Beyond these figures it seems that no new significant information will come forth.* Now if the preferred orientation pattern belongs to one of the

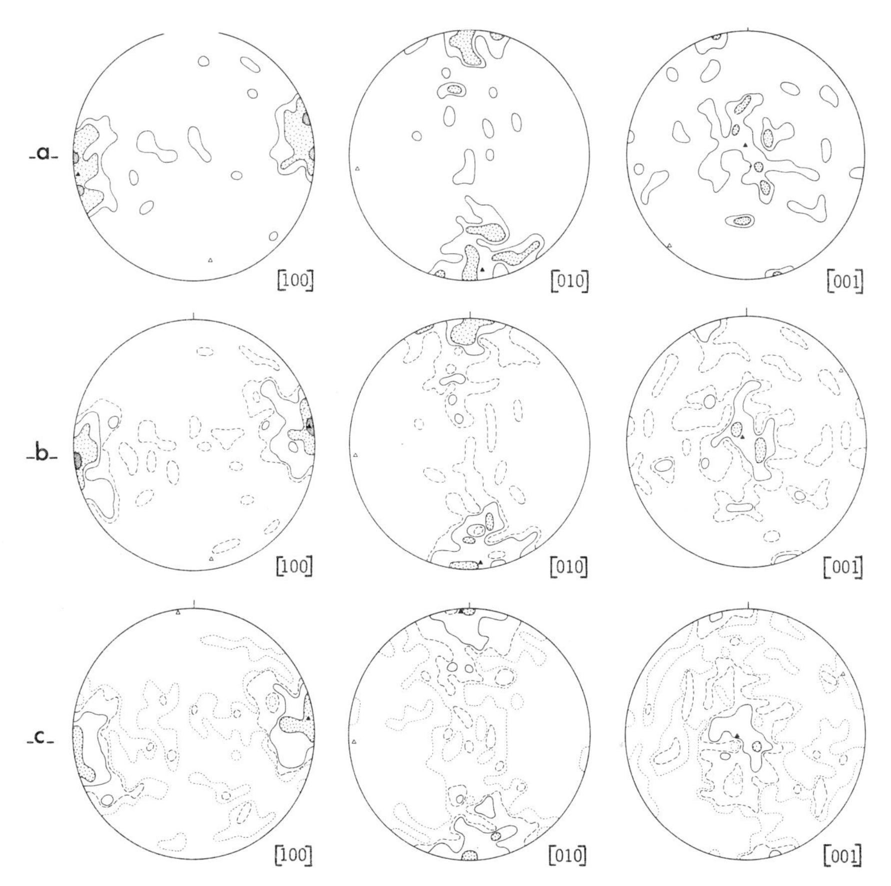

Fig. 8.2. Influence of the number of measurements on preferred orientation diagrams (olivine from a basalt xenolith with a porphyroclastic texture). Equal area projections; contours per 0·45 % area (see § 8.3.4); full triangle: best computed axis; open triangle pole of best plane (see § 8.3.5):
(a) 50 measurements; contours: 2, 4, 8 %
(b) 100 measurements; contours: 1, 2, 4, 8 %
(c) 200 measurements; contours: 0·5, 1, 2, 4 %

* This raises the interesting question of knowing what is to be expected from diagrams of preferred orientation. An example will illustrate our point. In the low temperature plastic deformation of quartzites studied by one of us (J.L.B.), 5 % of the deformed grains have their slip plane normal to the lineation and exhibit a globular habit. This contrasts with the general orientation parallel to the lineation and the flattening of the other grains. If the existence of this specific group is to be inferred from diagrams of orientation, several hundred grains would have to be measured. In fact it was identified by a close examination of the texture under the microscope and its remarkable orientation defined by specific measurements. Hence, generally speaking, diagrams of preferred orientation are of more use if they support a detailed textural study than if they are intended to be a substitute for it (see § 7.2).

simple types defined in § 8.3.5, the number of measurements necessary to characterize it may be in the range of 100, more or less depending on the degree of preferred orientation. This is what results, in the relatively simple fabrics of peridotite, from the use of Hopwood's test adapted for automatic processing by Bouchez (1971) (Figs. 8.2 and 8.3).

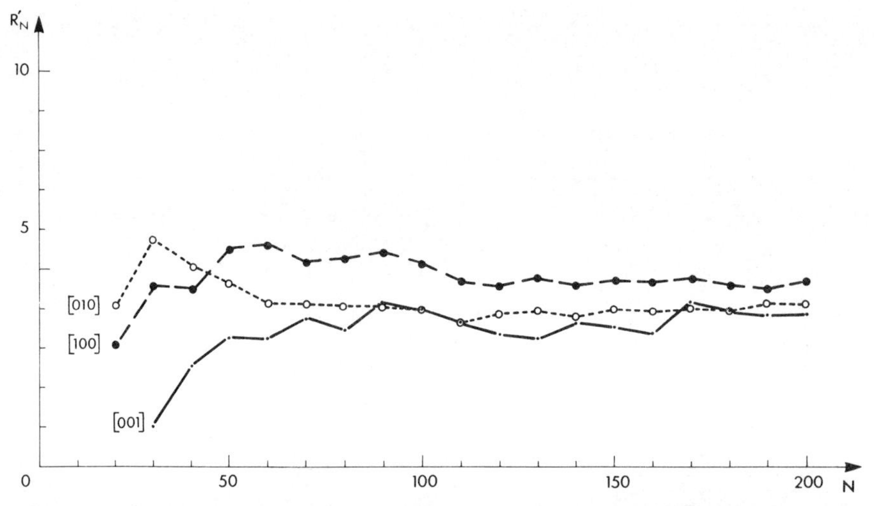

Fig. 8.3. Graph of the degree of preferred orientation coefficient R'_N (modified from Hopwood's) against the number of points N in the case of the olivine-preferred orientation shown on Fig. 8.2

8.3. DIRECTIONAL DATA WITHOUT TOPOGRAPHIC LOCATION

8.3.1. Nature of the directional data

The structural objects have to be expressed in a simple shape by planes and straight lines whose measurements constitute the directional data. Depending on its nature a structural object requires for its definition from one to three directional data. Since a plane is fully characterized by its normal, the directional data will be represented in their most general form by vectors in a fixed coordinate system. A vector is defined by the position of its origin, its orientation, sense and intensity. In structural geology the intensity and the sense are seldom used and henceforth we will refer to the orientation and position of straight lines. Examples involving the intensity of representing vectors would be the case of lineations due to various measurable amounts of stretching; an example involving sense would be the case of striae with a given polarity, on a fault plane; finally the complete representation of an ellipsoidal strain marker, like a deformed pebble, would require three vectors with intensity but without sense indications. In fact in this case, aspect ratios and orientations relative to two reference planes are usually reported (Ramsay, 1967; Dunnet, 1969; Elliott, 1970).

In this section we will only consider the orientation of the directional data. This simplification entitles us to translate all the lines to a common intersection. If, simultaneously, account is taken of the topographic position, the representation techniques are more complex and will be examined in § 8.4.

8.3.2. *Reference systems and principal modes of representation*

To define thoroughly a structural object one may need to use three directional data. They determine a reference system which is internal with respect to an external invariant reference system. Our purpose is now to introduce the main modes of representation of one system relative to the other. They are best illustrated by the case of the total orientation of a crystal when it is represented in a system of axes. Let us call X, Y, Z the crystallographic axes of the crystal which define an internal reference system. If many crystals were considered, there would be as many internal reference systems. The crystal system can be related in two ways to an external invariant reference system called ABC (for instance, the main structural elements in the sample containing the crystal or the geographic axes). Either the internal system is referred to the external invariant one (Fig. 8.4(a)) or vice versa (Fig. 8.4(b)). In the former case it is a *direct* representation, in the latter an *inverse* representation (see § 8.3.4).

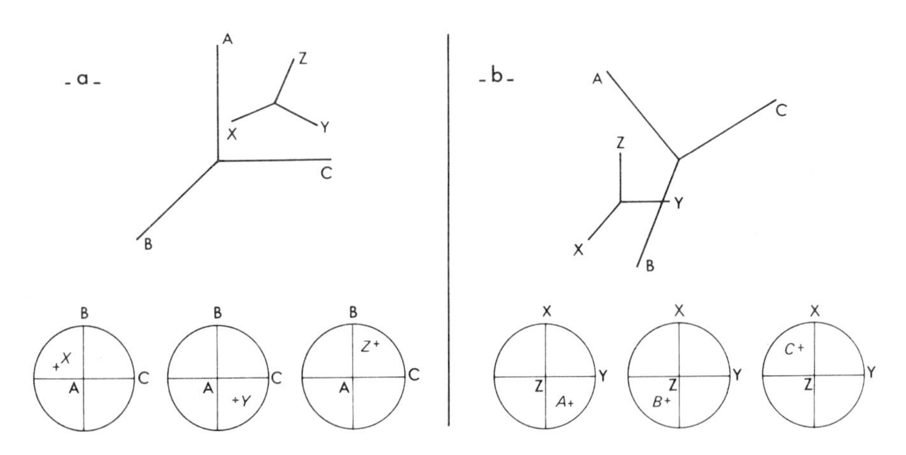

Fig. 8.4. Direct (a) and inverse (b) modes of representation. A, B, C: external invariant system, for instance geographic coordinate axes; X, Y, Z: internal system, for instance crystallographic axes in a given mineral

Two different modes of representation have been envisaged. The common one consists of planar equatorial projection of the equal area type (see below) of every one of the reference axes or other directional data of one system in the other system. As many diagrams are to be presented as reference axes and other directional data in the other system. The planar equatorial projection is described in the forthcoming sections and is the only one used in this volume. In the other mode of representation, one of the systems is fully represented with respect to the other by the three Euler angles. These angles constitute the coordinate axes of a three-dimensional diagram in which one system, for instance three representative crystallographic axes of a given crystal, is entirely represented by one point. The distribution of these representative points is studied by *the orientation distribution function or ODF*. This representation has been recently introduced in geology by Baker (1969), Baker et al. (1969), Baker and Wenk (1972) and Wenk and Wilde (1972). Baker (1969) gives a full account of how the ODF can be recovered from a suite of pole figures determined with the X-ray goniometer and, in his other papers, applies this representation to the analysis of quartz-bearing rocks. Wenk and Wilde (1972) show, from an example in marble, how to use the ODF when the preferred orientations are determined with the universal-stage technique. We do not share their enthusiasm about this way of presenting orientation information and wish to restrict our discussion on it to weighing up its advantages and drawbacks with respect to the planar projection diagrams.

Apart from our problem of a physical analysis of preferred orientations the ODF seems to be appropriate for mathematical analyses of preferred orientation as is the case in metallurgy. Its promoters claim that it must be used in calculating the elastic coefficients in anisotropic aggregates, but this is currently achieved by using the numerical data and conventional equatorial representation (Crosson and Lin, 1971; Baker and Carter, 1972; Peselnick et al., 1974). In our case, the main advantage of the ODF consists in giving the complete orientation of a crystal in a single diagram; this makes total symmetry and various correlations easier to detect. However, this advantage is only apparent because the representation is three-dimensional and can be more thoroughly analysed only through planar projections, each one recording only partial information (Fig. 8.5). Correlation problems are also solved in conventional equatorial projection diagrams by identifying individual crystals in each diagram. For instance, correlating a single Y maximum on the Y diagram with one of the two X maxima on the X diagram is achieved in this way. The main drawback of the ODF representation from our point of view lies in the difficulty in visualizing the preferred orientation as Wenk and Wilde also noted (Fig. 8.5). Another drawback, when the ODF is applied in using X-ray goniometer data, is that it requires a considerable number of measurements (8 days of goniometer operation in Baker et al., 1969) to bring the development of the harmonic analysis to a high enough order. Thus, $(12/n+1)$ pole figures are necessary to get a 12th order development if the crystal has an n-fold symmetry axis. Even at such an order, the pole figures regenerated by the ODF are quite smooth, losing information with respect to the original ones (Baker and Wenk, 1972).

Direct versus inverse representations

It is worth coming back to the question of direct and inverse representations to decide in which case one or the other is better adapted. In equal area projection, the inverse preferred orientation diagram, more commonly called

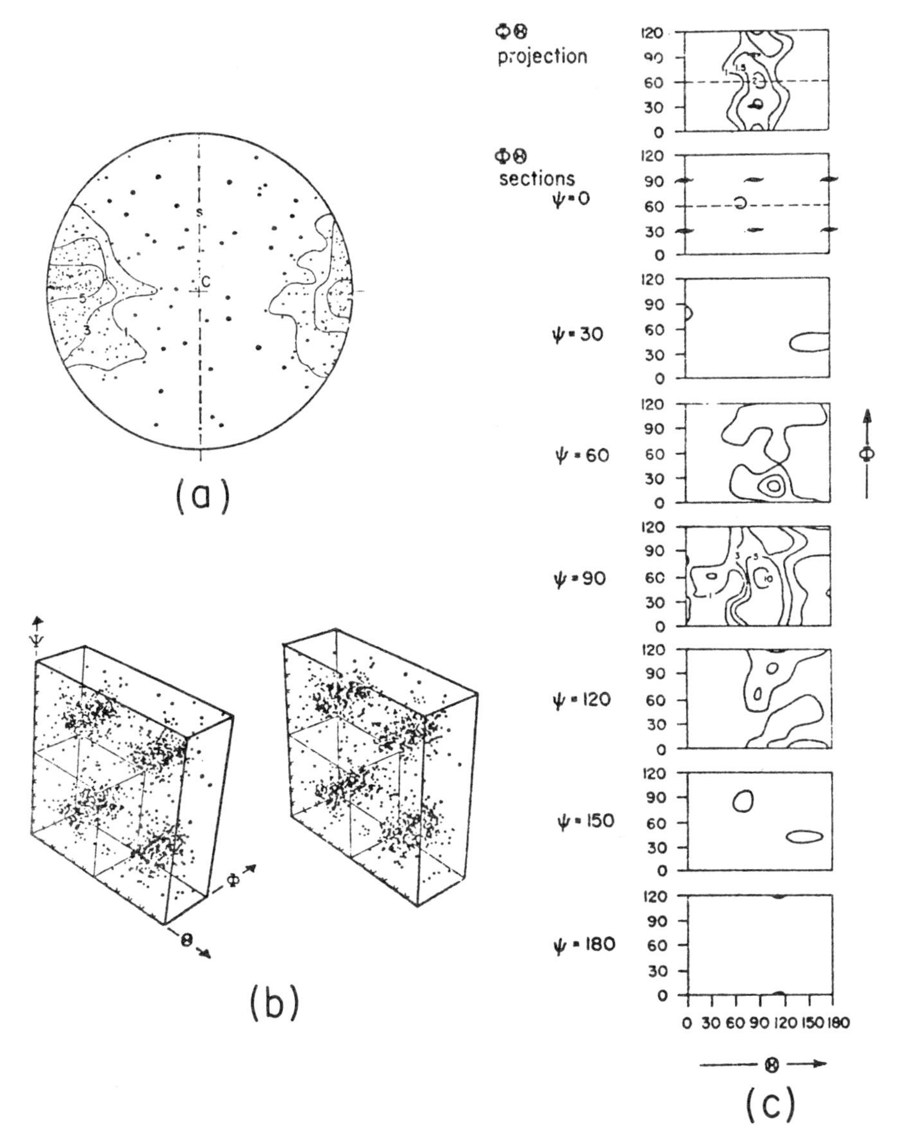

Fig. 8.5. Orientation distribution diagrams of the c-axis of calcite grains from a Yule marble specimen:

(a) Equal area projection of the c-axis preferred orientation; upper hemisphere; contours: 1, 3, 5 % per 1 % area

(b) Stereoscopic pair of orientation distribution diagrams; 343 points in the asymmetric unit

(c) Contoured $\Phi\Theta$ projections and sections of the asymmetric unit of the orientation distribution of (b); (top panel), contours: 0·5, 1, 2 and 3 % per 1 % volume for Ψ projections; (lower panels), contours: 1, 3, 5 % and 10 % per 1 % volume for sections. (Reproduced with permission from Wenk and Wilde, *Flow and Fracture of Rocks*, Griggs volume, **16**, 83 (1972), copyright by American Geophysical Union)

inverse pole figure, has become familiar due to the X-ray goniometer determination of the preferred orientations developed during axial deformation tests. In the numerous examples in Chapters 5 and 6, the density contouring is traced, after automatic processing, in multiples of a uniform distribution. In such cases, this representation offers the advantage of directly visualizing the orientations of the deviatoric stress relative to any crystallographic element in the deformed crystals. As also observed, all the information is contained in an elementary cell on the projection from which the whole diagram can be generated by the crystal symmetry operations. For instance, in the case of quartz fabrics, the diagram is reduced to a sector of one-sixth of the equal area net (Fig. 6.10).

The choice of a direct or inverse representation depends on the kind of problem under investigation. Let us suppose that we are studying the crystallographic orientation of crystals with respect to the orientation of many structural elements. The direct representation is more suitable because all the structural elements will be represented on each of the three density diagrams, one for each crystallographic axis, if the crystallographic orientation has been defined by three axes (see for example Fig. 8.4). On the contrary, if a few structural elements (for instance, deviatoric stress or foliation and lineation) are to be related to many crystallographic planes or axes in the crystals, the inverse pole figure makes it possible to represent on each one of a few diagrams, one for each structural element, all the necessary crystallographic information.

In short, the inverse representation is recommended in studies on flow mechanisms involving a large amount of crystallographic information and a limited amount of structural information; the direct representation, in studies on interpretation of field structures in relation to mineral preferred orientations involving many structural elements and a few crystallographic ones.

8.3.3. Techniques of planar representation

It has been seen in § 8.3.1 that non-located directional data may, most commonly, be restricted to straight lines intersecting at a single point. This point is the centre of a hemisphere and the study of the orientations of the directional data is now based on the distribution of their intersections with the hemisphere. As it is more convenient to work in a plane, these intersections are projected by various systems in the equatorial plane of the hemisphere. In geology, the usual two systems are the *equatorial equiangular* or *equatorial stereographic projection* (Fig. 8.6(a)) and the *Lambert* or *equatorial equal area projection* (Fig. 8.6(b)). The projected data are then analysed, using reference nets, which are respectively the *Wulff and the Schmidt nets* for the stereographic and the equal area projection. In particular cases, like those of axially symmetrical distributions, *polar nets* are also usual. They are derived from projection in the plane tangent to the sphere at the pole. These systems of projection are fully described by Vistelius (1966). The equal area projection is best adapted to structural geology studies where the density of points distribution is considered.

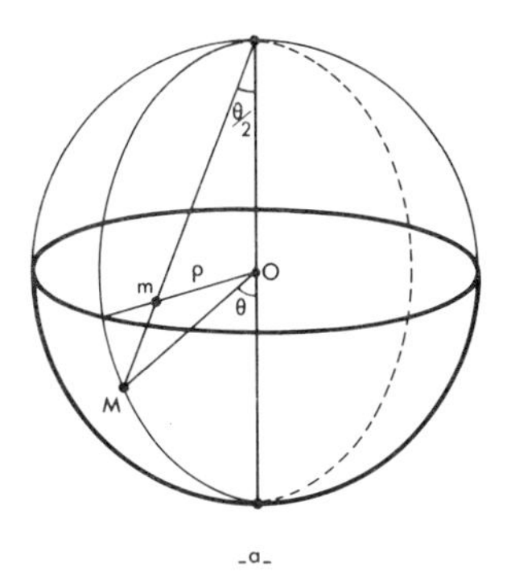

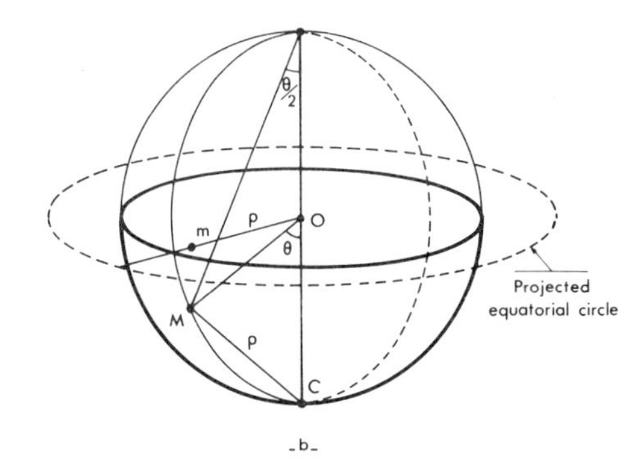

Fig. 8.6. Equatorial projections. OM is parallel to the projected straight line and intersects the reference hemisphere (radius R) in M; m is the projection in the equatorial plane of M following two principal modes:
(a) Equatorial stereographic (equiangular) projection: $\text{Om} = \rho = R \tan \theta/2$
(b) Lambert or equatorial equal area projection: $\text{Om} = \rho = \text{MC} = 2\,R \sin \theta/2$

The techniques of plotting directional data on the net, and the various geometrical operations which can be performed on them, have been described by many authors. We wish to avoid presenting them again and recommend Turner and Weiss' book (1963, p. 52) for the general treatment and Badgley's (1959, p. 187) and Ragan's (1968) for field problems. We will insist only on the techniques of construction of density diagrams which have been recently renewed by computer processing.

A particularly simple situation occurs when it comes to representing directional data entirely characterized by a single angular value either because they all lie in a single plane or because in $3D$-space they are determined by their angle with a single reference direction. The representation is then directly planar and graphs like histograms and rose diagrams are quite appropriate.

8.3.4. Construction of orientation density diagrams

The raw data are points on the equal area net, representing the directional data. Such a representation, although preserving all the elementary information, is not adapted to a visual study of the preferred orientations. A convenient substitute for it is a representation by areas of equal numbers of points per constant area of reference on the net, separated by contours. The subsequent drawing is that of an *orientation density diagram.* The general method consists in diluting the punctual information as homogeneously as possible over the entire area of the net. The various techniques have been reviewed by Stauffer (1966) and Bouchez and Mercier (1974).

Whatever the techniques employed, an attempt is always made to point out equal density areas by counting the number of points inside a constant area of reference. The constant area is either each cell of a grid laid on the point diagram (counting on the net), or a circular disc (counting on the net) or a spherical cap (counting on the hemisphere) which is centred on pre-determined points.

The grid techniques differ from each other by the shape of the grid-cell: triangular–hexagonal (Kalsbeek, 1963), square (Stauffer, 1966) or curvilinear (Denness, 1972). In the circular disc techniques, the counting disc is either centred at the mesh points of a squared grid (Schmidt, 1925) or on the data points themselves (Mellis, 1942); Mellis technique is simplified by the free counter technique (Turner and Weiss, 1963, p. 62). In the spherical cap techniques, the counting is carried our directly on the hemisphere by a computer, except in the variable ellipse technique (Strand, 1944) in which the counting is left up to the operator, using a special net (Dimitrijevic, 1956). This net is derived from the equal area projection of the spherical caps which are nearly elliptical discs of increasing eccentricity from the centre, where the cap is a circle, towards the periphery of the net.

All these manual techniques, except the variable ellipse and Denness', lead to important counting distortions towards the periphery due to the constant shape of the counting cell. For this reason and also because the procedure is easy and rapid, we recommend using for manual counting Dimitrijevic's or Denness' counting nets. More complete descriptions of manual counting and contouring procedures will be found in Turner and Weiss (1963, p. 58) and Vistelius (1966, p. 65).

Introduced in the sixties, the techniques involving computers make manual plotting and contouring obsolete, particularly when a change of reference system (rotation of data) is required. The power of computers is used in the most profitable conditions when the counting of the data points is carried out on the hemisphere and not in the projection plane, as in the methods of Tocher (1967), Starkey (1970) and Milnes (1971).* This seems to have been achieved for the first time by Noble and Eberly (1964). In these techniques, spherical caps upon the hemisphere are equivalently replaced by small cones of a given aperture, but for the sake of description it is easier to come back to the caps and hemisphere model. As in the manual techniques using discs, two sets of modes exist: one centres the spherical caps upon the data points themselves (Spencer and Clabaugh, 1967; Franssen and Kummert, 1971); the other centres them on the successive mesh points of a hemispherical counting grid (Noble and Eberly, 1964; Möckel, 1969; Bouchez and Mercier, 1974). The only drawback in the former set of techniques, which occurs also in Mellis' technique, is that it yields contours whose artificial complexity increases with the number of data points and would call for another smoothing. The counting grid techniques differ from each other by the position and number of mesh points. The number of mesh

* This author recommends for regional trend analysis an automatic counting method in the projection plane using 30° counting discs centred on a very limited number of counting points (25 mesh points). In such conditions, only the major structural features are brought out.

points must be fixed within certain limits: above these limits, the density diagram tends towards a point diagram; below them, information is lost by excessive smoothing. It seems that 500 points is a good figure; this also keeps down the processing time. The symmetry of the grid is then essential in ensuring a regular distribution of the mesh points over the hemisphere. Bouchez (1971) and Bouchez and Mercier (1974) have used 457 points and devised a grid of high symmetry resulting in a low distortion in the typical case of representing an ideal uniform distribution (Fig. 8.7). Their flowchart is presented in Table 8.1.

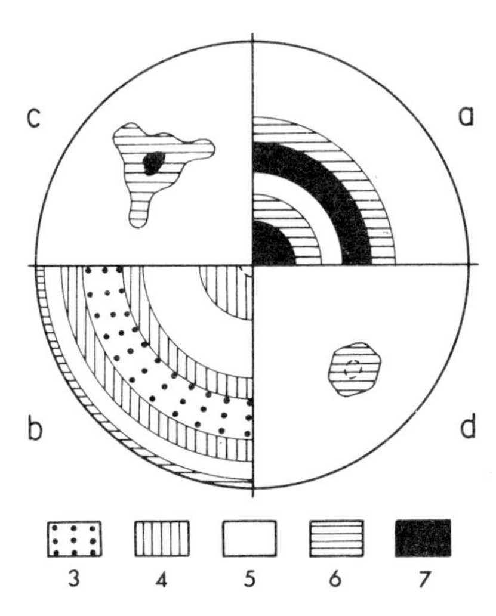

Fig. 8.7. Equal area projections of equal overlapping classes for spherical counting caps, 1 % of the hemisphere area, except in (d) where it is 0·45 %. The class of mean overlapping is taken as 5; the classes above have 1/5, 2/5, and more overlapping and the classes below, 1/5, 2/5 and less. The caps are centred on the mesh points of the following counting nets: (a) Dimitrijevic's, (b) Noble and Eberly's, (c) and (d) Bouchez'. (Reproduced by permission of Fondation Scientifique de la Géologie et de ses Applications)

In manual contouring of equal density areas on the equal area net, the counting circle represents 1 % of the net area and whatever the counting mode, the density is traditionally expressed in % for 1 % of the net area. In computer counting with the grid techniques this no longer holds true. The optimum radius of the spherical cap is determined so that, when the cap has been applied on the successive mesh points, all the hemisphere has been explored with minimum overlapping (Fig. 8.7). Therefore the radius of the cap depends on the number of mesh points on the grid and is such that the cap area does not necessarily represent 1 % of the hemisphere area. As an example, in Bouchez and Mercier's technique, the surface of the spherical cap is 1/220th that of the hemisphere, resulting in densities which are lower (2.2 times at most) and/or restricted to smaller areas than in the corresponding manual plotting. Although comparing densities in diagrams obtained by various methods is not straightforward and masks inherent difficulties, it is recommended to indicate the exact surface of the counting cell.

Table 8.1.

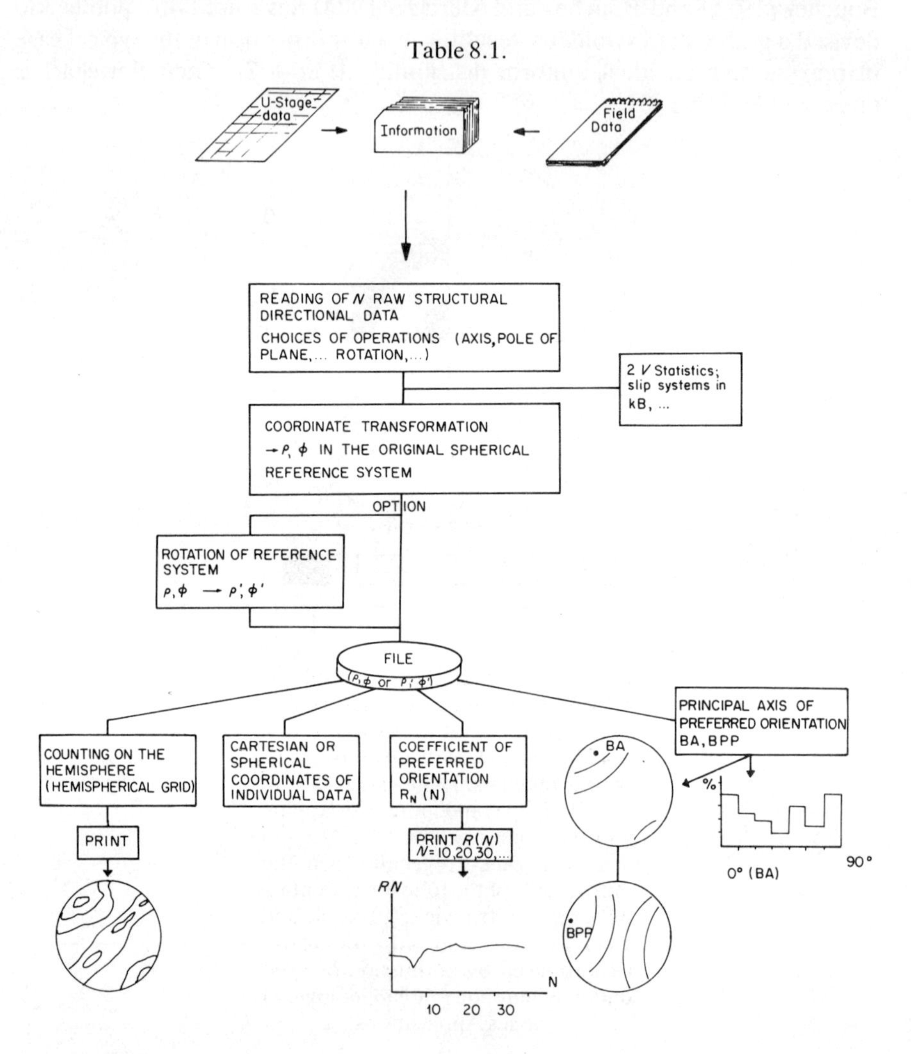

8.3.5. Analysis of orientation density diagrams

Although many of this section's results are valid for any type of directional data, it deals principally with mineral preferred orientation.

Symmetry

Symmetry arguments introduced by Sander (1930, 1948, 1950, 1970) have always been regarded as important in the analysis of mineral orientation density diagrams (Paterson and Weiss, 1961; Turner and Weiss, 1963 p. 42, p. 64 and p. 264; Den Tex, in Sherbon Hills, 1963; Whitten, 1966 p. 115). In analysis using direct modes of representation they seem to us mostly of academic interest and are possibly misleading in that they may bias the analysis towards irrelevant aspects. We think it far more important to relate the density diagram to other pieces of structural information, the main one being the shape preferred orientation defined by the penetrative shape fabric (see § 7.3). Symmetry arguments may then contribute to the analysis in so far as significance is attached only to the total symmetry but they possibly do not need explicit discussion. This point is illustrated by the examination of the peridotite fabric in § 10.2. In the case of inverse pole figures, since the reference system is the crystallographic orientation of the crystal, symmetry relations are useful, for instance they restrict the size of the representative figure.

The remarks about symmetry arguments and the point stressed above are trivial. Nevertheless, descriptions in terms of the symmetry of mineral preferred orientation diagrams without any reference to external structural information can be found in some of the studies mentioned above and even in more recent papers.

Types of preferred orientations

The patterns of density diagrams representing directional data of any nature can be very complex. If the data do not represent a mixed population such as two independent lineations or two successive generations of crystals, the density diagrams usually belong to one of the following three elementary types or a combination of them (Den Tex, in Sherbon Hills, 1963, p. 397; Ramsay, 1967, p. 14):

(i) *The circular or point maximum* in which the data points cluster around the best axis (B.A.). The directional data tend to be parallel to a line (Fig. 8.8(a) left).

(ii) *The girdle maximum* in which the data points are spread along a great circle of the equal area net. The directional data tend to lie in a plane which is defined in projection by the axis normal to its representative girdle (B.P.P. in Fig. 8.8(b) left).

(iii) *Annular or small circle maximum.* The directional data are distributed preferentially on a cone surface which is defined, in projection, by the centre (B.A.) and the radius of the corresponding best fit small circle (Fig. 8.8(c) left).

A combination of the three fundamental types of density diagram is common. For instance, the poles of mica cleavage planes lying in a layer which has been isoclinally folded, are plotted in a girdle within which there will be a strong point maximum representing the orientations in the limbs. Another instance is the common crossed girdle fabric of quartz in naturally deformed quartz-bearing rocks (Fig. 6.11), in which the *c*-axes tend to lie in two planes symmetrically inclined to the shape fabric elements.

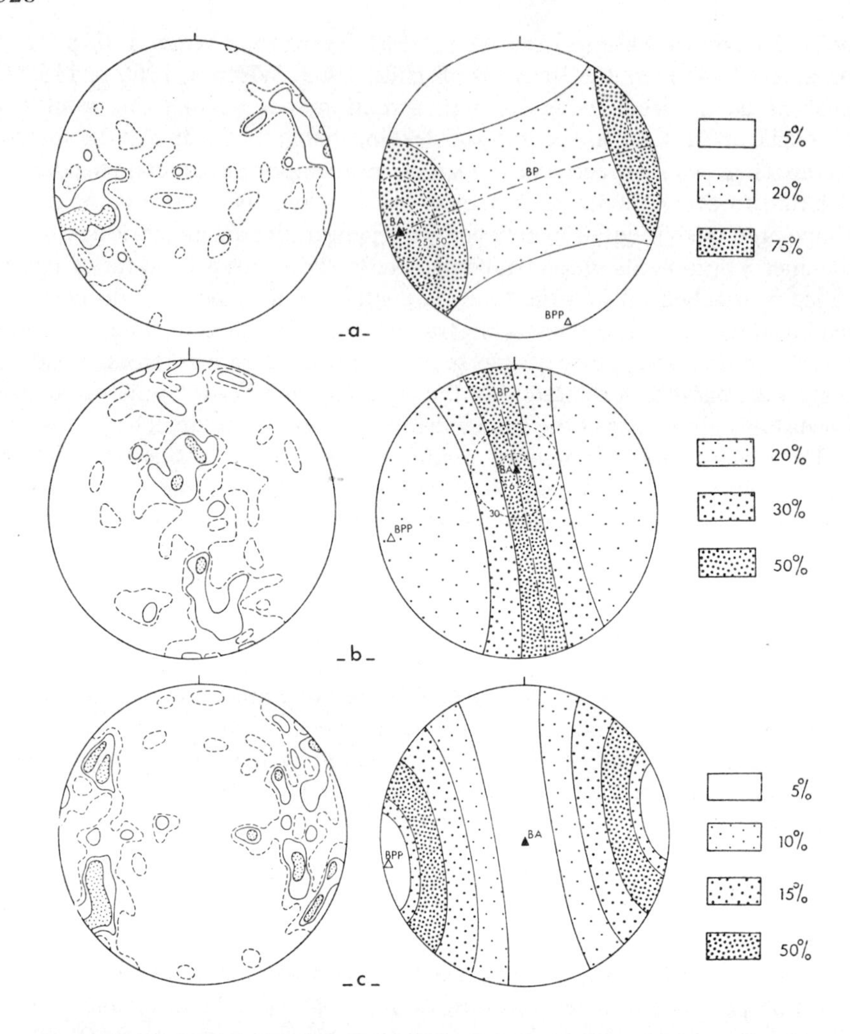

Fig. 8.8. Principal types of preferred orientations (left-hand diagrams) and corresponding analysis (right-hand diagrams). 100 points contours: 1, 2, 3, 4 % per 0·45 % area: (a) Point maximum; (b) girdle maximum; (c) small circle maximum (after Darot and Bouchez, in press, reproduced by permission of The University of Chicago Press, Chicago)

Analysis of the preferred orientations

Attention is restricted here to analysing the elementary types defined in the previous section. Two distinct problems are dealt with: (1) defining the nature and the orientation of the geometric elements (axes, planes) which fit the preferred orientation best; (2) Studying the density distribution about geometrical axes. The latter problem may be solved by using statistical tests which have been recently reviewed by Stauffer (1966). Hopwood's derivation of a coefficient of degree of preferred orientation (see § 8.2) illustrates a non-statistical method in that it is not based on distribution laws. Methods intended to describe the nature and orientation of the geometrical elements fitting the density diagram best have been introduced by Loudon (1964), Watson (1965, 1966), Scheidegger (1965), Ramsay (1967, p. 14), Kiralý (1969), Darot and Bouchez (in press). The last authors' method, which is not intrinsically original but is adapted to our purpose, calculates the best axis (B.A.), the best pole of a plane (B.P.P.) and automatically samples the real distribution in annuli around these axes. Thus it differs from Watson's method which refers to a parametric spherical distribution law. It is up to the operator to decide, in view of the density diagram pattern, which type of preferred orientation fits this pattern best (point, girdle, annular, combination) and to choose accordingly either a B.A. or a B.P.P. distribution. In cases where the diagram suggests an annular-preferred orientation, the density distribution is fitted by a B.A. type of preferred orientation, only if the distribution on the small circle is homogeneous. If this condition is not satisfied, the B.A. is not centred in respect of the small circle and does not reflect the fabric. Figure 8.8 also presents the diagrams corresponding respectively to these three types of distribution.

8.4. DATA WITH A TOPOGRAPHIC LOCATION

The data now under consideration have a topographic location in a two-dimension space which must be retained on the representative map.

8.4.1. Located dimensional data

All the systems of representing located dimensional data are based on a planar representation of a three-dimensional space: the horizontal coordinates give the topographic location and the vertical coordinate the value of the datum. The various graphic representations are described in Bertin (1967) and computed systems of visualization are presented in Mallet (1974).

We wish here to call attention briefly to the possibilities and limits of automatic mapping in structural geology. In its elaborate form it consists of a planar representation of a continuous surface known only by a limited number of points. This surface represents the value of the dimensional data everywhere in the plane. The most common visualization of this surface in the base plane is carried out by an equal value contouring (Fig. 8.9). Other than the general conditions introduced in § 8.2, the principal prerequisite condition in using automatic mapping successfully is to have a fairly regular pattern of data points inside the body under investigation.

In geological problems, different mathematical methods have been used to determine the representative surface. They have been reviewed by Laffitte (1972, p. 540) who classifies them into three groups according to their approximation system:

(1) Approximation of a polygonal surface by a triangular network directly built on the data points; it is a computer adaptation from manual map construction;

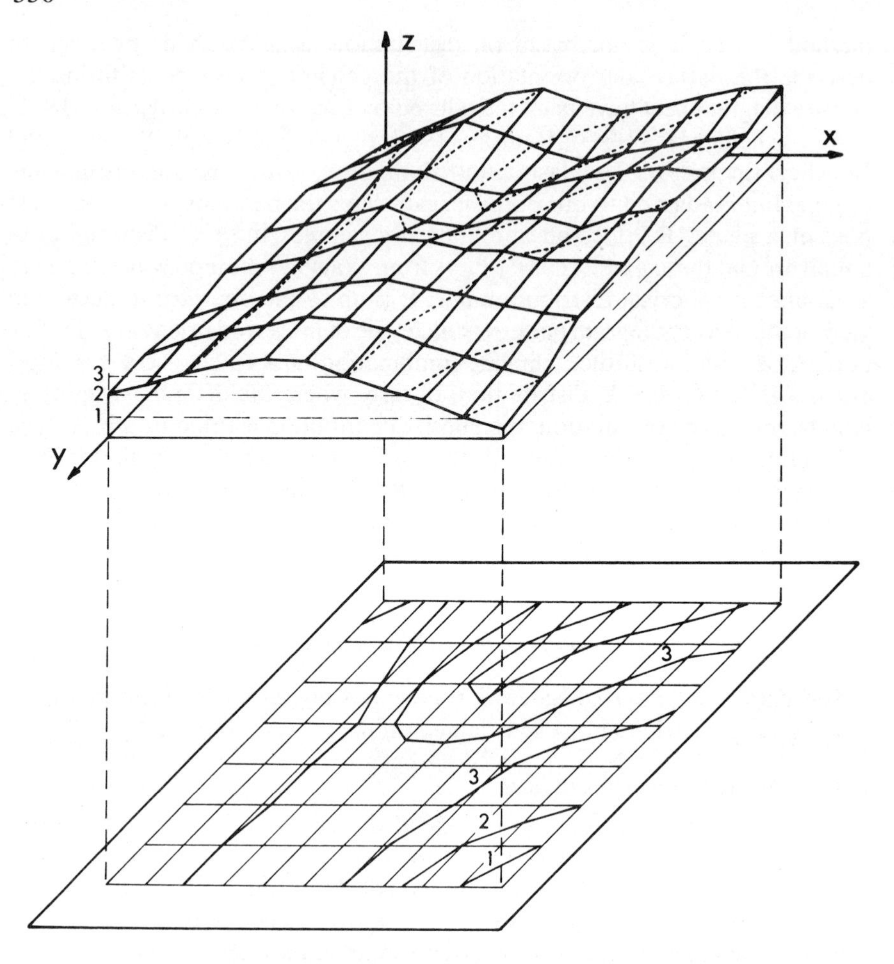

Fig. 8.9. Automatic mapping using a numerical grid. The Z values of dimensional data located in a (x, y) plane are projected as isolines (after Darot, 1973)

(2) Approximation of a surface of the type $z = f(x, y)$, usually calculated using the harmonic expansion of various polynoms (orthogonal, Fourier's, Tchebycheff's) (Harbaugh and Merriam, 1968; Mallet, 1974);

(3) Approximation of a polygonal surface by transposing the data points network into a regular network with square or hexagonal cells (Harbaugh and Merriam's numerical grid method). Matheron and Huijbregts' (1970) method of universal kriging belongs to this third group. It yields a representative map which is complemented by an isovariance map bearing indications concerning the precision obtained in any point. In this regard, the method is more satisfactory than any other one; this is obtained at the price of particularly complex mathematical treatment.

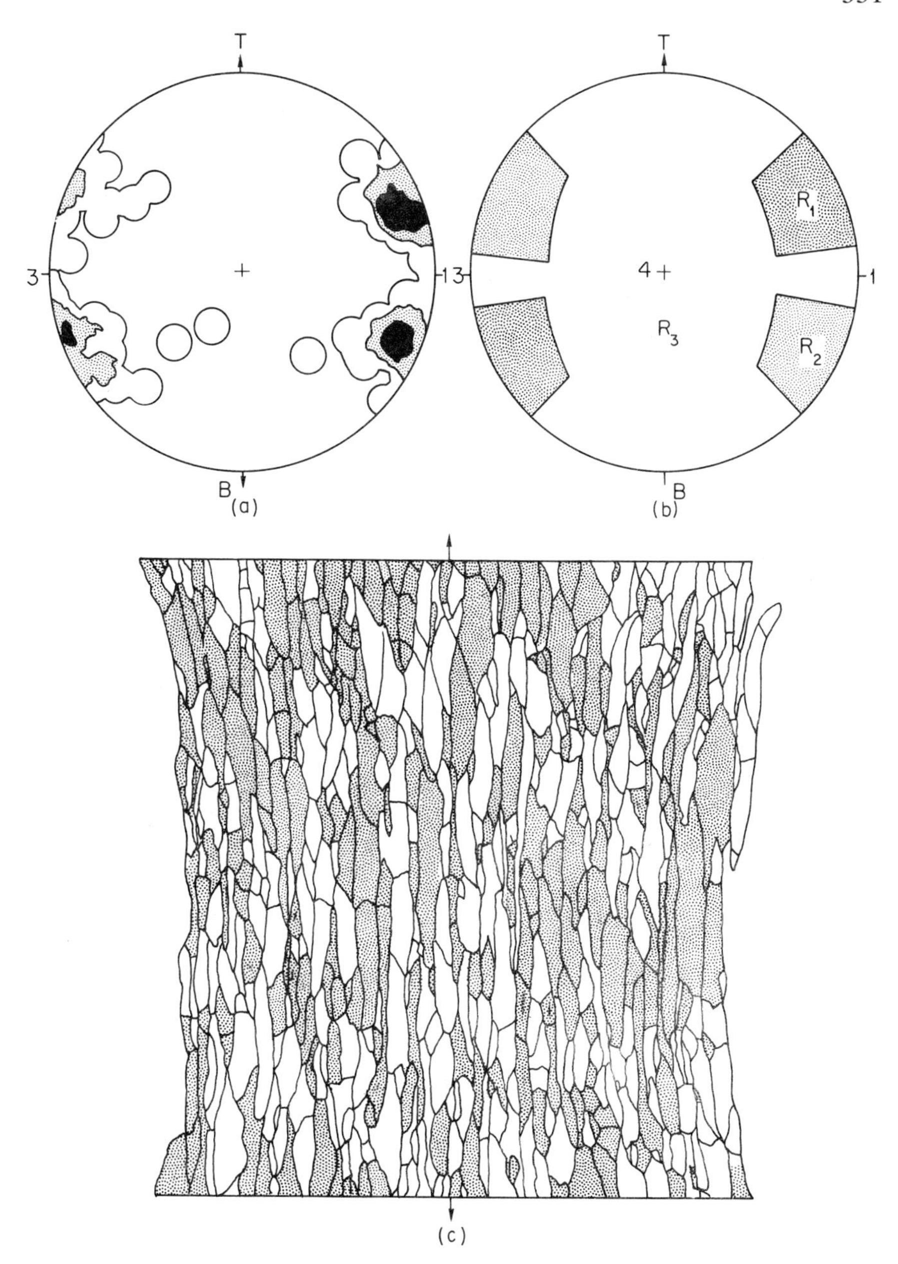

Fig. 8.10. Axial distribution analysis of Yule marble elongated 118 %. (a) Orientation diagram for [0001] in calcite, 100 grains in neck. Contours 10 %, 5 %, 1 % per 1 % area. (b) Division of diagram (a) into direction groups. (c) Axial distribution diagram, shaded to correspond to direction groups of diagram (b) (after Turner and Weiss, 1963, reproduced by permission of the McGraw-Hill Book Company, New York).

In deciding among the various methods which one is best adapted to representation of given dimensional data, the most important factor is the manner in which these data vary over the considered area. If the fluctuations in the value of the data are high, the method using a triangular network is certainly the most suitable, although the regular network method may be applied; fitting by a polynomial surface is not appropriate, whatever the nature of the polynoms: either the degree of the harmonic expansion is low and the surface too smoothed, or it is high and then the surface presents undulations without any physical meaning if they are not subsequently filtered (Robinson and Ellis, 1971). On the contrary, if the variable fluctuations are moderate, the fitting by a polynomial surface is altogether suitable and interesting due to its further developments (for instance, in hydrogeology, calculations of flow lines from the determination of a phreatic surface).

8.4.2. Located directional data

When the directional data are not evenly distributed within the investigated surface (massif, thin section, etc.) a representation taking into account their individual location may be required. In studies on mineral orientations this is important whenever the density diagrams yield evidence of a complex distribution; it may help to discover heterogeneity in the structure or in the nature of the directional data which were overlooked in the examination of the thin sections.

When the data are not too numerous, they need merely to be individually plotted in maps using the conventional signs for planes and lines. When they are numerous, synthetic modes of graphic representation are sought. They are not satisfactory (see Bertin, 1967) because two angular values being attached to every directional datum, part of the information has to be abandoned unless individual plotting at the mesh points of a regular grid is acceptable.

Of special interest here is the problem of the spatial representation of mineral orientations. This is achieved using the axial distribution analysis (A.D.A.), more often called A.V.A. after its German name (Achsenverteilungsanalyse). It has been developed in Sander's school (Sander, 1970, p. 385) and the technique is clearly defined by Turner and Weiss (1963, p. 247). It consists in dividing the preferred orientation patterns into groups of close orientations and in decorating accordingly the grains in the investigated area of the sample (Fig. 8.10). Information is lost and moreover it is a tedious task. Tenants of the A.V.A. have claimed that it was necessary to use it when the suspected inhomogeneity of the fabric could not be revealed otherwise. It seems to us doubtful that a significant fabric element with a specific location might be missed in the careful examination of the texture that we have recommended before any measurement is interpreted. Therefore we think it is more profitable to divide the thin section into homogeneous domains or to sort out the different categories of grains before studying preferred orientation and to present separate diagrams for the various subfabric elements.

CHAPTER 9

Laboratory Techniques

9.1. STUDIES WITH THE OPTICAL MICROSCOPE

As exposed in the chapter on methods of kinematic analysis, investigations on flow rely heavily on observations and measurements using the optical microscope, complemented by the universal stage. In this regard, some recommendations presented below may be trivial. They are nevertheless legitimated in that overlooking them cuts the investigator from an indispensable and cheap source of information. We shall restrict ourselves to describing only those technical operations and their sequence which are not exposed in recent textbooks.

9.1.1. Thin section preparation

The oriented sample collected in the field is sliced in the three orthogonal planes determined by the orientations of the foliation and the lineation (Fig. 7.1). To achieve this purpose, the sample must be reoriented in the laboratory using as a support plasticine or a sample holder built of non-ferromagnetic metals to avoid any disturbance of the compass used in the operation. The traces of foliation and lineation measured in the field are then reported. This procedure can be omitted if these traces have been directly reported on the sample in the field. The thin section slab is oriented in a given reference plane relative to the traces of the two other reference planes and its contour is drawn. A useful convention is to prepare always the thin section so that, when applied on its origin emplacement in the sliced sample, its base plate comes into contact with the surface of the sample (Fig. 9.3). This will avoid an inversion responsible, in particular, for inversion of the determined shear sense.

If it has not been possible to measure correctly the penetrative structural elements in the field, in some cases it is still feasible in the laboratory on sliced and lacquered sections of the oriented sample. Mineral elongation may be revealed by etching or colouration treatments of the surfaces (for instance, Laduron's colouration technique (1966) in quartz and feldspar-bearing rocks or treatment with diluted HCl, in peridotites). The shape fabric is then studied under a binocular microscope complemented by an orientometer device (Whitaker and Gatrall, 1969) (Fig. 9.1). A more tedious technique, necessary

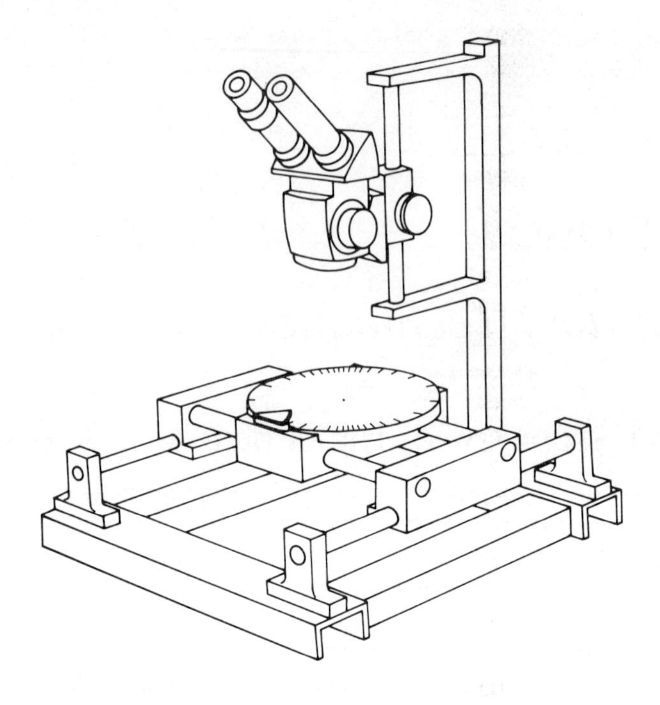

Fig. 9.1. Orientometer used for determining foliation traces and lineations on polished slabs (after Darot, 1973)

when the trace of shape preferred orientation needs to be determined with precision, consists in comparing the grains aspect ratios for successive orientations in a thin section observed under the microscope (Fig. 9.2, see also § 9.1.2).

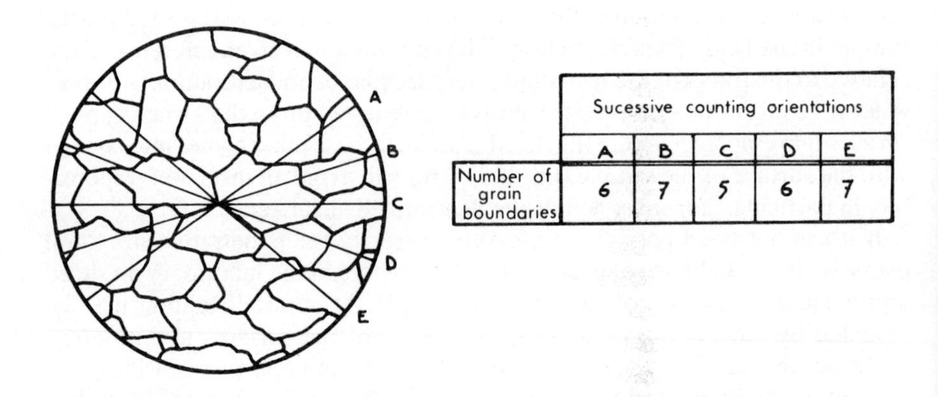

	Sucessive counting orientations				
	A	B	C	D	E
Number of grain boundaries	6	7	5	6	7

Fig. 9.2. Determination of shape-preferred orientation under the microscope from calculation of the grain-size index for successive orientations (after Darot, 1973)

With the determination of the penetrative elements on the oriented sample in view, the sample should be first sliced normal to all the visible elongation trends. This will result in a cut close to the *YZ* section of the finite strain ellipsoid (section normal both to the foliation and the stretching lineation, see Fig. 7.1). The techniques mentioned above should help in defining in this section the exact orientation of the foliation plane. The next slice is then parallel to the foliation plane in order to show up the lineation; finally the most important *XZ* section (normal to the foliation and parallel to the lineation) is cut.

In a petrofabric study, it may be found more convenient to section the sample in the horizontal plane and to orient the thin section relative to the geographic axes, since the data are usually presented in this plane and referred to these axes. In manual plotting, this saves the troublesome rotation of data, otherwise necessary. The now widespread computer processing of data eliminates this difficulty and the practice of sectioning in the horizontal plane should therefore be abandoned because this plane probably does not coincide with the remarkable *XZ* section, thus making relevant observations difficult.

To identify correctly a texture under the microscope it is evidently necessary to have a direct scope over areas large enough to be representative of this texture. In common textures, homogeneous and with a mean grain size in the range of the millimeter, this requires the observation at the same time of a large area if not the whole area of a standard thin section. This is impossible in most microscopes which have a narrow field even at the lowest magnification. It is easily achieved with a small stereomicroscope. When the grain size is coarse or the texture fairly inhomogeneous it may be necessary to carry out this important preliminary observation on large thin section (for instance, 60 × 45 mm mounts). Accurate observations on substructural features, like subgrains, are greatly facilitated by polishing the thin section on both sides and choosing a mounting epoxy with refractive index close to that of the studied minerals.

9.1.2. Principal studies in thin section

Lattice preferred orientations

When a petrofabric study is in view, the preliminary investigation is destined to sort out the various subfabric elements, for instance distinguishing two successive generations of grains in the same mineral species which ought to be studied independently. In an inhomogeneous texture investigated without the help of the AVA representation, it permits the delineation of the subdomains on which independent measurements will be carried out. In any case, it is recommended that an enlarged photograph be taken (40 × 30 cm) of the thin section upon which the grains oriented with the U-stage are spotted with a distinctive number.

The technique of measurements of mineral orientation using the universal stage is exposed by Emmons (1943) and can also be found in Winchell (1961, Part I, p. 239). It is applied to the 5-axes universal stage which constitutes an

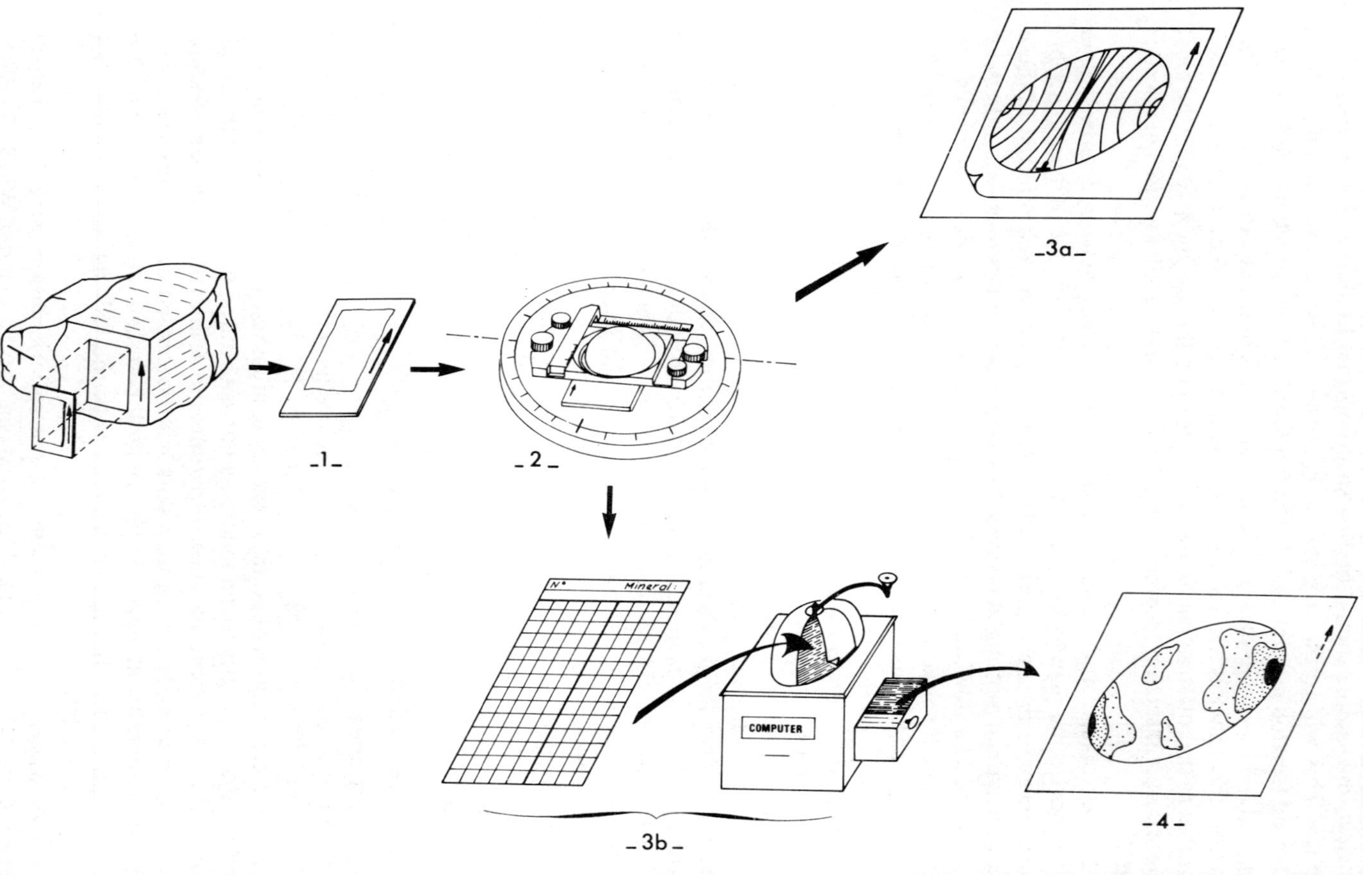

Fig. 9.3. Sketch of the sequence of operations involved in establishing diagrams of preferred orientations, (3a) manual plotting; (3b) and (4) computer plotting

important progress relative to the 4-axes U-stage, still used in some laboratories. Modes of representation and analysis of mineral orientation determined with the universal stage are discussed in the previous chapter.

At this point, it seems useful to follow the sequence of elementary operations leading to the density diagrams of mineral preferred orientations discussed in Chapter 8. The starting point is the thin section referred to the oriented sample as described in the previous section (Fig. 9.3 (1)). For the U-stage measurement the thin section must be maintained in a known orientation. This is achieved by keeping its sides in contact with a rectangular guide sledge moving in a slot of the upper hemisphere mount of the stage (Fig. 9.3 (2)). In the fundamental orientation on a Leitz 5-axes U-stage, the arrow on the thin section which is parallel to its long side is oriented NS, such that the index of the inner plate of the stage indicates 90° on the graduated rim. The sequence (2)–(3a) illustrates the corresponding orientation on a projection net of the overlay if the measurements obtained with the U-stage are to be hand-plotted; (3b) and (4) are the operations in computer processing, a flow-chart of which is presented in Table 8.1.

A rapid and global estimation of the trace of mineral preferred lattice orientation in a plane can be obtained using a photocell coupled with a microammeter (Fig. 9.4). The reading of the transmitted light intensity, which depends on the anisotropy of the lattices orientations, is reported on a polar net for successive orientations of the thin section. In the example of Fig. 9.4 from a peridotite (Darot, 1973), it is seen that there is a strong preferred

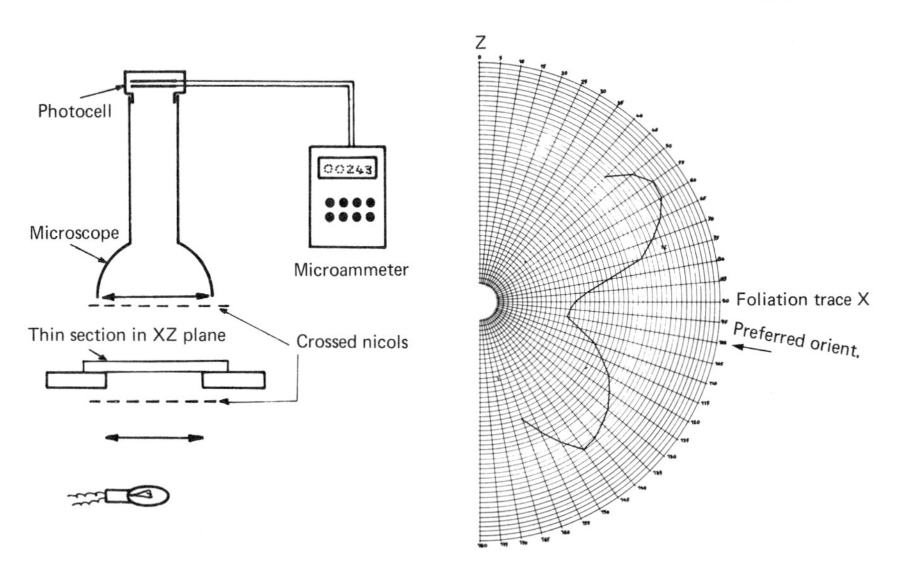

Fig. 9.4. Estimation of the trace of mineral preferred lattice orientations in a chosen thin section plane (usually the *XZ* section) using a photocell mounted on a low magnification microscope. The photocell output is plotted for varying orientations on both sides of the foliation trace (after Darot, 1973)

orientation of the minerals (olivine and enstatite) and that the minimum of transmitted light (100°) which coincides with a preferred orientation of the lattices depart from the foliation trace (90°) by about 10°. This kind of measurement obviously has meaning only if it is carried in a remarkable plane, usually *XZ*, if there is one mineral dominant in the rock or responsible for the fabric anisotropy and if it is complemented by preferred orientations measurements with the U-stage. In such conditions it presents the advantage of being rapid and accurate and therefore suitable in systematic studies.

Grain shape measurements

General methods for quantitative studies on the shape of minerals under the microscope are presented by De Hoff and Rhines (1968). The two principal parameters here are grain-size and grain shape anisotropy measured by aspect ratio in remarkable planes; statistically considered in the texture, the latter defines the shape preferred orientation, also expressed at the scale of the sample and above by the penetrative foliation and lineation. Grain-size and aspect ratios measurements are carried out in the *XZ* and *YZ* sections. If the minerals need not to be measured independently for each species or each generation in a given species, these parameters are statistically obtained by Snyder and Graff's technique (1938). It consists in counting all grain-boundaries crossed in traverses respectively parallel to X (n_x) and to Z (n_z) in the *XZ* plane and if necessary parallel to Y (n_y) in the *YZ* plane (Fig. 9.2). Equal distances or nearly so for cumulated traverses along each direction X, Y and Z (d_x, d_y and d_z) are measured on the vernier of the sliding device attached to the stage of the microscope. The grain size index and the aspect ratios are, respectively,

$$g = \frac{n_x + n_y + n_z}{d_x + d_y + d_z}$$

and

$$r_{x/z} = \frac{n_x}{d_x} \cdot \frac{d_z}{n_z}; \qquad r_{x/y} = \frac{n_x}{d_x} \cdot \frac{d_y}{n_y}; \qquad r_{y/z} = \frac{n_y}{d_y} \cdot \frac{d_z}{n_z}$$

The technique does not take into account textural factors like inclusions, serrated boundaries which would artificially reduce the grain size and the aspect ratios. Hence, it is preferably applicable in comparative studies on evenly textured rocks (for example, see § 11.2.2).

In the case of shape measurements limited to defining the direction of the long diameter of each inequidimensional grain in a remarkable plane, an interesting mode of simultaneous representation of lattice and shape fabric has been introduced by Collée (1963). It consists of superposing a rose diagram of the orientations of the long diameters upon the equatorial projection of lattice preferred orientations (Fig. 9.5).

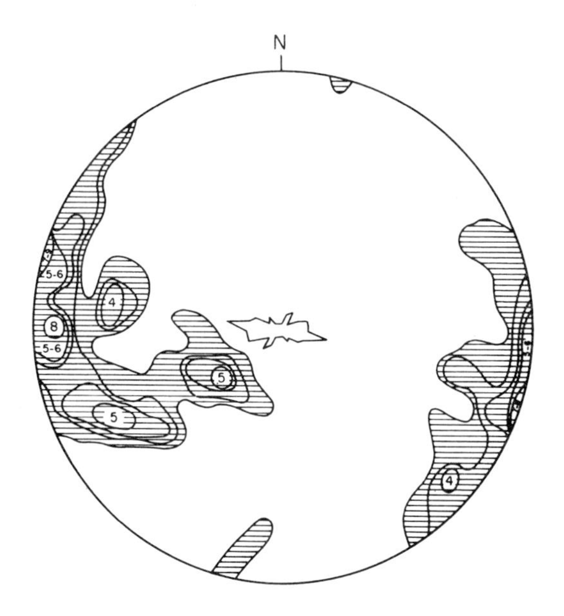

Fig. 9.5. Representation in the case of a mineral of the preferred orientation of a crystallographic axis along with the trace in the projection plane of its shape preferred orientation (after Collée, 1963, reproduced by permission of the Geological Institute, Leiden)

Optical determination of the slip systems

In naturally deformed rocks the determination of the slip systems relies mainly on the study of two substructural features: deformation lamellae and kink bands. The deformation lamellae (Plate 3(b)) are present only in low temperature deformation conditions. They were attributed by Christie et al. (1964) to planar arrays of locked-in slip dislocations creating a stress field responsible for the optical effects characterizing these structures. In this case, they represent the active slip plane, and the orientation measurements can contribute to determining the slip systems (Raleigh, 1968). TEM studies have shown that similar structures can be produced by elongated subgrains or twins (§ 5.9.2). Hence, we doubt whether any slip system determination can be seriously achieved, at present, using only optical observation of deformation lamellae.

The kink banding geometry is also used to determine the slip system. The method has been exposed by Christie et al. (1964) and Raleigh (1968). In short, a kinked structure implies slip in one or both bands (KB) separated by the kink band boundary (KBB) in directions which are roughly perpendicular to the KBB (Fig. 9.6) if the angle of tilt of the KB relative to the KBB is small. The slip plane is determined as containing the slip direction and the axis of external

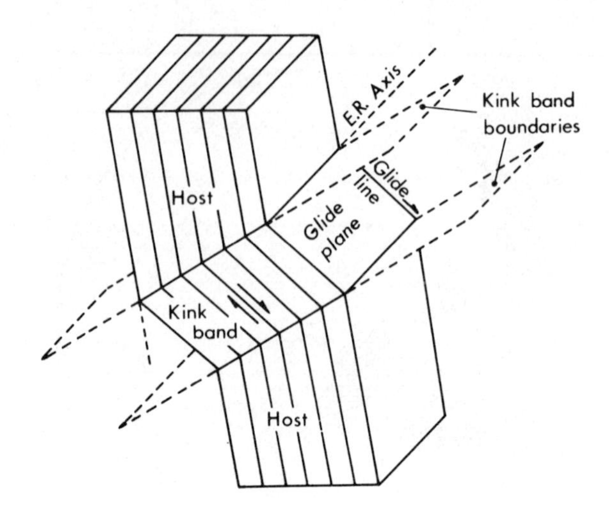

Fig. 9.6. Kink band structure: Geometry and associated slip system

rotation (E.R.) of the KB relative to the KBB. In fact, the bands on each side of the KBB are usually symmetrically inclined.* The method consists, then, in measuring with the U-stage the orientations of the KBB and of the crystallographic axes in each band, and in determining their axis of external rotation (E.R. axis) which is the common intersection of the planes bisecting the angle between each corresponding crystallographic axes and which is also constrained to be in the KBB plane. The determination is achieved graphically on a stereonet. The slip lines are the crystallographic direction coinciding with the KBB pole or symmetrically inclined on it at a small angle. In the case of olivine this is illustrated by Fig. 9.7.

Designed for kink banding structures, this construction is theoretically also applicable to polygonized subgrains separated by subgrain-boundary because they are geometrically identical. In the latter case, the E.R. axis cannot be determined with accuracy because the rotation from one band to the other is usually smaller than 10°, that is in the range of the experimental error, considering the large amplification of the measurement errors in this construction. The subgrain boundary has also to be a pure tilt wall. If it has a twist component, the axis of external rotation is no longer in the wall (in case of pure twist it is normal to the wall). This can be accounted for in considering only the measurements such that the E.R. axis is in the subgrain wall; then it must be realized that only partial information on the slip systems is recorded. A more serious drawback is that in high temperature creep it is not certain that the

* Otherwise the spacing between slip planes varies from one band to the other and must be accommodated by elastic strain, for instance Raleigh reports that departure from symmetry of no more than 2·5° in olivine would result in tensile stress of 2 kB.

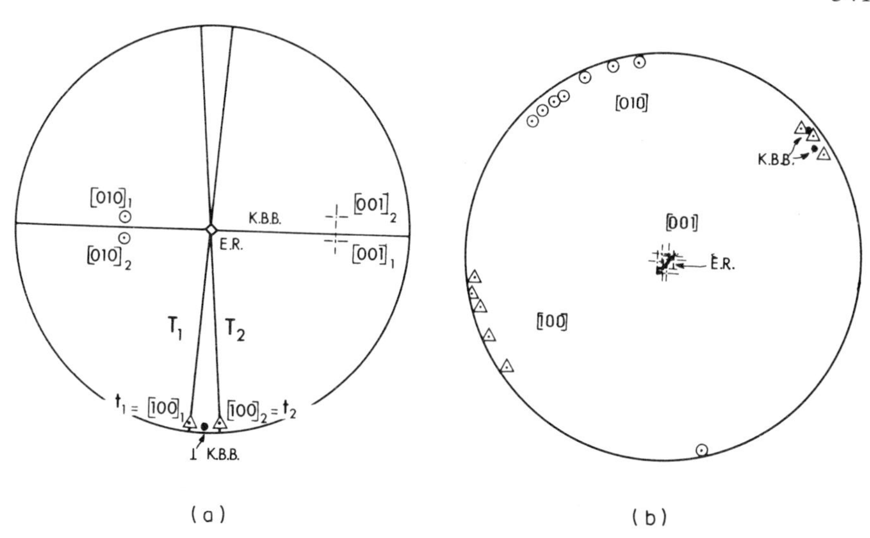

Fig. 9.7. (a) Relation between olivine crystal axes on either side of kink-band boundary (KBB) on (100) in naturally deformed crystal. External rotation of the axes symmetrically about ER brings [100] normal to the kink-band boundary. The slip plane, T containing $[100]=t$ and ER is a plane of indices {0kl}. Equal area projection. (b) Crystal axes of olivine with differing orientations in several (100) kink bands within a single naturally deformed grain show external rotation about a line near $Y=[001]$. Slip system deduced is (010), [100]. Equal area projection. (Reproduced from Raleigh, *Jour. Geophys. Res.*, **73**, 5391, Fig. 15 (1968), copyright by American Geophysical Union)

systems thus determined were *active*, since the subgrain formation is due to *climb* and not to slip: it is then a stable configuration of dislocations not necessarily involved in slip. The method is strictly valid only in low temperature creep in case of slip polygonization. We run then into a new difficulty: the substructure is wavy (undulatory extinction) and the angles of rotation between bands so small that the errors inherent in the method almost certainly preclude it. Provided the construction makes it possible to show that the subgrain-boundary is not a twist wall an important information is obtained: the orientation of the slip direction statistically coinciding with the normal to the subgrains-boundary. In quartzites naturally deformed at low temperature in the basal slip regime, Bouchez (in press) has used this indication to determine the flow direction in the rock.

Remark. It is recalled that the slip directions determined by the KB method as normal or at high angle to the KBB or subgrain-boundary have generally an opposed kinematic meaning in the case of kink banding *stricto sensu* and in that of a polygonized substructure as experimentally shown (Nicolas et al., 1973; § 6.2). In the former case, the deformation structure is the direct result of compression and the slip direction is nearly normal to the macroscopic flow direction; in the latter the substructure results from moderate bending and twisting of the crystal lattice during plastic flow and the slip direction is at a small angle to the flow direction in the rock.

9.2. X-RAY DETERMINATION OF PREFERRED ORIENTATION

Analysis of mineral preferred lattices orientations by X-ray diffraction, using a texture goniometer, is increasingly important in petrofabric studies. It is well suited in the case of fine-grained rocks (grain size $< 10\ \mu$, although rocking and translating the specimen under the beam makes larger grain size acceptable). In this regard it complements in a fortunate way the U-stage technique in which it is difficult to measure grains smaller than $30\ \mu$. It is also superior when dealing with uniaxial crystals (e.g. quartz, calcite) in which the U-stage provides only the optic axis orientation whereas the X-ray technique gives the full orientation. It has been introduced in geological sciences in the early sixties (Higgs et al., 1960; Starkey, 1964) and is fully described by Baker et al. (1969), by Lhote et al. (1969) and by Phillips and Bradshaw (1970) to whom we wish to refer in order to avoid presenting it here.

9.3. TECHNIQUES OF OBSERVATION OF SUBSTRUCTURES AND DISLOCATIONS

In § 4.2, we have seen that the strain rate and temperature regime of deformation and/or subsequent anneal can be, up to a certain point, deduced qualitatively from the observation of typical dislocation configurations and microstructural features (subgrains, KBB, etc.). Also, in some favourable cases, quantitative measurements of subgrain size or dislocation densities can provide a certain amount of information about the value of the prevailing shear stress during deformation. Last, but not least, the best evidence that can be given of the activity of a slip system, is the direct observation of dislocations in the slip planes and the determination of their Burgers vector.

In this paragraph, we will briefly describe and assess the most important experimental techniques allowing direct observation of dislocations and other microstructural features.

9.3.1. Dislocation decoration

Dislocations and dislocation groups such as subgrain-boundaries are preferential sites for the nucleation of precipitates of a second phase. Certain impurities, introduced by diffusion in a crystal during a high temperature anneal, may precipitate (exsolve) into a more stable phase when the temperature is lowered. Dislocations are preferential sites for precipitation since the free enthalpy of the crystal is lowered when a small volume of the matrix phase containing a portion of dislocation is replaced by a dislocation-free precipitate (the total length of dislocation line decreases). In the early stages of precipitation, dislocations may therefore be 'decorated' by small precipitates. It is possible, in a transparent crystal, to observe by optical microscopy decorated dislocation lines or sub-boundaries in the depth of a thin section.

Although decoration methods sometimes allow good visualization of the dislocation structures in three dimensions, they are not widely used, due to the following drawbacks:

It is not always possible to find a practical decoration method for a given crystal.

The high temperature diffusion anneal, necessary to introduce the decorating impurity, may alter, by recovery, the dislocation structure under investigation.

As the magnification is limited by the possibilities of the optical microscope, a clear picture can be obtained only for very low dislocation densities.

Table 9.1 lists the few decoration methods that have been found successful in some of the rock-forming minerals considered in Chapter 5.

Table 9.1. Decoration

Mineral	Procedure	Reference
Halite	Annealing in evacuated tube with $AuCl_3$ ($T=650$ °C, 4 h)	D. J. Barber, K. B. Harvey and J. W. Mitchell, *Phil. Mag.*, **2**, 704 (1957)
Olivine	Anneal in vacuum furnace with Mn ($T=1200$ °C)	C. Young, *Am. J. Sci.*, **267**, 841 (1969)
	Anneal with Mn or Mg prior to deformation ($T=750$–1200 °C, 12 h)	N. L. Carter and H. G. Avé Lallemant, *Geol. Soc. Am. Bull.*, **81**, 2181 (1970)
	Anneal in air ($T=900$ °C, 2 h)	D. L. Kohlstedt (pers. comm.)

It is interesting to notice that decoration of dislocations can sometimes occur naturally. For instance, in olivine, dislocations, being preferential sites for the oxidation of Fe are sometimes revealed optically by the presence of minute particles of Fe_2O_3 (Zeuch, 1974) (Plate 7(b)).

9.3.2. Etch pits

When a crystal is immersed in a suitable reagent, etch pits can develop at the emergence points of dislocations on the surfaces, thus materializing a section through the dislocation substructure, which can be observed by optical microscopy in reflexion.

Experimental data on etch-pitting and current theories have been reviewed by Amelinckx (1964) and by Robinson (1968). The reasons for the nucleation of etch pits at the emergence points of dislocations can be roughly exposed as follows.

In the vicinity of a dislocation line, the crystal has a higher stored elastic energy density than far from the dislocation and a core energy can be attributed to the line itself. The free-energy change necessary to the formation of a pit at the surface by dissolution of a volume element is lowered at a site where dislocation intersects the surface by the elastic strain energy of the dissolved volume plus the core energy of the length of dislocation disappearing. In other terms, the chemical potential of the crystal atoms is higher at the point of emergence of a dislocation which is therefore a preferred site for dissolution.

For an etch pit to be visible in optical microscopy the sloping sides of the pit should make an angle of at least 5 to 10° with the surface. For an etch pit to have a slope steep enough, the dissolution rate must therefore be higher at the bottom of the pit, along the dislocation, than on the sides (Fig. 9.8). Dissolution

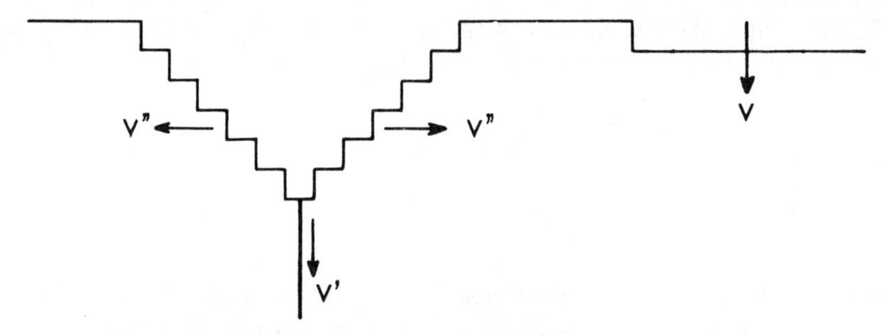

Fig. 9.8. Etch pit at the point of emergence of a dislocation:
V = Dissolution rate at the perfect surface
V' = Dissolution rate along the dislocation
V'' = Dissolution rate at lateral edges of etch pits
If $V' > V$ etch pits nucleate at points of emergence of dislocations
If $V' > V''$ the etch pits can grow in depth

on the sides occurs by migration of atomic ledges, deepening of the pit will be thus favoured by any factor decreasing the migration rate of the ledges or increasing the dissolution rate along the dislocation. The dissolution rate along the dislocation can sometimes be enhanced by the segregation of impurities around the dislocation in the crystal. The lateral dissolution rate can be decreased by adding specific impurities to the etchant, which 'poison' the ledges.

The crystallographic orientation of the surface usually has also a great influence on the shape and depth of pits, the close-packed surfaces being the most easily etched. Some reagents give etch pits only on certain crystallographic surfaces.

The etch pits technique has several advantages:

It is usually possible to develop a satisfactory etchant for any crystal.

The subgrains-boundaries are clearly revealed and this allows measurement of the subgrain size, even in opaque crystals.

It is quite easy to measure dislocation densities by counting the number of etch pits by unit area provided that the dislocation density is not high, so that the pits do not overlap and can be resolved individually. However, certain important precautions must be taken to check the correspondance between the emergence point of dislocations and the etch pits produced by a given reagent. A one-to-one correspondence must not be taken for granted for the following two reasons:

There may exist other sites of preferential dissolution at the surface (small heterogeneities, inclusions, etc.).

Not all dislocations may be revealed. There are cases when a certain reagent will reveal only 'fresh' dislocations without an impurity atmosphere, whereas another reagent will reveal only 'aged' dislocations where impurities have segregated.

The correspondence between etch pits and emergence point of dislocations can be checked by producing and/or observing typical structures known to be due to dislocations, such as slip bands or polygonization walls (Plates 11(a) and 14); or by etching the two matching surfaces of a cleaved crystal, the etch pits

Table 9.2. Etch Pits

Mineral	Plane	Etchant	Reference
Halite	(100)	(4 g $FeCl_3$–1 l glacial acetic acid) ($T \simeq 25$ °C, 30 s, moderate agitation)	S. Mendelson, *J. Appl. Phys.*, **32**, 1579 (1961)
Quartz	All	Concentrated HF vapour (a few seconds)	J. A. Christie, D. T. Griggs and N. L. Carter, *J. Geol.*, **72**, 734 (1964)
Calcite	All	Formic acid 90 % ($T \simeq 25$ °C, 15 s)	R. E. Keith and J. J. Gilman, *Acta Met.*, **8**, 1 (1960)
	All	Tartaric acid 2 % in water	V. I. Startsev, *J. Phys. Soc. Japan*, **18** Supp. III, 16 (1963)
Olivine	(010)	$NaOH + KOH + H_2O$ in equal parts ($T = 220$ °C)	C. Young, *Am. J. Sci.*, **267**, 841 (1969)
	(100)	Concentrated HF	
	sub (100)	(5 parts HF–2 parts HNO_3)	
	(010) (001) (100)	1 part vol. conc. HF 48 %, 1 part vol. conc. HCl 37 %, 1 part vol. solution 60 g citric acid monohydrate in 75 ml distilled water ($T = 24$ °C, 30 s–5 mn) Wash and rinse in deionized water Immerse 5 min. in boiling citric acid solution, rinse	M. Wegner, J. M. Christie, *Contrib Mineral Petrol.*, **43**, 195 (1974)

patterns should then be mirror images. However, the best method by far consists in etching the surface and examining the crystal by X-ray topographic methods (see below) to image the dislocation emerging at the etch pit.

Table 9.2 lists some etchant for the principal rock-forming minerals.

9.3.3. *X-ray topography*

(*a*) *Generalities*

X-ray topographic methods are based on the diffraction of X-rays by crystals. The principles and applications of the methods have been abundantly reviewed (see, e.g., Weissmann and Kalman, 1969; Lang, 1970) and it would be entirely out of the scope of this book to expose the theory of the contrast for the various defects and microstructural features. We will therefore limit ourselves to giving a very schematic and intuitive account of the principle of the methods and to assessing their value.

Bragg's law relates the angle θ at which a beam of X-rays of wavelength λ can be reflected on a family of planes of spacing d (Fig. 9.9):

$$2d \sin \theta = n\lambda$$

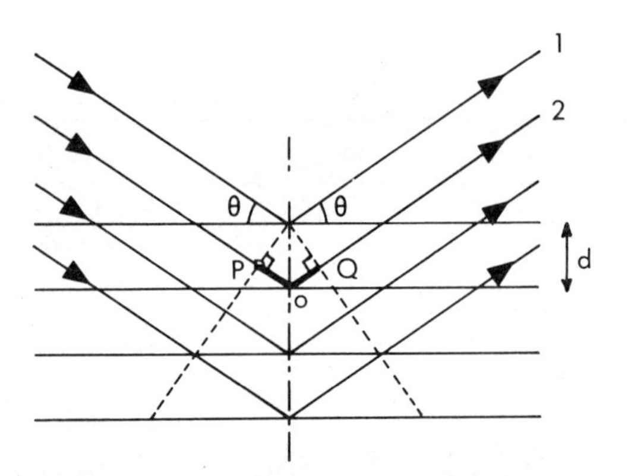

Fig. 9.9. Reflection of X-rays on lattice planes. The path length difference between rays 1 and 2 is:
$PO + OQ = 2OQ = 2\,d \sin \theta$
(d = distance between planes, θ = angle of incidence of the beam)
Constructive interference occurs for $2\,d \sin \theta = n\lambda$ (Bragg's law)
(n: integer, λ: wavelength of the rays)

If a finely collimated monochromatic beam of X-rays strikes the surface of a crystal, the reflected intensity will be strong if there is a family of lattice planes in Bragg's orientation (that is inclined at an angle θ determined by Bragg's law for a given λ and a given d), conversely the reflected intensity will be weak or nil if this is not the case. If a crystal is set to be in Bragg's orientation on one family of planes, local misorientation due to the presence of subgrains, for instance, will be revealed by variations in reflected intensity. This is called *orientation contrast.*

X-rays penetrate significant distances into crystals according to the absorption coefficient of the latter. If an ideally perfect crystal is in Bragg's orientation for an incident X-ray beam, the beam will suffer multiple reflections between successive planes giving rise to destructive interferences which will attentuate considerably the reflected intensity for penetrations of more than 1 μm. This is the phenomenon of *primary extinction.*

If the crystal departs locally from ideal perfection, that is, if a defect, like a dislocation for instance, distorts slightly the atomic planes, the once-diffracted beam is not in the exact Bragg's position to be diffracted again, the primary extinction is locally decreased and the reflected intensity increases. This *extinction contrast* may allow dislocations to be seen.

Topographic methods can be classified into two groups:

In *reflection methods*, the reflecting planes make a small angle with the surface of the crystal; the diffracted beam intensity can be recorded by a photographic emulsion located on the same side of the crystal as the X-ray source. These methods can be used on thick specimens and convey information on the superficial layers of the crystal, to a depth depending on the absorption of X-rays.

In *transmission methods*, the reflecting planes make a small angle with the normal to the surface of the crystal, the diffracted beam has therefore to go through the crystal and its intensity is recorded on a photographic emulsion situated on the other side of the crystal with respect to the X-ray source. These methods must be used with thin specimens, up to a few millimeters for crystals containing only light elements.

Many X-ray topographic methods exist but we will only describe shortly the two most widely used methods.

(b) Berg–Barrett method

This is a reflection method. The reflection from the crystal surface is recorded on a fine-grained photographic emulsion, situated as close to the specimen as possible for better resolution (Fig. 9.10).

The incident X-ray beam is usually a slit-collimated characteristic radiation beam (practically monochromatic).

If the reflecting planes are chosen so that Bragg's angle is small, the illuminated area on the specimen can be as large as a few square millimeters for a suitable cross-section of the incident beam. On the other hand, the choice of a long-wavelength radiation (Co K_α or Cr K_α) giving a large Bragg's angle allows the location of the photographic plate to within a fraction of a millimetre of the surface.

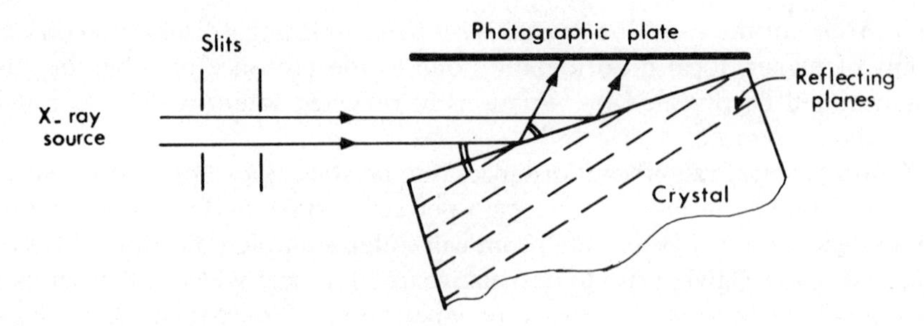

Fig. 9.10. Berg Barrett's X-ray topographic method

Due to the angle between the surface of the crystal and the plate, the topograms are slightly distorted. Their magnification is 1 and they must be examined under an optical microscope.

The Berg–Barrett method is very suitable to investigate the subgrain structure by orientation contrast (Plate 11(b)) (Fig. 9.11). It is also possible to obtain

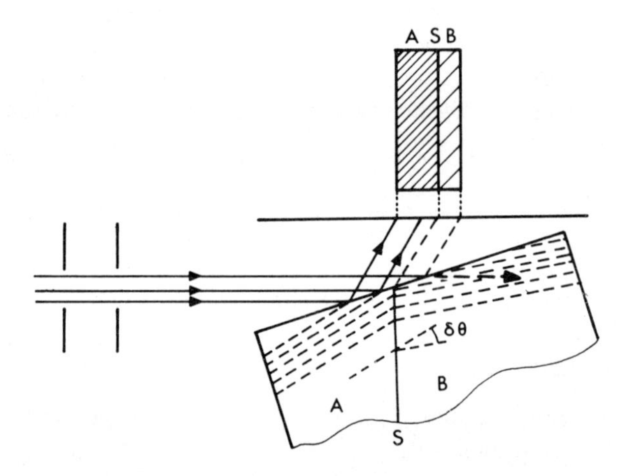

Fig. 9.11. Orientation contrast in Berg–Barrett's method. Subgrains A and B, misoriented by $\delta\theta$, are separated by a sub-boundary S. In A, reflecting planes are in Bragg's position, hence the reflected intensity on the photographic plate is strong; in B the reflecting planes are off Bragg's position by $\delta\theta$, the reflected intensity on the photographic plate is weak or very weak

isolated dislocation images by extinction contrast and analyse them quantitatively (Newkirk, 1959). Dislocation arrays or sub-boundaries can also be examined in extinction contrast and quantitative information derived by a method exposed by Wilkens (1967).

(c) Lang's method

This is a transmission method. A slit-collimated beam of nearly monochromatic characteristic radiation is reflected at Bragg's angle on planes nearly normal to the surface of the thin crystal. The incident and diffracted beam both go through the crystal (Fig. 9.12). The diffracted beam passes through a

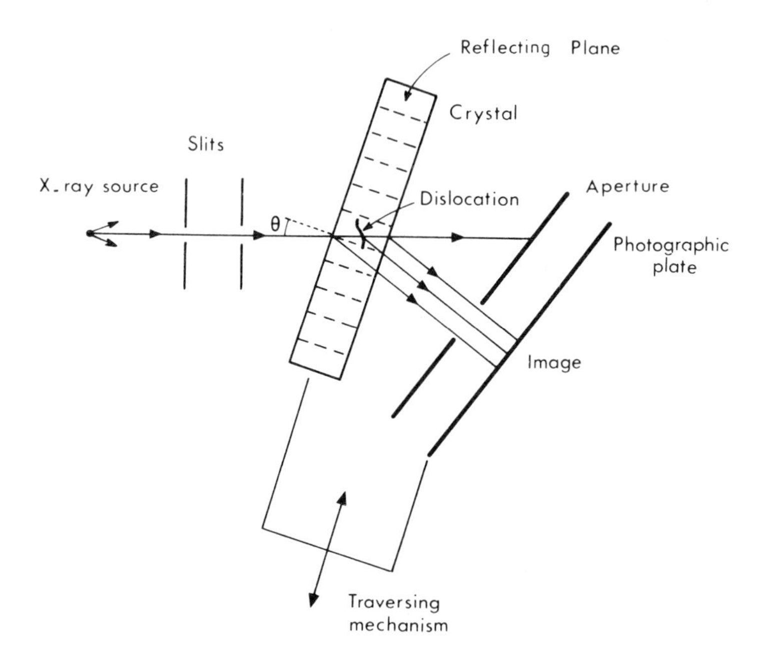

Fig. 9.12. Lang's X-ray topographic method. Only a narrow section of the thin crystal is irradiated by the beam, a topogram of the whole crystal may be obtained by translating the crystal and the photographic plate in front of the slit

selecting aperture and strikes a photographic plate, blackening a narrow strip on it. If a defect is present in the irradiated crystal portion its image will be recorded in extinction contrast. Only a thin slice of crystal is irradiated and the topogram records the position of the intersection of dislocations with the irradiated slice. This type of topogram is called *Section topogram.*

The whole picture of the defects in the crystal can be obtained by scanning the crystal by the irradiated slice. This is obtained by mounting the crystal and the plate together on a traversing mechanism oscillating back and forth in front of the incident beam during the time of an exposure; a *Projection topogram* is thus obtained (Plate 6(b)). Three-dimensional information about the distribution of dislocations in the specimen can be obtained by taking stereo pairs (the topograms are recorded by viewing the reflecting planes from one side and from the other, with the same Bragg's angle). The subgrains are seen in orientation contrast on projection topograms.

(d) Interest of the X-ray topographic methods

All X-ray topographic methods have in common certain advantages and drawbacks.

They are non-destructive in the sense that they can be used on large crystals without having to destroy them. This is obvious for the reflection methods, but even the transmission methods allow the examination of bulk specimens, depending on their absorption coefficient and the wavelength of X-rays. The examination of the dislocation structure can therefore be coupled if needed with experiments in bulk on the crystal: deformation, annealing treatments for instance.

The X-ray irradiation doses are low and the crystals do not suffer radiation damage. This is especially interesting for crystals which are heavily damaged by electron irradiation and cannot be observed for a long time in transmission electron microscopy. This is the case of quartz for instance: Lang (1970) reports that forty topograms, totalling 800 h of exposure, could be taken on the same quartz crystal without any detectable radiation damage resulting in loss of contrast.

The topograms have a magnification of 1 and the resolution of the X-ray topographic methods is much lower than the resolution of transmission electron microscopy: it is generally limited to about 1 μm. This precludes the observation of single dislocations by extinction contrast in crystals where the dislocation density is high. This is also the source of some doubt as to whether some features in topograms are the image of isolated dislocations or groups of dislocations. Nevertheless, in crystals with low dislocation density the image of dislocations can be analysed quantitatively as in TEM (see later) and the Burgers vectors determined.

The absence of magnification and the lack of resolution of the topograms can be sometimes more than compensated by the fact that it is possible to investigate quantitatively microstructural features such as glide bands or sub-boundaries at a scale of the order of 100 μm to 1 mm, too large for TEM.

9.3.4. Transmission electron microscopy (TEM)

(a) Generalities

Transmission electron microscopy is a very powerful technique for the observation of defects in thin crystalline foils. The theoretical and practical aspects of TEM have been well covered in many excellent books (Amelinckx, 1964; Hirsch et al., 1965, for instance). The theory of the defect contrast, based on the theory of electron diffraction by crystals, is much too complex to be treated here and we will give only a rough intuitive picture of it.

In first approximation the problem of the electron diffraction can be treated (as in the case of X-rays) as the problem of the reflection of a wave on the lattice planes of the crystal. The wavelength associated with the electron beam is inversely proportional to the square root of the energy of the electrons; in the usual case of monokinetic electrons accelerated under a 100 kV tension, the associated wavelength is $\lambda = 0{\cdot}037$ Å, the Bragg angles are therefore much smaller than in the case of X-rays (θ of the order of 10^{-3} radian). In an electron microscope, the electrons are produced by a heated filament and accelerated at a tension of 100 kV or more in an electron gun; the beam is focused by several electromagnetic lenses, it is collimated by appropriately located apertures and strikes the thin crystalline foil (thickness of the order of 1000 Å). The reflecting planes are almost normal to the foil surface.

It can be shown, and it is experimentally verified, that due to the extremely small wavelength of the beam and to the fact that the foil is very thin, the incident beam can be reflected on planes which deviate from Bragg's conditions; accordingly, several families of planes will give rise to several diffracted beams (the thinner the foil the more it will be possible for planes to deviate from the Bragg angle and still give a reflexion).

The transmitted beam and the diffracted beams go through more lenses and eventually form a pattern of spots: the electron diffraction pattern, on a fluorescent screen or a photographic plate. The crystalline orientation of the thin foil with respect to the incident beam can be deduced from the diffraction pattern.

The observation of images of defects is usually made in such conditions that only one diffracted beam is present or at least has a much higher intensity than the others. To achieve this result, the thin foil mounted on a tilting stage must be precisely oriented with respect to the incident beam. Two modes of operation are then possible.

—The transmitted beam is allowed to pass through an aperture which blocks the diffracted beam. If a defect is present and distorts slightly the lattice diffracting away more intensity than the perfect crystal, its image will appear in black on a brighter background; it is the *bright field* technique.

—The contrast can be reversed and the image appears in bright on a dark background, the transmitted beam is blocked and the diffracted beam only is allowed to pass through the aperture; it is the *dark field* technique.

(*b*) *Contrast of defects*

(*i*) *Dislocations.* Dislocations are 'seen' by the electrons through their strain field. If the reflecting lattice planes, far from an edge dislocation for instance (Fig. 9.13), make an angle with the incident beam deviating a little from Bragg's angle, the distorted planes close to the dislocation are rotated towards or away from Bragg's position. The intensity of the diffracted beam is therefore higher and lower than for the perfect crystal according to which side of the dislocation is considered. The dislocation will be imaged as a dark or bright line (in bright or dark field) whose position does not coincide exactly with the geometrical position of the dislocation.

It must be pointed out that the interpretation of a TEM micrograph requires a good familiarity with the physical theory of contrast and much experience, for many artifacts can arise from contrast effects. For instance,

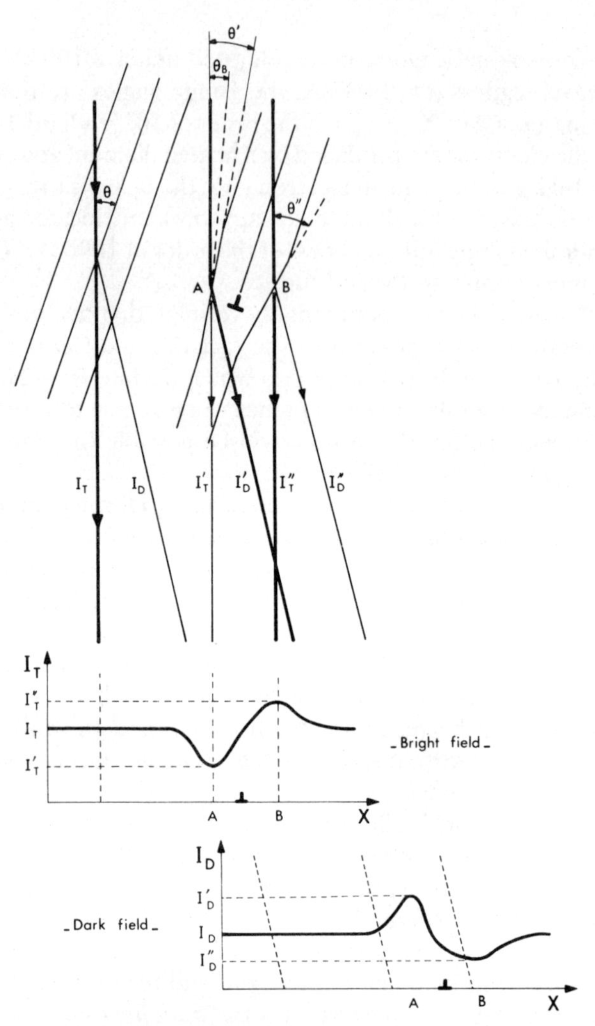

Fig. 9.13. Diffraction of an electron beam by a crystal containing an edge dislocation ($\perp$):

θ_B = Bragg's angle
θ = Angle of incidence of the beam on the undistorted lattice planes
θ', θ'' = Angles of incidence on distorted lattice planes close to the dislocation
I_T, I_D = Transmitted and diffracted intensities far from the dislocation (background)
I'_T, I'_D = Transmitted and diffracted intensity at A (angle of incidence θ')
I''_T, I''_D = Transmitted and diffracted intensity at B (angle of incidence θ'')

$$\text{A}: \theta_B < \theta' < \theta \begin{cases} I'_T < I_T \\ I'_D > I_D \end{cases} \qquad \text{B}: \theta'' > \theta > \theta_B \begin{cases} I''_T > I_T \\ I''_D < I_D \end{cases}$$

depending on the diffraction conditions, a single dislocation line may appear as a dotted line (not to be confused with a row of short segments of dislocations) or as a double line (not to be confused with a dissociated dislocation).

(*ii*) *Planar defects.* Planar defects, oblique with respect to the plane of the foil, usually give rise to a fringe contrast. The transmitted and diffracted waves having penetrated a distance ξ_g into the foil (extinction distance) interact to give standing waves. The intensities vary periodically with the depth in the foil, and if the thin foil is wedge-shaped the transmitted and diffracted intensities will vary according to the thickness of the foil giving rise to *thickness fringes.* The fringe spacing depends on the reflecting plane.

A planar grain-boundary, oblique with respect to the foil will give rise to such fringes if the first crystal is in Bragg's position and not the second (the first crystal is then wedge-shaped).

Other fringes may arise when a planar defect introduces a phase shift in the waves. Such is the case for stacking faults (Plates 5(b) and 9).

These two types of fringes must not be confused with moiré fringes occurring when there is a slight difference in lattice parameters between each side of a coherent interface or when there is slight rotation between two crystals.

We want to emphasize that it is often impossible, by simple inspection of a micrograph, to determine the nature of a feature presenting a fringe contrast. To make this point clear, it is enough to list a few of the planar defects presenting a fringe contrast: grain-boundaries, dislocation walls, dislocation multipoles, twin boundaries, microtwins, stacking faults, thin exsolution lamellae, etc.

However, a thorough analysis performed on the electron microscope and involving the taking of several micrographs under carefully selected diffraction conditions can allow not only the determination of the nature of the defect but also the provision of quantitative information about it.

(*c*) *Preparation of thin foils*

To be observable, the foils must be transparent to electrons; their thickness must therefore be inferior to a certain value which depends on the absorption coefficient of the material (higher for heavy elements) and on the energy of the electrons (proportional to the accelerating voltage). With 100 kV electrons, most crystals can be observed for foil thickness varying between 1000 Å and 3000 Å. High-voltage electron microscopy (HVEM) with electrons of 1 MeV or 3 MeV allows the observation of most materials under thicknesses of the order of 1 μm.

There are various means for preparing thin foils depending on the material. Most recipes are specific to one material and must be developed especially. Generally three kinds of methods can be distinguished in the case of minerals.

(*i*) *Cleavage* (*for layer minerals like micas*). Very thin flakes are cleaved by means of adhesive tape, loosened in an appropriate solvent and picked up on grids.

(*ii*) *Chemical thinning.* This method applies if the mineral is soluble in a chemical reagent. Slices are mechanically or chemically cut and thinned to a thickness of the order of 30 μm. The final thinning in chemical solution is pursued until desagregation into thin foils which are rinsed and picked up on grids. Other methods are of the jet polishing type: a jet of solution impinges on a dimple in the slice until a hole appears. The edges of the hole may be thin enough for observation.

In all cases the chemical attack must be slow and not give rise to etch pits.

(*iii*) *Ion bombardment.* This is now the most commonly used technique, and it is applicable to all minerals (Barber, 1970).

High-energy ions impinge at a grazing angle on the surface of the sample and knock out atoms from the surface. The process must be well controlled and is very slow. Reliable ion bombardment apparatus is now commercially available.

(*d*) *Quantitative measurements*

One of the main interests of TEM is to allow quantitative measurements on the defects. The various possibilities are detailed in specialized works such as Hirsch et al. (1965). We will mention only a few here.

(*i*) *Measurement of local misorientations* (between subgrains for instance) or of orientation of characteristic features (slip bands, twin boundaries, KBB, etc.). These measurements are made by taking selected area diffraction patterns of the area under consideration. The diffraction patterns contain all the information relative to the crystalline orientation and allow the corresponding micrograph to be indexed with the relevant orientations.

(*ii*) *Measurement of dislocation densities.* Several statistical methods can be used. The most commonly used consists of counting the number N of intersections with dislocation images made by random lines of total length L in a given area. It can be shown (see Hirsch et al., 1965) that the dislocation density ρ (in cm/cm^3) is given by:

$$\rho = \frac{2N}{Lt}$$

where t is the thickness of the foil.

(*iii*) *Determination of Burgers vectors of dislocations.* The methods for determining the direction and sense of the Burgers vector **b** of a dislocation are detailed in Hirsch (1965).

The principle of the method for the determination of the direction of **b** can be understood in Fig. 9.14, in the case of an edge dislocation. At first order, all lattice planes are distorted around a dislocation except the planes parallel to the Burger vector which remain flat. If these planes are chosen as reflecting planes the dislocation will give no diffraction contrast and will therefore be invisible. If by orienting the foil, by means of a tilting stage, one can find two

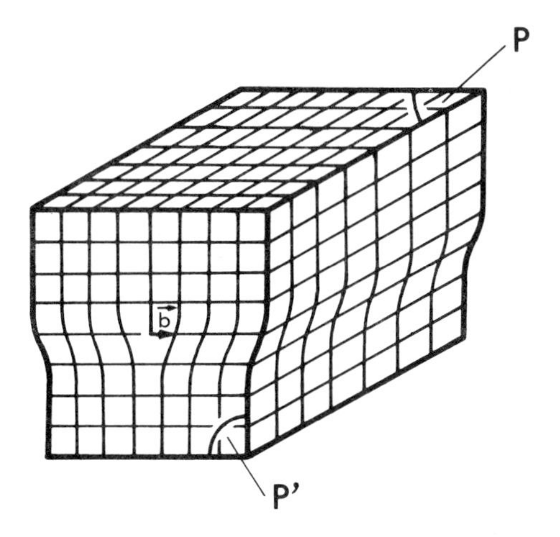

Fig. 9.14. Lattice planes P and P′ parallel to the Burgers vector **b** of the dislocation are undistorted. The dislocation is, in principle, invisible if these planes are used as reflecting planes for electrons

reflecting planes for which the image of the dislocation disappears, the direction of the Burgers vector will be the intersection of these planes (Plate 15). It must be noted that things are not always quite so simple in practice since in most cases a residual contrast may exist even though the extinction conditions are fulfilled, which makes it sometimes difficult to decide which planes contain **b**. Besides, it may be geometrically difficult or impossible to use a reflecting plane which would give an extinction. The method can therefore be supplemented by outside information about the Burgers vector. However, the best method, when possible, consists of finding two planes which extinguish the dislocation image.

(*e*) *Limitations*

The advantages and the interest of TEM are so obvious and so immense that it is more interesting to dwell a little on the limitations and difficulties of this method than extol its merits.

The scale of observation is the scale of the micron. Focusing only on observations in TEM may be dangerous in the sense that important and relevant features on a larger scale may be overlooked.

The usual methods of preparation of thin foils may introduce dislocations or cause a loss of dislocations at the surfaces. More generally, the structure observed in foils a few thousands Ångströms thick may not be representative of the bulk material. The observation of thicker specimens by HVEM has therefore a considerable advantage in this respect and the dislocation structure is much more typical of the bulk material.

Irradiation by high-energy electrons may damage the crystals: when the electrons possess an energy higher than a threshold energy which depends on the crystal, they knock atoms out of their lattice sites, thus creating vacancies and interstitials. These point-defects cluster, and the clusters collapse into dislocation loops which can grow by absorbing more point-defects and produce dense networks of dislocations. TEM and especially HVEM produce irradiation damage in most minerals (quartz, calcite, halite, etc.), it is therefore imperative to work at low illumination levels to decrease the flux of electrons and/or low temperatures (liquid N_2 or He) where the diffusion coefficients of point defects are lower. Irradiation may also induce a crystalline to amorphous transition as has been observed in quartz (Das and Mitchell, 1974) (see Plate 4(b)).

CHAPTER 10

Kinematic Interpretation of Structures, Textures and Preferred Orientations in Peridotites

10.1. IMPORTANCE OF FLOW STUDIES IN PERIDOTITES

The object of this chapter is to illustrate how the results of physical metallurgy developed in Chapters 2 to 4, bound together by the methodological concepts of Chapters 7 to 9, provide clues for deciphering the main flow properties in naturally deformed peridotites.

Peridotites have been selected because, as far as mechanical properties are concerned, they are the simplest and best known rocks. Moreover, studying their flowage in the mantle is one of the major geophysical issues in view of further developments of global tectonics. The next chapter will deal specifically with upper mantle flow.

Quartzites and marbles might have been also selected to illustrate kinematic interpretations. They are monomineralic rocks and the plastic flow properties of both the constituent minerals and their aggregates have been experimentally investigated in great depth. Unfortunately, the interpretation of the textures and mineral preferred orientations in naturally deformed specimens is more debatable than in peridotites. This is largely due to the higher crystallographic symmetry in quartz and calcite than in olivine. As a result more slip systems may have contributed to the development of the deformation fabric. thus making the interpretation more complex; the high symmetry also makes it difficult to have quartz and calcite optically oriented.

10.2. DIFFERENT GROUPS OF DEFORMED PERIDOTITES AND THEIR RESPECTIVE OCCURRENCES

10.2.1. Peridotites in massifs

With his classification of peridotite-gabbro complexes in stratiform and alpine types, Thayer (1960) was the first to make a clear distinction between peridotites in the stratiform complexes whose cumulate structure and texture result from the direct crystallization of a melt, and those belonging to the alpine stem whose tectonic structure and texture are indicative of high temperature solid state flow. Jackson and Thayer (1972) have since introduced a new type of complexes called concentric complexes displaying features common to both the stratiform and alpine types. Until now, they have been identified only in the Urals and the orogenic belts on the west coast of America from Alaska to Venezuela.

We are interested here mainly in the alpine type which is represented by massifs emplaced in orogenic belts, ranging in surface from a few square metres to over 1000 km^2. Den Tex (1969) has stated that this type, which he calls orogenic peridotites, should be divided into two subgroups depending upon whether the peridotites are parts of ophiolite complexes, i.e. ophiolitic peridotites, or tectonic fishes incorporated in 'root zones' of orogenic belts, constituted by high-grade polymetamorphic terrains. The latter 'root zones' peridotites can be equated with Green's 'high temperature' peridotites (1967). Jackson and Thayer (1972) divide the alpine type, using mineralogical and chemical criteria, into lherzolite and harzburgite subtypes. Nicolas and Jackson (1972) relate the two kinds of subdivisions by showing that the 'root zone' peridotites are essentially lherzolites and the ophiolitic peridotites, harzburgites. The 'root zone' lherzolite massifs are exceptionally numerous and well exposed in the alpine belt of the western Mediterranean; the ophiolitic harzburgite massifs occur much more extensively and are present, together with the other members of the ophiolitic complex, along most phanerozoic orogenic belts.

For Nicolas and Jackson, the lherzolite massifs originated inside the upper mantle beneath the continental crust from where they were intruded. The harzburgites originated inside the uppermost oceanic mantle; this mantle section and the overlying oceanic crust is then equated with an ophiolite section. If the lherzolites are considered as representative of a normal upper mantle, the difference with harzburgites in mineralogical and chemical composition is attributed to the partial melting at the oceanic ridges: through this process the lherzolites are depleted of their most fusible components and changed to harzburgites.

However, the case of the 'root zone' peridotites may be more complex. The lherzolite bodies with a tectonic fabric, found in suture zones of the alpine belt, like Beni Bouchera, Ronda, Lherz, Liguria, Lanzo, Baldissero and Balmuccia (Fig. 10.1), probably represent upper mantle slices intruded during the alpine orogenesis. Other bodies and small fishes inserted in high-grade metamorphic terrains with which they are texturally and mineralogically equilibrated may have been intruded in the same fashion during an older orogenesis and later deformed and metamorphosed along with their country rocks. They may also have originated in the exact place where they are found and, when associated with mafic granulitic facies as in the Ivrea zone, represent a transition zone between the deep continental crust and the underlying upper mantle (Lensch, 1971). Other examples from western Europe illustrating this type of 'root zone'

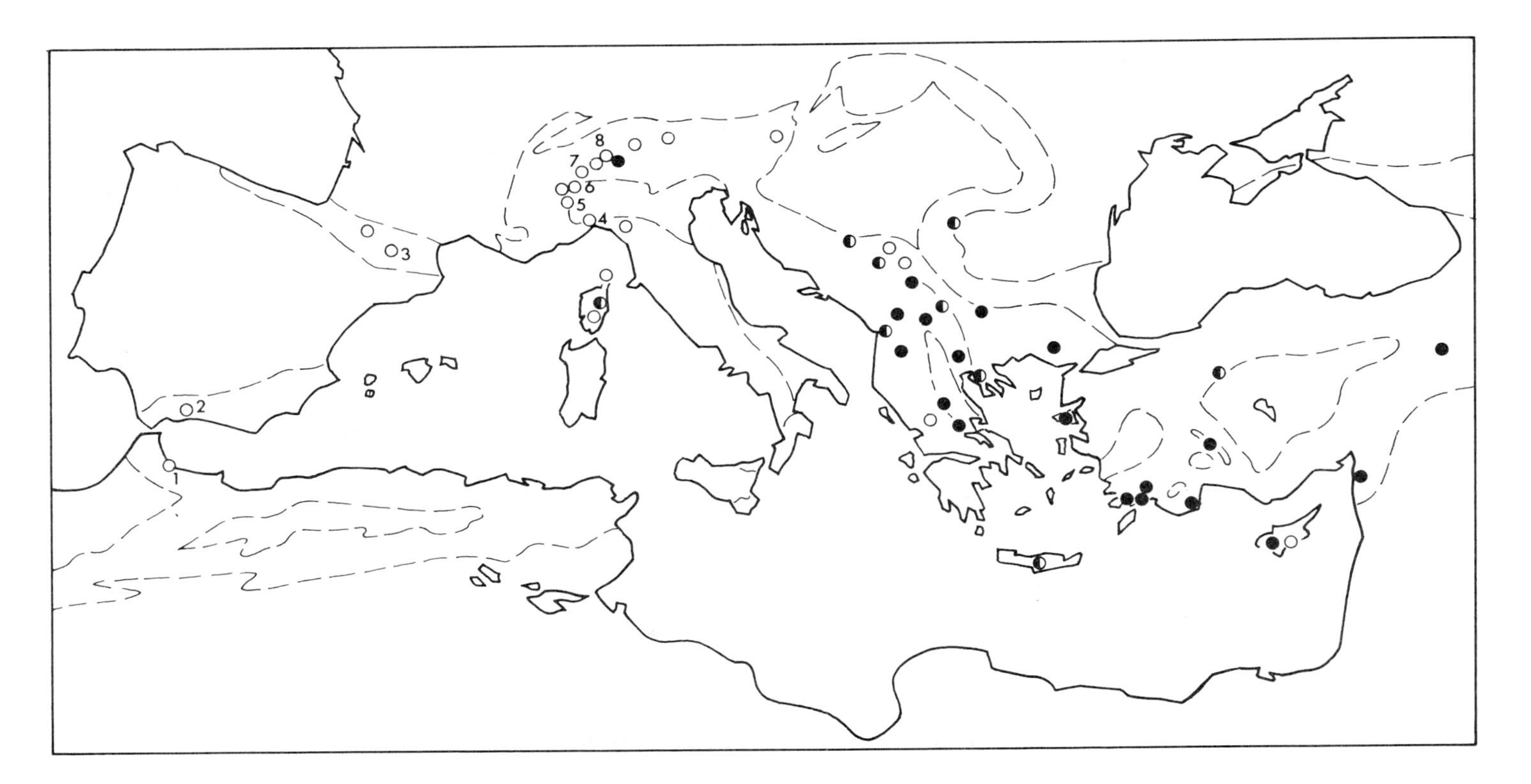

Fig. 10.1. Peridotite massifs in the Mediterranean alpine belts. Open circles: lherzolite massifs; full circles: harzburgite massifs; half-filled circles: intermediate massifs: 1: Beni Bousera; 2: Serrania di Ronda; 3: Lherz; 4: Liguria; 5: Lanzo; 6: Baldissero; 7: Balmuccia; 8: Alpe Arami (after Nicolas and Jackson, 1972)

peridotites include Cabo Ortegal in Galicia and the numerous massifs of Norway, Czechoslovakia and the Central Alps.

From experimental evidence and considering mineral equilibria, it is known that the deformation structures of alpine-type peridotites were developed during flowage at high temperature: possibly in the range of 700–1000 °C for cold-worked structures and above 1200 °C for typical hot-worked structures. These peridotites have often undergone later deformations during tectonic events at lower temperatures, usually in the presence of water. Their effects are concentrated in entirely serpentinized peripheral or narrow zones. Due to the weakness of serpentine which absorbs most of the strain, it is common to observe, outside these particular zones, well-preserved structures due to high temperature flow in the peridotites.

10.2.2. Peridotites in xenoliths

Peridotites are also found in xenoliths from alkali basalts and from kimberlites. The size of these xenoliths is commonly that of a hand specimen, although in kimberlites they may be a few tens of centimetres in diameter. In alkali basalts, most xenoliths are spinel lherzolites, whereas in kimberlites garnet lherzolites are dominant. These xenoliths represent mantle fragments which originated at depths of 30–100 km in alkali basalts and down to 200 km in kimberlites.

In a few occurrences it can be shown that the peridotite xenoliths from alkali basalts have a cumulate texture indicating that they were formed by magmatic settling in magma chambers; however, in the overwhelming majority of cases and also in the kimberlite occurrences, such textures are absent and on the contrary the peridotites have tectonic or recrystallized textures. It has been argued that the latter textures could have been superimposed on cumulate textures during the ascent in the basalt or the kimberlite. This hypothesis is rejected considering the relative viscosities of magmas and peridotites which are respectively of the order of 10^5 poises and 10^{20} poises.

From this it is concluded that the peridotite xenoliths from basalts and kimberlites represent a sampling of mantle rocks from between about 30 km down to 200 km. Since they are are not modified by the intrusion process, the study of their structure provides straightforward information on that of the upper mantle. However, these xenoliths only represent that part of the mantle where magma genesis has occurred and it might not be representative of the 'normal' upper mantle. It is possible that the plastic flow structures observed in xenoliths do not result from the flow directly responsible for plate motion but from that of local origin like diapiric intrusions to which Green and Gueguen (1974) attribute the origin of kimberlites.

On comparing the information on flow in the mantle obtained by studying peridotite massifs and xenoliths, one notices their complementarity: the xenoliths provide direct information on mantle flow but the flow structures are observed on a limited scale, whereas the massifs deformed in the upper mantle and crust afford the opportunity of large-scale continuous observations.

10.3. TEXTURES AND PREFERRED ORIENTATIONS

In this section, we will describe some typical textures and associated mineral preferred orientations from various peridotites. They have been selected to illustrate points of special interest in a kinematic interpretation using the physical metallurgy concepts. This is in no way a review of textures and preferred orientations in deformed peridotites, and a few remarkable types, such as the one described by Möckel (1969) in the Alpe Arami body (Switzerland; Fig. 10.1), are omitted because no interpretation can at present be proposed for them.

The preferred orientations presented here are only those of olivine and enstatite although these rocks usually contain some diopside too. Diopside is not considered because it is of minor importance: 5–10 % compared with 60–70 % for olivine and 20–25 % for enstatite, and also because it exhibits few signs of plastic flow and generally very poor preferred orientations.

10.3.1. Baldissero: cold-working—dominant irrotational shear flow

The Baldissero body outcrops over 3 km^2 along the Canavese line in the Ivrea zone of the western Alps (Fig. 10.2). Its petrology has been described by Lensch (1968, 1971) and its structure by Etienne (1971). It consists of a spinel lherzolite with spinel websterite layers and pyroxenite dikes. Its internal structure is fairly homogeneous. Nicolas et al. (1972) have interpreted it as an intrusion from the uppermost mantle underlying the Ivrea zone and ascribed the process to a strike slip and overthrust motion of the Southalpine plate against and over the European plate during the alpine orogenesis (Fig. 11.6).

The texture (Plate 22(b)) and the preferred orientations (Fig. 10.3) are those of a fairly deformed lherzolite. As other specimens in this chapter, the texture is observed in the (XY) strain ellipsoid section which will be also the equatorial plane in the density diagram. Features characterizing cold-working (Chapter 5) are distinctly visible:

Undulatory extinction due to slip polygonization. It appears in the olivine crystals whose slip direction, indicated by the normal to the sub-boundaries between bands, is close to the foliation trace. It is a consequence of the bending and twisting undergone by the crystal lattice during elongation by slip.

Kink banding with large rotations from one band to the next. It occurs in olivine and enstatite crystals whose slip direction is at a high angle to the foliation trace. As a result the crystal is shortened normal to the foliation.

Almost complete absence of recrystallization in olivine, complete in enstatite. It is only observed as trails of olivine neoblasts, 0·01–0·02 mm in diameter, aligned along some grain-boundaries at low angles on either side of the foliation trace (Plate 22(b)).

Inside the olivine porphyroclasts in the vicinity of the grain boundaries, the substructure is tighter and more contorted indicating concentrations of stress and strain probably responsible for this grain-boundary recrystallization.

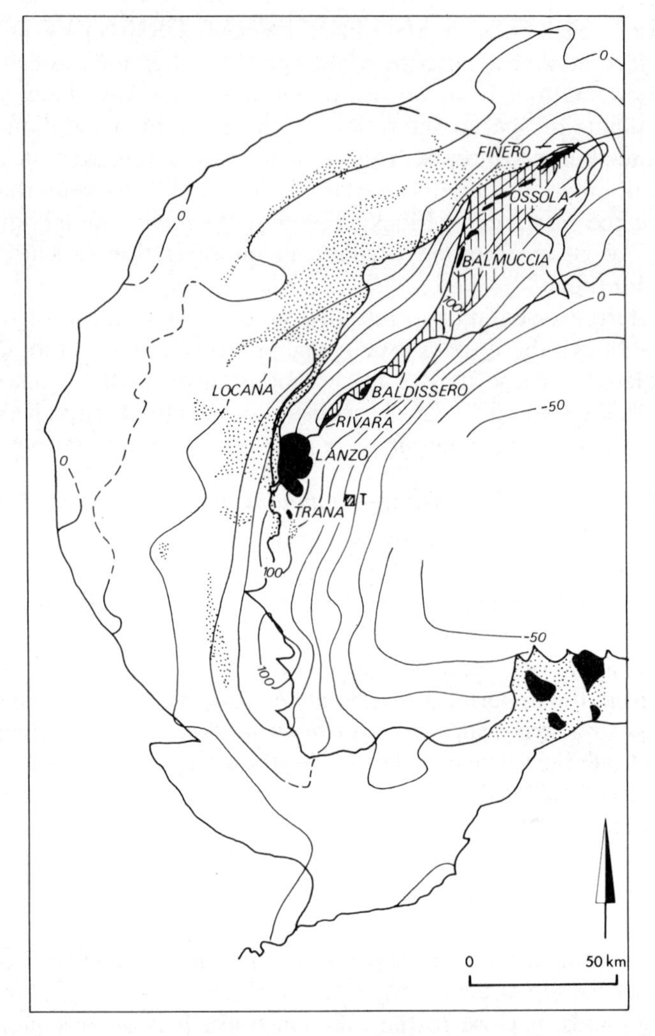

Fig. 10.2. Location map of the principal lherzolite massifs (black decoration) in the western Alps, superimposed on a gravimetry map. (Simplified from Vecchia, 1968). Stippled areas: metamorphic ophiolites. Hatched area: the Ivrea zone

Finally, irregularity of the grain-boundaries even when observed with high magnification. This denotes the minor importance of grain boundary migration.

The lattice preferred orientation has been measured in both the olivine and enstatite (Fig. 10.3). The enstatite has no significant preferred orientation. This is common in this massif (Etienne, 1971) and more generally in peridotites deformed in cold-working conditions. Avé Lallemant's (1967) thorough studies in Lherz body (Pyrénées), which resembles Baldissero, support this

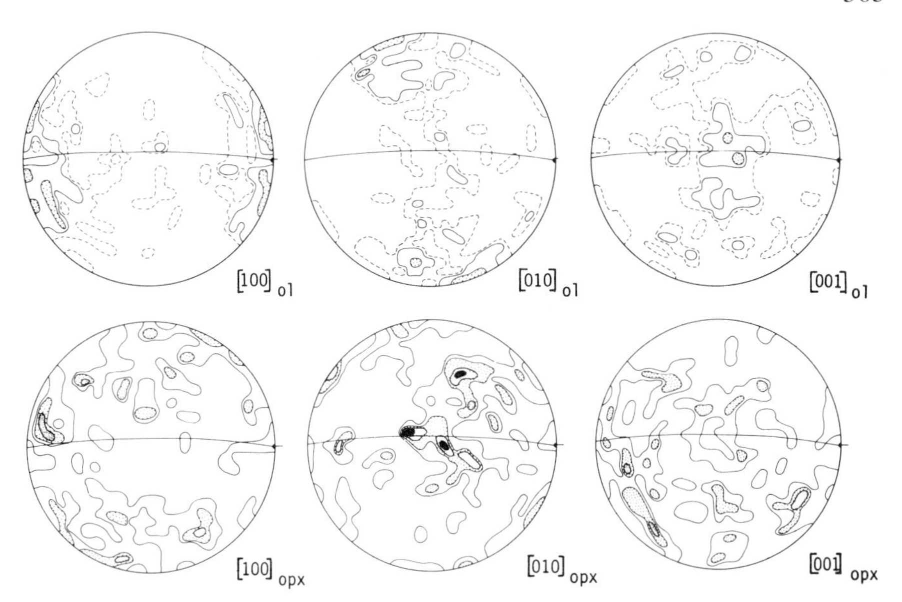

Fig. 10.3. Olivine (ol) and enstatite (opx) preferred orientations in the Baldissero specimen illustrating cold-working and a dominantly irrotational shear flow; 100 porphyroclasts of each species. Equal area projection, lower hemisphere; contours: 1, 2, 4, 8 % per 0·45 % area (see § 8.3.4). Full triangle: best computed axis; open triangle: pole of best plane (see % 8.3.5). Horizontal line: trace of the foliation plane; E. W. dots: trace of the mineral aggregate lineation

conclusion, although in two cases a good enstatite preferred orientation is described. Along with the absence of textural evidence of plastic deformation, this indicates that, in moderate to high strain and cold-working conditions, the enstatite does not participate significantly to the flow and behaves passively like diopside. The olivine porphyroclasts have a preferred orientation related to the orientation of the penetrative foliation and spinel lineation. Dimensional measurements in thin sections have confirmed that the olivine preferred orientation was parallel to that of the spinel visible in the field: [100] olivine density maximum is close to the lineation trend but symmetrically inclined to it by 20° in the plane containing the lineation and normal to the foliation, although more scattered; [001] olivine cluster in the foliation at right angles to the lineation; [010] olivine form crossed girdles at high angles to the foliation with point concentration nearly normal to it. The diagram is very consistent with what could be foreseen from observing Plate 22(b).

The origin of the olivine preferred orientation has been ascribed to large flowage with slip on the systems (010) [100] and {0kl}[100] as the dominant flow mechanism (Nicolas et al., 1971). Recrystallization trails in grain-boundaries weakly inclined to the foliation also suggest the contribution of grain-boundary sliding. In the least deformed facies of the Baldissero massif, the spinel-pyroxene relationship (Plate 23(a)) is characteristic of what has been called in xenoliths from basalts a protogranular texture (Mercier, 1972; Mercier and Nicolas, 1975; § 10.3.4). The olivine preferred orientation, although not measured, is obviously weak. If the rock originated with a protogranular texture, it can be stated by referring to the basalt xenoliths that there was no significant preferred orientation at the onset of the flow (see Fig. 10.11) and consequently that the flow is entirely responsible for what is now observed. The division of the olivine preferred orientation pattern into two almost equally important subfabrics which are symmetrical with respect to the penetrative elements has been interpreted, in the aforementioned study, as indicative of a non-rotational flow regime (flattening). This is in keeping with experimental results (§ 6.2.1.) and predictions from the theoretical model of § 6.9.

10.3.2. Lanzo core: hot-working—dominant rotational shear flow

The structural setting of Lanzo massif and its internal features are specifically considered in § 11.2.2 (see also Fig. 10.2). The samples under investigation come from the central area of the northern body of the massif, an area where the deformation is moderate compared with that found at the margins and illustrated in the next section. The rock is a feldspathic lherzolite and, because of this, the penetrative elements are remarkably visible in the field. From the

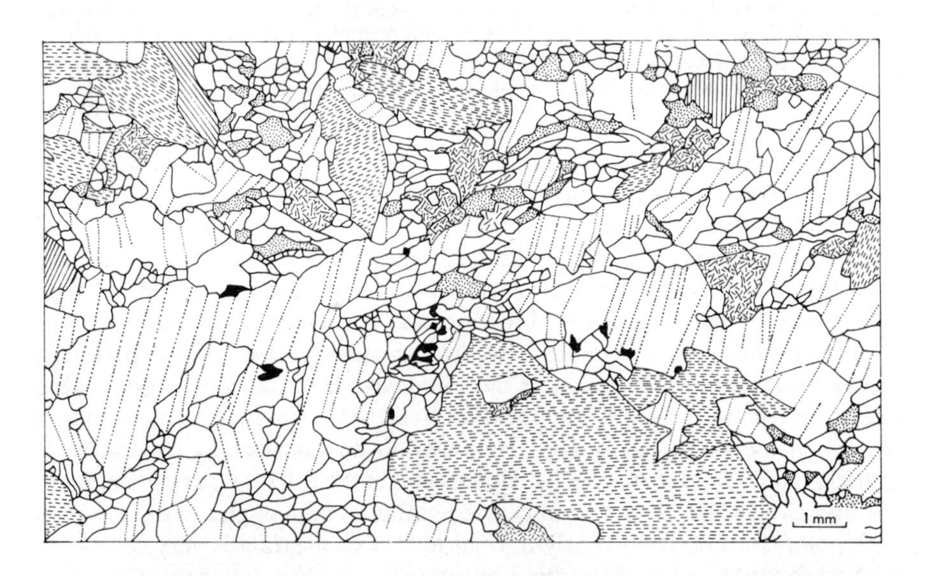

Fig. 10.4. Drawing after a thin section in a Lanzo specimen illustrating hot-working and dominant rotational shear flow. Observation plane: normal to the foliation and parallel to the aggregate lineation (same as in Fig. 10.5 presenting the mineral preferred orientations). Blanks with dotted lines: olivine with sub-boundary traces; aligned dashes: enstatite with (100) trace; disordered dashes: enstatite with other orientations; hatches: diopside with (100) trace; dots: feldspar; black: spinel (after Nicolas, 1974a, reproduced by permission of Societe Geologique de France, Paris)

evidence of partial melting at the time of the flowage, it has been concluded that the P, T conditions were 5–10 kB, 1200 °C. A geological mean strain rate (around $10^{-14}\,s^{-1}$) is postulated from the intrusion history (§ 11.2.2).

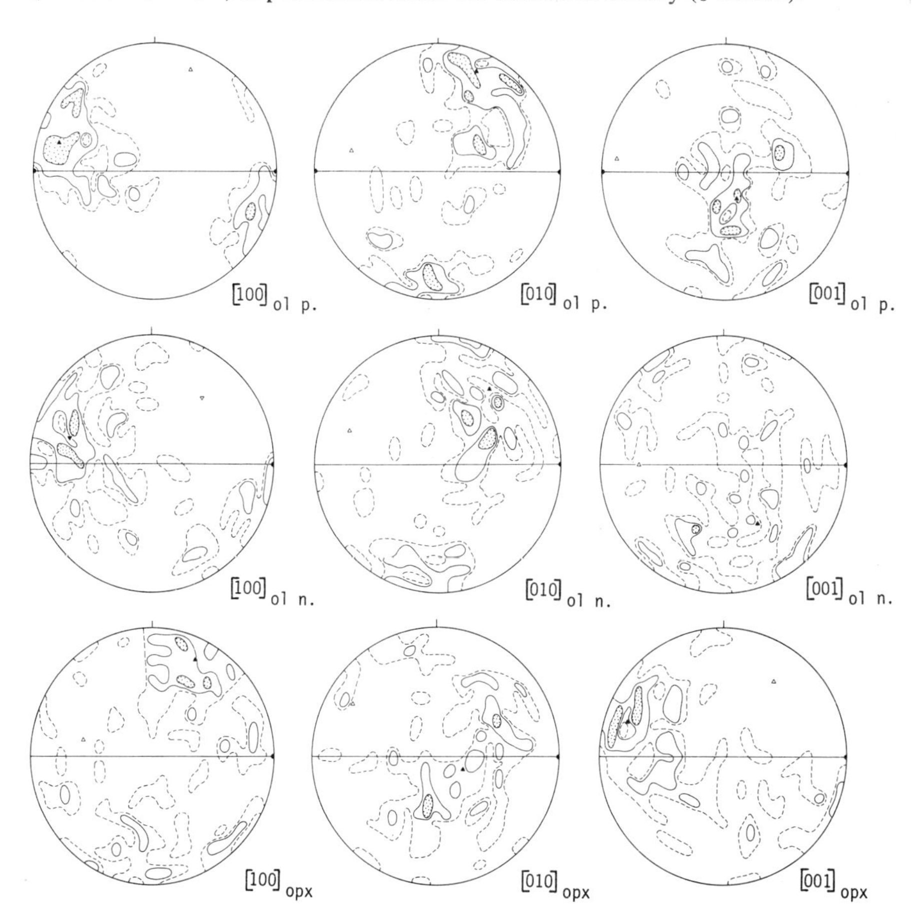

Fig. 10.5. Olivine (ol) and enstatite (opx) preferred orientations in the Lanzo specimen illustrating hot-working and dominant rotational shear flow (see also Fig. 10.4). Olivine: 100 porphyroclasts (ol p.) and 100 neoblasts (ol n.); enstatite: 100 prophyroclasts. For graphical arrangements see Fig. 10.3 (after Boudier, 1976)

The texture (Fig. 10.4) and the preferred orientations (Fig. 10.5) are observed in the (XZ) section of the strain ellipsoid which is deduced from the orientation of the penetrative structure.

The following features in olivine are typical of a hot-working regime (Chapter 5):

Strain-free subgrains separated by sharp subgrain-boundaries in the porphyroclasts;

Presence of numerous polygonal neoblasts.

Planar grain-boundaries. The ragged contour of the porphyroclasts, when observed on a finer scale can be resolved into small rectilinear segments which converge into triple points with 120° mean angles together with subgrain and neoblast boundaries. This observation proves that the substructure in porphyroclasts predates the recrystallization and therefore is not a late feature in the deformation course, formed independently of the texture and preferred orientation development.

A closer examination of the substructure reveals two distinct types, somewhat comparable with those described in the cold-working specimen just considered. The slip polygonization substructure is replaced here by a climb polygonization substructure. The segmentation into sharply defined subgrains is caused by dislocation climb during recovery (§ 4.1.2). This substructure is characterized by (100) subgrain-boundaries at high angle to the foliation trend (Fig. 10.4), in keeping with [100] as slip direction (§ 7.4.2). The misorientation angle between adjacent subgrains is small, with a mean value of 5° and never exceeding 15°. In contrast, the kink banding substructure, present here in only a few grains, is characterized by (001) KBBs at moderate angles to the foliation and larger misorientation angles between the KBs (mean value of 9°).* This indicates a shortening component normal to the foliation as in the cold-working case. These distinct interpretations of subgrain features have been experimentally verified (§ 6.2.1.).

The two elementary substructures are associated inside some olivine porphyroclasts which are thus divided into rectangular subgrains. The KBBs are responsible for the (001) subgrain-boundaries parallel to the crystal elongation and the climb polygonization walls for the (100) boundaries at right-angles to the former (Fig. 10.6). This is best observed on sections parallel to the (010) plane.

It is remarkable that in the enstatite, neither recovery structures nor any recrystallization are observed in contrast with the common occurrence of kink banding.

The olivine and enstatite porphyroclasts preferred orientations (Fig. 10.5), when related to the penetrative structural elements, conform to an origin by slip. [100] olivine, which is the dominant slip direction in high temperature flow, concentrates in a good point maximum at a small angle (20°) to the lineation, as expected. Unlike the previous case of Baldissero specimen, [010] olivine also concentrates in a single maximum at a high (60°) to the

* Due to the recovery conditions, beyond 15° the subgrains evolve into independent grains (§ 4.5.4; § 10.3.5). It is possible to trace clusters of recrystallized grains back to kinked structures. Misorientations between them, reflecting those between adjacent KBs, can be as high as 50°, with a mean value of 20°. This is certainly higher than what is measured in the case of recrystallized grains derived from climb polygonization substructures.

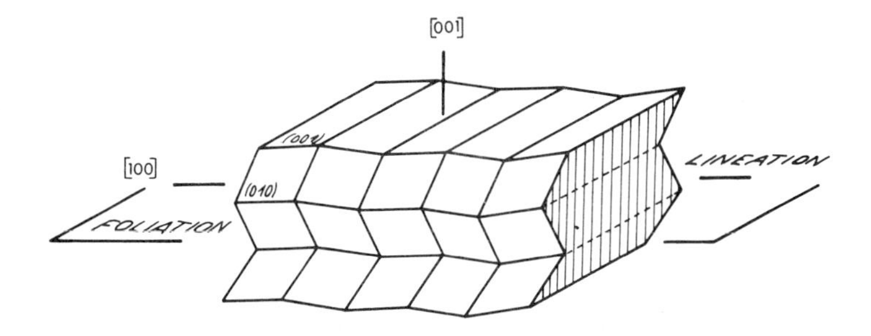

Fig. 10.6. Subgrains and kink bands in an olivine porphyroclast typically related to the foliation and lineation orientation. Misorientations across the (100) sub-boundaries are in the range of a few degrees and substantially larger across the (001) kink band boundaries

foliation. This suggests that in these higher temperature conditions, the dominant slip plane is (010) and not any (0kl) plane, in keeping with Carter and Avé Lallemant's predictions (1970; § 5.2.2). The enstatite preferred orientation, though more diffuse, also fits in with that interpretation: [001], the only slip direction, coincides with [100] olivine and [100], pole to the only slip plane, coincides with [010] olivine.

The obliquity of the lattice orientations relative to that of the penetrative elements has been already interpreted as due to a dominantly rotational shear flow (Nicolas et al., 1972). This conforms with the theoretical predictions of § 6.9. The 20° obliquity angle and its sense record here a dextral sense of shear and a shear angle in the range of 60° (for explanation see § 10.4.1). This was clearly apparent and could be measured directly on the thin section of Fig. 10.4 by considering the dominant inclination in one direction of the subgrain-boundaries.

The simultaneous occurrence in olivine porphyroclasts of two substructural elements (Fig. 10.6), one related to [100] slip and the other to [001] slip, makes it possible for a porphyroclast to be elongated by [100] slip, with subsidiary bending and twisting, at the same time that it is shortened by kinking in the [001] direction. In a three-dimensional flow consisting of a dominant simple shear component and a minor pure shear one, this may favour the olivine orientation with [100] in the shear direction and [001] in the pure shear shortening direction. As a result, [001] rather than [010] would be at a high angle to the foliation. This is not observed in the samples under investigation where, as a matter of fact, the [001] KBBs are uncommon. Simple shear (planar rotational flow), in this case, is concluded to be a good approximation of the flow. The situation is very different in the area in the massif where folding is ubiquitous (Fig. 11.2). It is interpreted as a consequence of a deviation from planar rotational flow with a pure shear component directly responsible for it (§ 10.4.2). This situation is reflected in olivine textures and preferred orientations which locally display an unusually large number of kinked crystals with (001) KBBs and a tendency for [001] to be at a high angle to the foliation.

The preferred orientation of the olivine neoblasts, also presented in Fig. 10.5, is clearly related to that of the porphyroclasts, although more diffuse. The polygonal grain-boundaries indicate that grain-boundary migration was operative during the recrystallization process. Recrystallization by progressive misorientation of subgrains (§ 4.5.4) is also suggested by the close crystallographic orientations of the new crystals and their host and by detailed textural observations like those developed for a comparable case in § 10.3.5.

10.3.3. Lanzo margin: extensive rotational shear flow

The samples now under consideration come from a shear zone at the North West margin of the Lanzo northern body (Fig. 11.2). In the vicinity of the shear zone, the foliation trend progressively rotates into parallelism with that of the zone; at the same time, the grain size is reduced and other signs of increasing deformation are reinforced. In the field, the outcrop exhibits the remarkable lamellar lineation of enstatite. The temperature during the flowage has been estimated as around 900 °C on the basis of the presence of hydrous phases in associated mylonitic gabbros and on the preliminary data on coexisting pyroxenes chemistry (G. M. Brown, oral comm.). The pressure corresponds to a crustal environment: 10 kB at most, strain rates are probably higher than in the previous case, since the flowage of the massif is now concentrated in the margin.

The texture is mylonitic with a matrix of recrystallized olivine forming bands of varied grain size, occasionally too minute for microscope resolution (Plate 19(a)). The coarsest bands contain some porphyroclast relics with undulatory extinction. The most striking feature, in the (XZ) strain ellipsoid section, is the occurrence of ribbon-shaped enstatite strips parallel to the foliation trend, with aspect ratios in the range of 25 to 1 and sometimes as high as 100 to 1. Locally, in these strips, the enstatite recrystallizes in neoblasts a few microns in diameter. In the very elongated ribbons, the trace of the (100) enstatite cleavage, which is at a small angle to the elongation of the ribbon, is oriented in a single direction with respect to this elongation pointing to a sinistral sense of shear. The mean angle between these two directions is 6°, indicating a shear angle of 84° or probably even higher since it can be assumed that the olivine has flown more than the enstatite. A pure shear component, exerted normal to the foliation, is also suggested by the boudinage figures observed in the enstatite strips (Plate 19(a)). Enstatite strips with aspect ratios of the order of 6/1, have glided in the opposite sense. This conforms to predictions made by the theoretical model presented in § 6.9. Another aspect is illustrated by retort shaped grains like those on Plate 19(a). This particular shape has been explained by a two-stage deformation in grains with a special initial orientation:

(1) Moderate slip in the opposite sense to that externally imposed. It is responsible for the body of the retort and operates until a lockage orientation is attained;

(2) External rotation in the externally imposed sense. The grain is then unlocked and undergoes considerable slip in that sense. In contrast with the theoretical model where a bodily rotation operates, here it operates within the lattice thus giving a recording in the grain shape of the deformation sequence.

The enstatite porphyroclasts and the olivine neoblasts preferred orientations are shown in Fig. 10.7. Due to the large amount of slip, the enstatite has

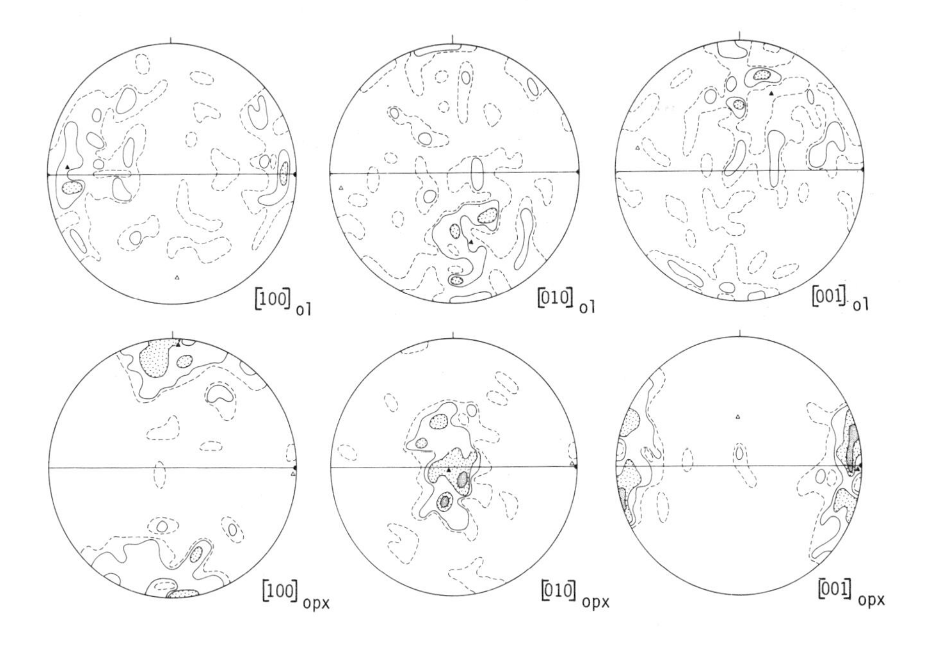

Fig. 10.7. Olivine (ol) and enstatite (opx) preferred orientations in a mylonitic sample from Lanzo margin illustrating an extensive rotational shear flow. 100 olivine neoblasts and 100 enstatite porphyroclasts. For graphical arrangements see Fig. 10.3 (after Boudier, 1976)

developed a very strong preferred orientation such that the slip direction [001] is parallel to the lineation and the slip plane (001) parallel to the foliation. In fact this mean orientation, as clearly seen in the [001] diagram, is the addition of the orientations of grains with opposed senses of slip. The olivine neoblasts fabric is probably an attenuated inheritance from former porphyroclasts oriented by the plastic flow.

Indeed, the dominant rotational shear flow regime is here best reflected in the direct observations mentioned above.

10.3.4. Basalt xenoliths sequence; increasing strain, recrystallization, deformation cycles

The main structural types met in lherzolite xenoliths from alkali basalts have been comprehensively studied and a classification proposed by Mercier (1972) and Mercier and Nicolas (1975). This investigation was based on xenoliths from Hawaii and Massif Central (France), although complemented by a few samples from other volcanic districts. Remarkably enough, so far as the structures are concerned, no essential difference has been observed according to the volcanic district of origin. Three main types were distinguished from textural and mineral preferred orientation criteria. Olivine which constitutes around 70 % in these nodules is principally considered in drawing limits between the types. They are now briefly described in view of a kinematic interpretation, starting with the least deformed type. The textures are presented in the plane normal to the foliation and parallel to the spinel lineation (*XZ* strain ellipsoid section) whenever these elements could be defined in the sample.

Protogranular texture (*Fig.* 10.8; *Plate* 23(a))

The olivine and enstatite grain size is typically coarse (4 mm). At a previous stage the grain size of olivine could even have been in the range of 10 mm because groups of crystals, with a close common orientation and straight lined mutual grain boundaries, suggest that they evolved from subgrains by the misorientation mechanism described in § 4.5.4. Otherwise the grain boundaries are curvilinear and the grain shape equant, preventing the occurrence of any foliation or lineation. The substructure is weakly developed and highly recovered with a mean subgrain dimension of 1 mm. The enstatite exhibits a very characteristic association with diopside and spinel. These two minerals

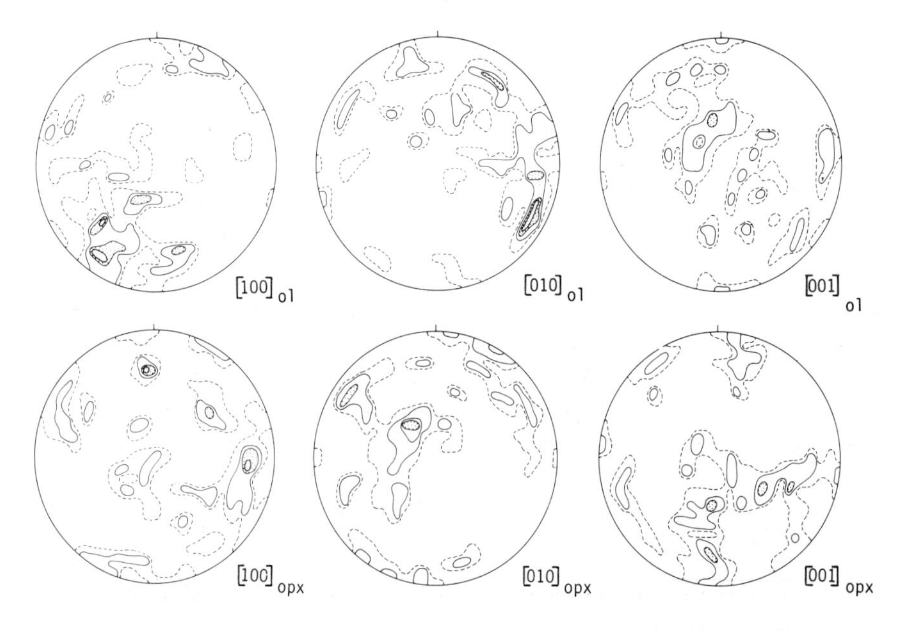

Fig. 10.8. Olivine (ol) and enstatite (opx) preferred orientations in a lherzolite xenolith from basalt with a protogranular texture. 100 grains of each species. For graphical arrangements see Fig. 10.3 (after Mercier and Nicolas, 1975)

form blebs or vermicules inside the enstatite which can recrystallize into clusters of polygonal-shaped grains of the three species (Plate 23(a)).

Typically the olivine and enstatite have no preferred orientations (Fig. 10.8).

Porphyroclastic texture (*Fig.* 10.9, *Plate* 23 (b) *and* (c))

Typical samples displaying this texture have well-defined foliations and spinel lineations. The olivine is present in two fashions:

Large porphyroclasts (8 × 2 mm) with numerous subgrains, 0·5 mm in diameter;

Later formed polygonal, strain free neoblasts (0·5 mm). The relative proportion of the two kinds of grains varies over the entire spectrum, the two other textural types being at each extremity of it.

The enstatite essentially constitutes porphyroclasts. The spinel and diopside are still preferentially associated with the enstatite, although scattered grains are also observed. The spinel tends to recrystallize into aggregates with a holly leaf shape.

The texture and preferred orientations of olivine and enstatite (Fig. 10.9) are very similar to the hot-working specimen from the Lanzo massif (§ 10.3.2) and call for the same interpretation by plastic flow. Here likewise, the porphyroclasts-preferred orientations are oblique relative to the foliation and lineation (angle of 18°) implying a shear regime.

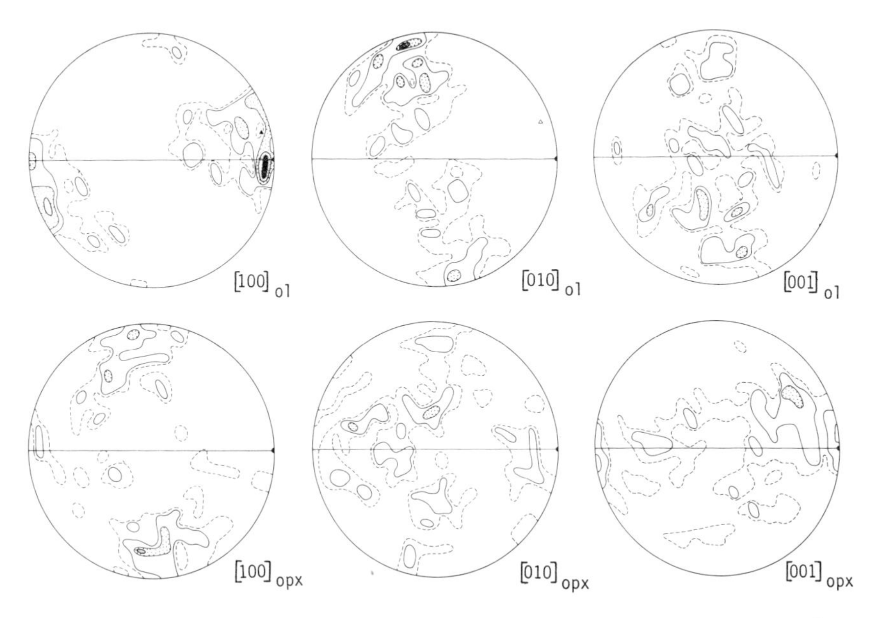

Fig. 10.9. Olivine (ol) and enstatite (opx) preferred orientations in a lherzolite xenolith from basalt with a porphyroclastic texture. 100 porphyroclasts of each species. For graphical arrangements see Fig. 10.3 (after Mercier and Nicolas, 1975)

The olivine neoblasts preferred orientation conforms to that already described in the Lanzo sample. Their relation to the host porphyroclasts has been more specifically investigated as an illustration of the recrystallization process by progressive misorientation of subgrains (Poirier and Nicolas, 1975, § 4.5.4). Figure 10.10(a) is a drawing of porphyroclasts and neoblasts

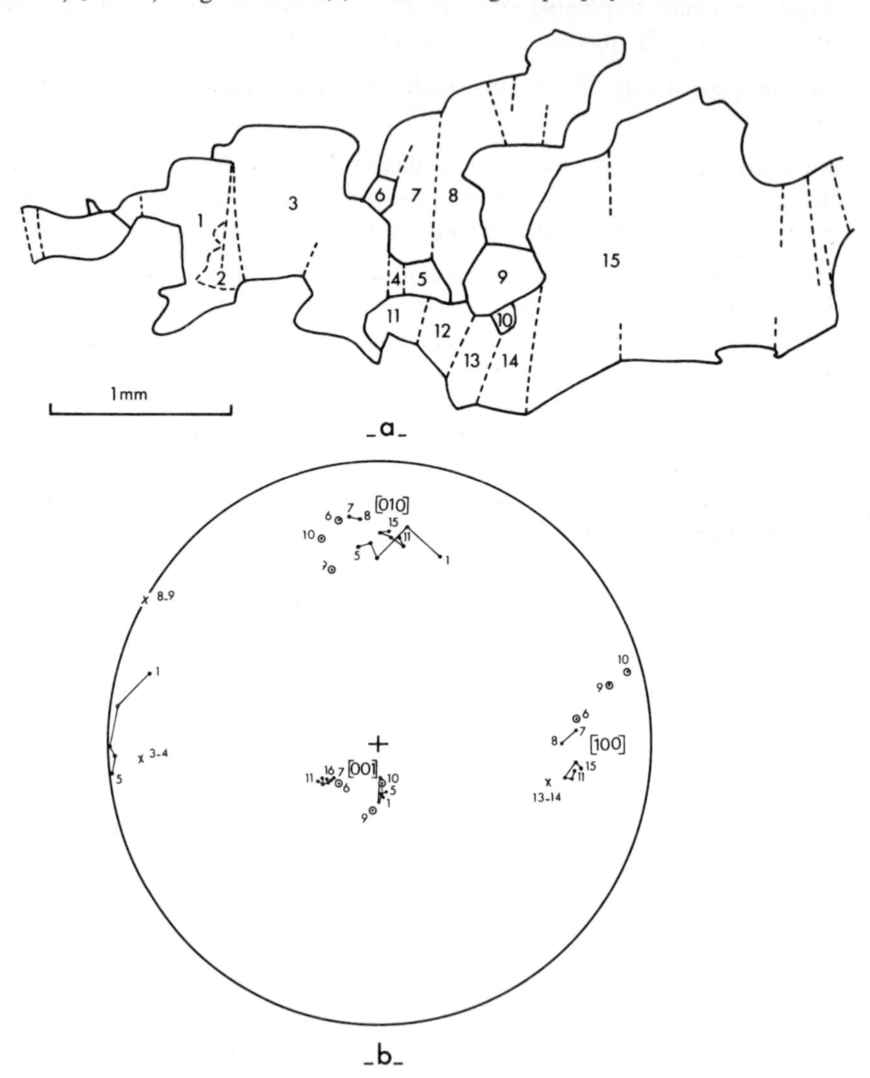

Fig. 10.10. Recrystallization by progressive misorientations of subgrains:
(a) Drawing of an olivine porphyroclast and related neoblasts from a porphyroclastic textured xenolith in basalt. Dashed lines: sub-boundaries
(b) Lattice orientation in the subgrains and associated neoblasts. Successive subgrains are connected by a line; circles: neoblasts: crosses: poles of sub-boundaries (after Poirier and Nicolas, 1975 reproduced by permission of The University of Chicago Press, Chicago)

which originated as a single crystal, as shown by their close orientations (Fig. 10.10(b)). The misorientation develops on either side of (100) subgrain walls. It is the largest in the [100] direction (slip direction) and smallest in the [001] direction suggesting that [001] is close to the axis of rotation and that the walls are therefore dominantly tilt walls. As in the Lanzo sample, the transition from subgrain to grain occurs for misorientation above 14°; it is sometimes observed in a single porphyroclast.

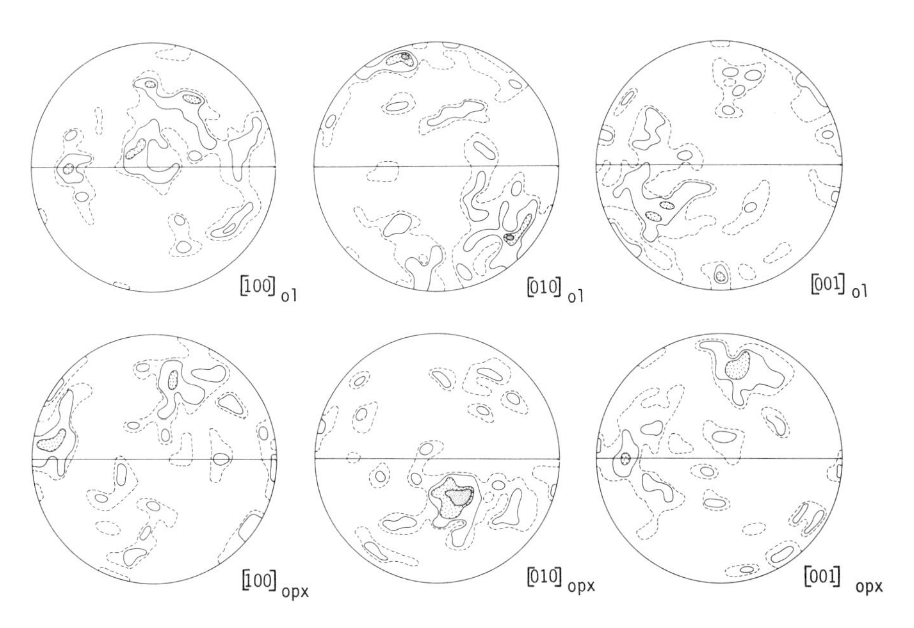

Fig. 10.11. Olivine (ol) and enstatite (opx) preferred orientations in a lherzolite xenolith from basalt with an equigranular mosaic texture. 100 neoblasts of each species. For graphical arrangements see Fig. 10.3 (after Mercier and Nicolas, 1975)

Equigranular textures

This structural type has been divided into two sub-types from grain shape and preferred orientation considerations: the mosaic equigranular texture (Fig. 10.11, Plate 24(b)) and the tabular equigranular texture (Fig. 10.12, Plate 24(a)). They both have a rather fine grain size (0·7 mm) and an even dimension for the minerals, principally the olivine. The grains are polygonal with straight boundaries and triple point at 120°. The pyroxenes and spinel are scattered and the spinel is recrystallized in blebs (0·05–0·2 mm) inside the olivine or at grain-boundaries, mainly at triple points.

In the tabular sub-type, the silicates form tabular or prismatic crystals with 1·4×0·5 mm ratios in the (XZ) section of the sample. They define a good foliation but a poor lineation. The olivine occurs as strain-free tablets and as few elongated porphyroclasts with a ragged contour and (100) and (010) subgrain boundaries. At a higher magnification, the ragged contour is resolved

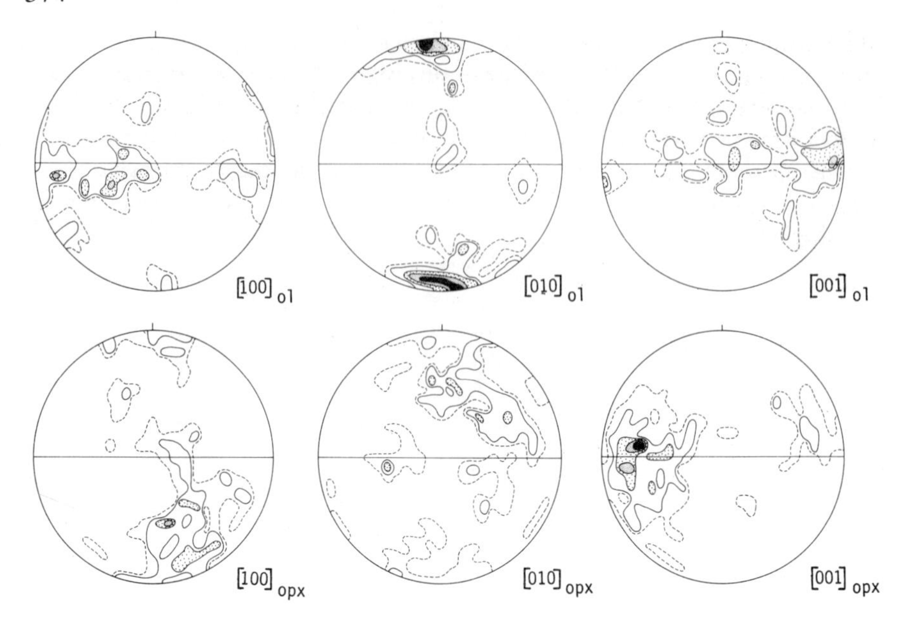

Fig. 10.12. Olivine (ol) and enstatite (opx) preferred orientations in a lherzolite xenolith from basalt with an equigranular tabular texture. 100 grains of each species. For graphical arrangements see Fig. 10.3 (after Mercier and Nicolas, 1975)

in straight segments. The olivine preferred orientation is dominated by a good [010] point maximum normal to the foliation; [100] defines a weak maximum at some angle to the lineation, which in fact could not be measured accurately. The enstatite presents a [001] maximum close to [100] of olivine.

In the mosaic sub-type, the silicates are equant (0·8 mm) or slightly elongated in only one direction, thus defining a foliation. The olivine and enstatite preferred orientations are typically weak.

Transitional textures

This three-fold classification is somewhat arbitrary since every transitional stage exists between the main types, clearly shown in the prophyroclastic type by the complete range of proportions between porphyroclast and the two kinds of neoblasts. This is not surprising considering that the sequence arises from increasing strain. The strain is estimated, in a qualitative fashion, by the degree of spinel scattering away from the enstatite. The scattering and the spinel habit associated with its successive stages are illustrated on plates.

With increasing strain, the equant and strain-free olivine grains in the protogranular texture are elongated, divided into subgrains and their outline is ragged (Plate 23(b)). A weak preferred orientation is developed with Z oblique to the foliation, in a single direction. There is yet no recrystallization. The enstatite and the associated diopside and spinel are also slightly elongated. Gradation into the porphyroclastic texture is marked by:

Neoblast recrystallization in olivine, generally along porphyroclast boundaries.

Definite elongation and strong preferred orientation in the porphyroclasts.

Increase in the number of subgrains in the porphyroclasts and decrease in their dimension.

Spinel recrystallization in spindle-shaped aggregates which are largely responsible for the tectonite structure, visible in hand specimen.

As deformation still proceeds, the minerals increasingly recrystallize and the equigranular texture is reached when the olivine porphyroclasts have almost entirely disappeared.

Plastic flow and recrystallization

All the specimens illustrating this deformation sequence display evidence of hot-working flow. If the flow took place during the magma generation the temperature was certainly high, in the range of 1200–1300 °, at a pressure of 15–30 kB. Nothing precise is known about the strain rates.

Assuming the sequence represents a continuous process, the textural variations are attributed to increasing strain. For moderate strain, possibly imposed at low stress considering the large dimension of the subgrains (§4.2.3), plastic flow is dominant as in the case of hot-working flow in Lanzo (§ 10.3.2). It is responsible for the development of the preferred orientations, whose obliquity with respect to the penetrative elements indicates a rotational regime. Though the relation between the angle of obliquity and the shear angle is not well assessed (see § 10.4.1) even in the case of simple shear, it is suggested that the 15° angle, which is a mean value in the porphyroclastic textures, indicates shear angles around 75°.

Beyond this angle recrystallization becomes dominant. It is associated with very large strain as shown by the scattering of spinel and diopside in isolated grains. Larger stress than in the earliest stages is likely because of the smaller subgrain size in porphyroclasts. The increasing importance of recrystallization with greater strains has already been deduced from experiments (§ 6.2.1).

Preliminary studies in TEM on the dislocation density noted in the recrystallized grains of the equigranular mosaic texture give higher figures than would be expected if the recrystallization had proceeded during annealing (H. W. Green, oral comm.). It is thought that it occurred during the flow for this particular structure. It is not necessarily the same for the tabular texture. This important matter of syntectonic versus post-tectonic recrystallization needs to be settled by more TEM studies. There is little hope of an univocal conclusion being reached from textural and preferred orientation data only. Illustrating this point, one could cite the remarkable similarity between the equigranular mosaic texture described above and the one developed during annealing below 1000 °C in the peridotites incorporated in the granulitic terrains of the Ivrea zone (Lensch, 1971).

Evidence has been given that the mechanism of recrystallization by progressive misorientation of subgrains with increasing strain was important. A further indication is provided by observing that both the mean and the maximum misorientation angles between adjacent subgrains increase from the protogranular to the most deformed porphyroclastic textures: thus the mean misorientation angle changes from 2° to nearly 4° and the maximum misorientation angle from 5° to 14° (Poirier and Nicolas, 1975). Recrystallization involving grain-boundary migration is also witnessed by the fact that in the equigranular texture many spinel blebs are totally enclosed in olivine neoblasts and by the observation of the polygonal outline of grains. The occurrence of olivine tablets, devoid of any strain features and directly superimposed upon the substructure of porphyroclasts, also proves that the classical strain-induced recrystallization mechanism (§ 4.5) is present in these rocks. Indeed it is more representative in the porphyroclastic textures of peridotite xenoliths from kimberlites where these tablets are idiomorphic (Boullier and Nicolas, 1973). It appears that recrystallization of olivine in tablet-shaped crystals is due to different processes which are not thoroughly understood.

Secondary deformation cycles

In a few volcanic vents like Puy Beaunit (Massif Central), a suite of xenoliths illustrating a new deformational cycle has been found. The starting point is a secondary protogranular texture which in many ways resembles the primary one described above. The striking difference is that the typical pyroxene spinel associations are absent here and that the spinel is scattered in minute grains inside the olivine crystals (Plate 24(c)). Since the xenoliths are found together with equigranular textured ones, it is believed that they are derived from them by grain growth. The absence of strain features in the large olivine grains indicates that the process was annealing recrystallization.

A secondary porphyroclastic texture can be recognized provided that the olivine porphyroclasts have spinel grain inclusions (Plate 24(c)). In the few cases where this was clearly observed (Puy Beaunit, Pali in Hawaii), the character of the deformation was cold-working (wavy substructure, peripheral recrystallization in tiny grains with irregular outlines). The olivine preferred orientation was also found to be symmetrically inclined with respect to the structural elements, suggesting development during a non-rotational flowage.

10.3.5. Kimberlite xenoliths: superplastic flow

A comprehensive study on the textures and mineral preferred orientations in peridotite xenoliths from kimberlites has been conducted by Boullier (1975) and Boullier and Nicolas (1973, 1975) in a way very similar to that followed in basalt xenoliths. However, the structural classification is based only on xenoliths from South Africa, apart from a few from Siberia. It compares well with that of basalt xenoliths with two coarse-grained textures (coarse granular and coarse tabular) which are probably corresponding to the protogranular texture in basalt xenoliths, a porphyroclastic texture similar to that in basalt xenoliths and a mosaic texture which is equated with the equigranular mosaic one. The possibility of successive deformation cycles separated by anneals has also been considered in kimberlite xenoliths (Harte et al., 1974). Emphasis will be laid here on particular features only met in the mosaic textured kimberlite xenoliths displaying evidence of an extensive flowage.

The mosaic-textured xenoliths have equilibrated at temperatures up to 1400 °C and pressures of 60 kB (Boyd and Nixon, 1972; MacGregor, 1974). The strain rates are unknown. Transitional facies show that they derive, through increased deformation, from the porphyroclastic texture. As in the corresponding equigranular mosaic group of basalt xenoliths, the olivine is entirely, or nearly, recrystallized into a mosaic of small polygonal grains (0·07 mm). The other minerals, enstatite, diopside and garnet, first remain as porphyroclasts. Eventually these minerals will be scattered in strings of small grains. The enstatite, along with slip features, presents some peripheral recrystallization in tiny polygonal grains (0·01 mm).

The olivine lattice orientations are almost random when individual grains are picked throughout the thin section. This contrasts with local strong preferred orientations when juxtaposed grains are measured in different areas of the thin section. It has been interpreted as evidence of recrystallization by subgrain misorientation in former large porphyroclasts (§ 4.5.4) and, in this case, as an indication that this texture developed from a porphyroclastic one (Boullier, 1975).

The mosaic texture progressively grades into a fluidal mosaic texture. This occurs when the recrystallized grains around the enstatite porphyroclasts (Plate 22(b)) begin to stretch into the foliation. They finally constitute narrow stripes (0·01–0·3 mm thick) connecting together several enstatite porphyroclasts before ending in the olivine matrix where they cut through the grains as microfaults (Plate 18(a)). Referring the mean length of the stripes to that of the porphyroclasts makes it possible to estimate strain in the order of 800 to 900 %. This spectacular texture has been ascribed to superplastic flow (Boullier and Nicolas, 1975; see § 4.3.5), an interpretation which has received great support from measurements on lattice orientations and density of dislocations in the enstatite grains from the stripes (Gueguen and Boullier, in press).

It is remarkable to note in these particular flow conditions a ductility inversion between olivine and enstatite. In the plastic flow regime examined up to now, the olivine is more ductile than the enstatite, which is confirmed by experimental creep. Here the deformation is controlled by the enstatite since stripes end in fractures through the olivine crystals. This means that olivine could not keep up with the flow in enstatite and that it responded by fracturing.

10.4. FIELD STRUCTURES

10.4.1. Foliation and lineations

The relations established in the preceding section between the flow direction and that of the penetrative structural elements (foliation, lineation), make it possible, by measuring their orientation in the field, to extend the kinematic analysis to the scale of a massif, as illustrated in the cases of Lanzo and Ronda (Nicolas et al., 1972; Darot, 1973a and b; see also next chapter).

The foliation plane is defined in the field by alignment and/or flattening of spinel grains. It is enhanced in feldspathic lherzolites by the feldspar rims around the spinel. In strongly deformed facies, like the one described in § 10.3.3, the enstatite elongation also contributes to defining this plane. In this case, parallel shear zones form a minute banding at a small angle to the foliation which is no longer visible in the field (Plate 17).

Several types of lineations are related to the flowage. Darot and Boudier (1975) have comprehensively described them and stated their respective conditions of occurrence. The three main types in spinel peridotites are the tabular enstatite lineation, the spinel or spinel feldspar aggregate lineation and the lamellar enstatite lineation. The tabular enstatite lineation is a variety of pull apart lineations as were first described in chromitite pods within dunites by Thayer (1943). It has been examined in the general section on lineations (§ 7.3.2). The axis of the roughly disc-shaped enstatite grains is, within measurement errors in the field, parallel to the spinel lineation and coincides with the stretching direction in the foliation (X-axis of finite strain ellipsoid). It is observed principally in peridotites moderately deformed at high temperature. The spinel lineation and the corresponding spinel feldspar aggregate lineation in feldspathic lherzolites are due to stretching of these minerals in the foliation and are parallel to the X-axis of the strain ellipsoid. They are ubiquitous but sometimes difficult to observe in the field and need to be brought out by special laboratory treatments (§ 9.1.1). Finally, the lamellar enstatite lineation results from extensive slip in this mineral, as described in the mylonite of § 10.3.3. It is only found in such highly deformed peridotites and its trend indicates the X-direction of the strain ellipsoid.

The precise kinematic meaning of these structural elements depends on the nature of the flow. If the laboratory investigations have shown that the flow was irrotational they can be directly equated with the main flow directions. If it has been concluded that the flow was dominantly rotational from the observation of an angle between the shape and lattice fabric elements, the flow plane and flow line are at some angle from the foliation and the stretching lineation. In previous studies, we have assimilated the flow elements to the lattice fabric elements, that is the flow line to the point maximum of the orientations of the crystallographic axis coinciding with the operative slip direction and the flow plane to the preferred orientation of the crystallographic planes regarded as slip planes. Etchecopar's shear flow simulation (§ 6.9) casts some doubt on this assimilation. If the model is correct the flow elements are found at a still higher angle to the orientation of the penetrative structural elements. There is at present no way to deduce precisely this angle from the available information (Fig. 10.13); it seems to be lower than 10° for shear angles in the range of 70°. For these reasons, the assimilation made above is still accepted, making it possible to deduce the kinematic elements from the foliation and lineation

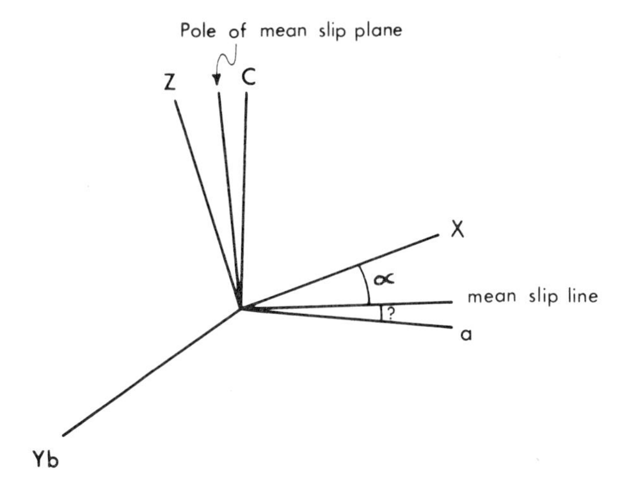

Fig. 10.13. Expected relative orientation of the finite strain principal axes (X, Y, Z), maxima of the lattice-preferred orientation equated with the mean slip orientation and the kinematic axes (a, b, c) in the case of a rotational shear flow

orientations. The amplitude and the sense of the smaller rotation leading from the latter to the former represent the necessary information. As seen in the previous section and in § 9.1.2, they are deduced from microscopic studies on oriented samples. The sense of this rotation corresponds to the sense of shear. It is dextral if the smaller rotation leading from the penetrative elements to the kinematic ones is clockwise and sinistral if it is counterclockwise (Fig. 10.14).

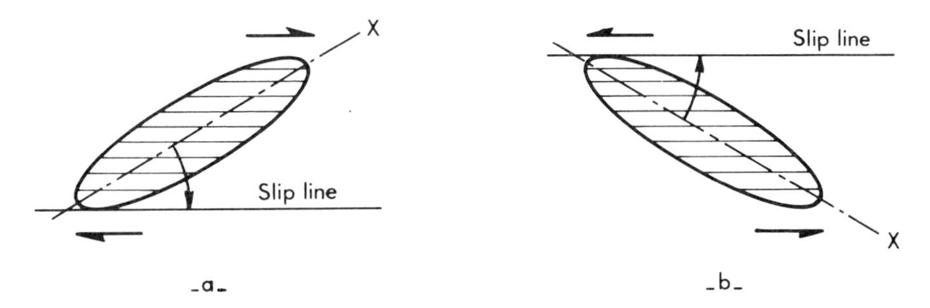

Fig. 10.14. Relationship between the dextral (a) or sinistral (b) sense of shear and the clockwise (a) or counterclockwise (b) rotation leading from the stretch direction (X) to that of the slip line

Relations in the field between the penetrative and the non-penetrative structural elements may also provide the means of directly determining the shear sense (§ 7.3.3). In the Lanzo massif, this was achieved considering the orientation, with respect to the penetrative structure, of feldspathic gashes formed during syntectonic partial melting (Fig. 7.8) and the small-scale offsetting of layering by shear surfaces associated with the foliation (Fig. 7.9).

10.4.2. Folds

Folding of pyroxenitic layers is fairly common in peridotite massifs. This matter has been reviewed and more specifically investigated, in the Lanzo massif, for its kinematic bearing by Nicolas and Boudier (1975).

The folding is observed on any scale in the Northern body of this massif. Figure 11.2 is an interpretative map constructed on 600 measurements of foliation and pyroxenitic layering. Letting aside the peripheral folding which deforms both the foliation and the layering, and therefore post-dates the development of the foliation, attention is called to the large-scale fold present in the core of the massif. In the hinge area, a great number of folds are visible on the scale of the outcrop. They are open and display a parallel style in the precise hinge zone; they are tight and isoclinal away from it (Plate 26). Refolded folds are common (Fig. 10.15). Petrofabric studies on the rotation of the mineral lattices in hinges show that the folding mechanism is flexural flow and not shearing.

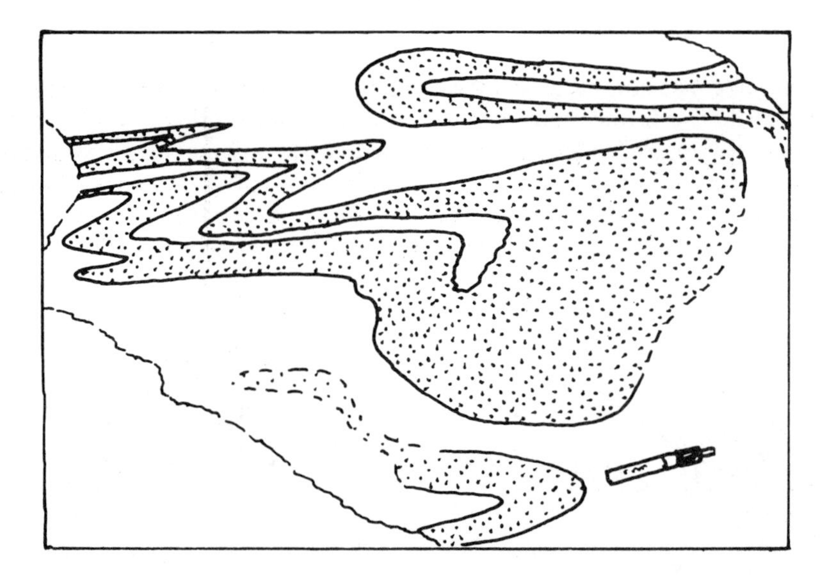

Fig. 10.15. Refolded fold in the hinge area of the large-scale fold, northern body of Lanzo massif (after Nicolas and Boudier, 1975)

The foliation usually coincides with the axial plane of these folds. A permanent character is that the spinel feldspar aggregate lineation always coincides with the fold axis. This is verified by measurements on individual folds as well as by statistical constructions like the one presented for the boxed area in Fig. 11.2. The density diagrams of Fig. 10.16 demonstrate that the

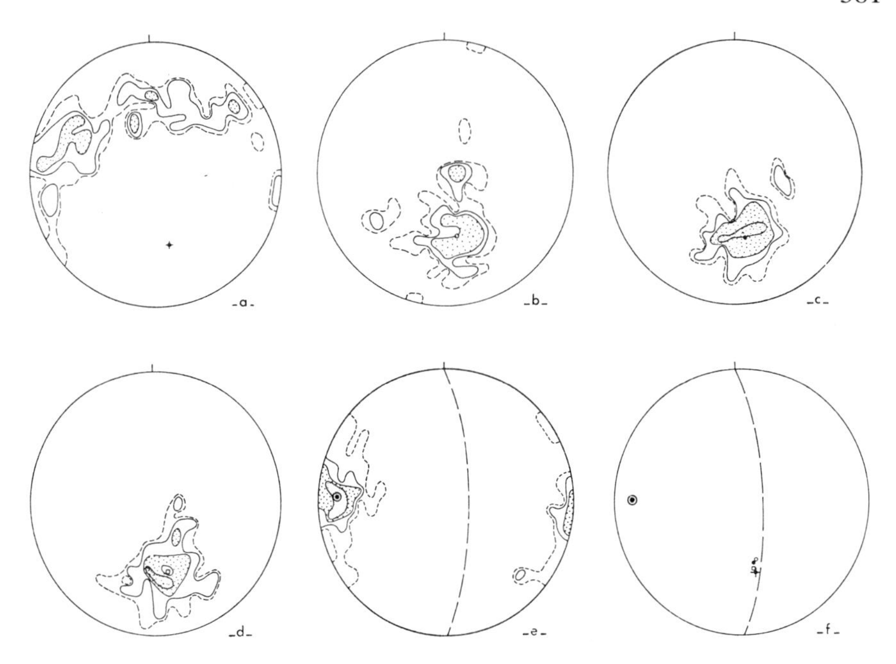

Fig. 10.16. Lower hemisphere projections of the field measurements in the large-scale fold of Lanzo massif (boxed area on Fig. 11.2). Equal area projections; contours: 1, 2, 4, 8 % per 0·45 % area; each sign indicates the computed best axis in diagrams (b), (c), (d), (e) and the pole of the best plane in (a). These best axes and pole of best plane are plotted together on diagram (f).
(a) 98 poles of layering; (b) 81 intersections layering-foliation; (c) 81 mineral aggregates lineations; (d) 74 directions calculated as perpendicular to the pull-apart enstatite lineations in the foliation plane; (e) 86 poles of foliation and trace of mean foliation plane (after Nicolas and Boudier, 1975)

large-fold constructed axis is nearly contained in the mean local foliation plane and parallel to the mean local orientation of: (1) the spinel feldspar aggregate lineation, (2) the normal in the foliation plane to the pull-apart enstatite lineation and (3) the layering foliation intersection.

These remarkable relationships together with considerations on mineral equilibrium in hinges of folds indicate that the folding was produced during plastic flow. They also make it possible to carry further the kinematic analysis. In fold analysis it is usually accepted, without much justification, that the fold axes are normal to the direction of the flowage responsible for their development. In the case of deformed peridotite it is shown that the flow direction, which forms a small angle with the mineral lineation, is close to the direction of the fold axis. From this it has been concluded that in these rocks the flowage directly responsible for the folding is a minor flow component compared with that recorded in the mineral preferred orientation. A descriptive model is to consider the fold as developing in the pyroxenitic layer by flexuring and flattening in response to a compression exerted parallel or close to the layer. The olivine-rich matrix around the folding layer is meanwhile flowing with a direction close to the fold axis and a flow plane close to the axial plane. The finite strain associated with this flowage is far larger than the one involved in the fold formation and is consequently the one recorded in the final fabric*. Another equivalent analysis is proposed in Fig. 10.17. The three-dimensional

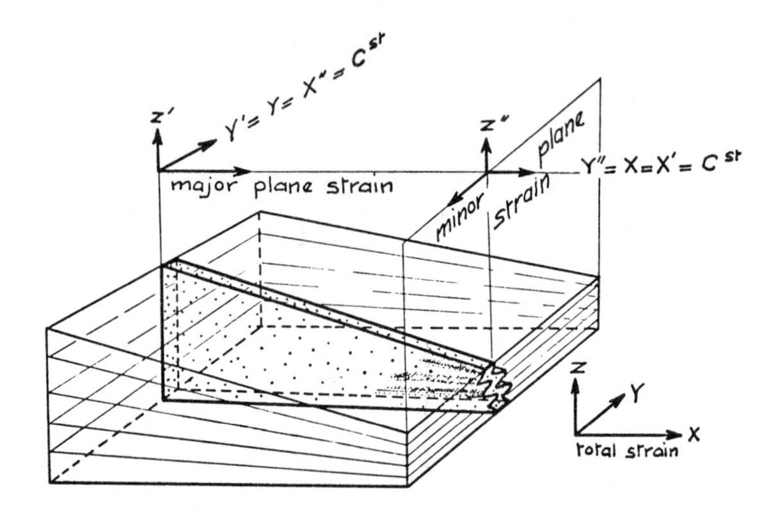

Fig. 10.17. Theoretical model for development of folds in a layer (stippled on the figure) in response to a minor plane strain (X'', Z''). The flow structures (foliation and lineation) in the matrix are mainly due to a major plane strain (X', Z'). The total strain (X, Y, Z) combines the two (after Nicolas and Boudier, 1975)

finite strain (X, Y, Z) characterizing the folded structure can for the sake of description be divided into two elementary planar strains. The major one (X', Y', Z') is rotational and it is responsible for the imprinting in the rock of the penetrative elements (foliation, lineations). The minor one (X'', Y'', Z'') is non-rotational and it is responsible for the folding. This minor component of flow is certainly more easily derived from the study of folds than from that of the texture and preferred orientations.

* However, a more sophisticated analysis of the mineral preferred orientation and of the associated substructure features in these folded areas has succeeded in bringing into evidence the minor flow component linked to the fold generation (see § 10.3.2.).

CHAPTER 11

Large-scale Flow in Peridotites, Upper Mantle Geodynamics

11.1. INTRODUCTION

In this chapter we wish to describe geometrical models of large-scale flow incorporating the results of the analysis of peridotites presented in the preceding chapter. We will also call attention to another important aspect of how a physical metallurgy approach can contribute to understanding geophysical and geological problems, i.e. how the creep laws established for olivine crystals and aggregates may be used together with other geophysical data to construct rheological models for the upper mantle.

11.2. GEOMETRICAL MODELS OF FLOW IN PERIDOTITES

11.2.1. Flow studies in the different peridotite groups

The principal occurrences of peridotites displaying a tectonite fabric were reviewed in the preceding chapter. Two groups were described: the peridotites from massifs and those from basalt and kimberlite xenoliths. The latter yield direct information on mantle flowage; in this regard, both sets are complementary: the xenoliths are small fragments of the flowing mantle whereas the massifs exhibit large-scale exposure of flow structures possibly developed in the mantle and subsequently altered in crustal conditions. A common origin in the mantle is inferred when observations in weakly deformed specimens from massifs (§ 10.3.1) show remnants of the protogranular texture, typical of plastically undeformed xenoliths from basalts (§ 10.3.4).

In the peridotite massifs we distinguished (§10.2.1) between the harzburgite massifs enclosed in the ophiolite complexes and attributed to the uppermost mantle underneath the oceans and the lherzolite massifs, attributed to the upper mantle underneath the continental crust. Complex local situations can modify this general sketch (Nicolas and Jackson, 1972). The cause of large-scale flowage in the two environments are distinctly different, even though their structures, textures and preferred orientations are comparable.

11.2.2. Flow related to the intrusion in lherzolite bodies

In the lherzolite bodies inserted within high-grade metamorphic terrains, the flow can only be studied where metamorphic structures have not obliterated those presumably due to plastic flow at high temperature (see § 10.2.1). Examining the flow structures in bodies which have not been re-equilibrated in crustal conditions produces information on the kinematics of their intrusion through the upper mantle and the deep continental crust. Typical in this regard are the following massifs in the Alpine ranges of the western Mediterranean: Beni Bouchera (Morocco), Serrania de Ronda (Andalusia), those in the Pyrenées and some in the Ivrea zone (Fig. 10.1).

The geometry of the large-scale flowage is analysed here principally in the case of Lanzo massif (Fig. 10.2). It is one of the largest and best exposed of the

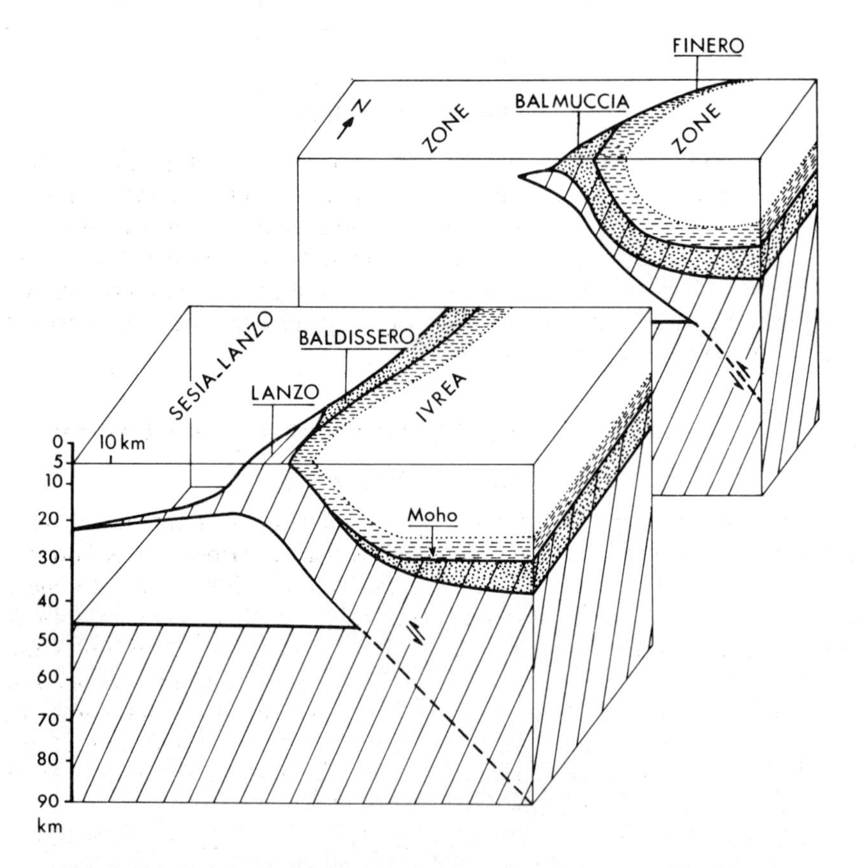

Fig. 11.1. Sketch of the assumed relationship between the peridotite massifs and the Ivrea zone in the western Alps; data concerning the deep structure from Berckhemer's (1968) geophysical model. Hatches: mantle; hatches and dots: Baldissero and Balmuccia massifs and their inferred zones of origin in the mantle; dashes: granulites from the Ivrea zone (after Nicolas et al., 1972, reproduced from *Tectonophysics* by permission of Elsevier Scientific Publishing Company, Amsterdam)

known lherzolite massifs; it has also received the most attention for its internal structure and structural setting though it is entirely surrounded by a serpentinite rim (Nicolas et al., 1971, 1972; Boudier, 1972; Boudier and Nicolas, 1972). For this reason the geophysical results of the detailed campaigns over the western Alps in the 60s are of paramount importance. The model combining seismological and gravimetrical evidence (Giese, 1968; Berckhemer, 1968, 1969) strongly suggests that the peridotite massif is rooted deep in the mantle (Fig. 11.1). On this basis, we thus have a unique opportunity of relating the internal large scale flow in the massif to the external orogenetic displacements best analysed in terms of relative plate movements (references and review in Frey et al., 1974 and Nicolas, 1974).

Internal structures of Lanzo massif

Lanzo massif is divided into two bodies by a strike-slip fault (Fig. 11.2). The southern body has a larger extension than is exposed as shown by the occurrence of similar facies 6 km south of the massif on the other side of the Susa Valley. The deformation structures are less conspicuous in the southern body than in the northern one. The forthcoming analysis deals principally with the northern one.

In § 10.3.2 and § 10.4.2, it was shown that the flow in the northern body was rotational although a minor irrotational component was deduced from considering the folds. The sense of shearing is dominantly sinistral (Fig. 11.3), a conclusion which is derived from independent and concordant methods: obliquity between lattice and shape fabric elements, feldspathic tension gashes and sense of layer rotation during discontinuous shear (§ 7.3.3).

The principal flow directions were equated in § 10.4.1 with those of the lattice preferred orientations in the two dominant minerals, olivine and enstatite. The angle between the principal directions of the penetrative elements and those of the mineral preferred orientations is 16° on the average throughout the massif. Since in Figs. 11.2 and 11.4 all the data relate to the field-penetrative structures, to deduce the flow directions from them a counter clockwise rotation of 16° must be operated around a nearly vertical axis. This crude analysis does not account for the fact that the flow has an irrotational component which lowers the rotation angle by unknown quantity; moreover, the nearly vertical axis of rotation is an approximation based on the moderate plunges of the lineation in the massif (Fig. 11.4). In the southern body, where no definite sense of shear can be ascertained, it is assumed that the flow was irrotational and therefore that its principal directions coincide with those of the foliation and lineation.

Fig. 11.3 is a computed map of the grain size in the massif established from 110 measurements carried out by the technique presented in § 9.1.2. It is meant to represent the plastic deformation inside the massif. The inverse relation between grain size and strain on which this point is based, although it is not a universal rule, is observed in the continuous suites of experimentally (§ 6.2.1) and naturally (§ 10.3.5) deformed peridotites; in Lanzo it is locally confirmed by field and laboratory evidence.

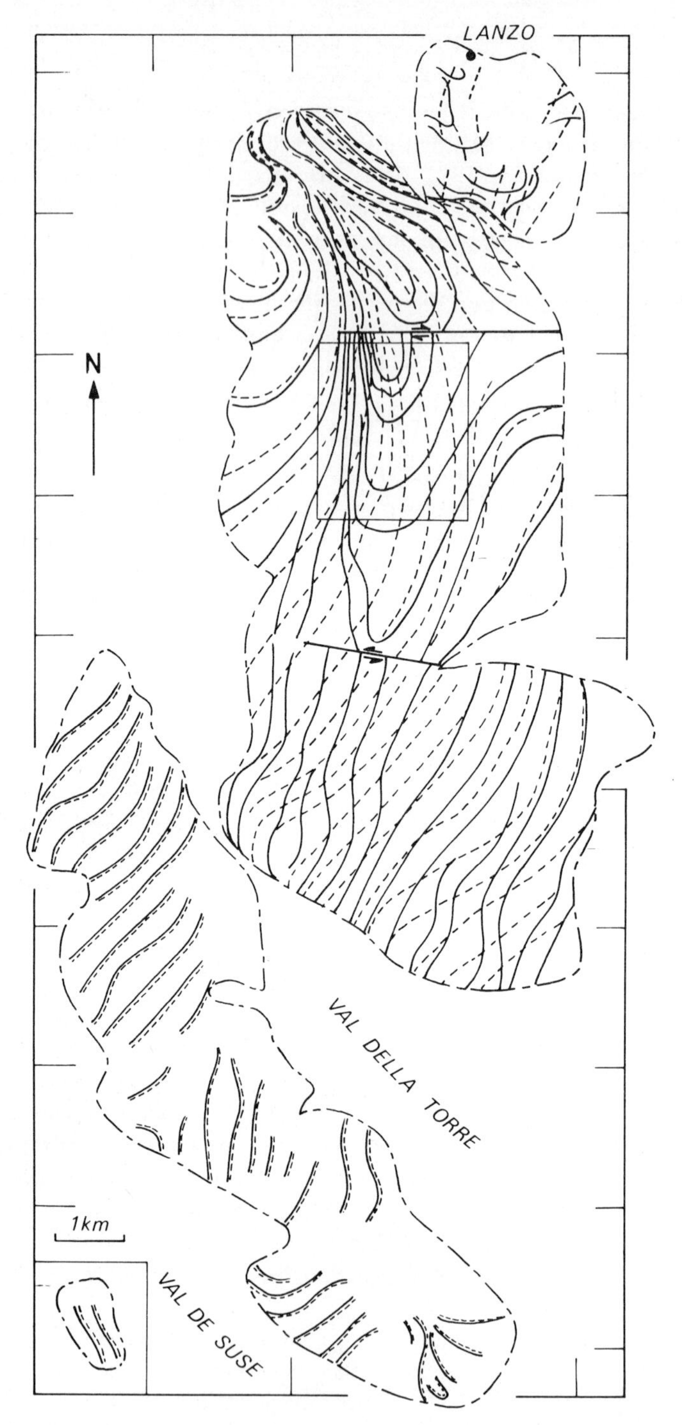

Fig. 11.2. Trends of layering (full lines) and foliation (dashed lines) in the Lanzo peridotite massif. The orientations in the large-scale fold of the boxed area are presented on Fig. 10.16. The two strike-slip faults shown in the northern body are deduced from the discontinuous change in the layering and foliation relative directions (after Nicolas and Boudier, 1975)

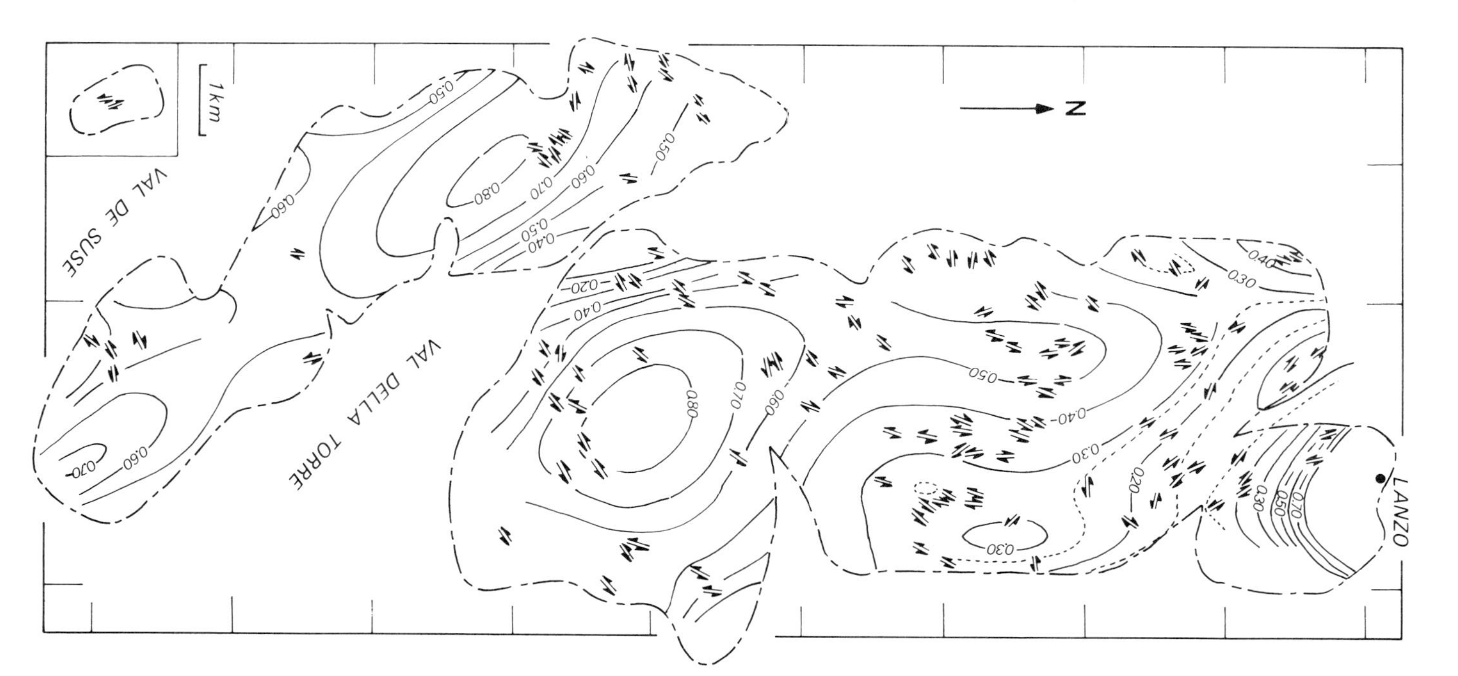

Fig. 11.3. Senses of shear during plastic flow and grain size distribution in the Lanzo massif

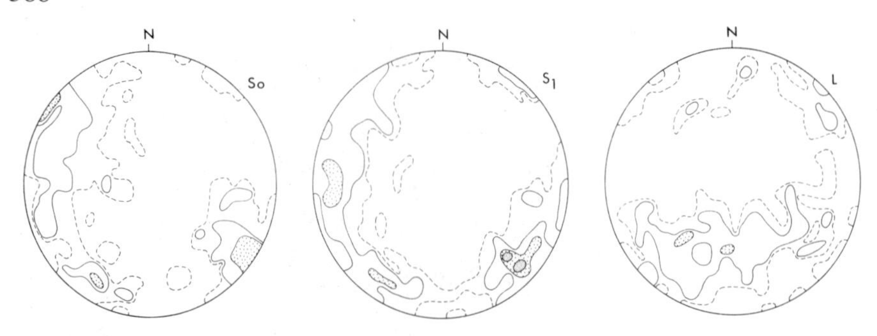

Fig. 11.4. Equal area projections (lower hemisphere) of layering (S_0, 966 measurements) foliation (S_1, 851 measurements) and flow lineations (L, 533 measurements) over the Lanzo massif. Contours: 0·5, 1, 2, 3 % per 0·45 % area

Mylonitic bands

The foliation and lineation trends are remarkably continuous and regular in the large central area of the northern body. At its north and south edges, they progressively rotate and the strain simultaneously increases, producing the mylonites described in § 10.3.4 with a shear angle in the range of 85° or more. Their dominant shear sense is confirmed by the observation on the map of the counterclockwise rotation of the foliation in the shear zones. These two shear zones separate the central area of the northern body from the southern body and from the area located around the town of Lanzo. In both cases, this separation corresponds to noticeable structural and petrological differences. This suggests that the separation was initiated inside the mantle and therefore that the shear zones were active early in the intrusion of the whole mass. On the other hand, the mylonites parageneses grade into serpentinite and blueschist assemblages, indicating that the shearing lasted longer than the plastic flow in the rest of the massif and eventually continued during the alpine events recorded in the surrounding blueschists and *schistes lustrés* (Nicolas, 1974).

Geodynamic interpretation

The direction and sense of flow in Lanzo massif can be correlated with the orogenetic displacements acknowledged in this part of the western Alps provided two conditions are fulfilled: (1) the massif must be rooted in the mantle, (2) the flow structures have to be developed during its intrusion. The geophysical evidence for a direct deep burying in the mantle is accepted here; it is indeed the only piece of evidence since the massif is surrounded by a continuous rim of serpentinites inside which it could have been translated and rotated.

Due to the high temperatures the flow structures indicate, they cannot have been imposed during the low temperature alpine events recorded in the surrounding formation and in the serpentinite rim. The flowage could have operated at great depth and the massif broken off then rigidly intruded into its present setting. The distribution of the strain deduced from that of the grain size (Fig. 11.3) shows that the cores of the two bodies of the massif are less deformed than their margins. Such strain gradients are more in keeping with the interpretation of the flowage occurring during the intrusion. They would result from the velocity gradients existing between the core and the margins of the flowing mass; at the margins the flowage would continue until it was superseded by flow in the serpentinite rim. This interpretation is also supported by evidence from partial melting in the lherzolites, producing thin feldspathic lenses and gabbro dikes (Boudier, 1972; Boudier and Nicolas, 1972). Both correspond to a melting taking place around 7 kB, that is in a crustal environment, while the peridotite mass was flowing. The latter point is demonstrated by the geometrical relations between the feldspathic lenses and the flow structures (§ 7.3.3) and by detailed observations made in the dikes linking their location or nature to whether or not they were deformed. The emerging picture is that of a peridotite mass intruding at high temperature, which in turn triggers off melting once the lherzolite solidus is crossed due to a lowering in pressure (Fig. 11.5). This sketch is speculative, relying on unwarranted assumptions

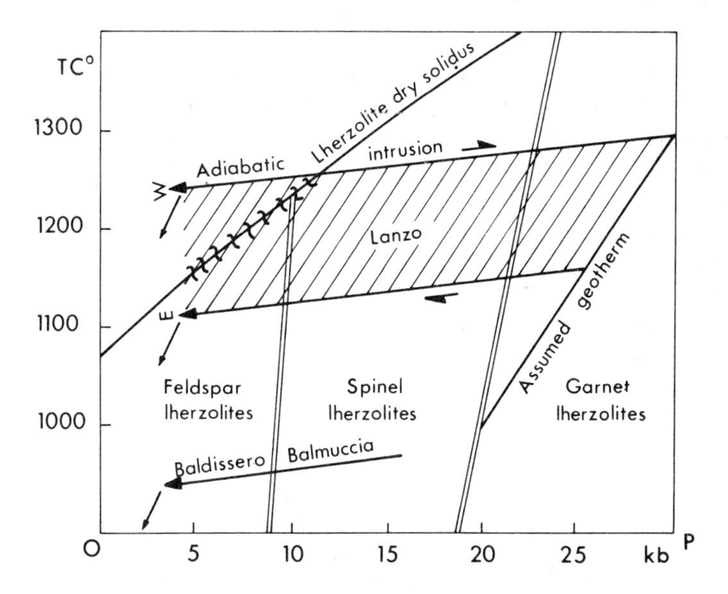

Fig. 11.5. *P*, *T* diagram illustrating the approximate conditions of emplacement of the peridotite massifs shown on Fig. 11.1. Lanzo is considered as a sliver about 15 km thick (hatched area) initially in equilibrium in the mantle on a geotherm and adiabatically intruded by a dominantly rotational shear flow. Crossing of the dry solidus in a crustal environment produces partial melting (*en echelon* decoration) over the western part of the massif. Intrusion temperature in Baldissero and Balmuccia is only inferred. Shallow intrusions are attributed to flow in serpentinized matrix (after Nicolas et al., 1972, reproduced from *Tectonophysics* by permission of Elsevier Scientific Publishing Company, Amsterdam)

regarding the geotherm and the adiabatic character of the intrusion. Geographically, only the southern body and the western margin of the northern one are pervaded by lenses and dikes formed *in situ*. This indicates that the eastern part of the northern body represented the top of the mass, initially intruding in the mantle at a temperature below the one required for crossing the solidus during the ascent. Among the numerous assumptions on which the sketch on Fig. 11.5 is based, is the absence of a geotherm gradient inversion at the onset of the intrusion. Such an inversion has been postulated in a subduction zone (Minear and Toksöz, 1970; Griggs, 1972; Toksöz et al., 1971) but it is located at the top of the downgoing plate while we regard Lanzo as belonging to the mantle in the overlying plate.

This geometrical outlay fits the model of Fig. 11.1, combining both the geophysical and geological information. It shows the Southalpine plate overriding the European plate. Lanzo and the other lherzolite bodies with tectonite fabric (chiefly, Baldissero and Balmuccia) are located in the plates boundary where they are believed to have intruded. Lanzo corresponds to the deepest intruded mantle from the Southalpine plate, initially surmounted by the Baldissero–Balmuccia lherzolites which could have been located close to the Mohorovičić discontinuity beneath the ultramafic–mafic continental crust of the Ivrea zone. Due to their initial location, the Baldissero–Balmuccia lherzolites were intruded in cold-working conditions (§ 10.3.1) contrasting with the hot-working conditions prevailing in Lanzo. The Lanzo lherzolites, which are equilibrated in the feldspar lherzolite facies stable at lower pressure, originate deeper down than the Baldissero ones, equilibrated in the spinel lherzolite facies stable at higher pressure (O'Hara, 1967). This apparent paradox is explained if temperatures close to the dry solidus are assumed to have promoted a continuous re-equilibration during the ascent in Lanzo, whereas lower temperatures in Baldissero precluded any subsolidus re-equilibration.

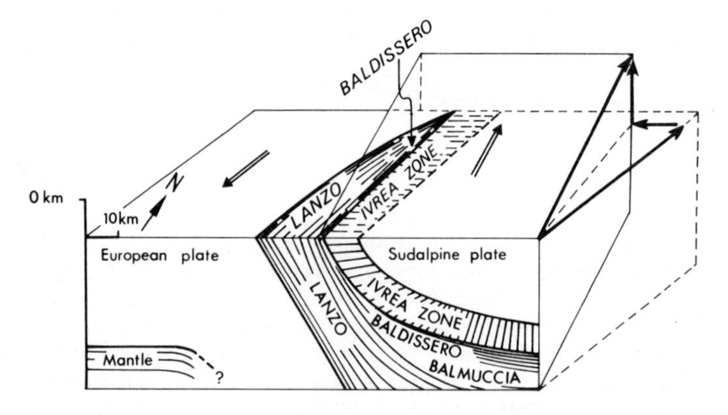

Fig. 11.6. Block diagram illustrating the relative displacement of the South alpine and European plates in the vicinity of the Ivrea zone as inferred from the flow structure in Lanzo massif (after Nicolas, 1974)

The causes of the intrusion are sought, during the alpine orogenesis, in a relative motion of the European and Southalpine plates, both topped by continental crusts, after the complete subduction of the Piedmont basin. The generally accepted sketch in this part of the western Alps shows a relative northward motion of the Southalpine plate along the north–south trending plates boundary, probably with minor westward and upward components (Fig. 11.6). The westward component accounts for east–west compressive structures and thrusting and the upward component for the major north–south thrusting of the Southalpine plate over the European one; this is observed, north–east of the area under consideration, in the Swiss and Austrian Alps.

The orientation of the foliation and lineation in Lanzo massif (Figs. 11.2 and 11.4), indicative of the main flow orientation, records the major relative displacement of the plates, parallel to their boundary. This is roughly north–south in direction with a moderate vertical component, along a steeply dipping plane and with a sinistral sense of shear. The flattening responsible for the folds formation (§ 10.4.2) records the east–west convergence of the two plates (Fig. 11.7). One inference from this highly speculative picture is that the 25° mean

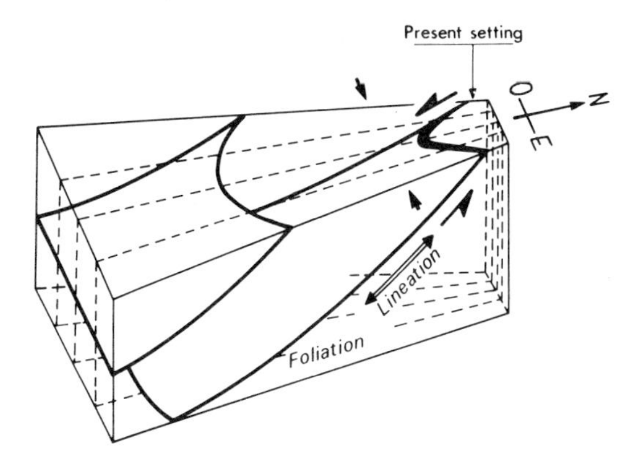

Fig. 11.7. Block diagram illustrating the overall deformation in Lanzo massif as a result of a composite flow during the intrusion of a mantle slab. One component is simple shear, the other pure shear. The latter is directly responsible for the folding of the layering (heavy lines). The foliation is represented by the dashed lines (after Nicolas and Boudier, 1975)

plunge of flow directions in the massif may lead to overestimating the vertical component of the plates' relative motion. This motion may be essentially horizontal and the plunge of flow directions in the massif progressively accentuated during the intrusion as a result of the flattening component (Nicolas and Boudier, 1975).

11.2.3. Flow studies in harzburgite massifs within ophiolite complexes

It is now widely accepted that ophiolites are fragments of the oceanic crust formed at a ridge and of the underlying uppermantle, tectonically emplaced into their present setting, during low temperature conditions of deformation. The harzburgite massifs within ophiolite complexes are then equated with the uppermost oceanic mantle. In these massifs, once the deformations assigned to tectonic emplacements by thrusting have been differentiated, studying the high temperature plastic flow structures ameliorates the structural and kinematic model of an expanding oceanic ridge. It also contributes to understanding the seismic anisotropy measured in the oceans (see next section).

In oceanic harzburgite,* plastic flow structures, comparable with those observed in ophiolitic harzburgites have been recorded although the dredged samples are usually serpentinized. The unaltered specimens dredged from the Tonga Trench display a remarkably similar structure to the freshest specimens from ophiolites. This structure is clearly due to plastic flow at high temperature contrary to Engel and Fisher's opinion (1969).

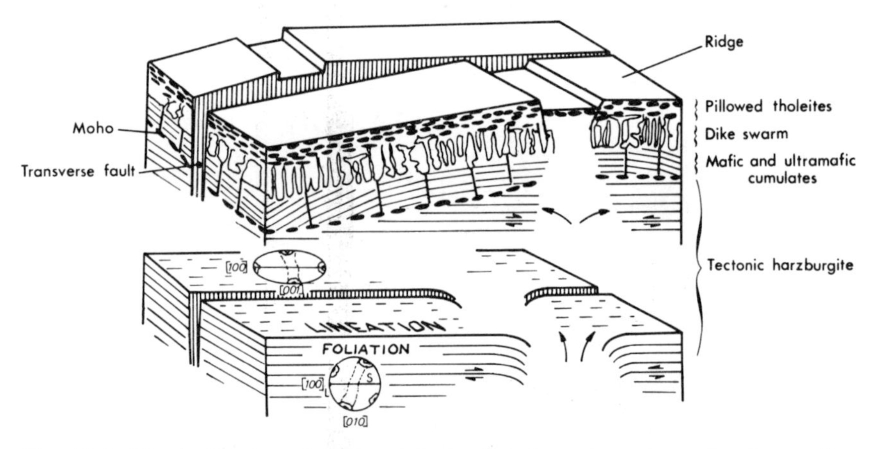

Fig. 11.8. Block diagram of the oceanic crust and uppermost mantle at an active ridge, illustrating the theoretical relationship between structural features in the crust and flow structures in the uppermost mantle

Structural models can be established in the ophiolite sections which were not dismembered during the actual emplacement with a view to comparing them with the structure at a ridge (Fig. 11.8). Valuable pieces of information in such sections are:

Depositional plane in the overlying sediments and in the magmatic cumulates (gabbros, pyroxenites, peridotites). They indicate more or less accurately the horizontal plane at the time of their deposition. The planar structures in the pillowed and flowed lavae can also be considered.

* References to oceanic harzburgites can be found in Aumento and Loubat's paper (1971) and to ophiolitic harzburgites in Juteau et al. (1973).

Orientations of dikes in the dike swarms which may be in zone with the direction of the presumed ridge.

Magmatic flow directions in lavas and in cumulates; statistically they may be normal to the ridge direction.

The large-scale structures of high temperature solid-state flow in the associated harzburgites do not allow further comparison. By considering the penetrative elements, i.e. foliation and lineation, the sense of shear can be oriented relative to the other elements of the model. Thus it has been observed that the flow plane was often at a low angle to the magmatic depositional plane in gabbro cumulates which are locally deformed at these high temperatures. Hence, below a ridge, the harzburgites, after their ascent, flow horizontally immediately beneath the crust, until they are frozen due to the lateral lowering in the isotherms away from the ridge. Mylonitic zones at a high angle to the foliation plane in the flow direction are to be sought in harzburgite massifs from ophiolite complexes and would represent transverse faults. Seismic evidence shows that they cut into the uppermost mantle and, owing to a large shear, harzburgites mylonites are bound to be expected.

11.2.4. Uppermost mantle seismic anisotropy

The existence of a seismic velocity anisotropy for compressional waves in the uppermost mantle beneath the Pacific ocean was discovered by Raitt (1963) and Shor and Pollard (1964). Recently, Bamford (1973) reported evidence of a comparable anisotropy in the mantle beneath the continental crust of Western Germany. The highest seismic velocities in the oceans have been measured parallel to transverse faults or normal to ridge trend; the difference with the velocities normal to transverse faults, or parallel to ridge trend, varies from 0·2 to 0·7 km s^{-1} (Raitt et al., 1969; Morris et al., 1969; Keen and Barrett, 1971).

Olivine possesses a remarkable velocity anisotropy with a maximum value of 9·87 km s^{-1} along the [100] direction and a minimum value of 7·73 km s^{-1} in the [010] direction (Verma, 1960). In a peridotite where this mineral ranges from 60% to nearly 100%, a lattice-preferred orientation can readily explain the velocity anisotropy observed in the oceans, as shown by experiments and computations. The velocity of ultrasonic waves has been measured at room temperature and pressures up to 10 kB on peridotite specimens cut in several orientations by Birch (1960, 1961), Christensen (1966, 1971), Kasahara et al. (1968). Christensen and Ramananantoandro (1971) took their measurements in the directions parallel to the olivine lattice fabric maxima. Peselnick et al. (1974) took theirs in the directions of the foliation-lineation reference system. From the elastic stiffness coefficients of olivine and other silicates present in peridotites it is possible to compute the velocity surfaces for both compressional and shear waves in a peridotite of any composition, provided the minerals mode and preferred orientations are known (Crosson and Lin, 1971; Baker and Carter, 1972; Peselnick et al., 1974). These calculations show the

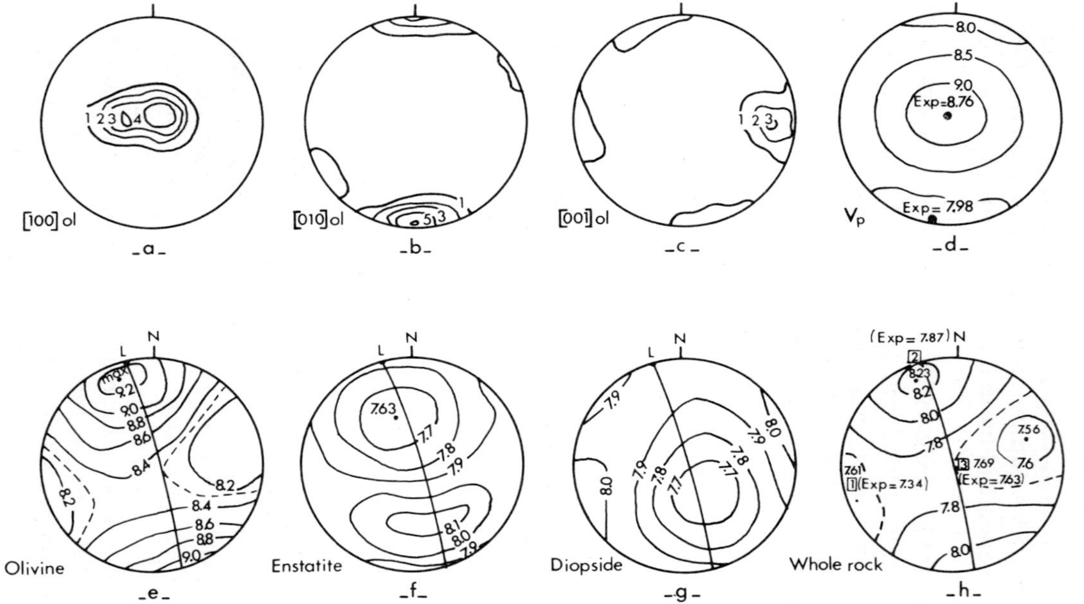

Fig. 11.9. (a), (b), (c) Pole figures of olivine preferred orientations in a specimen from the Twin Sisters dunite; (d) calculated and measured (Exp) P seismic velocities (in km s^{-1}) in the same specimen. Lower hemisphere, equal area projections; 100 grains in (a), (b), (c) with contours in multiples of the expected number for a random distribution. (After Baker and Carter, 1972; experimental data from Christensen and Ramananantoandro (1971) at 10 kB, 25 °C)
(e), (f), (g), (h) Calculated and measured P velocities in a Lanzo lherzolite specimen in relation with the structural elements (foliation: full line; aggregate lineation L: dot). (e) Velocities in hypothetical olivine aggregate having the olivine preferred orientation of the specimen; (f) the same in enstatite; (g) the same in diopside; (h) whole rock calculated and measured (Exp) velocities. The velocities are calculated from the results of (e), (f), (g) and the modes of the corresponding minerals. The cores for the measurements (2 kB, 25 °C) were oriented with respect to the foliation and lineation (square symbols). (Reproduced from Peselnick et al., *Jour. Geophys. Res.*, **79**, 1175 (1974), copyright by American Geophysical Union)

major contribution of olivine to velocity anisotropy. This is due to its predominance in the peridotite and to its intrinsic anisotropy and preferred orientation which are both stronger than those of pyroxenes. The computed velocities are higher than the measured ones, because in the models no account is taken of the serpentine usually present in peridotites. Both measured and computed anisotropies are found to be higher in a dunite than in a lherzolite (Fig. 11.9). Typical values illustrated by that figure would be a maximum measured anisotropy at 10 kB of $\Delta V_P = 1{\cdot}32$ km s^{-1} in the Twin Sisters dunite (Christensen and Ramananantoandro, 1971) and a computed one of $\Delta V_P = 1{\cdot}32$ km s^{-1} (Baker and Carter, 1972), a measured anisotropy at 2 kB of $\Delta V_P = 0{\cdot}60$ km s^{-1} in the Lanzo Iherzolite and a computed one of $\Delta V_P = 0{\cdot}62$ km s^{-1} (Peselnick et al., 1974).

Since Hess' publication (1964), many interpretations have been proposed to account for the seismic anisotropy measured in the oceanic mantle. They all suggest that solid-state flow produced the necessary preferred orientations in the mantle peridotites. This is not surprising, since sea-floor spreading at ridges entails flow in the upper mantle and since flow is an efficient process for developing mineral preferred orientations. We shall not report here the majority of the interpretations (Sugimara and Uyeda, 1967; Avé Lallemant and Carter, 1970; Lappin, 1971; Carter et al., 1972) which rely on diffusion creep as the flow mechanism and on Kamb's theory for the development of preferred orientations, because this theory has been shown to be inadequate (§ 4.4.3). Other serious drawbacks are discussed in Nicolas et al. (1973) and Peselnick et al. (1974). Hess attributed the seismic anisotropy to plastic flow in the transverse faults. Considering (010) [100] as the most probable slip system in olivine, he admitted that the shear would orient the [100] axes along the flow direction, that is along the trend of the faults and the (010) planes in the flow plane, close to the fault plane, that is vertically. This preferred orientation would produce the maximum of seismic anisotropy in the horizontal plane, where actually it is recorded (Fig. 11.8). Francis (1969), aware of Raleigh's recent discovery (1968) of the {0kl}[100] slip system in olivine, considers that plastic flow would develop essentially a [100] axes maximum in the flow direction, a conclusion well supported by the evidence in the previous chapter. His major line of disagreement with Hess consists in his attributing the anisotropy to the mantle flowage at an expanding ridge, during its actual ascent and turnover. Francis discarded Hess' interpretation because shear along transverse faults seemed insufficient to produce preferred orientations in peridotites over wide areas of the uppermost mantle; his other argument that in the shear zone the temperature would be too low (200 °C) for plastic flow has been questioned by Avé Lallemant and Carter (1970) (see also below). Francis' model is supported by Nicolas et al. (1973) who bring observations from ophiolite sections into the picture. The observations, summarized on Fig. 11.8, indicate that the flowage in harzburgites, after the ascent, continued with a horizontal orientation before the temperature away from the ridge dropped below that compatible with plastic flow. In this regard, vertical shear flow can also operate along transverse faults near the ridge. In these interpretations the recorded seismic anisotropy is not the highest possible since the flow plane is horizontal or nearly so; it corresponds to the difference between the velocity along the [100] axes mean orientation and that along a [010] and [001] axes mixed orientation. It is nevertheless compatible with the recorded anisotropy (Fig. 11.9).

From the measurements of foliation and lineation over Lanzo massif consisting of lherzolites (Fig. 11.4), and the velocity measurements at 2 kB in a representative sample along those structural directions, the principal velocities of the triaxial ellipsoid assumed to represent the whole massif have been computed (Fig. 11.10) (Peselnick et al., 1974).

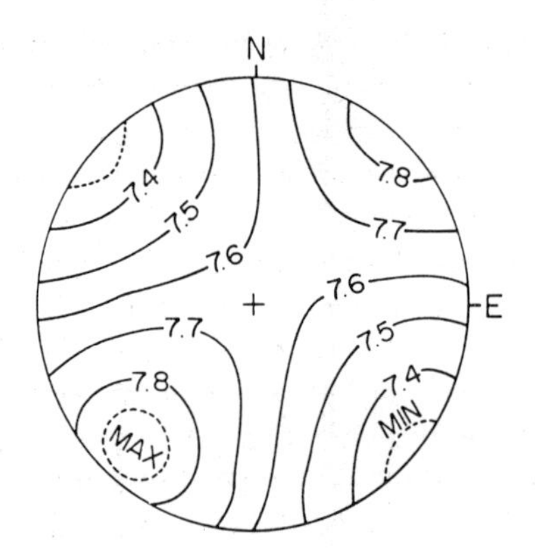

Fig. 11.10. *P* velocity contours (in km s^{-1}) for the Lanzo massif as calculated from the velocity measurements in the remarkable structural directions on the sample on Fig. 11.9 and correlation of these measurements with the orientation of the structural directions in the massif (about 500 field measurements). (Reproduced from Peselnick et al., *Jour. Geophys. Res.*, **79**, 1175 (1974), copyright by American Geophysical Union)

11.3. CREEP OF THE UPPER MANTLE

11.3.1. Plate tectonics and creep of the upper mantle

The essential concept on which plate tectonics is founded is the division of the Earth's outermost layer or lithosphere in about 10 rigid plates in relative motion with respect to one another (Le Pichon et al., 1973). It has recently been possible to define absolute velocities of the plates with respect to the lower mantle considered as quasistationary. The velocity of one lithosphere plate is of order of 1 to 10 cm/yr. This motion can obviously take place only if there is a viscous layer of the upper mantle, undergoing shear: the asthenosphere. From the evidence of peridotite xenoliths brought up by basalts and kimberlites it is now generally agreed that the upper mantle consists of peridotite whose main constituent is olivine (see § 10.2.2). Seismological data point to existence of two phase transitions from olivine to denser structures: one at 400 km to the spinel structure and one at 700 km to denser oxides (periclase and stishovite). The rate of shear (creep rate), in the asthenosphere $\dot{\varepsilon} = u/t$ obviously depends on the adopted thickness t, for a given absolute plate velocity u; estimates for $\dot{\varepsilon}$ range between 10^{-15} and 10^{-13} s^{-1} according to whether shear is thought to take place in the whole upper mantle or in the upper layer above the 400 km phase transition.

The viscosity of the flowing material of the asthenosphere can be defined by $\eta = \sigma/\dot{\varepsilon}$, where σ is the shear stress.

If η is constant, the flow is Newtonian, and it is non-Newtonian if η depends on the shear stress (see § 2.5). Values of the viscosity of the upper mantle, assuming Newtonian flow, have been derived from the analysis of post-glacial rebound. Current estimates range between 10^{19} and 10^{22} poises.

The question of knowing whether the flow in the asthenosphere is Newtonian or not is an important one and it bears strongly on the validity of the kinematic and dynamic analyses of plate motion. Furthermore, thermomechanical models of the asthenosphere can be made only if the dependence of the effective viscosity on temperature and pressure is reasonably known. Thus, the most needed information is the nature of the flow mechanism of the mantle material, along with the values of the physical parameters in the corresponding equation $\dot{\varepsilon} = f(\sigma, T, P)$.

The possible high temperature flow mechanisms and creep laws for crystalline materials were reviewed in Chapter 4 and the results of laboratory experiments on flow of peridotites and olivine were given in § 5.2. In recent years many authors have tried to assess the applicability of such creep laws and data to large-scale flow in the asthenosphere; the next paragraphs will be devoted to the exposition and discussion of their results.

At this point, it must be mentioned that the asthenosphere plasticity has sometimes been more or less explicitly ascribed to partial melting of the mantle material. Partial melting has been introduced to account for the existence of a high seismic attenuation, low seismic velocity zone (LVZ) below about 100 km. This problem will be discussed in § 11.3.4; we will merely note here that the identification of the asthenosphere with a partially melted zone does not appear to be necessary and the mechanisms considered in what follows will refer only to high temperature flow in the solid state.

The flow mechanisms which will be considered below are diffusion-controlled creep mechanisms, well documented in materials at high temperature; however, it must be borne in mind that there is a phase transformation boundary olivine-spinel at 400 km, and it is not impossible that shear flow could occur there by transformational 'superplasticity' (Gordon, 1971; Sammis and Dein, 1974) (see § 4.3.6).

11.3.2. Viscosity of the upper mantle: assessment of various flow mechanisms

This problem has generally been approached along the following lines:

(i) Several flow mechanisms are considered, giving various creep laws. Usually they are the following (for more details see Chapter 4):

(a) Newtonian creep mechanisms, by diffusional transport of matter through the grains (Nabarro–Herring) or along grain-boundaries (Coble).

The creep law has the form:

$$\dot{\varepsilon} \propto \frac{D\sigma\Omega}{kTd^2} \quad \text{for Nabarro–Herring creep}$$

and

$$\dot{\varepsilon} \propto \frac{D_{GB}\delta\sigma\Omega}{kTd^3} \quad \text{for Coble creep}$$

where

D is the bulk self-diffusion coefficient,
D_{GB} is the grain-boundary self-diffusion coefficient,
δ is the grain-boundary width,
d is the grain diameter.

Here, the viscosity $\eta = \sigma/\dot{\varepsilon}$ is independent of σ.

Weertman (1970) and Green (1970) have remarked that this flow should not be Newtonian for large strains since the grains suffer elongation and the diffusion path length increases with time.

(b) Non-Newtonian creep mechanisms by diffusion controlled dislocation glide or climb.

The creep laws have the form:

$$\dot{\varepsilon} \propto D\sigma^n$$

with $3 \leqslant n \leqslant 5$.

In this case the viscosity must be considered for a given strain rate $\dot{\varepsilon}$ or a given stress σ.

In both the diffusion controlled classes of models (a) and (b) the diffusion coefficient D is given by its theoretical expression:

$$D = D_0 \exp -\left(\frac{Q^* + PV^*}{kT}\right)$$

or a semi-empirical expression:

$$D = D_0 \exp -\left(\frac{gT_M(P)}{T}\right)$$

where

Q^* is the activation energy for diffusion,
V^* is the activation volume for diffusion,
g is an empirical constant,
$T_M(P)$ is the actual melting temperature under hydrostatic pressure P.

(ii) The physical parameters are chosen as follows:

The grain size d is often taken of the same order of magnitude as the one of peridotite xenoliths. The diffusion parameters Q^* and V^* (or g) are not directly known for olivine. They may be extrapolated from data compiled for ceramic materials whose characteristics are thought to be similar, or from experimental values of the activation energy for laboratory creep of olivine or peridotites.

The pressure dependence on depth z is roughly known from seismic data.

A hypothetical geotherm $T(z)$ is adopted.

(iii) Viscosity η is calculated as a function of depth z for various flow mechanisms at given stresses or strain rates compatible with the observations (seafloor spreading rate, post-glacial rebound, etc).

If the viscosity calculated for one mechanism is too high to allow flow in reasonable stress or strain rate conditions throughout the assumed thickness of the asthenosphere, it is thought that this particular mechanism does not operate. Potentially active mechanism are then compared. The dominant flow mechanism is the one with the lowest viscosity (or, equivalently, the one that yields the given strain rate for the lowest shear stress).

The hypotheses and conclusions of recent studies are gathered in Table 11.1 (Gordon, 1965; Raleigh and Kirby, 1970; Weertman, 1970; Carter, Baker and George, 1972; Kirby and Raleigh, 1973; Stocker and Ashby, 1973).

The importance of the choice of the geotherm can be seen on Figs. 11.11 and 11.12 (Kirby and Raleigh, 1973). Especially interesting is the study of Stocker

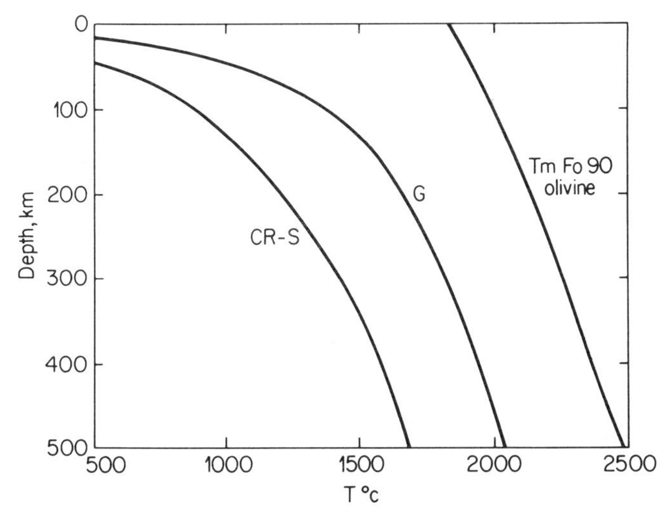

Fig. 11.11. Temperature of melting (T_m) for Fo_{90} olivine and selected geotherms:
G: Griggs (1972) oceanic geotherm
CR-S: Clark and Ringwood (1964) geotherm for shield areas (after Stocker and Ashby, 1973, reproduced from *Tectonophysics* by permission of Elsevier Scientific Publishing Company, Amsterdam)

Table 11.1.

Reference	Creep laws, parameters	Geotherm	(z), Dominant flow mechanism
Gordon, 1965	Periclase Nabarro–Herring $d = 0{\cdot}1$ cm $Q^* = 138$ kcal/mole $V^* = 0{\cdot}5\ \Omega$ (oxygen)	Mean geotherm (Gutenberg, 1951; Lubimova, 1958)	Nabarro–Herring creep possible at all depths down to $z \simeq 1000-1500$ km ($\eta_{NH} \simeq 10^{22}$)
Raleigh and Kirby, 1970	Olivine $D = D_0 \exp -\left(\frac{26{\cdot}4\,T_M(P)}{T}\right)$ Nabarro–Herring $d = 1$ cm Dislocation creep $n = 5$	Clark, Ringwood, 1964 (Low) Lee, 1968 (High)	For $\dot{\varepsilon} = 10^{-14}\ s^{-1}$ Dislocation creep dominant for $z < 200$ km (low geotherm) or $z < 130$ km (high geotherm) Nabarro–Herring creep dominant for $z > 130$ km (high geotherm)
Weertman, 1970	Olivine $D = D_0 \exp -\left(\frac{18\,T_M(P)}{T}\right)$ Nabarro–Herring $d = 22$ cm (For $\sigma = 10^{-2}$ b and $T = T_M$) Dislocation creep $n = 3$	Jacobs, 1956	For $\dot{\varepsilon} = 10^{-16}\ s^{-1}$ Dislocation creep dominant at all depths
Carter, Baker and George, 1972	Olivine $D = D_0 \exp -\left(\frac{28{\cdot}7\,T_M(P)}{T}\right)$ ($Q^* = 120$ kcal/mole) Nabarro–Herring $d = 0{\cdot}5$–5 cm Dislocation creep $n = 3$–$4{\cdot}8$	Oceanic Geotherm (Ringwood, 1969)	For $\dot{\varepsilon} = 10^{-14}\ s^{-1}$ Nabarro–Herring creep and dislocation creep both possible for $z < 400$ km but the viscosity increases more rapidly with depth (for $z > 200$ km) for Nabarro–Herring than for dislocation creep

Kirby and Raleigh, 1973	Olivine $D = D_0 \exp -\left(\frac{23\,T_M(P)}{T}\right)$ ($Q^* = 96$ kcal/mole $V^* = \Omega$ (oxygen) Nabarro–Herring $d = 0{\cdot}1$ cm Dislocation creep $n = 3$ $\mu = f(z)$	Clark, Ringwood, 1964 (Low) Griggs, 1972 (High)	Dislocation creep dominant at all z for $10^{-16} < \dot{\varepsilon} < 10^{-13}\ s^{-1}$ Nabarro–Herring creep dominant for $z > 150$ km only if $\dot{\varepsilon} < 10^{-16}\ s^{-1}$
Stocker and Ashby, 1973	Olivine $Q^* = 158$ kcal/mole $V^* = 11\ cm^3/mole = \Omega$ (oxygen) $V^* = 37\ cm^3/mole = \Omega$ (SiO_4) Coble creep $0{\cdot}01 < d < 1$ cm Dislocation creep $n = 3$–$4{\cdot}2$	'Reasonable' geotherm anchored at intersection with wet pyrolite solidus and at olivine spinel transition	Dislocation creep dominant for $1 < \sigma < 500$ b Coble creep dominant only if $n = 4{\cdot}2$, $V = 20\ cm^3/mole$ and $z < 400$ km, $\dot{\varepsilon} < 10^{-14}\ s^{-1}$

d = grain diameter

D = diffusion coefficient $D = D_0 \exp -\left(\frac{Q^* + PV^*}{RT}\right)$ Q^* = activation energy, V^* = activation volume.

or

$D = D_0 \exp -\left(\frac{gT_M(P)}{T}\right)$ $T_M(P)$ = actual melting temperature under pressure P.

$\dot{\varepsilon}$ = creep rate

Nabarro–Herring formula (NH): $\dot{\varepsilon}_{NH} \propto \frac{D\sigma\Omega}{kTd^2}$ σ = shear stress, Ω = atomic volume.

Dislocation creep (DC) formula: $\dot{\varepsilon}_{DC} \propto \frac{D\mu b}{kT}\left(\frac{\sigma}{\mu}\right)^n$ μ = shear modulus.

z = depth

$\eta = \frac{\sigma}{\dot{\varepsilon}}$ = viscosity The dominant mechanism is the one for which the viscosity is lowest (or the shear stress lowest for a given creep rate).

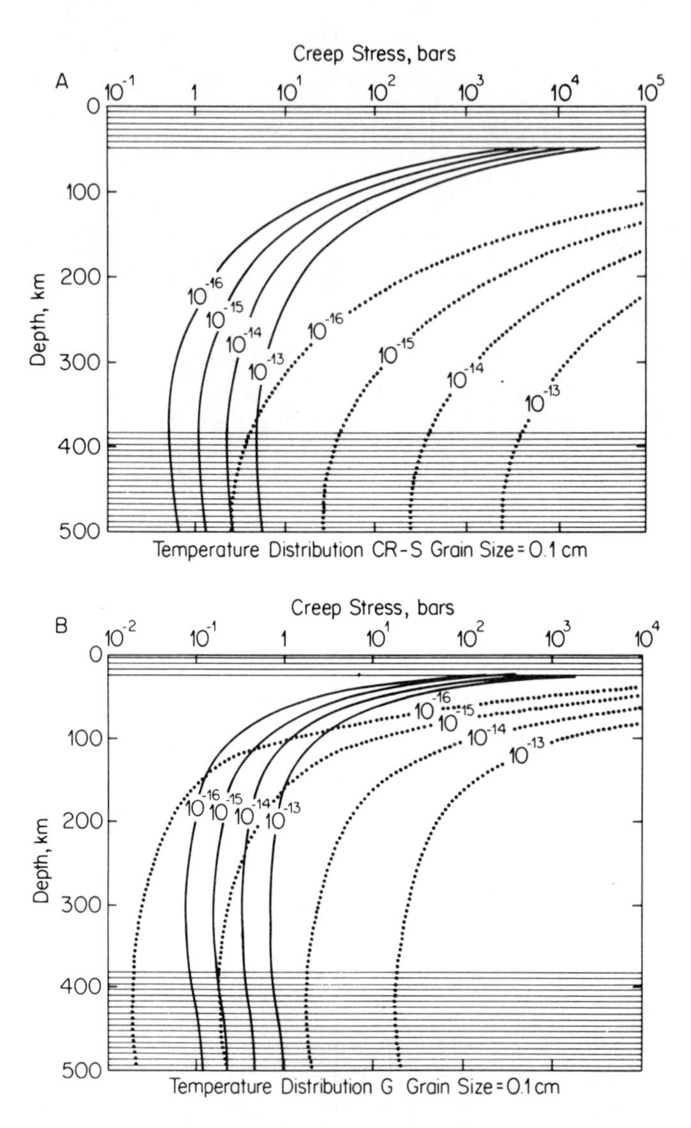

Fig. 11.12. Calculated creep-stress distribution in the upper mantle based on empirical power law (solid lines) and on theoretical Nabarro–Herring law (dotted lines). Strain-rates given in units of s^{-1} for each curve. Regions over which predicted creep stresses are invalid are shown as light horizontal ruled areas
(A): Griggs (1972) oceanic geotherm
(B): Clark and Ringwood shield geotherm
(after Stocker and Ashby, 1973, reproduced from *Tectonophysics* by permission of Elsevier Scientific Publishing Company, Amsterdam)

and Ashby (1973) since it considers most of the potential flow mechanisms with a spectrum of rheological constants derived from experimental data and empirical compilation. The relative activity of the flow mechanisms in olivine can be explored as a function of stress and temperature by the construction of deformation maps (Fig. 11.13); the chosen geotherm (Fig. 11.14) yields

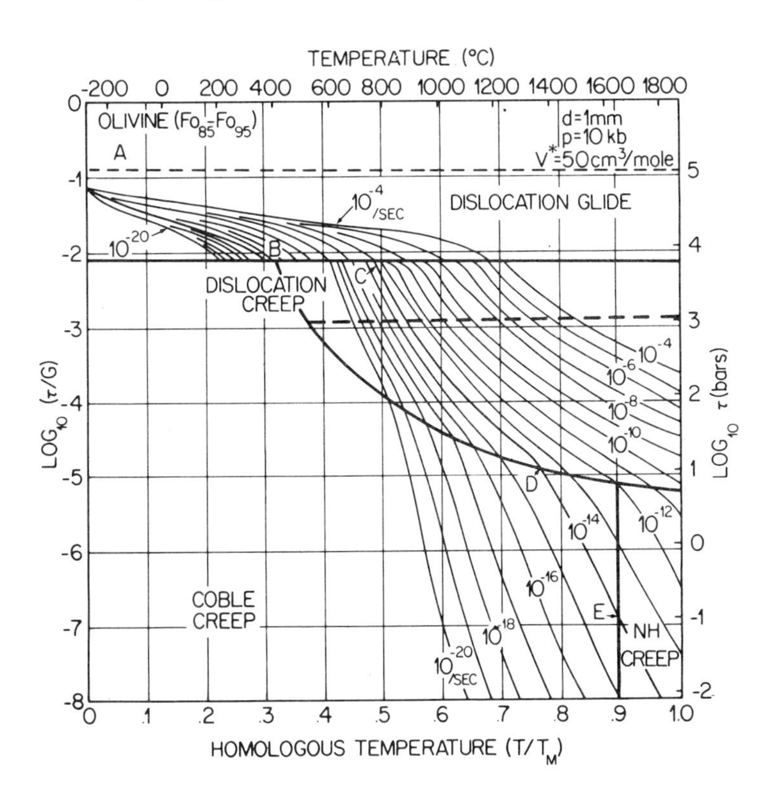

Fig. 11.13. A deformation map for olivine with strain-rate contours. (Reproduced from Stocker and Ashby, *Rev. Geophys. Space Phys.*, **11**, 391 (1973), copyright by American Geophysical Union)

deformation maps of the upper mantle where the domains of the active mechanisms are delineated as a function of depth for various strain rates.

Mass transport through a fluid phase at the grain-boundaries is considered as a possible flow mechanism (see § 4.4) but it is superseded by other creep mechanisms when a realistic set of parameters is chosen (Figs. 11.15, 11.16).

11.3.3. Discussion

Before starting to discuss the large-scale flow mechanisms in the mantle, it should perhaps be pointed out that the existence of a minimum in the viscosity depth curve is in no way a criterion of the activity of a specific flow mechanism.

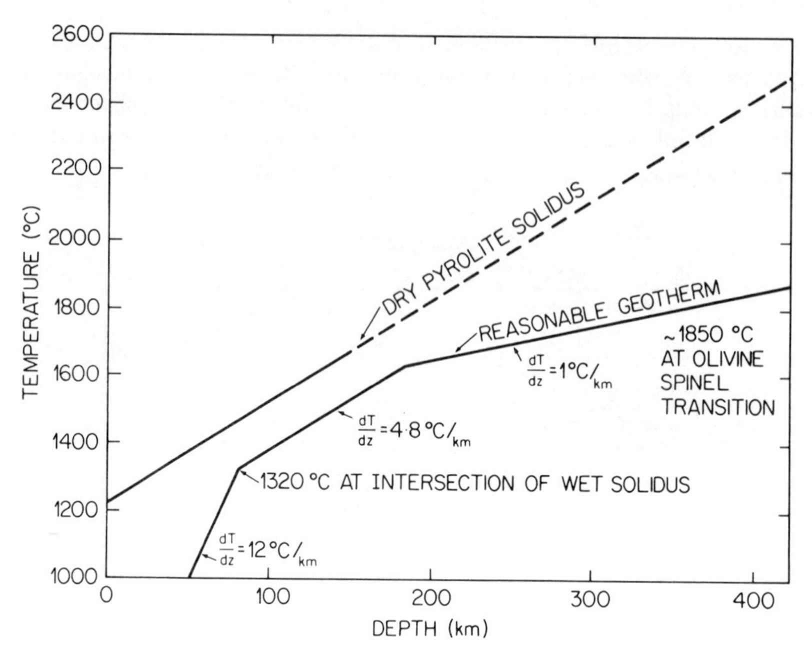

Fig. 11.14. A reasonable geotherm. (Reproduced from Stocker and Ashby, *Rev. Geophys. Space Phys.*, **11**, 391 (1973), copyright by American Geophysical Union)

Any thermally activated flow model with an activation energy increasing with pressure (as is the case with diffusion-controlled flow mechanisms), taken in conjunction with a geotherm whose slope decreases as depth increases, must necessarily yield a $\eta(z)$ curve possessing a minimum, i.e. must predict the existence of an asthenosphere. This obviously reflects the fact that viscosity varies as the diffusion coefficient, i.e. it decreases at first as the influence of temperature is dominant and then increases when the geotherm becomes less steep and pressure becomes important.

The thickness and depth of the asthenosphere and whether the minimum viscosity is low enough to allow significant flow at reasonable strain rates, depend mostly on the hypotheses made concerning the shape of the geotherm and the diffusion parameters (activation energy and volume). All models are especially sensitive to the latter parameters whose values reflect physical conditions which are largely unknown. The activation energy for diffusion may be widely different if one assumes that diffusion takes place in the bulk or along grain-boundaries, or if there is water present ('wet' or 'dry' mantle) or impurities which may alter the vacancy concentration (intrinsic or extrinsic diffusion) (see § 3.2). The activation volume will clearly depend on the nature of the diffusing species (oxygen ions or SiO_4 groups). The comparison between different flow mechanisms and the determination of the one which gives the lowest viscosity is also dependent on the hypotheses made, as can be seen by

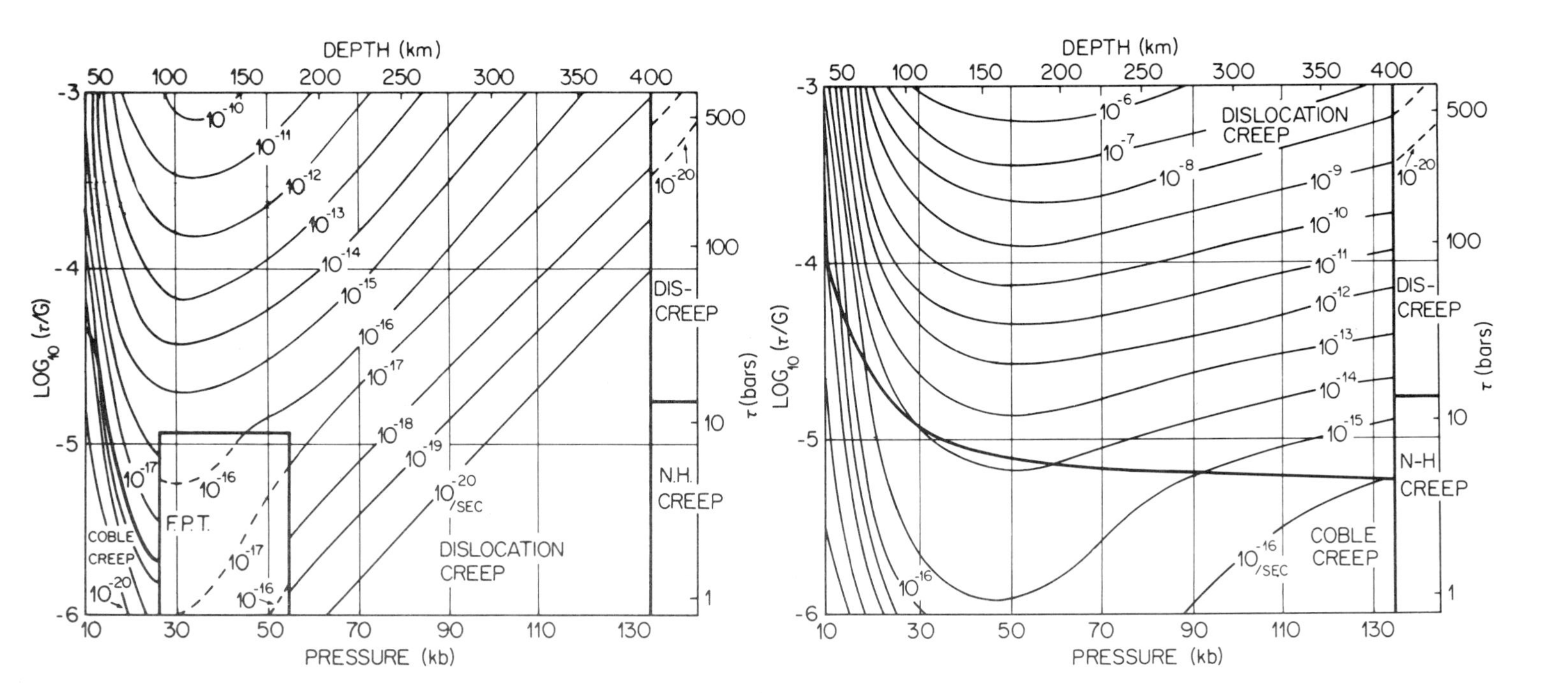

Fig. 11.15. Deformation map for the upper mantle with stress exponent $n = 3 \cdot 0$, pre-exponential constant $A = 0 \cdot 7$, activation volume $V^* = 50\ \text{cm}^3/\text{mole}$ and grain size $d = 1$ mm. Temperatures from the reasonable geotherm of Fig. 11.14. The strain-rate contours in the fluid phase transport (FPT) field and the fields below 400 km are schematic. (Reproduced from Stocker and Ashby, *Rev. Geophys. Space Phys.*, **11**, 391 (1973), copyright by American Geophysical Union)

Fig. 11.16. Deformation map for the upper mantle with $n = 4 \cdot 25$, $A = 1 \cdot 2 \times 10^4$, $V^* = 20\ \text{cm}^3/\text{mole}$ and $d = 1$ mm. Temperatures from the reasonable goetherm of Fig. 11.14. The FPT field is missing because Coble creep produces a larger strain-rate. Compare with Fig. 11.15. (Reproduced from Stocker and Ashby, *Rev. Geophys. Space Phys.*, **11**, 391 (1973), copyright by American Geophysical Union)

mere inspection of Table 11.1. The choice between diffusion creep and dislocation creep as the most probable flow mechanism is in a large measure determined by the adopted grain size for the mantle material. Nabarro–Herring creep is obviously at a disadvantage at the outset if the grain size is assumed to be 20 cm or more.

Thus, the $\eta(z)$ curve calculated for a flow model is worth only as much as the physical parameters that have been used, and it is interesting at this stage to try to evaluate these parameters from the experimental evidence available.

The best evidence that can be obtained on the deformation of the mantle material comes from the peridotite xenoliths from basalts and kimberlites which constitute a sampling from the upper mantle (see § 10.2.2). From phase equilibria studies between minerals, it has recently been possible to assess the temperature and pressure (i.e. depth) at which xenoliths were equilibrated before their ascent to the surface. (Boyd, 1973; Nixon et al., 1973; MacGregor and Basu, 1974). It appears that xenoliths originate from depths between 30 and 200 km and were equilibrated at temperatures between 900 °C and 1400 °C. Thus, they are probably representative of the upper mantle material.

In particular, the grain size of the protogranular nodules (§ 10.3.5) is very probably typical of the grain size in the upper mantle; it ranges between 1 and 10 mm, which does not support the idea that the grain size should be very large (up to a few metres or more) as proposed by Orowan (1967).

Another very important piece of information is the fact that basalt xenoliths are deformed by dislocation creep in hot-working conditions (temperature ranging from 950 ° to 1120 °C, J. Dickey, written communication).

The obliquity of lattice preferred orientation on the foliation points to a rotational shear flow (§ 10.3.5, § 10.3.6) quite similar to the flow regime of the massif peridotites (§ 11.2.2).*

The dislocation structures of naturally deformed xenoliths and experimentally deformed peridotite cores at high temperatures have been compared by several experimenters by optical or electron petrography (Raleigh and Kirby, 1970; Green and Radcliffe, 1972; Phakey et al., 1972; Nicolas et al., 1973), and analogous structures (see § 4.2) have been repeatedly observed.

It is therefore possible to draw the conclusions that:

The mode of flow responsible for the natural shear of the xenoliths, hence the mode of flow in the asthenosphere, is probably climb-controlled dislocation creep.

Although the minimum creep rate attainable in laboratory experiments ($\sim 10^{-8}\ s^{-1}$) is much higher than the probable creep rate in the asthenosphere, the non-Newtonian creep law $\dot{\varepsilon} \propto \sigma^n$ with $3 \leqslant n \leqslant 5$ obtained in laboratory experiments may be reasonably thought to be valid for the asthenosphere.

* Green and Gueguen (1974) suggested, however, that the deformation structures of kimberlite xenoliths (and possibly basalt xenoliths) may not be directly associated with the flow in the asthenosphere but may be representative of flow in diapirs.

Even though post glacial rebound has been previously analysed with a Newtonian creep law, it has been shown recently that it can be accounted for by non-Newtonian creep (Post and Griggs, 1973; Brennen, 1974). Another argument, possibly less compelling, in favour of non-Newtonian creep has been put forward by Stocker and Ashby (1973). The observed seismic anisotropy near the oceanic ridges (see § 11) is thought to be due to a flow-induced lattice preferred orientation of peridotite and no diffusion creep mechanism gives rise to such a lattice fabric.

An interesting one-dimensional model of the upper mantle has been recently proposed by Froidevaux and Schubert (1975) making use of a non-Newtonian ($n = 3$), diffusion-controlled creep law in the shear zone. The corresponding equation is coupled with a temperature equation expressing the balance between heat accumulation, radiogenic heat generation and viscous heat dissipation. The resulting thermomechanical equation, integrated with known boundary conditions, yields the geotherm $T(z)$ together with the viscosity profile $\eta(z)$. The model predicts the existence of a narrow asthenosphere which is not too far from the surface, only if the activation volume for diffusion is large and the activation energy small (experimental data suggest that the activation energy may be small if the mantle contains water) (see § 5.2.4). If the geotherm is constrained to match the 'pyroxene geotherm' obtained by phase equilibria measurements on kimberlite xenoliths (Boyd, 1973), the asthenosphere depth in a 'wet' mantle is found to be compatible with the depth from which the sheared xenoliths are presumed to originate. Calculations with a Newtonian diffusion creep law yield an unrealistically high geotherm.

Although this model has yet to be refined, its main interest lies in the fact that it needs no *a priori* assumptions about the shape of the geotherm. Also it allows a clear view to be obtained of the physical influence of the rheological parameters. The existence of a thermally activated viscosity (depending exponentially on temperature) makes possible a thermal feedback effect. Together with the fact that viscosity is non-Newtonian ($n > 1$), this leads to a narrow asthenosphere where shear is concentrated. The depth at which this asthenosphere lies depends on the effect of pressure on viscosity (activation volume).

In conclusion, there seems to be rather convincing evidence (especially from the comparison between naturally deformed xenoliths and laboratory experiments) that the flow in the asthenosphere is a diffusion-controlled dislocation creep, but too little is known as yet about the activation energies and volumes to predict the width and depth of the asthenosphere with reasonable confidence, whatever model is used.

11.3.4. Asthenosphere and low velocity zone

Seismological data have shown the existence of a zone below the lithosphere (roughly between 80 and 180 km) where the velocity of the seismic waves drop by 3 to 6% and where these waves suffer an attenuation that can reach 10 %. The occurrence of the low velocity zone (LVZ) has been tentatively explained by two main categories of mechanisms ascribing the attenuation and low velocity, either to the presence of a small amount of fluid phase resulting from partial melting of the mantle material or to the absorption of energy by dislocations vibrating in the presence of a friction force due to their interaction with point defects (see Gueguen and Mercier, 1973). It is not our purpose here to discuss the validity of these mechanisms but to consider briefly the relations between the LVZ and the asthenosphere (defined as the plastic shear zone). The LVZ coincides approximately with the region of the asthenosphere where a minimum in the viscosity profile $\eta(z)$, occurs, i.e. the zone where most of the plastic flow is expected to take place. As one of the proposed models for the LVZ involves the existence of a fluid phase or partial melting of the mantle material, it has been thought that the plasticity of the asthenosphere was due to the presence of this fluid phase at grain-boundaries. Although one feels intuitively that the cohesion of a polycrystalline aggregate vanishes when the grains are entirely wetted by a fluid film, this is true only in uniaxial tension, but not necessarily in shear (Weertman, 1970). Creep experiments were performed on rock samples of granite and gabbro at 900 °C after they suffered a few percent of partial melting, and their creep strength was found to be much closer to that of the solid rock than to that of the melt (Arzi, 1972; see also Murrell and Chakravarty, 1973). More recently Auten et al. (1974) investigated the creep behaviour of a model alloy of aluminium 2 at. % gallium which contains about 1 % melt above 29 °C (the melting temperature of Al is 660 °C). After some time during which the solid sample was crept at 20 °C a rapid increase of temperature to 33 °C resulted in no increase in the steady-state creep rate (Fig. 11.17), although a significant attenuation and drop of velocity of ultrasonic waves occurred. The authors remark that except if the deformation rate is controlled by diffusion in the fluid phase, the creep strength of a

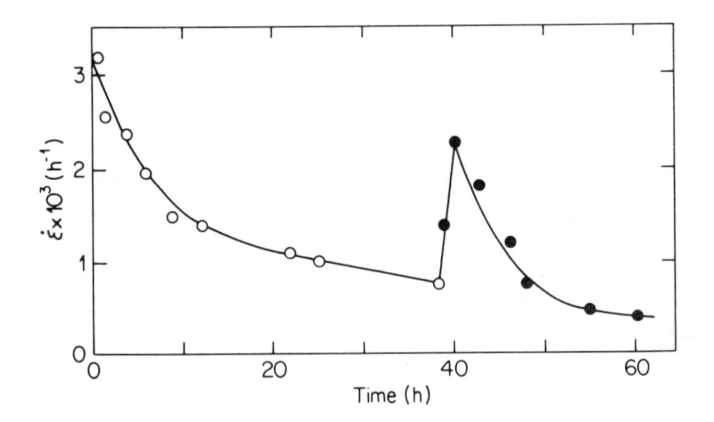

Fig. 11.17. Creep rate of an Al–Ga alloy under constant compressive load, initially at 20 °C (open circles) and after 38 h at 33 °C (solid circles). (Reproduced from Auten et al., *Nature*, **250**, 317 (1974), by permission of Macmillan (Journals) Ltd., London)

partially melted polycrystal does not depend on the presence of a fluid phase at the grain-boundaries but on the strength of the individual grains; thus it depends on T/T_m where T_m is the melting temperature of the solid grains (here 660 °C) and not solidus temperature (here 29 °C). Low sound velocity and high acoustic attenuation can be found together with a low creep rate in a system where T_m is high and the solidus temperature low. *In conclusion, the asthenosphere must not be equated with a partially melted zone only because partial melting would explain the existence of a LVZ.*

CHAPTER 12

Conjectures on Flow Interpretation in Metamorphic Rocks

12.1. GENERAL STATEMENT

We wish here to assess the general relevance of the interpretations proposed for peridotites concerning the flow structures: foliation, lineations and folds. The kinematic interpretation of some non-penetrative features such as crenulation or slip cleavage and tectonic banding is less controversial and the content of § 7.3.3. reflects a wide consensus on this matter. In the case of the interpretation of the former and more controversial structures in other metamorphic rocks than peridotites, we are calling upon a unifying concept and are encouraged by the remarkable similarity on the scale of the crystalline plasticity between metals, oxides and silicates.

Our conjecture is evidently tentative and often leads to queries since, due to the various causes exposed earlier, little is known at present about the flow mechanisms in most naturally deformed metamorphic rocks. It is hoped that these questions will promote relevant studies in the field of flow interpretation.

We will first examine the case where only penetrative planar and linear structures without any simultaneous flexural folding were induced by the flowage, the flow being laminar but not necessarily homogeneous, thus including the possibility of developing shear folds. We will next question the origin and kinematic interpretations of flexural folds associated with the deformation-induced penetrative elements. Folds related to subsequent tectonic events are excluded from the discussion. Deciding whether folds were formed during the flowage or subsequently and whether their mechanism is flexuring or shearing may rely on sophisticated investigations (for instance in the latter case, petrofabrics in fold hinges) and straightforward answers will not always be obtained.

12.2. KINEMATIC INTERPRETATION OF PLANAR AND LINEAR PENETRATIVE STRUCTURES

The planar penetrative structures briefly described in § 7.3.1 are the slaty cleavage in non or weakly metamorphic rocks and the foliation or schistosity in crystalline metamorphic rocks like gneisses. Both cases should be considered independently since the flow and mineral-orientating mechanisms are probably distinct. Although intermediate cases are evidently important, the present discussion applies to a layer silicates-rich phyllite displaying a slaty cleavage on the one hand and a mica poor, quartz and feldspar-rich gneiss on the other.

12.2.1. Foliation

In § 7.3.1, foliation or schistosity was defined as the plane of mineral flattening, containing the principal axis of stretching and the intermediate axis of the finite strain ellipsoid (XY section). Whether it is enhanced by some mineral segregation (tectonic banding) or by non-penetrative surfaces (strain-slip cleavage) is not specifically dealt with here. As mentioned in § 1.1, in gneisses plastic flow is probably important only after the completion of the prograde metamorphism, when deformation proceeds essentially in the solid state. Earlier, the crystallization–recrystallization of minerals in the presence of fluids or melts cause the intervening of other flow mechanisms. The way in which foliation is formed during flowage depends on the flow mechanisms in the dominant minerals. Conversely, considering the shape and lattice fabrics in these minerals, provided the flow mechanisms are known, may be the clue to elucidating the kinematic role of foliation.

The following questions should be asked regarding a gneiss whose principal minerals are quartz, feldspar and mica (or amphibole):

What are the flow mechanisms in each of these minerals?
How effectively do they contribute to the bulk flowage?
How are the minerals fabrics related to their flowage?

The second question is particularly important since the minerals mentioned above have obviously very different ductilities. Feldspars, which are strong, often constitute large augen which persist even in intensely deformed gneisses (Plate 17(a)). They introduce an element of heterogeneity in the deformation which is unknown in peridotites. This sets a limit to the applicability of the method advocated in this book. It seems that a complete kinematic analysis of a gneiss deformation should combine this method with the continuum mechanics one, that is incorporate the results of the study of plastic properties to those obtained considering the flow of a continuous medium around rigid objects. The rigid feldspar augen often have a shape anisotropy; their alignment, the way in which the flow line swirls around them and their pressure shadows provide means to interpreting the flow regime (see the brief review in § 7.3). In the forthcoming discussion, attention is paid only to the plastic flow analysis, obviously fitting better the case of mylonitic gneisses.

The experiments reported in Chapter 6 provide some of the answers in the case of quartz and mica aggregates. Such experiments are critically needed in the case of feldspars and of polyphased aggregates. It is suggested that in a gneiss, quartz and to a lesser extent feldspars contribute to the rock flowage, deforming by plastic flow (dislocations slip and climb, helped by twinning in feldspar) with strain-induced recrystallization. When plastic flow operates on one dominant slip system, on obliquity develops between the elongation direction of the deforming grain and that of its slip direction (Fig. 10.14). The subsequent fabric will reflect this, in an attenuated way when recrystallization is ubiquitous. Micas can be plastically deformed with kinking and often an important strain-induced recrystallization rendering difficult any interpretation; during prograde metamorphism they grow and recrystallize in large strain-free plates, the orientation of which are controlled by the medium and the micas own anisotropy. In this case no obliquity is observed between the shape and lattice orientations in the grains. Late oriented growth of micas may also alter a former foliation developed in the other minerals subsequent to the flowage. Hornblende, which seems to be harder than most other rock-forming minerals (§ 5.6), may grow or recrystallize during the flowage or, if it predates it, behave as a passive marker.

Finally, in a plastically deformed gneiss the general situation is more complex than in the peridotites studied in Chapter 10, but nevertheless may not be intrinsically different. The foliation can be equated with the flow plane in an irrotational shear flow. It is oblique to the flow plane in a rotational shear flow. If the preferred orientation of the slip planes and slip directions in minerals are respectively equated with the flow plane and flow direction (see discussion § 6.9.3), their obliquity with respect to the foliation indicates the sense of shear and, with the additional constraint of a plane strain regime, the angle of shear.

Rotational shear flow with some component of irrotational flow is probably important in the crust; petrofabric studies carried on the different mineral species present in the gneiss may bring it out by a careful and systematic quest of the characteristic obliquity.

12.2.2. Lineations

The types of lineations defined in § 7.3.2 and observed in gneisses along with the foliation are important for the full characterization of the finite strain ellipsoid orientation. Depending on their nature, they are either parallel to the X elongation axis of the strain ellipsoid or to the Y intermediate axis. The lineations oriented in the X-direction are those formed by: (1) plastically flowing minerals, (2) strain markers and (3) hard minerals with a prismatic habit like amphiboles which are passively oriented in their flowing matrix. However, if the flow was turbulent with generations of small folds, the axes of which were at some angle to the flow direction, then locally the prismatic hard minerals can be oriented accordingly. Minerals which are pulled apart during the flowage may yield lineations in the Y direction (Fig. 7.5).

This expresses again the age-old argument about the kinematic meaning of lineations which can be greatly clarified by examining whether they are parallel to the X- or to the Y-direction of the finite-strain ellipsoid, leaving aside other unwarranted relations with stress or flow directions deduced from fold consideration. The solution depends on the plastic properties of the mineral forming the lineation compared with those of the other minerals, on their respective behaviours and on the flow regime.

12.2.3. Slaty cleavage

In § 7.3.1, the slaty cleavage was equated with the *XY* section of the finite-strain ellipsoid, a conclusion which is largely valid but needs further discussion now. The relation of cleavage with the flow orientation is obscure because observations and measurements on minerals shape and lattice-preferred orientations are difficult to make and because the flow mechanisms are not well understood in the dominant rock-forming minerals, quartz and layer silicates. Also, due to orientation mechanisms directly controlled by the *stress* intervening in micas, the way the strain increments were added is not insignificant here.

At low temperatures, quartz may contribute to the flowage by basal slip as experimentally shown (§ 6.4) and verified in the natural deformation of quartz-rich layers in slates (Bouchez, 1975). Another mode of flow is diffusion, favoured here by the presence of a fluid transporting the elements (§ 4.4.1). This mechanism involving pressure-solution, solution-transfer and redeposition is certainly important in slates (Durney, 1972). It is very dubious whether Kamb's theory can be invoked to explain in quartz the simultaneous development of a lattice preferred orientation parallel to the shape one (§ 4.4.2).

In the experimental section on micas (§ 6.3), the conclusion reached was that slip was not an important mechanism in the development of preferred orientations during flow. Rotations of the lamellar particles, crystallization and recrystallization intervene in various proportions depending mainly on the temperature conditions. In slates where direct crystallization is possible in the presence of large amounts of water, these different processes are probably associated. Crystallization and recrystallization are controlled by the growth anisotropy of layer silicates, by the pore anisotropy of the medium and by the pressure solution mechanism.

If the dominant flow mechanism in quartz is pressure solution and if the flow was irrotational (strain increments added co-axially), the slaty cleavage has to be normal to the *principal normal compressive stress* σ_1, and the lineation parallel to σ_3. A strict parallelism is to be expected between the finite strain axes, the minerals shape preferred orientations and the micas lattice orientation. Since we reject Kamb's model for the development of preferred orientations in connection with pressure solution, no *lattice preferred orientation should be expected in quartz.* A more complex situation would occur if the strain increments were not added co-axially (as on the limb of a fold or in the course of a rotational shear flow). With only this mechanism operating in micas, the effects are not clearly predicted and it is uncertain whether the cleavage would still be parallel to the *XY* section of the strain ellipsoid (Etheridge et al., 1974).

If the flow mechanisms are slip in quartz and rotation and crystallization–recrystallization guided by a strain-induced pore anisotropy in layer silicates, then the slaty cleavage and the micas preferred orientation coincide with the *XY section of the finite strain ellipsoid,* whatever the way the strain increments are added; the cleavage may also be at any angle to σ_1. The flow can be rotational or irrotational. *Quartz develops a preferred lattice orientation* which, in a rotational shear flow regime, will necessarily be reflected by the characteristic obliquity between its lattice orientation and that of the slaty cleavage.

Deciding on one of these possibilities depends on careful investigation of minerals lattice-preferred orientation, mainly quartz, and of their exact relations with the orientation of the finite strain axes. Such investigation is now possible with the X-ray goniometer technique.

12.3. KINEMATIC INTERPRETATION OF FOLDS

In metamorphic terrains, when the folds associated with the penetrative elements of the deformed rocks are considered, a recurring observation is that the fold axis is parallel or close to the mineral and/or stretching lineations (pebbles, etc.), whereas the axial plane coincides with the foliation. The observation of this stretching parallel to the fold axis goes against the classical view that folds develop with their axis normal to the motion direction. Before examining the various interpretations intended to reconcile these conflicting points, two general comments may clarify the debate:

There is no univocal connection between the motion direction which is determined with respect to coordinates outside the studied structure and the flowage direction which refers to internal coordinates (see § 2.1). Often the motion direction is merely inferred and the only indisputable information bears on finite strain (and sometimes flow) to which the analysis should exclusively refer.

The fact that folds form at right-angles to the motion direction is a concept inherited from the study of folding in superficial terrains. These are close to the major discontinuity represented by the surface of the crust and possess large viscosity contrasts between the successive layers. In deeper terrains with attenuated viscosity contrasts this is not necessarily valid. *Both cases must be treated separately.* Bearing this in mind it is clear that the origin of folds and their kinematic meaning cannot be regarded as unique. The various interpretations mentioned above can be grouped under five headings.

(1) Two independent deformations have been superimposed: for instance, folds which were formed first went on later to control the flow direction.

(2) Both the motion and flow directions are normal to the fold axis, and the stretching parallel to the axis is only apparent. It can result from boudinage scalloping layers into long stripes or from parting along two shear planes intersecting in the fold axis direction.

(3) The motion is normal to the fold axis but the flowage is parallel to the axis. This situation arises locally during divergent flowage. Rolling, much like a piece of putty assuming a cigar shape when rolled between the hands has also been invoked to explain this situation. Mechanically, this rolling hypothesis possesses hidden assumptions which modify its actual meaning: if, due to

rolling, particles are elongated parallel to the fold axis, the surrounding matrix will be elongated in a similar or larger proportion. Now, whether the major flow direction is normal to the fold axis (rotational shear flow component) or parallel to the fold axis should be decided by studying the flow structures.

(4) The motion and the flow directions are parallel and they both make a small angle with the fold axis. This occurs in the case of non-cylindrical folds possibly as an original feature, later exaggerated by differential flow in the fold axial plane along the motion direction. As a result the axis trend is locally rotated toward the flow direction.

(5) The motion and flow directions coincide with the fold axis. This results from convergent flowage as a consequence of motion in a narrowing channel (§ 10.4.2).

With a view to deciding between these interpretations, plastic flow studies as advocated in this book and finite strain estimations in connection with the fold orientation provide objective information. For instance, Cloos (1947) showed that if the mean stretching is appreciably higher than 10 %, it is no longer compatible with the divergent flowage observed in arcuate belts; 75 % to 100 % stretchings which are fairly common figures, call for another interpretation. In the metamorphic rocks, displaying evidence of large penetrative flowage, we are inclined to think that, as in the case of peridotites, the dominant picture is that of fold axes parallel or close to the major flow direction yielding information on the general flow pattern which complements that obtained from the study of structures, textures and preferred orientations.

'Haüy de bonne humeur sort de l'Institut . . .'

Unknown author, 1822?

References

Adda, Y., and Philibert, J., 1966. '*La diffusion dans les solides*', Presses Universitaires de France, Paris.

Aladag, E., Davis, L. A., and Gordon, R. B., 1970. 'Cross slip and the plastic deformation of NaCl single and polycrystals at high pressure', *Phil. Mag.*, **21**, 469–478.

Alexander, H., and Haasen, P., 1968. 'Dislocations and plastic flow in the diamond structure', *Sol. State Physics*, **22**, 27–158.

Alexandrov, K. S., and Ryzhova, T. V., 1961. 'Elastic properties of rock-forming minerals, 2, Layered Silicates', *Bull. Acad. Sci. USSR, Geophys. Ser.* English Trans. No. 22, 1165–1168.

Amelinckx, S., 1964. *The Direct Observation of Dislocations*, Academic Press, New York.

Amelinckx, S., and Delavignette, P., 1962. 'Dislocations in layer structures.' In *Direct Observation of Imperfections in Crystals*, J. B. Newkirk and J. H. Wernick Eds., Interscience, New York.

Anderson, O. L., 1965. 'Determination and some uses of isotropic elastic constants.' In *Physical Acoustics*, Vol. 3.B, W. P. Mason Ed., Academic Press, New York, pp. 43–95.

Ardell, A. J., Christie, J. M., and McCormick, J. W., 1974. 'Dislocation images in quartz and the determination of Burgers vector', *Phil. Mag.*, **29**, 1399–1411.

Arzi, A. A., 1972. 'Experimental study of partial melting in natural rocks and subsequent creep under low stress', *Trans. Amer. Geophys. Union*, **53**, 513.

Ashby, M. F., 1970. 'The deformation of plastically non-homogeneous crystals', *Phil. Mag.*, **21**, 399–424.

Ashby, M. F., and Verrall, R. A., 1973. 'Diffusion accommodated flow and superplasticity', *Acta Metall.*, **21**, 149–163.

Aumento, F., and Loubat, H., 1971. 'The Mid-Atlantic ridge near 45° N. XVI serpentinized ultramafic intrusions', *Canadian Jour. Earth Sc.*, **8**, 631–663.

Auten, T. A., Gordon, R. B., and Stocker, R. L., 1974. 'Q^{-1} and mantle creep', *Nature*, **250**, 317–318.

Avé Lallemant, H. G., 1967. 'Structural and petrofabric analysis of an "Alpine type" peridotite: the lherzolite of the French Pyrénées', *Leidse Geol. Meded*, **42**, 1–57.

Avé Lallemant, H. G., and Carter, N. L., 1970. 'Syntectonic recrystallization of olivine and modes of flow in the upper mantle', *Geol. Soc. Am. Bull.*, **81**, 2203–2220.

Avé Lallemant, H. G., and Carter, N. L., 1971. 'Pressure dependence of quartz deformation lamellae orientations', *Am. Jour. Sci.*, **270**, 218–235.

Ayrton, S., 1969. 'On the origin of gneissic banding', *Eclogae Geol. Helv.*, **62**, 567–570.

Ayrton, S. N., and Ramsay, J. G., 1974. 'Tectonic and Metamorphic Events in the Alps', *Bull. Suisse Min. Petrogr.*, **54**, 609–639.

Badgley, P. C., 1959. *Structural Methods for the Exploration Geologist*, Harper, New York, 280 pp.
Baëta, R. D., and Ashbee, K. H. G., 1967. 'Plastic deformation of quartz at atmospheric pressure', *Phil. Mag.*, **15**, 931–938.
Baëta, R. D., and Ashbee, K. H. G., 1968. 'Evidence of plastic deformation in quartz at atmospheric pressure', *Proc. 4th Eur. Conf. Electr. Microscopy, Rome*, 427–428.
Baëta, R. D., and Ashbee, K. H. G., 1969. 'Slip systems in quartz, I—Experiments; II—Interpretation', *Am. Mineral.*, **54**/11–12, 1551–1582.
Baëta, R. D., and Ashbee, K. H. G., 1970. 'Mechanical deformation of quartz, I—Constant strain rate compression experiments; II—Stress relaxation and thermal activation parameters', *Phil. Mag.*, **22**, 604–624 and 625–635.
Baker, D. W., 1969. 'X-ray analysis and representation of preferred orientations in crystal aggregates', *Ph. D. Dissert. Univ. California, Los Angeles* (Univ. Microfilms, Ann Arbor, Mich., No. 70.14257), 160 pp.
Baker, D. W., and Carter, N. L., 1972. 'Seismic velocity and anisotropy calculated for ultramafic minerals and aggregates.' In, *Flow and fracture of rocks*, Geophys. Monogr. Ser., **16**, 157–166.
Baker, D. W., and Wenk, H. R., 1972. 'Preferred orientation in a low symmetry quartz mylonite', *Jour. Geol.*, **80**, 81–105.
Baker, D. W., Wenk, H. R., and Christie, J. M., 1969. 'X-ray analysis of preferred orientations in fine grained quartz aggregates', *Jour. Geol.*, **77**, 144–172.
Balderman, M. A., 1974. 'The effect of strain rate and temperature on the yield point of hydrolytically weakened synthetic quartz', *J. Geophys. Res.*, **79**, 1647–1652.
Bamford, D., 1973. 'Refraction data in Western Germany—a time term interpretation', *Zeitschrift für Geophys.*, **39**, 907–927.
Barber, D. J., 1970. 'Thin foils of non-metals made for electron microscopy by Sputter-Etching', *J. Mater. Sci.*, **5**, 1–8.
Bayly, B., 1974, 'Cleavage not parallel to finite strain ellipsoid's *XY* plane: discussion', *Tectonophysics*, **23**, 205–207.
Bearth, P., 1959. 'Uber eklogite, Glaukophanschiefer und metamorphe pillowlaven', *Schweiz. Min. Petr. Mitt.*, **39**, 267–286.
Bengus, V. Z., 1963. 'Dislocation interaction in calcite twinning', *Sov. Phys. Cryst.*, **8**, 322–326.
Bengus, V. Z., Kommik, S. N., and Startsev, V. I., 1961. 'Some phenomena observed on the boundaries of a twin lamellae in calcite', *Sov. Phys. Cryst.*, **6**, 491–496.
Benioff, H., 1964. 'Earthquake source mechanisms', *Science*, **143**, 1399–1406.
Berckhemer, H., 1968. 'Topographie des "Ivrea Körpers" abgeleitet aus seismischen und gravimetrischen daten', *Schweiz Mineral. Petrogr. Mitt.*, **48**, 235–254.
Berckhemer, H., 1969. 'Direct evidence for the composition of the lower crust and the Moho', *Tectonophysics*, **8**, 97–105.
Bertin, J., 1967. *Sémiologie graphique*, Gauthier-Villars Mouton, Paris, 431 pp.
Bhattacharyya, D. S., and Pasayat, S., 1968. 'Deformation texture in quartz: a theoretical approach', *Tectonophysics*, **5**, 303–314.
Biot, M. A., 1961. 'Theory of folding of stratified viscoelastic media and its implication in tectonics and orogenesis', *Geol. Soc. Am. Bull.*, **72**, 1595–1620.
Biot, M. A., Ode, H., and Roever, W. L., 1961. 'Experimental verification of the theory of folding of stratified viscoelastic media', *Geol. Soc. Am. Bull.*, **72**, 1621–1632.
Birch, F., 1960. 'The velocity of compressional waves in rocks to 10 Kilobars, *1*', *Jour. Geophys. Res.*, **65**, 1083–1102.
Birch, F., 1961. 'The velocity of compressional waves in rocks to 10 Kilobars, *2*', *Jour. Geophys. Res.*, **66**, 2199–2224.
Blacic, J., 1971. 'Hydrolytic weakening of quartz and olivine', *Ph.D. Thesis, Univ. of California, Los Angeles*, 205 pp.

Blacic, J. D., and Christie, J. M., 1973. 'Dislocation substructure of experimentally deformed olivine', *Contr. Mineral and Petrol*, **42**, 141–146.
Blacic, J. D., and Griggs, D. T., 1965. 'New phenomena in experimental deformation of quartz at low strain rates', *Am. Geophys. Union. Trans.*, **46**, 541.
Blum, W., and Ilschner, B., 1967. 'Uber das kriechverhalten von NaCl einkristallen', *Phys. Stat. Sol.*, **20**, 629–642.
Boas, W., and Schmid, E., 1931. 'Zur deutung der deformations texturen von metallen', *Tech. Physik*, **12**, 71–75.
Bocquet, J., Delaloye, M., Hunziker, J. C., and Krummenacher, D., 1974. 'K–Ar and Rb–Sr dating of blue amphiboles, Micas and associated minerals from the Western Alps', *Contrib. Mineral. Petrol*, **47**, 7–26.
Boland, J. N., and McLaren, A. C., 1971. 'Dislocations associated with optical features in naturally deformed olivine', *Contr. Min. Petr.*, **30**, 53–63.
Bollmann, W., 1970. *Crystal Defects and Crystalline Interfaces*, Springer, Berlin.
Borg, I., and Turner, F. J., 1953. 'Deformation of Yule marble: Part VI', *Geol. Soc. Amer. Bull.*, **64**, 1343–1352.
Borg, I., Friedman, M., Handin, J. W., and Higgs, D. V., 1960. 'Experimental deformation of St Peter Sand: a study of cataclastic flow; *Rock deformation*', Griggs and Handin Eds., *Geol. Soc. Am. Memoir*, **79**, 133–191.
Borg, I., and Handin, J. W., 1966. 'Experimental deformation of crystalline rocks', *Tectonophysics*, **314**, 249–368.
Borg, I., and Handin, J. W., 1967. 'Torsion of calcite single crystals', *Jour. Geophys. Res.*, **72**, 641–669.
Borg, I. Y., and Heard, H. C., 1969. 'Mechanical twinning and slip in experimentally deformed plagioclase', *Contr. Mineral and Petrol*, **23**, 128–135.
Borg, I. Y., and Heard, H. C., 1970. 'Experimental deformation of plagioclase.' In, *Experimental and natural rock deformation*, Paulitsch Ed., Springer, Berlin, 375–403.
Borradaile, G. J., 1974. 'Bulk finite tectonic strain estimates from the deformation of neptunian dykes', *Tectonophysics*, **22**, 127–139.
Borradaile, G. J., 1974. 'Contribution to discussion concerning the relationship between slaty cleavage and the *XY* plane of the strain ellipsoid', *Tectonophysics*, **23**, 208.
Borradaile, G. J., and Johnson, H. D., 1973. 'Finite strain estimates from the Dalradian dolomitic formation, Islay, Argyll, Scotland', *Tectonophysics*, **18**, 249–259.
Bouchez, J. L., 1971. 'Exemples de traitement automatique des données numériques en géologie structurale et en pétrologie', *Thèse 3ème Cycle Nantes*, 117 pp.
Bouchez, J. L., 1975. 'Plasticité et morphologie du quartz dans la déformation progressive d'horizons gréseux associés aux schistes d'Angers', *3° Réunion Sc. Terre*, Montpellier., 61.
Bouchez, J. L., and Mercier, J. C., 1974. 'Construction automatique des diagrammes de densité d'orientation—Présentation d'un réseau de comptage', *Sciences de la Terre, Nancy*, XIX, 55–64.
Boudier, F., 1972. 'Relations lherzolite-gabbro-dunite dans le massif de Lanzo (Alpes piémontaises): exemple de fusion partielle', *Thèse 3ème Cycle, Nantes*, 106 pp.
Boudier, F., 1976. 'Le Massif lherzolitique de Lanzo, étude structurale et pétrologique', *Thèse Doctorat d'Etat, Nantes*, 167 pp.
Boudier, F., and Nicolas, A., 1972. 'Fusion partielle gabbroïque dans la lherzolite de Lanzo (Alpes piémontaises)', *Bull. Suisse Mineral. Pétrogr.*, **52**, 39–56.
Boullier, A. M., 1975. 'Structure des peridotites en enclaves dans les Kimberlites d'Afrique de Sud', *Thesis 3rd Cycle*, University of Nantes, 122 pp.
Boullier, A. M., and Gueguen, Y., 1974. 'SP-mylonites. Origin of some mylonites by superplastic flow', *Contrib. Mineral. Petrol.*, **50**, 93–104.
Boullier, A. M., and Nicolas, A., 1973. 'Texture and fabric of peridotite nodules from

kimberlite at Mothae, Thaba Putsoa and Kimberley.' In *Lesotho Kimberlites*, P. H. Nixon Ed., Lesotho National Devel. Corpn., 57–66.
Boullier, A. M., and Nicolas, A., 1975. 'Classification of textures and fabrics of peridotites xenoliths from South African Kimberlites.' In *Physics and Chemistry of the Earth*, Proceedings of the 1st International Conference on Kimberlites, L. M. Ahrens Ed., Pergamon, Oxford, **9**, 97–105.
Boyd, F. R., 1973. 'A pyroxene geotherm', *Geochim. et Cosmochim. Acta*, **37**, 2533–2546.
Boyd, F. R., and Nixon, P. H., 1972. 'Ultramafic nodules from the Thaba Putsoa Kimberlite pipe', *Carnegie Inst. Annual Rep. Yearbook*, **71**, 362–373.
Braillon, P., Mugnier, J., and Serughetti, J., 1972. 'Deformation plastique des cristaux de calcite en compression suivant [111]', *C. R. Acad. Sci. Paris*, **275 B**, 605–608.
Braillon, P., Mugnier, J., and Serughetti, J., 1974. 'Transmission electron microscopy observations of dislocations in calcite single crystals', *Crystal Lattice Defects*, **5**, 73–78.
Brennen, C., 1974. 'Isostatic recovery and the Strain-Rate dependent viscosity of the Earth's mantle', *Jour. Geophys. Res.*, **79**, 3993.
Bridgman, P. W., 1964. *Studies in Large Plastic Flow and Fracture*, Harvard University Press, Cambridge, Mass.
Brothers, R. N., 1959. 'Flow orientation of olivine', *Am. Jour. Sci.*, **257**, 574–584.
Brown, W. L., Morimoto, N., and Smith, J. V., 1961. 'A structural explanation of the polymorphism and transition of $MgSiO_3$', *Jour. Geol.*, **69**, 609–611.
Buck, P., 1970. 'Verformung von Hornblende Einkristallen bei Drucken bis 21kB', *Contr. Mineral. Petrol.*, **28**, 62–71.
Buck, P., and Paulitsch, P., 1969. 'Experimentelle verformung von Glimmer und Hornblende Einkristallen', *Die Naturwissenschaften*, **9**, 460.
Burke, P. M., 1968. 'High temperature creep of polycrystalline sodium chloride', *Thesis, Stanford University*, 122 pp.
Burns, K. L., and Spry, A. H., 1969. 'Analysis of the shape of deformed pebbles', *Tectonophysics*, **713**, 177–196.
Calnan, E. A., and Clews, C. J. B., 1950. 'Deformation textures in face centered cubic metals', *Phil. Mag.*, **41**, 1085–1100.
Calnan, E. A., and Clews, C. J. B., 1951a. 'The development of deformation textures in metals. Part. II, body centered cubic metals', *Phil. Mag.*, **42**, 616–635.
Calnan, E. A., and Clews, C. J. B., 1951b. 'The development of deformation textures in metals. Part III, hexagonal structures', *Phil. Mag.*, **42**, 919–931.
Carter, N. L., and Avé Lallemant, H. G., 1970. 'High temperature flow of dunite and peridotite', *Geol. Soc. Am. Bull.*, **81**, 2181–2202.
Carter, N. L., Baker, D. W., and George, R. P., 1972. 'Seismic anisotropy, flow and constitution of the upper mantle.' In *Flow and fracture of rocks*, Geophys. Monograph. Ser., **16**, 167–190.
Carter, N. L., Christie, J. M., and Griggs, D. T., 1961. 'Experimentally produced deformation lamellae and other structures in quartz sand' (Abstract), *Trans. Am. Geophys. Union*, **66**, 2518.
Carter, N. L., Griggs, D. T., and Christie, J. M., 1964. 'Experimental deformation and recrystallization of quartz', *J. Geol.*, **72**, 687–733.
Carter, N. L., and Heard, H. C., 1970. 'Temperature and rate dependant deformation of halite', *American Jour. Sci.*, **269**, 193–249.
Carter, N. L., Raleigh, C. B., and Decarli, P. S., 1968. 'Deformation of olivine in stony meteorites', *Jour. Geophys. Res.*, **73/16**, 5439–5461.
Carter, N. L., and Raleigh, C. B., 1969. 'Principal stress directions from plastic flow in crystals', *Geol. Soc. Am. Bull.*, **80**, 1231–1264.
Caslavsky, J. L., and Vedam, K., 1970. 'The study of dislocations in Muscovite mica by X-ray transmission topography', *Phil. Mag.*, **22**, 255–268.

Challenger, K. D., and Moteff, J., 1973. 'Quantitative characterization of the substructure of AISI 316 stainless steel, resulting from creep', *Metall. Trans.*, **4**, 749–755.
Champness, P. E., and Lorimer, G. W., 1973. 'Precipitation (exsolution) in an orthopyroxene', *J. Mater. Sci.*, **8**, 467–474.
Chapple, W. M., 1968. 'A mathematical theory of finite amplitude folding', *Geol. Soc. Am. Bull.*, **79**, 47–68.
Choudhury, A., Palmer, D. W., Amsel, G., Curien, H., and Baruch, P., 1965. 'Study of oxygen diffusion in quartz by using the nuclear reaction O^{18} (p, α) N^{15}', *Sol. State. Comm.*, **3**, 119–122.
Choukroune, P., 1971. 'Contribution à l'étude des mécanismes de la déformation avec schistosité grâce aux cristallisations syncinématiques dans les "zones abritées," "pressure shadows" ', *Bull. Soc. Geol. France*, **3–4**, 257–271.
Christensen, N. I., 1966. 'Elasticity of ultrabasic rocks', *Jour. Geophys. Res.*, **71**, 5921–5931.
Christensen, N. I., 1971. 'Fabric seismic anisotropy and tectonic history of the Twin Sisters dunite, Washington', *Geol. Soc. America Bull.*, **82**, 1681–1694.
Christensen, N. I., and Ramananantoandro, R., 1971. 'Elastic moduli and anisotropy of dunite to 10 kilobars', *Jour. Geophys. Res.*, **76**, 4003–4010.
Christie, O. M. J., Ed., 1962. *Feldspar volume, Norsk Geol. Tidsskrift*, **42**, 606 pp.
Christie, J. M., and Ardell, A. J., 1974. 'Substructure of deformation lamellae in quartz', *Geology*, **August**, 105–408.
Christie, J. M., and Green, H. W., 1964. 'Several new slip systems in quartz' (Abstract), *Am. Geophys. Union Trans.*, **45**, 102.
Christie, J. M., Griggs, D. T., and Carter, N. L., 1964. 'Experimental evidence of basal slip in quartz', *Jour. Geol.*, **72**, 734–756.
Chudoba, F. K., and Frechen, J., 1950. 'Ueber die plastische verformung von olivine', *N. Jahrb. Min. Abh. Abt. A.*, **81**, 183–200.
Clark, S. P. (Ed.), 1966. 'Handbook of Physical Constants', rev. edn., *Geol. Soc. Amer. Mem.*, **97**.
Clark, S. P., and Ringwood, A. E., 1964. 'The density distribution and constitution of the mantle', *Rev. Geophys.*, **2**, 35–88.
Cloos, E., 1947. 'Oolite deformation in the South Mountain fold, Maryland', *Bull. Geol. Soc. America*, **58**, 843–918.
Cloos, E., 1962. 'Lineation, a critical review and annoted bibliography', *Geol. Soc. America Mem.*, **18**, 122 pp., with Supplement.
Coble, R. L., 1963. 'A model for boundary diffusion controlled creep in polycrystalline materials', *J. Appl. Phys.*, **34**, 1679–1682.
Coe, R. S., 1970. 'The thermodynamic effect of shear stress on the ortho-clino inversion in enstatite and other coherent phase transitions characterized by a finite simple shear', *Contr. Mineral. Petrol.*, **26**, 247–264.
Coe, R. S., and Muller, W. F., 1973. 'Crystallographic orientation of clinoenstatite produced by deformation of ortho enstatite', *Science*, **180**, 64–66.
Collée, A. L. G., 1963. 'A fabric study of lherzolites with special reference to ultrabasic nodular inclusions in the lavas of Auvergne (France)', *Leidse Geol. Meded.*, **28**, 1–102.
Cottrell, A. H., 1964. *The Mechanical Properties of Matter*, Wiley, New York, 223 pp.
Cropper, D. R., and Pask, J. A., 1973. 'Creep of LiF single crystals at elevated temperatures', *Phil. Mag.*, **27**, 1105–1124.
Crosson, R. S., and Lin, J. W., 1971. 'Voigt and Reuss prediction of anisotropic elasticity of dunite', *Jour. Geophys. Res.*, **76**, 570–578.
Dana, E. S., 1911. *System of Mineralogy*, Wiley, New York, 1134 pp.
Darot, M., 1973. 'Methodes d'analyse structurale et cinematique. Application à l'étude du massif ultrabasique de la Sierra Bermeja (Serrania de Ronda—Andalousie, Espagne)', *Thése 3° Cycle, Nantes*, 120 pp.

Darot, M., 1973. 'Cinematique de l'extrusion à partir du manteau, des péridotites de la Sierra Bermeja (Serrania de Ronda, Espagne)', *Comptes Rendus Ac. Sc. Paris*, **278**, 1673–1676.

Darot, M., and Bouchez, J. L., 'Study of directional data distributions from principal preferred orientation axes'. *Jour. Geology*, in press.

Darot, M., and Boudier, F., in press. 'Mineral lineations in deformed peridotites: Kinematic meaning', *Lithos*,

Das, G., and Mitchell, T. E., 1974. 'Electron irradiation damage in quartz', *Rad. Effects*, **23**, 49–52.

Daubrée, A., 1879. *Etudes Synthétiques de Géologie Expérimentale*, Librairie des corps des Ponts et Chaussées, des Mines et des Télégraphes, Paris, 828 pp.

Davidge, R. W., and Pratt, P. L., 1964. 'Plastic deformation and work hardening in NaCl', *Phys. Stat. Sol.*, **6**, 759–776.

Davis, G. J., Edington, J. W., Cutler, C. P., and Padmanabhan, K. A., 1970. 'Superplasticity: a review', *J. Mater. Sci.*, **5**, 1091–1102.

Davis, L. A., and Gordon, R. B., 1969. 'On the deformation of alkali halide single crystals at high pressure', *Phys. Stat. Sol.*, **36**, K 133–K 135.

Deer, W. A., Howie, R. A., and Zussman, J., 1963. *Rock Forming Minerals*, Longmans, London, 4 vols.

DeHoff, R. T., and Rhines, F. N., 1968. *Quantitative Microscopy*, McGraw Hill, New York.

Demny, J., 1963. 'Elektronenmikroskopische Untersuchungen an sehr dünnen Glimmerfolien', *Zeit. Natürforsch*, **A 18**, 1088–1096.

Denness, B., 1972. 'A revised method of contouring stereograms using curvilinear cells', *Geol. Mag.*, **109**, 157–163.

Den Tex, E., 1969. 'Origin of ultramafic rocks, their tectonic setting and history': a contribution to the discussion of the paper 'The origin of ultramafic and ultrabasic rocks' by P. J. Wyllie, *Tectonophysics*, **7**, 457–488.

Den Tex, E., 1970. 'Principal γ-olivine fabrics: Their tectonic and metamorphic significance.' In *Experimental and Natural Rock Deformation*, P. Paulitsch Ed., pp. 486–495, Springer, New York.

De Vore, G. W., 1966. 'Elastic strain energy and mineral recrystallization: a commentary on rock deformation', *Contr. to Geology*, **5/2**, 19–44.

De Vore, G. W., 1969. 'Elastic strain and preferred orientation in monoclinic crystals', *Lithos*, **2/1**, 9–24.

Dieterich, J. H., 1969. 'Origin of cleavage in folded rocks', *Am. Jour. Sci.*, **267**, 155–165.

Dieterich, J. H., and Carter, N. L., 1969. 'Stress history of folding', *Am. Jour. Sci.*, **267**, 129–154.

Dillamore, I. L., and Roberts, W. T., 1964. 'Rolling textures in face centered cubic and body centered cubic metals', *Acta Metall.*, **12**, 281–293.

Dimitrijevic, M., 1956. 'Jedna nova mreza za izradu konturnih dijagrama', *Trans. Mining Geol. Fac. Univ. Beograd*, 69–70.

Dimitrijevic, M. D., 1967. 'Sur la systematique des surfaces S et des éléments linéaires L dans les tectonites', *Bull. Soc. Geol. de France*, **7/IX**, 153–157.

Dingley, D. J., 1973. 'Enhanced ductility in metals (superplasticity)', *Physik 1972* (2nd General conference of the European Physical Society), E.P.S., Geneva, 319–356.

Doherty, R. D., 1974. 'The deformed state and nucleation of recrystallization', *Metal Sci.*, **8**, 132–142.

Dollinger, G., and Blacic, J. D., 1974. 'New glide system in experimentally deformed hornblende', *Trans. American Geophys. Union*, **56**, 1194.

Dorizzi, P., 1974. 'Etude physique du fluage à haute température de monocristaux d'or, d'argent et de solution solide or-argent 50–50', *Report CEA-R*-4536, 57 pp.

Dunnet, D., 1969. 'A technique of finite strain analysis using elliptical particles', *Tectonophysics*, **7/2**, 117–136.

Dunnet, D., and Siddans, A. W. B., 1971. 'Non-random sedimentary fabrics and their modification by strain', *Tectonophysics*, **12**, 307–325.

Durney, D. W., 1972. 'Solution transfer, an important geological deformation mechanism', *Nature*, **235**, 315–317.

Durney, D. W., and Ramsay, J. G., 1973. 'Incremental strains measured by syntectonic crystal growths.' In *Gravity and tectonics*, Van Bemmelen volume, Dejong and Scholten Eds., Wiley, New York, 67–96.

Ellenberger, F., 1958. *Etude Géologique du Pays de Vanoise*, Imprimerie Nationale, Paris, 561 pp.

Elliott, D., 1970. 'Determination of finite strain and initial shape from deformed elliptical objects', *Geol. Soc. America Bull.*, **81**, 2221–2236.

Elliott, D., 1972. 'Deformation paths in structural geology', *Geol. Soc. America Bull.*, **83**, 2621–2638.

Elliott, D., 1973. 'Diffusion flow laws in metamorphic rocks', *Geol. Soc. Amer. Bull.*, **84**, 2645–2664.

Emmons, R. C., 1943. 'The universal stage', *Geol. Soc. America Mem.*, **8**, 205 p.

Engel, C. G., and Fisher, R. L., 1969. 'Ultramafic and basaltic rocks dredged from the nearshore flank of the Tonga trench', *Geol. Soc. America Bull.*, **80**, 1373–1378.

Eshelby, J. D., Newey, G. W. A., Pratt, P. L., and Lidiard, A. B., 1958. 'Charged dislocations and the strength of ionic crystals', *Phil. Mag.*, **3**, 75–89.

Etchecopar, A., 1974. 'Simulation par ordinateur de la déformation progressive d'un agrégat polycristallin. Etude du développement de structures orientées par écrasement et cisaillement', *Thèse 3ème Cycle, Nantes*, 115 pp.

Etheridge, M. A., and Hobbs, B. E., 1974. 'Chemical and deformational controls on recrystallization of mica', *Contr. Mineral. Petrol.*, **43**, 111–124.

Etheridge, M. A., Hobbs, B. E., and Paterson, M. S., 1973. 'Experimental deformation of single crystals of biotite', *Contr. Mineral. Petrol.*, **38**, 21–36.

Etienne, F., 1971. 'La lherzolite rubanée de Baldissero-Canavese', *Thèse 3ème Cycle, Nancy*, 150 pp.

Fairbairn, H. W., and Hawkes, H. E. Jr., 1941. 'Dolomite orientation in deformed rocks', *Am. Jour. Sci.*, **239**, 617–632.

Ferreira, M. P., and Turner, F. J., 1964. 'Microscopic structure and fabric of Yule marble experimentally deformed at different strain rates', *Jour. Geol.*, **72**, 861–875.

Flinn, D., 1962. 'On folding during three-dimensional progressive deformation', *Quart. Jour. Geol. Soc. London*, **68**, 385–433.

Fontaine, G., 1968. 'Dissociation des dislocations sur les plans (110) dans les cristaux ioniques de type NaCl', *J. Phys. Chem. Sol.*, **29**, 209–214.

Fontaine, G., and Haasen, P., 1969. 'Hydrostatic pressure and plastic deformation of the alkali halides', *Phys. Stat. Sol.*, **31**, K 67–K 70.

Francis, T. J. G., 1969. 'Generation of seismic anisotropy in the upper mantle along the mid oceanic ridges', *Nature*, **221**, 162–165.

Frank, W., 1970. 'Theorie der verfestigung von alkalihalogenid Einkristallen', *Mater. Sci. and Eng.*, **6**, 110–148.

Franssen, L., and Kummert, P., 1971. 'Présentation d'un programme de traitement des données en géologie structurale', *Ann. Soc. Geol. Belgique*, **94**, 39–43.

Frey, M., Hunziker, J. C., Frank, W., Bocquet, J., Dal Piaz, G. V., Jager, E., and Niggli, E., 1974. 'Alpine metamorphism of the Alps.' A review. *Bull. Suisse Min. Petrogr.*, **54**, 247–290.

Friedel, G., 1964. *Leçons de Cristallographie*, Albert Blanchard, Paris, 2nd Edn., 602 pp.

Friedel, J., 1964. *Dislocations*, Pergamon, London.

Froidevaux, C., and Schubert, G., 1975. 'Plate motion and structure of the continental asthenosphere. A realistic model of the upper mantle', *Jour. Geophys. Res.*, **80**, 2553–2564.

Gay, N. C., 1968a. 'Pure shear and simple shear deformation of inhomogeneous viscous fluids. I—Theory', *Tectonophysics*, **5**, 211–234.

Gay, N. C., 1968b. 'Pure shear and simple shear deformation of inhomogeneous viscous fluids. 2—The determination of the total finite strain in a rock from objects such as deformed pebbles', *Tectonophysics*, **5**, 295–302.

Gibbs, J. W., 1876. 'On the equilibrium of heterogeneous substances.' In *The Scientific Papers of J. Willard Gibbs*, Dover, New York, Vol. I, 55–371.

Giese, P., 1968. 'Die struktur der Erdkruste im bereich der Ivrea zone', *Schweiz Mineral. Petrogr.*, **48**, 261–284.

Giletti, B. J., Semet, M. P., and Yund, R. A., 1975. 'Oxygen self-diffusion measured in silicate minerals using an ion microprobe', *Geological Society of America annual meeting.*

Gilman, J. J., 1969. *Micromechanics of Flow in Solids*, McGraw Hill, New York.

Glover, G., and Sellars, C. M., 1973. 'Recovery and recrystallization during high temperature deformation of α-iron', *Metall. Trans.*, **4**, 765–775.

Goetze, C., and Brace, W. F., 1972. 'Laboratory observations of high temperature rheology of rocks', *Tectonophysics*, **13**, 583–600.

Goetze, C., and Kohlstedt, D. L., 1973. 'Laboratory study of dislocation climb and diffusion in olivine', *Jour. Geophys. Res.*, **78/26**, 5961–5971.

Goguel, J., 1965. 'La cause de l'orientation des minéraux dans les roches métamorphiques', *Bull. Soc. Geol. France*, **VII**, 747–752.

Goguel, J., 1965. *Traité de Tectonique*, Masson, Paris, 457 pp.

Goguel, J., 1967. 'L'orientation des minéraux des roches sous l'influence de la contrainte. II—Minéraux mono cliniques et micas', *Bull. Soc. Geol. France*, **IX**, 481–489.

Gordon, R. B., 1965. 'Diffusion creep in the Earth's mantle', *J. Geophys. Res.*, **70**, 2413–2428.

Gordon, R. B., 1971. 'Observation of crystal plasticity under high pressures with applications to the earth's mantle', *J. Geophys. Res.*, **76**, 1248–1254.

Gosh, S. K., and Sengupta, S., 1973. 'Compression and simple shear of test models with rigid and deformed inclusions', *Tectonophysics*, **17**, 133–175.

Green, D. H., 1967. 'High temperature peridotite intrusions.' In *Ultramafic and Related Rocks*, P. I. Wyllie Ed., John Wiley, New York, 212–222.

Green, H. W., 1966. 'Preferred orientation of quartz due to recrystallization during deformation', *Am. Geophys. Union Trans.*, **47**, 491.

Green, H. W., 1967. 'Quartz: extreme preferred orientation produced by annealing', *Science*, **157**, 1444–1447.

Green, H. W., 1968. 'Syntectonic and annealing recrystallization of fine grained quartz aggregates', *Ph.D. Thesis, Univ. of California, Los Angeles*, 203 pp.

Green, H. W., 1970. 'Diffusional flow in polycristalline materials', *J. Appl. Phys.*, **41**, 3899–3902.

Green, H. W., 1972. 'Metastable growth of coesite in highly strained quartz', *Jour. Geophys. Res.*, **77/14**, 2478–2482.

Green, H. W., Griggs, D. T., and Christie, J. M., 1970. 'Syntectonic and annealing recrystallization of fine grained quartz aggregates.' In *Experimental and Natural Rock Deformation*, Paulitsch Ed., Springer Verlag, 272–335.

Green, H. W., and Gueguen, Y., 1974. 'Origin of kimberlite pipes by diapiric upwelling in the upper mantle', *Nature*, **249**, 617–620.

Green, H. W., and Radcliffe, S. V., 1972a. 'Dislocation mechanisms in olivine and flow in the upper mantle', *Earth planet. Sci. Letters*, **15**, 239–247.

Green, H. W., and Radcliffe, S. V., 1972b. 'Deformation processes in the upper mantle.' In *Flow and fracture of rocks*, Geophys. Monograph, **16**, Am. Geophys. Union, 139–156.
Green, H. W., and Radcliffe, S. V., 1972c. 'The nature of deformation lamellae in silicates', *Geol. Soc. Amer. Bull.*, **83**, 847–852.
Greenwood, G. W., and Johnson, R. H., 1965. 'The deformation of metals under small stresses during phase transformations', *Proc. Roy. Soc. London*, **A 283**, 403–422.
Griggs, D. T., 1936. 'Deformation of rocks under high confining pressures', *Jour. Geol.*, **44**, 541–577.
Griggs, D. T., 1938. 'Deformation of single calcite crystals under high confining pressures', *Am. Mineral.*, **23/1**, 28–33.
Griggs, D. T., 1940. 'Experimental flow of rocks under conditions favouring recrystallization', *Geol. Soc. Am. Bull.*, **51**, 1001–1022.
Griggs, D. T., 1967. 'Hydrolytic weakening of quartz and other silicates', *Geophys. Jour. Roy. Astron. Soc.*, **14**, 19–31.
Griggs, D. T., 1972. 'The sinking lithosphere and the focal mechanism of deep earthquakes.' In *The nature of the solid earth*, Robertson Ed., McGraw Hill, New York, 361–384.
Griggs, D. T., 1974. 'A model of hydrolytic weakening in quartz', *J. Geophys. Res.*, **79**, 1653–1661.
Griggs, D. T., and Blacic, J. D., 1964. 'The strength of quartz in the ductile regime' (Abstract), *Trans. Am. Geophys. Union*, **45**, 102–103.
Griggs, D. T., and Miller, W. B., 1951. 'Deformation of Yule marble. Part I: Compression and extension experiments in dry Yule marble at 10.000 atmospheres confining pressure, room temperature', *Geol. Soc. Am. Bull.*, **62**, 853–862.
Griggs, D. T., Paterson, M. S., Heard, H. C., and Turner, F. J., 1960a. 'Annealing recrystallization in calcite crystals and aggregates.' In *Rock deformation*, Griggs and Handin Ed., *Geol. Soc. America Mem.*, **79**, 21–37.
Griggs, D. T., Turner, F. J., and Borg, I., 1953. 'Deformation of Yule marble. Part V: Effects at 300 °C', *Geol. Soc. Am. Bull.*, **64**, 1327–1342.
Griggs, D. T., Turner, F. J., Borg. I., and Sosoka, J., 1951. 'Deformation of Yule marble. Part IV: Effects at 150 °C', *Geol. Soc. Am. Bull.*, **62**, 1385–1406.
Griggs, D. T., Turner, F. J., and Heard, H. C., 1960b. 'Deformation of rocks at 500° to 800 °C.' In *Rock deformation*, Griggs and Handin Ed., *Geol. Soc. Am. Mem.*, **79**, 39–104.
Gross, K. A., 1967. 'X-ray line broadening and stored energy in deformed and annealed calcite', *Phil. Mag.*, **12**, 801–813.
Groves, G. W., and Kelly, A., 1963. 'Independent slip systems in crystals', *Phil. Mag.*, **8**, 877–887.
Gueguen, Y., and Boullier, A. M., 'Evidence of superplasticity in mantle peridotites', *NATO Petrophysics Proceedings*, Wiley & Academic Press, in press.
Gueguen, Y., and Mercier, J. M., 1973. 'High attenuation and the low velocity zone', *Physics of the Earth and Planet. Interiors*, **7**, 39–46.
Guggenheim, E. A., 1967. *Thermodynamics*, North Holland, Amsterdam.
Gutenberg, B., 1951. 'The cooling of the earth and the temperature of its interior.' In *Internal constitution of the Earth*, 2nd Edn., Dover, New York, 150–166.
Gutmanas, E. Yu., and Nadgornyi, E. M., 1970. 'Dislocation motion in secondary slips planes in alkali halide crystals at room temperature', *Phys. Stat. Sol.*, **38**, 777–782.
Haasen, P., Davis, L. A., Aladag, E., and Gordon, R. B., 1970. 'On the mechanism of stage III deformation in NaCl single crystals', *Scripta Metall.*, **4**, 55–56.
Handin, J. W., and Fairbairn, H. W., 1955. 'Experimental deformation of Hasmark dolomite', *Geol. Soc. Am. Bull.*, **66**, 1257–1273.
Handin, J. W., and Griggs, D. T., 1951. 'Deformation of Yule marble. Part II: Predicted fabric changes', *Geol. Soc. Amer. Bull.*, **62**, 863–886.

Handin, J. W., and Hager, R. V., 1958. 'Experimental deformation of sedimentary rocks under confining pressure: tests at high temperature', *Bull. Am. Assoc. Pet. Geol.*, **42**, 2892–2934.

Hara, I., and Paulitsch, P., 1971. 'C-Axis fabrics of quartz in buckled quartz Veins', *N. Jb. Miner. Abh.*, **155/1**, 31–53.

Harbaugh, J. W., and Merriam, D. F., 1968. *Computer applications in stratigraphic analysis*, John Wiley, New York, 282 pp.

Hardwicke, D., Sellars, C. M., and Tegart, W. J. McG., 1961. 'The occurrence of recrystallization during high temperature creep', *J. Inst. Met.*, **90**, 21–22.

Harker, A., 1932. *Metamorphism—a Study of the Transformations of Rocks Masses*, Methuen, London, 362 pp.

Harris, J. E., and Jones, R. B., 1963. 'Directional diffusion in magnesium alloys', *J. Nucl. Mater.*, **10**, 360–362.

Hart, E. W., 1970. 'A phenomenological theory for plastic deformation of polycrystalline metals', *Acta Metall.*, **18**, 599–610.

Harte, B., Cox, K. G., and Gurney, J. J., 1974. 'Petrography and geological history of upper mantle xenoliths from the Matsoku kimberlite pipe.' In *Physics and chemistry of the earth*, Proceedings of the 1st International Conference on Kimberlites, L. H. Ahrens Ed., Pergamon Press, Vol. 9, 617–646.

Hartman, P., and Den Tex, E., 1964. 'Piezocrystalline fabrics of olivine in theory and nature', *XXII Int. Geol. Congr. India*, **VI/4**, 84–113.

Hatch, F. H., Wells, A. K., and Wells, M. K., 1972. *Petrology of the Igneous Rocks*, Thomas Murby, London, 515 pp.

Haul, R., and Dümbgen, G., 1962. 'Untersuchung der Sauerstoffbeweglichkeit in titandioxyd, quarz und quarzglas mit hilfe des heterogen isotopenaustausches', *Zeit. für Elektrochem.*, **66**, 636–641.

Heard, H. C., 1963. 'Effect of large changes in strain rate in the experimental deformation of Yule marble', *J. Geol.*, **71**, 162–195.

Heard, H. C., 1972. 'Steady state flow in polycristalline halite at pressure of 2 Kilobars.' In *Flow and fracture of rocks*, H. C. Heard, I. Y. Borg, N. L. Carter, C. B. Raleigh Eds., American Geophysical Union, Washington, D.C., 191–209.

Heard, H. C., and Carter, N. L., 1968. 'Experimentally induced "natural" intragranular flow in quartz and quartzite', *Am. Jour. Sci.*, **266**, 1–42.

Heard, H. C., and Raleigh, C. B., 1972. 'Steady state flow in marble at 500° to 800 °C', *Geol. Soc. Am. Bull.*, **83**, 935–956.

Helm, D. G., and Siddans, A. W. B., 1971. 'Deformation of a slaty, lapillar tuff in the English lake district: Discussion', *Geol. Soc. Am. Bull.*, **82**, 523–531.

Herring, C., 1950. 'Diffusional viscosity of a polycrystalline solid', *J. Appl. Phys.*, **21**, 437–445.

Hess, H. H., 1964. 'Seismic anisotropy of the uppermost mantle under oceans', *Nature*, **203**, 629–631.

Hesse, J., 1965. 'Die plastische verformung von natriumchlorid', *Phys. Stat. Sol.*, **9**, 209–230.

Higgs, D. V., Friedman, M., and Gebhart, J. E., 1960. 'Petrofabric analysis by means of the X-ray diffractometer', *Geol. Soc. America Mem.*, **79**, 275–292.

Higgs, D. V., and Handin, J. W., 1959. 'Experimental deformation of dolomite single crystals', *Geol. Soc. Am. Bull.*, **70**, 245–278.

Hirsch, P. B., Howie, A., Nicholson, R. B., and Pashley, D. W., 1965. *Electron Microscopy of Thin Crystals*, Butterworths, London.

Hirth, J. P., 1972. 'The influence of grain boundaries on mechanical properties', *Metall. Trans.*, **3**, 3047–3067.

Hirth, J. P., and Lothe, J., 1967. *Theory of Dislocations*, McGraw Hill, New York.

Hobbs, B. E., 1968. 'Recrystallization of single crystals of quartz', *Tectonophysics*, **6/5**, 353–401.

Hobbs, L. W., and Goringe, M. J., 1970. 'Direct electron microscopical observations of dislocation arrangements in pure deformed alkali halide crystals.' In *Microscopie electronique 1970*, Société Française de Microscopie Electronique, Paris, 289–290.

Hobbs, B. E., McLaren, A. C., and Paterson, M. S., 1972. 'Plasticity of single crystals of synthetic quartz.' In *Flow and fracture of rocks*, the Griggs Vol., Am. Geophys. Union, **16**, 29–54.

Honeycombe, R. W. K., and Pethen, R. W., 1972. 'Dynamic recrystallization', *J. Less Common Metals*, **28**, 201–212.

Hopwood, T., 1968. 'Derivation of a coefficient of degree of preferred orientation from contoured fabric diagrams', *Geol. Soc. Am. Bull.*, **79**, 1651–1654.

Hörz, F., 1970. 'Static and dynamic origin of kink bands in micas', *Jour. Geophys. Res.*, **75/5**, 965–977.

Hörz, F., and Ahrens, T. J., 1969. 'Deformation of experimentally shocked biotite', *Amer. Jour. Sci.*, **267**, 1213–1229.

Hunziker, J. C., 1974. 'Rb–Sr and K–Ar age determination and the Alpine tectonic history of the western Alps', *Mem. Ist. Geol. Min. Univ. Padova*, **31**, 1–54.

Hurm, M., and Escaig, B., 1973. 'Propriétés mécaniques et structurales en fluage très haute température de magnesie polycristalline dopée', *J. de Phys.*, **34**, C 9, 347–358.

Hüther, W., and Reppich, B., 1973. 'Dislocation structure during creep of MgO single crystals', *Phil. Mag.*, **28**, 363–371.

Jackson, E. D., 1961. 'Primary textures and mineral associations in the ultramafic zone of the Stillwater complex, Montana', *Geol. Surv. Prof. Pap.*, **358**, 106 p.

Jackson, E. D., and Thayer, T. P., 1972. 'Some criteria for distinguishing between stratiform concentric and alpine peridotite gabbro complexes', *24th Inter. Geol. Congress, Sect. 2*, 289–296.

Jacobs, J. A., 1956. 'The Earth's interior.' In *Handbuch der Physik Geophysik I*, Ed. J. Bartels, Springer Verlag, Berlin, 364 pp.

Jeffery, G. B., 1922. 'The motion of ellipsoid particles immersed in a viscous fluid', *Royal Soc. London Proc.*, **A/102**, 161–179.

Johnsen, A., 1902. 'Biegungen und translationen', *Neues Jahrb. Mineral. Geol. Paleontol. Monatsh.*, **2**, 133–153.

Johnsen, A., 1918. 'Künstliche schiebungen und translationen nach untersuchungen von K. Veit', *Centralbl. Geol.*, **19**, 265–266.

Jonas, J. J., Sellars, C. M., and Tegart, W. J. McG., 1969. 'Strength and structure under hot working conditions', *Metals and Materials*, **3**, Met. Rev. No. 130, 1–14.

Jones, R. B., 1973. 'Creep by mass transport mechanisms', *J. of the Sheffield Univ. Metall. Soc.*, **12**, 34–40.

de Jong, M., and Rathenau, G. W., 1959. 'Mechanical properties of iron and some iron alloys while undergoing allotropic transformations', *Acta Metall.*, **7**, 246–253.

de Jong, M., and Rathenau, G. W., 1961. 'Mechanical properties of an iron carbon alloy during allotropic transformation', *Acta Metall.*, **9**, 714–720.

Juteau, T., Lapierre, H., Nicolas, A., Parrot, J. F., Ricou, L. E., Rocci, G., and Rollet, M., 1973. 'Idées actuelles sur la constitution, l'origine et l'évolution des assemblages ophiolitiques mésogéens', *Bull. Soc. Geol. France*, **5-6**, 478–493.

Kalsbeek, F., 1963. 'A hexagonal net for the counting out and testing of fabric diagrams', *Neues Jahr. für Mineral.*, **7**, 173–176.

Kamb, W. B., 1959. 'Theory of preferred orientation developed by crystallization under stress', *J. Geology*, **67**, 153–170.

Kamb, W. B., 1961. 'The thermodynamical theory of non-hydrostatically stressed solids', *J. Geophys. Res.*, **66**, 259–271.

Karl, F., and Kern, H., 1968. 'Uber beanspruchung und verformung von Gesteinen. II, Rotationssymmetrische und echt dreiachsige verformungen Marmoren', *Contr. Mineral. and Petrol.*, **18**, 199–224.

Kasahara, J., Suzuki, I., Kumazawa, M., Kobayashi, Y., and Tida, K., 1968. 'Anisotropism of P wave in dunite', *Jour. Seismol. Soc. Japan*, **21**, 222–228.

Kats, A., Haven, Y., and Stevels, J. M., 1962. 'Hydroxyl groups in α-quartz', *Phys. and Chem. of Glasses*, **3**, 69–75.

Keen, C. E., and Barrett, D. L., 1971. 'A measurement of seismic anisotropy in the northeast Pacific', *Canadian Jour. Earth Sc.*, **8**, 1056–1064.

Keith, R. E., and Gilman, J. J., 1960. 'Dislocation etch-pits and plastic deformation in calcite', *Acta Metall.*, **8**, 1–10.

Kern, H., 1971. 'Dreiaxiale verformungen an solnhofener Kalkstein in temperaturbereich von 20–650 °C. Röntgenographishe gefügeuntersuchungen mit dem texturgoniometer', *Contr. Mineral. and Petrol.*, **31**, 39–66.

Kern, H., and Braun, G., 1973. 'Deformation und gefügeregelung von steinsalz im temperaturbereich 20–200 °C', *Contr. Mineral. and Petrol.*, **40**, 169–181.

Kern, H., and Karl, F., 1968. 'Uber beanspruchung und verformung von Gesteinen. III: Synkristalline verformungen an Auerbach-Marmoren bei axial symmetrischer und echt dreiachsig wirkender beanspruchung', *Contr. Mineral. Petrol.*, **18**, 225–240.

Kern, R., and Weisbrod, A., 1964. *Thermodynamique de base pour minéralogistes, pétrographes et géologues*, Masson, Paris, 243 pp.

Kingery, W. D., and Montrone, E. D., 1965. 'Diffusional creep in polycrystalline NaCl', *J. Appl. Phys.*, **36**, 2412–2413.

Kiralý, L., 1969. 'Analyse statistique des fractures (orientation et densité)', *Geologische Rundschau*, **59**, 125–151.

Kirby, S. H., 1975. 'The role of crystal defects in the shear induced transformation of orthoenstatite to clinoenstatite.' (To be published.)

Kirby, S. H., and Coe, R. S., 1974. 'The role of crystal defects in the enstatite inversion' (Abstract), *Trans. Am. Geophys. Union*, **55/4**, 419.

Kirby, S. H., and Raleigh, C. B., 1973. 'Mechanisms of high temperature, Solid state flow in minerals and ceramics and their bearing on the creep behaviour of the mantle', *Tectonophysics*, **19**, 165–194.

Kirby, S. H., and Wegner, M. W., 1973. 'Dislocation substructure of mantle derived olivine as revealed by selective chemical etching', *Trans. Amer. Geophys. Union*, **54**, 452.

Klassen-Neklyudova, M. V., 1964. *Mechanical Twinning in Crystals*, Consultants Bureau, New York.

Klosterman, M. J., and Buseck, P. R., 1973. 'Structural analysis of olivine in pallasitic meteorites: deformation in planetary interiors', *Jour. Geophys. Res.*, **78/32**, 7581–7588.

Kohlstedt, D. L., and Goetze, C., 1974. 'Low stress, high temperature creep in olivine single crystals', *J. Geophys. Res.*, **79**, 2045–2051.

Kohlstedt, D. L., and Van Der Sande, J. B., 1973. 'Transmission electron microscopy investigation of the defect microstructure of four natural orthopyroxenes', *Contr. Mineral. and Petrol.*, **42**, 169–180.

Kretz, R., 1966. 'Interpretation of the shape of mineral grains in metamorphic rocks', *J. of Petrology*, **7**, 68–94.

Kumazawa, M., 1969. 'The elastic constants of single crystal orthopyroxene', *J. Geophys. Res.*, **74**, 5973–5980.

Kumazawa, M., and Anderson, O. L., 1969. 'Elastic moduli, Pressure derivatives and temperature derivatives of single crystal olivine and single crystal forsterite', *J. Geophys. Res.*, **74**, 5961–5972.

Laduron, D., 1966. 'Sur les procédés de coloration sélective des feldspaths en lame mince', *Ann. Soc. Geol. Belgique*, **89**, 281–294.

Laffitte, P., 1972. *Traité d'informatique géologique*, Masson, Paris, 624 pp.

Lang, A. R., 1970. 'Recent applications of X-ray topography.' In *Modern diffraction and imaging techniques in materials science*, S. Amelinckx, R. Gevers, G. Remaut, J. Van Landuyt Eds., North Holland, Amsterdam, 407–479.
Lappin, M. A., 1971. 'The petrofabric orientation of olivine and seismic anisotropy of the mantle', *Jour. Geol.*, **79**, 730–740.
Lasnier, B., Leyreloup, A., and Marchand, J., 1973. 'Découverte d'un granite "charnockitique" au sein des "gneiss oeillés", Perspectives nouvelles sur l'origine de certaines leptynites du Massif Armoricain Méridional (France)', *Contr. Mineral. and Petrol.*, **41**, 131–144.
Le Comte, P., 1965. 'Creep in rock salt', *J. Geology*, **73**, 469–484.
Lee, D., 1969. 'The nature of superplastic deformation in the Mg–Al eutectic', *Acta Metall.*, **17**, 1057–1069.
Lee, W. H. K., 1968. 'Effects of selective fusion on the thermal history of the earth's mantle', *Earth Planet. Sci. Lett.*, **4**, 270–276.
Lensch, G., 1968. 'Die ultramafitite der zone von Ivrea und ihre geologische interpretation', *Schweiz Mineral. Petr. Mitt.*, **48**, 91–102.
Lensch, G., 1971. 'Die ultramafitite der zone von Ivrea', *Ann. Univers. Saraviensis*, **9**, 5–146.
Le Pichon, X., Francheteau, J., and Bonnin, J., 1973. *Plate Tectonics*, Elsevier, Amsterdam, 300 pp.
Lhote, F., Leymarie, P., and Hetier, J. M., 1969. 'Utilisation du diffractomètre de texture pour la détermination des orientations cristallines des roches microgrenues. Application à la texture de quelques laves du Massif Central', *Bull. Soc. Fr. Mineral. Cristallogr.*, **92**, 299–307.
Loney, R. A., Himmelberg, G. R., and Coleman, R. G., 1971. 'Structure and petrology of the alpine type peridotite at Burro Mountain, California, U.S.A.', *Jour. Petr.*, **12/2**, 245–309.
Loudon, T. V., 1964. 'Computer analysis of orientation data in structural geology', *Off. Nav. Res., Geogr. Branch, Tech. Rep.*, **13**, 130 pp.
Lubimova, H. A., 1958. 'Thermal history of the earth with consideration of the variable thermal conductivity of its mantle', *Geophys. J.*, **1**, 115–134.
Luton, M. J., and Sellars, C. M., 1969. 'Dynamic recrystallization in nickel and nickel iron alloys during high temperature deformation', *Acta Metall.*, **17**, 1033–1043.
Macdonald, G. J. F., 1960. 'Orientation of anisotropic minerals in a stress field.' In *Rock deformation*, Griggs and Handin Ed., *Geol. Soc. America Mem.*, **79**, 1–8.
Macgregor, I. D., 1974. 'The system $MgO–Al_2O_3–SiO_2$: solubility of Al_2O_3 in enstatite for spinel and garnet compositions', *Amer. Mineral*, **59**, 110–119.
Macgregor, I. D., and Basu, A. R., 1974. 'Thermal structure of the lithosphere: A petrologic model', *Science*, **185**, 1007–1011.
McLaren, A. C., and Hobbs, B. E., 1972. 'Transmission electron microscope investigation of some naturally deformed quartzites.' In *Flow and fracture of rocks*, the Griggs Vol., Am. Geophys. Union, **16**, 55–66.
McLaren, A. C., and Phakey, P. P., 1965. 'Dislocations in quartz observed by transmission electron microscopy', *J. Appl. Phys.*, **36**, 3244–3246.
McLaren, A. C., and Retchford, J. A., 1969. 'Transmission electron microscopy study of the dislocation in plastically deformed synthetic quartz', *Phys. Stat. Sol.*, **33**, 657–668.
McLaren, A. C., Retchford, J. A., Griggs, D. T., and Christie, J. M., 1967. 'Transmission electron microscopy study of Brazil twins and dislocations experimentally produced in natural quartz', *Phys. Stat. Sol.*, **19**, 631–644.
McLaren, A. C., Turner, R. G., Boland, J. N., and Hobbs, B. E., 1970. 'Dislocation structure of the deformation lamellae in synthetic quartz: a study by electron and optical microscopy', *Contr. Mineral Petrol.*, **29**, 104–115.

McLean, D., 1965. 'The science of metamorphism in metals'. In *Controls of Metamorphism*, W. S. Pitcher and G. W. Flinn Eds. Oliver and Boyd, London, 103–118.

Mallet, J. L., 1974. 'Présentation d'un ensemble de méthodes et techniques de la cartographie automatique numérique, Sciences de la Terre', *Informatique Géologique*, **4**, 213 p.

March, A., 1932. 'Mathematische theorie der regelung nach der korngestalt bei affiner deformation', *Zeitschrift krist*, **81**, 285–298.

Marfunin, A. S., 1966. 'The feldspars, phase relations, optical properties and geological distribution', *Israel Program for Sc. Translations*, 317 pp.

Masset, J. M., 1973. 'Un système de visualisation des variations géographiques d'un paramètre géologique, Sciences de la Terre', *Informatique Géologique*, **1**, 171 p.

Matheron, G., 1962. *Traité de Géostatistique appliquée*, Tome 1, Editions Technip, Paris, 333 pp.

Matheron, G., 1963. *Traité de Géostatistique Appliquée*, Tome 2, Editions Technip, Paris.

Matheron, G., 1965. *Les variables régionalisées et leur estimation* (*Thèse*), Masson, Paris, 306 pp.

Matheron, G., and Huijbregts, C., 1970. 'Universal kriging.' In *Optimal method for estimating and contouring in trend surface analysis*, Transactions of the 1970 International Symposium on Computer applications and operation research in the mineral industry, Montréal.

Mattauer, M., 1973. *Les déformations des matériaux de l'écorce terrestre*, Hermann, Paris, 493 pp.

Means, W. D., and Paterson, M. S., 1966. 'Experiments on preferred orientation of platy minerals', *Contr. Mineral. Petrol.*, **13**, 108–133.

Means, W. D., and Rogers, J., 1964. 'Orientation of pyrophyllite synthetized in slowly strained material', *Nature*, **204**, 244–246.

Mellis, O., 1942. 'Gefügediagramme in stereographischer projektion', *Tscherm. Mineral. Petr. Mitt.*, **53**, 330–353.

Mendelson, S., 1962. 'Dislocations and plastic flow in NaCl single crystals', *J. Appl. Phys.*, **33**, 2175–2186.

Mercier, J. C., 1972. 'Structures des peridotites en enclaves dans quelques basaltes d'Europe et d'Hawaï—Regards sur la constitution du manteau supérieur', *Thèse* 3° *Cycle, Nantes*, 229 pp.

Mercier, J. C., and Nicolas, A., 1975. 'Textures and fabrics of upper mantle peridotites as illustrated by xenoliths from basalts', *Jour. Petrol.*, **16**, 454–487.

Milnes, J. W., and Toksöz, M. N., 1970. 'Thermal regime of a downgoing slab and new global tectonics', *Jour. Geophys. Res.*, **75**, 1397–1419.

Milnes, A. G., 1971. 'Regional variations in quartz *c*-axis preferred orientations in the Central Alps', *Nature*, **231**, 23, 117–122.

Minear, J. W., and Toksöz, M. N., 1970. 'Thermal regime of a downgoing slab and new global tectonics', *J. Geophys. Res.*, **75**, 1397–1419.

Möckel, J. R., 1969. 'Structural petrology of the garnet peridotite of Alpe Arami (Ticino, Switzerland)', *Leidse Geol. Medelelingen*, **42**, 61–130.

Moorbath, S., and Park, R. G., 1971. 'The Lewisian chronology of the southern region of the Scottish main land', *Scott. J. Geol.*, **8**, 51–74.

Morris, G. B., Raitt, R. W., and Shor, G. G., 1969. 'Velocity anisotropy and delay time maps of the mantle near Hawai', *Jour. Geophys. Res.*, **74**, 4300–4316.

Mügge, O., 1883. 'Beiträge für kentniss der strukturflächen des kalkspathes und über die beziechungen derselben untereinander und der Zwillingsbilden an kalkspath und einegen anderen mineralien', *Neues Jb. Miner. Geol. Paläont*, **1**, 32.

Mügge, O., 1898. 'Uber translation und verwandte erscheinungen in kristallen', *Neues Jahrb. Min. Geol. Pal.*, **1**, 71–158.

Mügge, O., and Heide, F., 1931. 'Einfache scheibungen am anorthit', *Neues Jahrb. Mineral. Geol. Paläontol.*, **I/64**, 163–170.
Mugnier, J., 1973. 'Deformation plastique de monocristaux de calcite', *Thesis 3rd Cycle*, University of Lyon 1, 61 pp.
Müller, P., and Siemes, M., 1974. 'Festigkeit verformbarkeit und gefügeregehung von anhydrit, experimentelle stauchverformung unter manteldrucken bis 5 Kbar bei temperaturen bis 300 °C', *Tectonophysics*, **23**, 105–127.
Murrell, A., A., F., and Charkravarty, S., 1973. 'Some new rheological experiments on Igneous rocks at temperatures up to 1120 °C'. *Geophys. J.R. Astr. Soc.*, **34**, 211–250.
Nabarro, F. R. N., 1967. 'Steady state diffusional creep', *Phil. Mag.*, **16**, 231–237.
Neumann, E. R., 1969. 'Experimental recrystallization of dolomite and comparison of preferred orientations of calcite and dolomite in deformed rocks', *Jour. Geology*, **77**, 426–438.
Newkirk, J. B., 1959. 'The observation of dislocations and other imperfections by X-ray extinction contrast', *Trans. Met. Soc. AIME*, **215**, 483–497.
Nicolas, A., 1969. 'Tectonique et metamorphisme dans les Stura di Lanzo (Alpes piémontaises)', *Bull. Suisse Min. Petrogr.*, **49**, 359–377.
Nicolas, A., 1973. 'Ecoulement des péridotites dans les déformations naturelles et expérimentales', *Bull. Soc. Geol. France*, **XV**, 587–599.
Nicolas, A., 1974. 'Mise en place des péridotites de Lanzo (Alpes piémontaises). Relation avec tectonique et metamorphisme alpins. Conséquences géodynamiques', *Bull. Suisse Min. Petrogr.*, **54**, 449–460.
Nicolas, A., Bouchez, J. L., Boudier, F., and Mercier, J. C., 1971. 'Textures, structures and fabrics due to solid state flow in some European lherzolites', *Tectonophysics*, **12**, 55–85.
Nicolas, A., Bouchez, J. L., and Boudier, F., 1972. 'Interpretation cinématique des déformations plastiques dans le massif de lherzolites de Lanzo (Alpes piémontaises)', *Tectonophysics*, **14**, 143–171.
Nicolas, A., and Boudier, F., 1975. 'Kinematic interpretation of folds in Alpine type peridotites', *Tectonophysics*, **25**, 233–260.
Nicolas, A., Boudier, F., and Boullier, A. M., 1973. 'Mechanisms of flow in naturally and experimentally deformed peridotites', *Am. Jour. Sci.*, **273**, 853–876.
Nicolas, A., and Jackson, E. D., 1972. 'Répartition en deux provinces des péridotites des chaînes alpines longeant la Méditerranée: implications géotectoniques', *Bull. Suisse Miner. Petrogr.*, **52/3**, 479–495.
Nixon, P. H., Boyd, F. R., and Boullier, A. M., 1973. 'The evidence of kimberlite and its inclusions on the constitution of the outer part of the Earth.' In *Lesotho kimberlites*, Nixon Ed., Lesotho National Development Corp., 312–318.
Noble, D. C., and Eberly, S. W., 1964. 'A digital computer procedure for preparing beta diagrams', *Am. Jour. Sc.*, **262**, 1124–1129.
Nye, J. F., 1949. 'Plastic deformation of silver chloride, I—Internal stresses and the glide mechanism', *Proc. Roy. Soc. London*, **A 198**, 190–204.
Nye, J. F., 1953. 'Some geometrical relations in disclocated crystals', *Acta Metall.*, **1**, 153–162.
Nye, J. F., 1957. *Physical Properties of Crystals*, Oxford University Press, 322 pp.
Oëlschlagel, D., and Weiss, V., 1966. 'Superplasticity of steels during the ferrite–austenite transformation', *Trans. American Soc. Metals*, **59**, 143–154.
Oertel, G., 1970. 'Deformation of a slaty lapillar tuff in the English lake district', *Geol. Soc. Am. Bull.*, **81**, 1173–1188.
Oertel, G., 1971. 'Deformation of a slaty, lapillar tuff in the English lake district'; reply, *Geol. Soc. Am. Bull.*, **82**, 533–536.
Oertel, G., 1974. 'Finite strain measurement: a comparison of methods', *Trans. Amer. Geophys. Union*, **55/7**, 695.

O'Hara, M. J., 1967. 'Mineral facies in ultrabasic rocks.' In *Ultramafic and Related Rocks*, Wyllie Ed., John Wiley, New York, 7–18.
Olsen, A., and Birkeland, T., 1973. 'Electron microscope study of peridotite xenoliths in kimberlites', *Contr. Mineral. and Petrol.*, **42**, 147–157.
Orowan, E., 1967. 'Seismic damping and creep in the mantle', *Geophys. J. Roy. Astron. Soc.*, **14**, 191.
Park, R. G., and Tarney, J., 1973. *The Early Precambrian of Scotland and related Rocks of Greenland*, University of Keele.
Parrish, D. K., 1973. 'A non-linear finite element fold model', *Am. Jour. Sci.*, **273**, 318–334.
Paterson, M. S., 1959. 'X-ray line broadening in plastically deformed calcite', *Phil. Mag.*, **4**, 451–466.
Paterson, M. S., 1969. 'Ductility of rocks.' In *Physics of Strength and Plasticity*, A. S. Argon Ed., M.I.T. Press, Cambridge, Mass., 377–392.
Paterson, M. S., 1970. 'A high pressure, high temperature apparatus for rock deformation', *Int. Jour. Rock Mech. Min. Sci.*, **7**, 517–526.
Paterson, M. S., 1973. 'Non-hydrostatic thermodynamics and its geologic applications', *Rev. Geophys. Space Phys.*, **11**, 355–389.
Paterson, M. S., and Turner, F. J., 1970. 'Experimental deformation of constrained crystals of calcite in extension.' In *Experimental and natural rock deformation*, Paulitsch Ed., Springer Verlag, 109–141.
Paterson, M. S., and Weiss, 1961. 'Symmetry concepts in the structural analysis of deformed rocks', *Geol. Soc. America Bull.*, **72**, 841–882.
Persoz, F., 1967. 'Le rôle des filons basiques dans l'élucidation des étages tectoniques', *Etages tectoniques*, Ed. de la Baconnière, Neuchatel, 151–159.
Peselnick, L., Nicolas, A., and Stevenson, P. R., 1974. 'Velocity anisotropy in a mantle peridotite from the Ivrea zone: application to upper mantle anisotropy', *Jour. Geophys. Res.*, **79**, 1175–1182.
Phakey, P., Dollinger, G., and Christie, J. M., 1972. 'Transmission electron microscopy of experimentally deformed olivine crystals.' In *Flow and Fracture of Rocks*, Geophys. Monograph., **16**, Am. Geophys. Union, 139–156.
Phillipps, F. C., and Bradshaw, R., 1970. 'The use of X-rays in petrofabric studies.' In *Experimental and Natural Rock Deformation*, P. Paulitsch Ed., Springer Verlag, Berlin, 75–97.
Pontikis, V., and Poirier, J. P., 1974. 'Déformation à haute température de AgCl monocristallin', *Scripta Met.*, **8**, 1427–1434.
Poirier, J. P., 1972. 'High temperature creep of single crystalline sodium chloride', *Phil. Mag.*, **26**, 701–725.
Poirier, J. P., 1974. 'Quelques remarques sur la sous-structure de fluage', Analogies hydrodynamiques, *Mater. Sci. and Eng.*, **13**, 191–193.
Poirier, J. P., 1975. 'On the slip systems of olivine', *J. Geophys. Res.*, **80**, 4059–4061.
Poirier, J. P., 1976. *Plasticité à haute température des solides cristallins*, Eyrolles, Paris, 322 pp.
Poirier, J. P., and Nicolas, A., 1975. 'Deformation-induced recrystallization by progressive misorientation of subgrain-boundaries, with special reference to mantle peridotites', *J. Geology*, **83**, 707–720.
Post, R. L. Jr., and Griggs, D. T., 1973. 'The Earth's mantle: evidence of non-Newtonian flow', *Science*, **181**, 1242–1244.
Pratt, P. L., 1953. 'Similar glide processes in metallic and ionic crystals', *Acta Metall.*, **1**, 103–104.
Ragan, D. M., 1968. *Structural Geology. An Introduction to Geometrical Techniques*, John Wiley, London, 208 pp.
Raitt, R. W., 1963. 'Seismic refraction studies of the Mendocino fracture zone', *Mar. Phys. Lab., Scripps Inst. Oceanogr.*, Univ. California, Los Angeles, Rep. MPL-U-23/63.

Raitt, R. W., Shorr, G. G., Francis, T. J. G., and Morris, G. B., 1969. 'Anisotropy of the Pacific upper mantle', *Jour. Geophys. Res.*, **74**, 3095–3109.

Raleigh, C. B., 1963. 'Fabrics of naturally and experimentally deformed olivine', *Ph.D. Thesis, California Univers., Los Angeles*, 215 pp.

Raleigh, C. B., 1965a. 'Glide mechanisms in experimentally deformed minerals', *Science*, **150**, 739–741.

Raleigh, C. B., 1965b. 'Crystallization and recrystallization of quartz in a simple piston cylinder device', *Jour. Geol.*, **73**, 369–377.

Raleigh, C. B., 1967. 'Experimental deformation of ultramafic rocks and minerals.' In *Ultramafic and related rocks*, Wyllie Ed., John Wiley, New York, 191–199.

Raleigh, C. B., 1968. 'Mechanisms of plastic deformation of olivine', *Jour. Geophys. Res.*, **73/14**, 5391–5406.

Raleigh, C. B., and Kirby, S. H., 1970. 'Creep in the upper mantle', *Min. Soc. Am. Spec. Pap.*, **3**, 113–121.

Raleigh, C. B., Kirby, S. H., Carter, N. L., and Avé Lallemant, H. G., 1971. 'Slip and the clinoenstatite transformation as competing rate processes in enstatite', *Jour. Geophys. Res.*, **76/17**, 4011–4022.

Raleigh, C. B., and Talbot, J. L., 1967. 'Mechanical twinning in naturally and experimentally deformed diopside', *Amer. J. Sci.*, **265**, 151–165.

Ramberg, H., 1952. *Origin of Igneous and Metasomatic Rocks*, University of Chicago Press, 317 pp.

Ramberg, H., 1963. 'Fluid dynamics of viscous buckling applicable to folding of layered rocks', *Bull. Am. Assoc. Petrol. Geol.*, **47**, 484–515.

Ramsay, J. G., 1960. 'The deformation of early linear structures in areas of repeated folding', *Jour. Geol.*, **68**, 75–93.

Ramsay, J. G., 1967a. *Folding and Fracturing of Rocks*, McGraw Hill, New York, p. 568.

Ramsay, J. G., 1967b. *A Geologist's Approach to Rock Deformation*, Inaugural lecture, Imperial College of Science and Technology, University of London, 131–143.

Ramsay, J. G., 1969. 'The measurement of strain and displacement in orogenic belts', *Geol. Soc. London, Sp. Publ.*, **3**, 43–79.

Ramsay, J. G., and Graham, R. H., 1970. 'Strain variation in shear belts', *Canadian Jour. of Earth Sc.*, **7**, 786–813.

Ransom, D. M., 1971. 'Host control of recrystallized quartz grains', *Mineralogical Mag.*, **38**, 83–88.

Reed-Hill, R. E., Hirth, J. P., and Rogers, H. C. (Eds.), 1964. *Deformation Twinning*, Gordon and Breach, New York.

Riecker, R. E., and Rooney, T. P., 1967. 'Deformation and polymorphism of enstatite under shear stress', *Geol. Soc. Am. Bull.*, **78**, 1045–1054.

Riecker, R. E., and Rooney, T. P., 1969. 'Water induced weakening of hornblende and amphibolite', *Nature*, **224**, 1299.

Ringwood, A. E., 1969. 'Composition and evolution of the upper mantle.' In *The Earth's Crust and Upper Mantle*, P. J. Hart Ed., Am. Geophys. Union, Washington, D.C., 1–17.

Roberts, D., and Stromgard, K. E., 1972. 'A comparison of natural and experimental strain patterns around fold hinges zones', *Tectonophysics*, **14**, 105–120.

Robinson, W. H., 1968. 'Dislocation etch-pit techniques.' In *Techniques of Metals Research*, R. F. Bunshah Ed., Interscience, New York, Vol. II, 291–340.

Robinson, J., and Ellis, M., 1971. Geocom programs (1) 1971. 'Spatial filters and Fortran IV program for filtering geologic maps', *Geocom. Bull.*, **4/1**, 1–21.

Roering, C., 1968. 'The geometrical significance of natural en-echelon crack arrays', *Tectonophysics*, **5**, 107–123.

Rooney, T. P., and Riecker, R. E., 1969. 'Experimental deformation of hornblende and amphibolite', *AF Cambridge Res. Lab. Envir. Res. Pap.*, **299**, 1–24.

Rooney, T. P., and Riecker, R. E., 1973. 'Constant strain rate deformation of amphibole minerals', *AF Cambridge Res. Lab. Envir. Res. Pap.*, **430**, 1–35.
Rooney, T. P., Riecker, R. E., and Ross, M., 1970. 'Deformation twins in hornblende', *Science*, **169**, 173–175.
Ruoff, A. L., 1965. 'Mass transfer problems in ionic crystals with charge neutrality', *J. Appl. Phys.*, **36**, 2903–2907.
Rutter, J. W., and Aust, K. T., 1965. 'Migration of ⟨100⟩ tilt grain boundaries in high purity lead', *Acta Metall.*, **13**, 181–186.
Sammis, C. G., and Dein, J. L., 1974. 'On the possibility of transformational superplasticity in the Earth's mantle', *J. Geophys. Res.*, **79**, 2961–2965.
Sander, B., 1930. *Gefügekunde der Gesteine*, Springer, Wien, 352 pp.
Sander, B., 1948. *Einführung in die Gefügekunde der Geologischen Körper*; I, Springer, Vienna, 215 pp.
Sander, B., 1950. *Einführung in die Gefügekunde der Geologischen Körper*; II, Springer, Vienna, 409 pp.
Sander, B., 1970. *An Introduction to the Study of Fabrics of Geological Bodies*, Pergamon Press, Oxford, 641 pp.
Sauvage, M., and Authier, A., 1965. 'Etude des bandes de croissance et des dislocations de macle dans la calcite', *Bull. Soc. Franç. Min. Crist.*, **88**, 379–388.
Sauvage, M., and Authier, A., 1965. 'Observations topographiques de dislocations de macle dans la calcite', *Phys. Stat. Sol.*, **13**, K 72–75.
Scheidegger, A. E., 1965. 'On the statistics of the orientation of bedding planes, grain axes and similar sedimentological data', *U.S. Geol. Serv. Prof. Pap.*, **525 C**, 164–167.
Schieferdecker, A. A. G., 1959. *Geological nomenclature*, Royal Geol. and Min. Soc. of the Netherlands. Gorinchem J. Noorduijr en Zoon N.V., 523 pp.
Schmid, E., and Boas, W., 1950. *Plasticity of Crystals*, F. A. Hughes, London.
Schmidt, W., 1925. 'Gefungestatistik', *tscherm. Min. und Petr. Mitt.*, **38**, 392–423.
Schneider, M., 1972. 'Shock induced mechanical deformations in biotites from crystalline rocks of the Ries crater (Southern Germany)', *Contr. Miner. Petrol.*, **37**, 75–85.
Schuh, F., Blum, W., and Ilschner, B., 1970. 'Steady state creep rate, impurities and diffusion in rock salt structure', *Proc. Brit. Ceram. Soc.*, **15**, 143–156.
Schwerdtner, W. M., 1964. 'Preferred orientation of hornblende in a banded hornblende gneiss', *Amer. Jour. Sci.*, **262**, 1212–1229.
Schwerdtner, W. M., 1968. 'Intragranular gliding in domal salt', *Tectonophysics*, **5**, 353–380.
Schwerdtner, W. M., 1970. 'Lattice orienting mechanisms in schistose anhydrite.' In *Experimental and natural rock deformation*, Paulitsch Ed., Springer Verlag, Berlin, 142–164.
Schwerdtner, W. M., 1973. 'A scale problem in paleo strain analysis', *Tectonophysics*, **16**, 47–54.
Schwerdtner, W. M., and Morrison, M. J., 1973. 'Internal-flow mechanism of salt and sylvinite in Anagance diapiric anticline near Sussex, New Brunswick.' In *Fourth Symposium on Salt*, The Northern Ohio Geol. Soc., **2**, 241–248.
Seifert, K. E., 1965. 'Deformation bands in albite', *Am. Miner.*, **50**, 1469–1472.
Šesták, B., 1972. 'Modern aspects of plastic deformation in metals', *Czech. J. Phys.*, **B 22**, 270–285.
Shelley, D., 1974. 'Mechanical production of metamorphic banding, a critical appraisal', *Geol. Mag.*, **III**, 287–292.
Sherbon Hills, E., 1963. *Elements of Structural Geology*, Sci. Paperbacks, Methuen, London, 483 pp.
Shor, G. G., and Pollard, D. D., 1964. 'Mohole site selection studies north of Maui', *Jour. Geophys. Res.*, **69**, 1627–1637.

Siddans, A. W. B., 1972. 'Slaty cleavage—A review of research since 1815', *Earth Science Rev.*, 205–232.
Silk, E. C. H., and Barnes, R. S., 1961. 'The observation of dislocations in mica', *Acta Metall.*, **9**, 558–562.
Smoluchowski, R., 1966. 'Dislocations in ionic crystals', *J. de Phys.*, **27**, C 3, 3–11.
Snyder, R. W., and Graff, H. F., 1938. 'Study of grain size in hardened high speed metals', *Metal Progress*, 377.
Sorby, H. C., 1853. 'On the origin of slaty cleavage', *Edinburgh New Philos. Jour.*, **55**, 137–148.
Spencer, A. B., and Clabaugh, P. S., 1967. 'Computer program for fabric diagrams', *Am. Jour. Sc.*, **265**, 166–172.
Spitzig, W. A., and Keh, A. S., 1970. 'Orientation and temperature dependence of slip in iron single crystals', *Metall. Trans.*, **1**, 2751–2757.
Spry, A., 1969. *Metamorphic Textures*, Pergamon Press, Oxford, 350 pp.
Squires, R. L., Weiner, R. T., and Phillips, M., 1963. 'Grain boundary denuded zones in a magnesium, $\frac{1}{2}$ Wt % zirconium alloy', *J. Nucl. Mat.*, **8**, 77–80.
Staker, M. R., and Holt, D. L., 1972. 'The dislocation cell size and dislocation density in copper deformed at temperatures between 25 and 700 °C', *Acta Metall.*, **20**, 569–579.
Starkey, J., 1964. 'Glide twinning in the plagioclase feldspars.' In *Deformation Twinning*, R. E. Reed Hill, J. P. Hirth, H. C. Rogers Eds., Gordon and Breach, New York, 177–191.
Starkey, J., 1964. 'An X-ray method for determining the orientation of selected crystal planes in polycrystalline aggregates', *Am. Jour. Sc.*, **262**, 735–752.
Starkey, J., 1968. 'The geometry of kink bands in crystals, a simple model', *Contr. Mineral. Petrol.*, **19**, 133–141.
Starkey, J., 1970. 'A computer programme to prepare orientation diagrams.' In *Experimental and Natural Rock Deformation*, Paulitsch Ed., Springer Verlag, Berlin, pp. 51–74.
Startsev, V. I., 1963. 'The formation of defects in crystal lattice by twinning', *J. Phys. Soc. Japan*, **18**, Supp. III, 16–20.
Stauffer, M. R., 1966. 'An empirical statistical study of three dimensional fabric diagrams as used in structural analysis', *Canadian Jour. Earth Sc.*, **3**, 473–498.
Steiger, R. H., 1964. 'Dating of orogenic phases in the central alpes by K–Ar ages of hornblende', *Jour. Geophys. Res.*, **69/24**, 5407–5421.
Steketee, J. A., 1958. 'Some geophysical applications of the elasticity theory of dislocations', *Can. J. Phys.*, **36**, 1168–1198.
Stephanson, O., and Berner, H., 1971. 'The finite element method in tectonic processes', *Phys. Earth Planet. Interiors*, **4**, 301–321.
Stocker, R. L., and Ashby, M. F., 1973. 'On the empirical constants in the Dorn equation', *Scripta Met.*, **7**, 115–120.
Stocker, R. L., and Ashby, M. F., 1973. 'On the rheology of the upper mantle', *Rev. Geophys. and Space Phys.*, **11**, 391–426.
Stokes, R. J., 1966. 'Mechanical properties of polycrystalline sodium chloride', *Brit. Ceram. Soc. Proc.*, **6**, 189–207.
Strand, T., 1944. 'A method of counting out petrofabric diagrams', *Norsk Geol. TidssKr.*, **24**, 112–113.
Streb, G., and Reppich, B., 1973. 'Steady state deformation and dislocation structure of pure and Mg doped LiF single crystals', *Phys. Stat. Sol.* (*a*), **16**, 493–505.
Strunk, H., 1972. 'Investigation of plastically deformed NaCl by X-ray topography and electron microscopy', *Phys. Stat. Sol.* (*a*), **11**, K 105–K 108.
Sugimura, A., and Uyeda, S., 1967. 'A possible anisotropy of the upper mantle accounting for deep earthquake faulting', *Tectonophysics*, **5**, 25–33.
Sun, R. C., and Bauer, C. L., 1970. 'Tilt boundary migration in NaCl bicrystals', *Acta Metall.*, **18**, 639–647.

Talbot, J. L., and Hobbs, B. E., 1968. 'The relationship of metamorphic differentiation to other structural features at three localities', *Jour. Geol.*, **76**, 581–587.

Talbot, C. J., 1970. 'The minimum strain ellipsoid using deformed quartz veines', *Tectonophysics*, **9**, 47–76.

Taylor, G. I., 1938. 'Plastic strain in metals', *Inst. Metals Jour.*, **62**, 307–324.

Taylor, G. I., and Elam, C. F., 1926. 'The distorsion of iron crystals', *Proc. Roy. Soc. London*, **A 112**, 337–361.

Thayer, T. P., 1943. 'Chrome resources of Cuba', *U.S., Geol. Survey Bull.*, **935 A**, 1–74.

Thayer, T. P., 1960. 'Some critical differences between alpine type and stratiform peridotite gabbro complexes', 21*th Inter. Geol. Congress, Sect. 13*, 247–259.

Thomas, J. M., and Renshaw, G. D., 1967. 'Influence of dislocations on the thermal decomposition of calcium carbonate', *Jour. Chem. Soc.* (*A*), 2058–2061.

Tocher, J. E., 1967. 'A point counting computer program for petrofabric analysis of uniaxial mineral orientations', *Min. Mag.*, **36**, 456–457.

Toksöz, M. N., Minear, J. W., and Julian, B. R., 1971. 'Temperature field and geophysical effects of a downgoing slab', *Jour. Geophys. Res.*, **76**, 1113–1138.

Trommsdorff, V., and Wenk, H. R., 1968. 'Terrestrial metamorphic clinoenstatite in kinks of bronzite crystals', *Contr. Mineral. Petrol.*, **19**, 158–168.

Tullis, J. A., 1970. 'Quartz: Preferred orientation in rocks produced by Dauphiné twinning', *Science*, **168**, 1342–1344.

Tullis, J. A., 1971. 'Preferred orientations in experimentally deformed quartzites', *Ph.D. Thesis, Univ. California, Los Angeles*, 344 pp.

Tullis, T., 1971. 'Experimental development of preferred orientations of mica during recrystallization', *Ph.D. Thesis, Univ. of California, Los Angeles*, 262 pp.

Tullis, J., Christie, J. M., and Griggs, D. T., 1973. 'Microstructures and preferred orientations of experimentally deformed quartzites', *Geol. Soc. Am. Bull.*, **84**, 297–314.

Tullis, T. E., and Wood, D. S., 1975. 'Correlation of finite strain from both reduction bodies and preferred orientation of mica in slates from Wales', *Geol. Soc. Amer. Bull.*, **86**, 632–638.

Tullis, J. A., and Tullis, T., 1972. 'Preferred orientation of quartz produced by mechanical Dauphiné twinning: thermodynamics and axial experiments.' In *Flow and fracture of rocks*, the Griggs Vol., Am. Geophys. Union, **16**, 67–82.

Turner, F. J., 1942. 'Preferred orientation of olivine crystals in peridotites with special reference to New Zealand examples', *Roy. Soc. New Zealand Proc. Trans.*, **72**, 280–300.

Turner, F. J., 1953. 'Nature and dynamic interpretation of deformation lamellae in calcite of three marbles', *Am. Jour. Sc.*, **251**, 276–298.

Turner, F. J., and Verhoogen, J., 1960. *Igneous and metamorphic Petrology*, McGraw Hill, New York, 694 pp.

Turner, F. J., and Chi'h, C. S., 1951. 'Deformation of Yule marble. Part III: Observed fabric changes due to deformation at 10.000 atmospheres confining pressure, room temperature, dry', *Geol. Soc. Am. Bull.*, **62**, 887–906.

Turner, F. J., Griggs, D. T., Clark, R. H., and Dixon, R. H., 1956. 'Deformation of Yule marble. Part VII: Development of oriented fabrics at 300 °C–500 °C', *Geol. Soc. Am. Bull.*, **67**, 1259–1294.

Turner, F. J., Griggs, D. T., and Heard, H., 1954a. 'Experimental deformation of calcite crystals', *Geol. Soc. Am. Bull.*, **65**, 883–934.

Turner, F. J., Griggs, D. T., and Weiss, L. E., 1954b. 'Plastic deformation of dolomite rock at 380 °C', *Am. Jour. Sc.*, **252**, 477–488.

Turner, F. J., Heard, H. C., and Griggs, D. T., 1960. 'Experimental deformation of enstatite and accompanying inversion to clinoenstatite', *Int. Geol. Congress, XXI Sess. Copenhagen*, **XVIII**, 399–408.

Turner, F. J., and Weiss, L. E., 1963. *Structural Analysis of Metamorphic Tectonites*, McGraw Hill, New York, 545 pp.

Van Der Plas, 1959. 'Petrology of the Northern Adula region, Switzerland (with particular reference to the glaucophane bearing rocks)', *Leidse Geol. Meded.*, **24/2**, 415–598.

Vecchia, O., 1968. 'La zone Cuneo-Ivrea-Locarno, élément fondamental des Alpes—Géophysique et Géologie', *Bull. Suisse Min. Petrogr.*, **48**, 215–226.

Veit, K., 1922. 'Künstliche schiebungen und translationen in mineralien', *N. Jahrb. J. Min.*, **45**, 121–148.

Verma, R. K., 1960. 'Elasticity of some high density crystals', *Jour. Geophys. Res.*, **65**, 757–766.

Vialon, P., 1974. 'L'importance du changement de comportement mécanique du matériau au cours d'une déformation unique, dans la définition des superpositions de structures et de phases tectoniques—Exemple du clivage schisteux', *2° Réunion Sc. Terre, Nancy*, 385.

Vistelius, A. B., 1966. *Structural Diagrams*, Pergamon Press, Oxford, 178 pp.

Voll, G., 1960. 'New work in petrofabrics', *Liverpool Manchester Geol. J.*, **2**, 503–567.

Von Plessman, W., 1964. 'Gesteinslösung ein haupt faktor beim schieferungsprozess', *Geol. Mitt. Aachen.*, **4**, 69–82.

Wager, L. R., and Brown, G. M., 1967. *Layered Igneous Rocks*, Oliver and Boyd, 588 pp.

Watson, G. S., 1965. 'Equatorial distribution on a sphere', *Biometrika*, **52**, 193–201.

Watson, G. S., 1966. 'The statistics of orientation data', *Jour. Geol.*, **74**, 786–797.

Weertman, J., 1968. 'Dislocation climb theory of steady state creep', *Trans. Amer. Soc. Metals*, **61**, 681–694.

Weertman, J., 1970. 'The creep strength of the earth's mantle', *Rev. Geophys. and Space Phys.*, **8**, 145–168.

Weertman, J., 1972. 'High temperature creep produced by dislocation motion.' In *J. E. Dorn memorial symposium, Cleveland, Ohio.*

Weertman, J., and Weertman, J. R., 1964. *Elementary Dislocation Theory*, McMillan, New York.

Wegmann, C. E., and Schaer, J. P., 1962. 'Chronologie et deformations des filons basiques dans les formations précambriennes du Sud de la Norvège', *Norsk. Geol. Tidsskrift*, **42**, 371–387.

Wegner, M. W., and Christie, J. M., 1974. 'Preferential chemical etching of terrestrial and lunar olivines', *Contr. Min. Petr.*, **43**, 195–212.

Weiss, L. E., and Turner, F. J., 1972. 'Some observations on translation gliding and kinking in experimentally deformed calcite and dolomite.' In *Flow and Fracture of Rocks*, the Griggs Vol., Am. Geophys. Union, **16**, 95–108.

Weissman, S., and Kalman, Z. H., 1969. 'X-ray diffraction topographic methods.' In *Techniques of Metals Research*, R. F. Bunshah Ed., Interscience, New York, Vol. II, 839–873.

Wenk, H. R., and Shore, J., 1974. 'Preferred orientation in experimentally deformed dolomite', *Trans. Am. Geophys. Union*, **55**, 419.

Wenk, H. R., and Shore, I., 1975. 'Preferred orientation in experimentally deformed dolomite', *Contr. Mineral. Petrol.*, **50**, 115–126.

Wenk, H. R., Venkitasubramanyan, C. S., and Baker, D. W., 1973. 'Preferred orientation in experimentally deformed limestone', *Contr. Mineral. Petrol.*, **38**, 81–114.

Wenk, H. R., and Wilde, W. R., 1972. 'Orientation distribution diagrams for three Yule marble fabrics.' In *Flow and Fracture of Rocks*, Geophys. Monog. Series Am. Geophys. Union, **16**, 83–95.

Whitaker, J. M. McD., and Gatrall, M., 1969. 'Two "orientometers" for measuring lineations on hand specimens', *Jour. Geol.*, **77**, 710–714.

White, S., 1973a. 'Syntectonic recrystallization and texture development in quartz', *Nature*, **244**, 276–278.

White, S., 1973b. 'The dislocation structures responsible for the optical effects in some naturally deformed quartzites', *Jour. Mat. Sc.*, **8**, 490–499.

Whitten, E. H. T., 1966. *Structural Geology of Folded Rocks*, Rand McNally, Chicago, 678 pp.

Wilkens, M., 1967. 'Application of X-ray topography to the analysis of the dislocation arrangement in deformed copper single crystals', *Can. J. Phys.*, **45**, 567–579.

Williams, P. F., 1972. 'Development of metamorphic layering and cleavage in low grade metamorphic rocks at Bermagni, Australia', *Am. Jour. Sc.*, **272**, 1–47.

Winchell, A. N., 1961. *Elements of Optical Mineralogy*; Part I: 'Principles and Methods', J. Wiley, New York, 263 pp.

Wood, D. S., 1974. 'Current views of the development of slaty cleavage', *Annual Review of Earth and Planetary Science*, F. A. Donath, F. G. Stehli, G. W. Wetherill (Eds.), Annual Reviews Inc., Palo Alto, Calif.

Young, C. III, 1969. 'Dislocations in the deformation of olivine', *Am. Jour. Sc.*, **267**, 841–852.

Zeuch, D. H., 1974. 'Naturally decorated dislocations in olivine', *Trans. Amer. Geophys. Union*, **55**, 418–419.

Zwart, H. J., 1963. 'Metamorphic history of the central Pyrenées', *Leidse Geol. Meded.*, **28**, 321–376.

Subject Index